International Conference on Mechanical, Material and Aerospace Engineering (2MAE 2017)

MATEC Web of Conferences Volume 114 (2017)

Beijing, China
12 - 14 May 2017

Editors:

Peter Liu
Xiaowei Tu

ISBN: 978-1-5108-4431-5

Printed from e-media with permission by:

Curran Associates, Inc.
57 Morehouse Lane
Red Hook, NY 12571

Some format issues inherent in the e-media version may also appear in this print version.

Printed by Curran Associates, Inc. (2017)

For additional information, please contact EDP Sciences – Web of Conferences
at the address below.

EDP Sciences – Web of Conferences
17, Avenue du Hoggar
Parc d'Activité de Courtabœuf
BP 112
F-91944 Les Ulis Cedex A
France

Phone: +33 (0) 1 69 18 75 75
Fax: +33 (0) 1 69 28 84 91

contact-edps@webofconferences.org

Additional copies of this publication are available from:

Curran Associates, Inc.
57 Morehouse Lane
Red Hook, NY 12571 USA
Phone: 845-758-0400
Fax: 845-758-2633
Email: curran@proceedings.com
Web: www.proceedings.com

TABLE OF CONTENTS

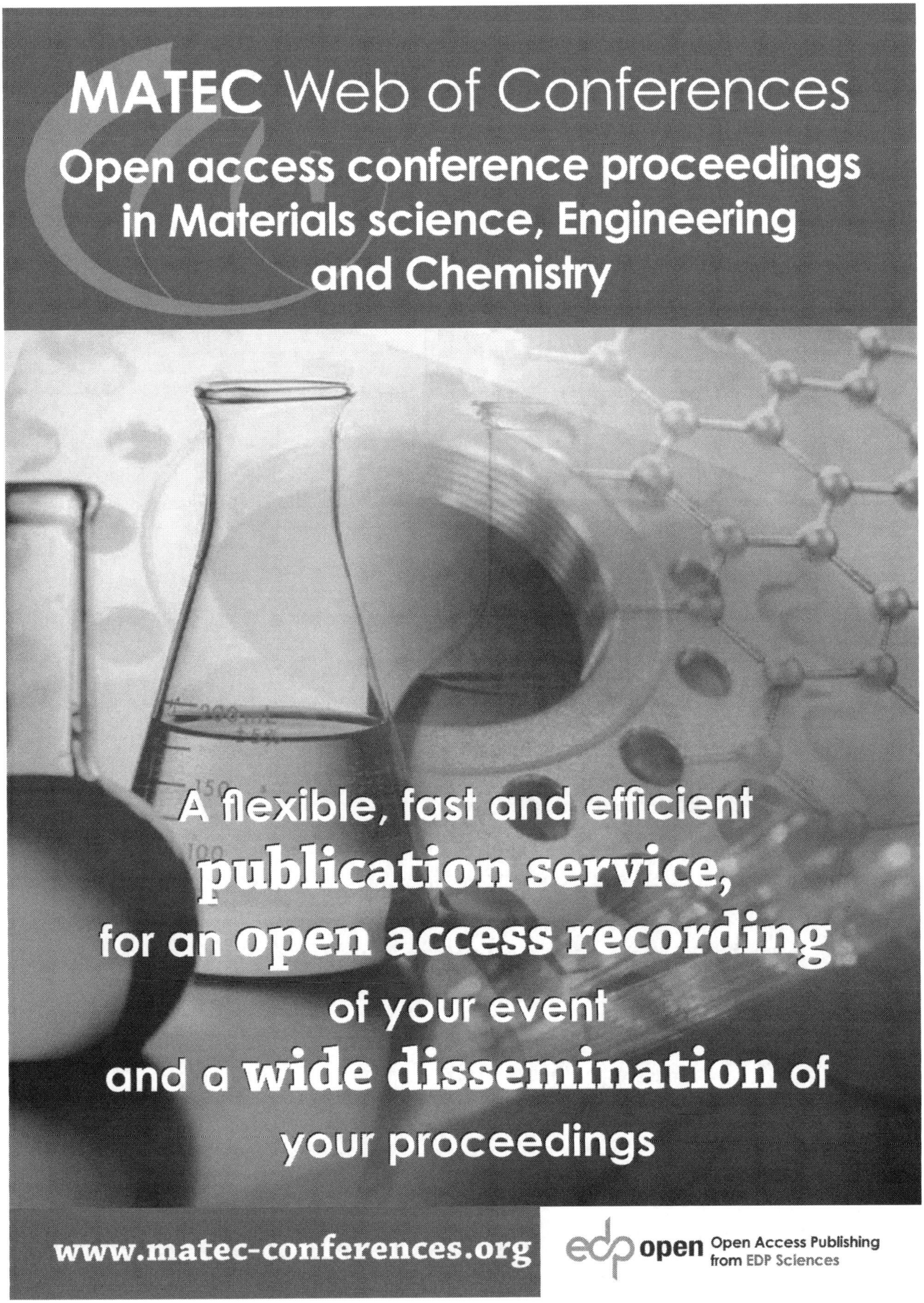
MATEC Web of Conferences
Open access conference proceedings in Materials science, Engineering and Chemistry
A flexible, fast and efficient
publication service,
for an open access recording
of your event
and a wide dissemination of
your proceedings
www.matec-conferences.org
edp open Open Access Publishing from EDP Sciences

PREFACE

2017 International Conference on Mechanical, Material and Aerospace Engineering (2MAE 2017) was held in Beijing during May 12-14, 2017. 2MAE 2017 was organized by Hong Kong Society of Mechanical Engineers (HKSME) and Technical Sponsored by Beijing Jiaotong University, China. The conference provides a useful and wide platform both for display the latest research and for exchange of research results and thoughts in broad Mechanical, Material and Aerospace Engineering. The participants of the conference were from almost every part of the world, with background of either academia or industry, even well-known enterprise. The success and prosperity of the conference is reflected high level of the papers received.

The proceedings are a compilation of the accepted papers and represent an interesting outcome of the conference. This book covers 4 chapters: Mechanical; Materials; Aerospace; Interdisciplinary.

We would like to acknowledge all of those who supported 2MAE 2017. Each individual and institutional help were very important for the success of this conference. Especially we would like to thank the organizing committee for their valuable advices in the organization and helpful peer review of the papers.

We sincerely hope that 2MAE 2017 will be a forum for excellent discussions that will put forward new ideas and promote collaborative researches. We are sure that the proceedings will serve as an important research source of references and the knowledge, which will lead to not only scientific and engineering progress but also other new products and processes.

Prof. Peter X. Liu, Carleton University, Canada
Prof. Xiaowei Tu, Shanghai University, China

COMMITTEES

General Chair

Prof. Peter X. Liu, Carleton University, Canada

Prof. Linmao Qian, Southwest Jiaotong University, China

Advisory Committees

Prof. Xin Zhang, Hong Kong University of Science and Technology, Hong Kong

Prof. Xiaowei Tu, Shanghai University, China

Conference Committee Chair

Prof. Christopher Y H CHAO, Hong Kong University of Science and Technology, Hong Kong

Conference Program Committee Chair

Prof. Liu Qiang, Beihang University, China

Local Committees

Prof. Liu Shichao, Beijing Jiaotong University, China

Prof. Wang Huanqing, Beijing Jiaotong University, China

Prof. Shen Haikuo, Beijing Jiaotong University, China

Lecturer Zheng Minhua, Beijing Jiaotong University, China

International Technical Committees

Dr. Alex Ong Zhi Chao, University of Malaya, Malaysia

Dr. B. F. Yousif, University of Southern Queensland, Australia

Prof. M. Chandrasekaran, Vels University, India

Prof. Shinn-Liang Chang, National Formosa University, Taiwan

Prof. Tang Sai Hong, Universiti Putra Malaysia, Malaysia

Dr. Darwin Lau, The Chinese University, Hong Kong

Dr. Sayeed Hossain, University of Bristol, UK

Dr. Laurent Tranchant, Kyushu Institute of Technology, Japan

Dr. AloK Jha, Kyushu Institute of Technology, Japan

Dr. Dehong Huo, Newcastle University, UK

Dr. Jie Zhang, University of Bristol, UK

edp sciences

MATEC Web of Conferences

Statement of Peer review

In submitting conference proceedings to MATEC Web of Conferences, I certify to the Publisher that I adhere to the **Policy on Publishing Integrity** of the journal in order to safeguard good scientific practice in publishing.

1. All articles have been subjected to peer review administered by the proceedings editors.
2. Reviews have been conducted by expert referees, who have been requested to provide unbiased and constructive comments aimed, whenever possible, at improving the work.
3. Proceedings editors have taken all reasonable steps to ensure the quality of the materials they publish and their decision to accept or reject a paper for publication has been based only on the merits of the work and the relevance to the journal.

Title, date and place of the conference

2017 International Conference on Mechanical, Material and Aerospace Engineering (2MAE 2017)

May 12-14, 2017 Beijing Jiaotong University,Beijing,China

Proceedings editor(s):

Prof. Peter Liu, Prof. Xiaowei Tu

Date and editor's signature 15 th juin 2017

DOI: 10.1051/matecconf/201711401001

Design and Construction of High-speed Dynamic Balance for High Load

Junli Sun[a], Nong Chen and Jin Liu

China Academy of Aerospace Aerodynamics, Beijing 100074, P.R. China

Abstract. There are high longitudinal loads and mismatched longitudinal/lateral aerodynamic characteristics of unconventional aerodynamic configurations, which present a new challenge to high-speed dynamic balance technology. The wind tunnel dimension limitation makes a new command in miniaturization design of dynamic balance. The article describes a solution method which is based on the electromechanical motion modulation testing principle and mechanical structure of saddle-type dynamic balance. There is an II-type structure assembled by oil film bearing and elastic pivot beam on the outer tube to achieve motion modulation and measurement. The double variable cross-section vertical beams are placed in front of inner shaft to measure the five components of high load. The double saddle-type structure is applied to connect the intermediate part of inner shaft with outer tube together. A Scotch yoke mechanism is composed of the transmission eccentric cam of sting and a mullion structure which is designed at the rear part of inner shaft. According to the redesign and alteration, the effective lateral load signal output could be obtained from the miniaturized dynamic balance, in which the maximum longitudinal load bearing capacity is extended more than twice. A great agreement is also obtained between the measurement accuracy and design requirements.

1 Introduction

The unconventional configuration design requirements are increased and the flight envelopes are expended gradually with the development of aerospace aircraft. It becomes much more serious about the unsteady and asymmetrical separation of the flow around aircraft, which raises a crucial point to make accurate prediction about characterizes of unsteady and dynamic [1-4]. The dynamic stability technology which based on the weak signal detection technique is a solution often used in aerodynamics. The direct, cross and cross-couple dynamic derivative parameters on three degrees of freedom of pitch, yaw and roll are obtained by motion modulation of the aircraft. The dynamic balance is a kernel part in this special experiment technology. The dynamic balance could achieve the motion modulation and measurement by its high flexibility, and its high stiffness ensures to measure aerodynamic force response and avoid interference of noise, while the disadvantage is the limitation of dimension and aerodynamic loads. The high longitudinal loads and mismatched longitudinal/lateral aerodynamic characteristics of unconventional aerodynamic configuration aircrafts lead to the redesign and alteration of dynamic balance.

The international researchers mainly adopt split shape and integrated shape of high-speed wind

[a] Corresponding author: sunjunli@zju.edu.cn

tunnel dynamic balances structure to achieve the two major functions: motion modulation and aerodynamic forces response measurement [5-15]. The split shape is front-rear series structure, whose front part often adopt the high stiffness conventional five-component balance, and the rear part is composed of an elastic bearing / ball bearing acted as vibration center and a Scotch yoke mechanism which is a reciprocating motion mechanism to converting rotational motion into linear motion [5-9], which results to high axial dimension and space occupation, so that it is suitable for high slenderness ratio aircraft. The dynamic balance suffers a large inertial force by the large space among calibration center, model gravity center and vibration center, which leads to the limitation on sensitivity level of torque components. It raises the difficulty of the weak physical parameters measurement [5,6,10,11]. The low signal-to-noise ratio and the weak suppression ability on noise decrease the test accuracy. Particularly, some unreasonable design of split shape dynamic balances lead to the resonance with wind tunnel sting, which will destroy the wake of aircraft and mistake measurement results. Two kinds of integrated shapes are in common use of dynamic test. One of them is a saddle-type structure. The inner shaft is designed as a ball bearing to associate with longitudinal moment beam to achieve motion modulation and measurement. The front part of outer shaft is designed as a rectangle beam to measure the five components. The immediate part of outer shaft which sits at the vibration center of inner shaft structure seems like a saddle structure. The rear part is a mullion structure which makes up a Scotch yoke mechanics with the transmission eccentric cam of wind tunnel sting. The front, immediate and rear part of outer shaft are assembled together before the assembly with inner shaft [10,11,12]. The whole structure is compactness and the space is much tinier among calibration center, model gravity center and vibration center. For the reason of assembled structure, it is hard to overcome the disadvantages of assembled structure, such as creep deformation, internal friction, hysteresis, non-liner interference and so on. The other one is a can-style structure. Its inner shaft is composed of cross-type elastic bearing used as a vibration center and the moment beam attached at the transmission mechanic in the wind tunnel sting. The outer structure is designed to be a wheel hub structure to obtain the five components [13,14], or vice versa [15]. The limited carrying capacity exists in some parts of the dynamic balance, like the inner shaft elastic bearing, torque beam and wheel hub structure of outer shaft which could not support the high aerodynamic loads. Furthermore, the integrated shape structure has higher measurement accuracy, while the radial dimension limitation of aircraft model allows it only fit for the 2 meters degree wind tunnel or larger. The split shape dynamic balance does not own high measurement accuracy, which is allowed to be used in the 2 meters degree wind tunnel or less for the reason of the axis dimension limitation of aircraft model, even just in high slenderness ratio aircraft.

The unconventional aerodynamic configuration aircraft is being focused in the article. For example, the normal aerodynamic force of flying-wing aircraft increases apparently with the raise of angle of attack, even reaches to 1000 kilograms ($Ma=0.9$, $\alpha=28°$) and the longitudinal/lateral aerodynamic characteristics are extremely mismatched (ratio of longitudinal load to lateral load equals to 20:1), which could bring measurement problems in wind tunnel experiment about the unsteady and dynamic characterizes [16,17]. It is necessary to redesign the split shape dynamic balances to satisfied the measurement accuracy requirements of dynamic characterizes [3,9]. In addition, the miniaturization design should be implemented for the reason that the test section dimension limitation of 1.2 meters degree wind tunnel makes a large constraint to the dynamic balance. The article is based on the saddle-type structure in order to solve the problems above. There is an II-type beam structure assembled by oil film bearing and elastic pivot on the outer tube to achieve motion modulation and measurement. The double variable cross-section vertical beams are placed in front of inner shaft to measure the five components of high loads. The double saddle-type structure is applied to connect the intermediate part of inner shaft with outer tube together. The inner shaft and outer tube could be welded together through the oil film bearing by the electron beam, and elastic pivot could be replaced according to the aerodynamic trim torque. A Scotch yoke mechanism is composed of the transmission eccentric cam of sting and a mullion structure which is designed at rear part of inner shaft. According to the redesign and alteration, the sensitivity of lateral load could be improved to achieve the five-component sensitivity measurement requirements from the miniaturized dynamic balance, in which the maximum longitudinal load bearing capacity is extended more than twice. The measurement accuracy could reach the design requirements.

DOI: 10.1051/matecconf/201711401001

2 Basic principles of the apparatus

The flow structure changed rapidly along with the raise of angle of attack and aerodynamic loads present the obvious unsteady features during the flight. The physical variables, like pressure, velocity, density, have changed with time since the interaction of various physical phenomena such as flow separation, shock wave boundary layer interference. The unsteady aerodynamic loads are much smaller than the static aerodynamic loads. Since most of the aerodynamic load signals are slowly varying over time or are approximated of a non-time-dependent direct-current (DC) component, which means the static aerodynamic load component is distributed in the low frequency region of signal. Suppose a conventional DC-type amplifier (such as a conventional balance) is used to amplify and detect a weak unsteady aerodynamic load signal directly, the 1/f noise and drift will be existed with seriously influence on the detection sensitivity.

The unsteady aerodynamic load measurement should adopt the modulation frequency technology to translate the signal in the frequency axis to avoid the 1/f noise and DC drift, which also ensures to keep the Dynamic Similarity Criteria between the dynamic frequency of the aircraft and its wind tunnel model, which is conducive to the unsteady aerodynamic load detection. It is necessary to suppress the noise by increasing the amplitude in the low test frequency (schematic diagram shown in figure 1) to improve signal-to-noise ratio. A modulation amplitude technique is required when the noise power spectrum of an unsteady aerodynamic load signal approaches to a low frequency. In general, the amplitude in low-speed wind tunnels is much larger than high-speed wind tunnel dynamic tests.

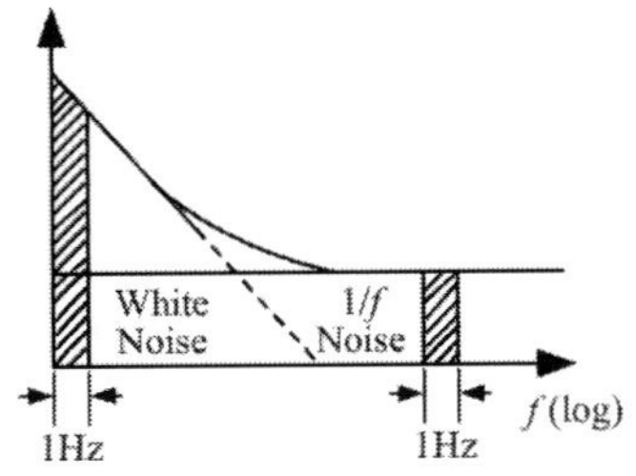

Figure 1. Amplitude modulation of the signal can reduce the output noise.

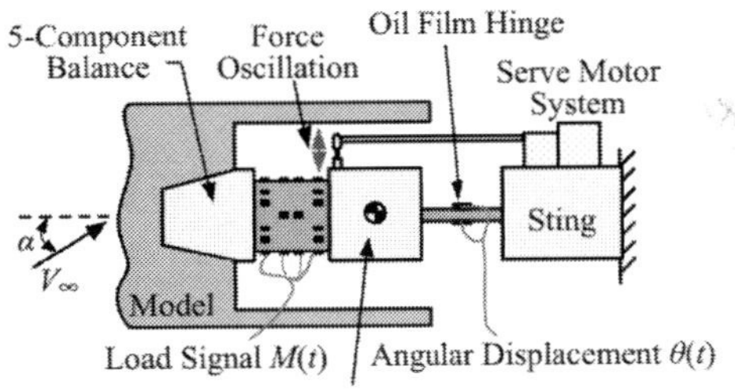

Figure 2. Schematic diagram of dynamic balance mechanical mechanism.

Figure 2 presents a schematic diagram about the electromechanical motion modulation testing principle and the mechanical structure of the saddle-type dynamic balance. Taking pitch/yaw dynamic balance as an example, the maximum frequency of the aerodynamic load signal $f(x)$ of the dynamic balance is assumed to ω_0 and the rotary switch modulation shown in figure 3 is adopted, a reasonable switch modulation function $S_k(t)$ shown in figure 4 is designed to simulate the pitching motion of the aircraft in figure 5. The output of the modulator (i.e. the output of the dynamic balance five-component) is the convolution of aerodynamic load signal $f(x)$ with switching function $S_k(t)$.

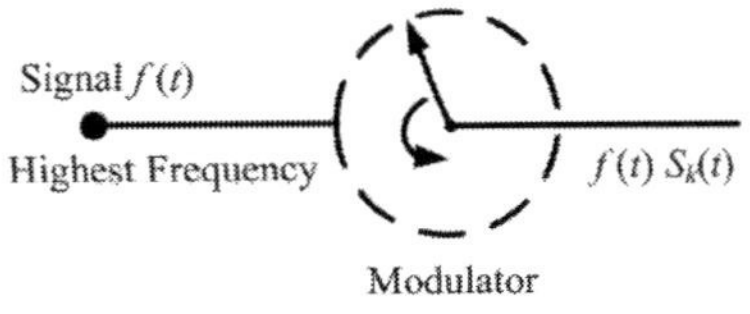

Figure 3. Pneumatic load signal modulation.

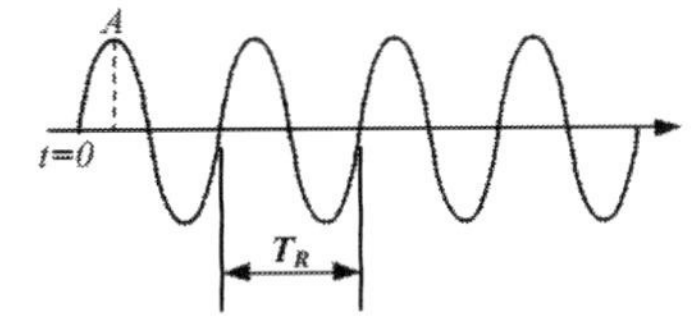

Figure 4. Switch modulation function S_k(t).

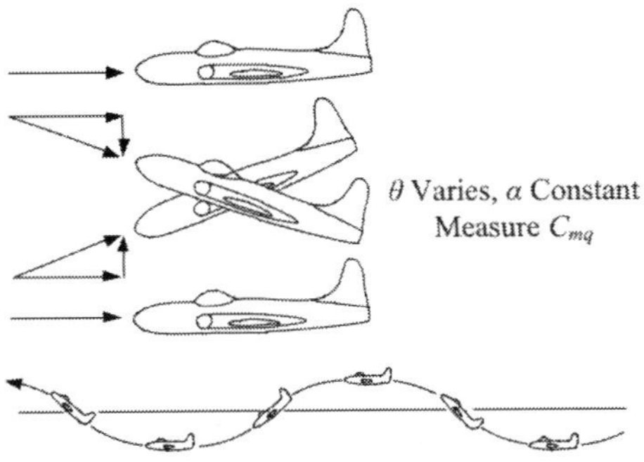

Figure 5. Aircraft pitching motion diagram.

DOI: 10.1051/matecconf/201711401001

The spectrum of the aerodynamic load signal $f(x)$ is represented by the Fourier transform $F(t)$. The modulated aerodynamic load signal could be expressed as $X(t) = f(x) * S_k(t)$, and its spectrum can also be represented by the Fourier Transform $X(f)$:

$$X(f) = \int_{-\infty}^{\infty} S_k(t) f(t) e^{-j\omega t} dt \tag{1}$$

The switch modulation function $S_k(t)$ is accomplished by next expressions:

$$S_k(t) = Ae^{j\omega_R t} \tag{2}$$

Where:
$\omega_R = 2\pi / T_R$ is the fundamental frequency of the modulator. A, TR are the amplitude and period of the switching modulation function shown in figure 4.
From the formula (1) and formula (2), it could form:

$$S_k(t) f(t) = f(t) \cdot Ae^{j\omega_R t} \tag{3}$$

And

$$X(f) = \int_{-\infty}^{\infty} f(t) \cdot Ae^{-j(\omega - \omega_R)t} dt \tag{4}$$

Due to the Fourier transform of the function $f(t)$, it could obtains:

$$F(f) = \int_{-\infty}^{\infty} f(t) e^{-j\omega t} dt \tag{5}$$

The equation (5) could be written as:

$$F(f - nf_R) = \int_{-\infty}^{\infty} f(t) e^{-j(\omega - \omega_R)t} dt \tag{6}$$

Substituting equation (6) into equation (4):

$$X(f) = A \cdot F(f - nf_R) \tag{7}$$

Getting the spectrum of signal after modulation, the discussion of frequency spectrum is as follows:
For $n=0$, where

$$X(f) = A \cdot F(f) \tag{8}$$

For $n=i$, where

$$X(f) = A \cdot F(f - if_R) \tag{9}$$

The corresponding demodulation algorithm [2, 3, 5-16] could be used to restore the original aerodynamic load signal $f(t)$ from the modulated aerodynamic load signal $X(t)$.

The weak signal detection technology based on the principle above is different from the general detection technology. The weak signal detection focuses on how to suppress noise and improve the signal-to-noise ratio rather than the physical model, sensing principle, the signal conversion circuit and instrument implementation. The dynamic balance is the special electromechanical motion modulation sensor which has been applied of weak signal detection technology. It could achieve the motion modulation and measurement by high flexibility, measure aerodynamic force response and avoid extraneous effects by high stiffness.

3 Apparatus description

3.1 Design loads and condition

The wind tunnel experiment is to be carried out in the wind tunnel of 1.2 meters. The internal space dimension is Φ55mm (maximum diameter) × 280mm (axial distance) for the flying wing model. According to the static test results [16], the lateral load of the dynamic balance is enlarged by considering the experimental scheme, and the design load of the dynamic balance is determined as shown in Table 1.

Table 1. The design scale and precision of the dynamic balance.

Load Component	Range (N, Nm)	Accuracy (%)	Precision (%)
Y	10000	≤0.5	≤0.2
Z	300	≤0.5	≤0.2
M_x	30	≤0.5	≤0.2
M_y	18	≤0.5	≤0.2
M_z	80	≤0.5	≤0.2

3.2 Whole structure of high-load dynamic balance

The developed dynamic balance based on the saddle-type dynamic balance structure is constructed of inner shaft and outer concentric tube, as shown in figure 6.

First of all, the inner shaft is machined strain-gauged beams for force measurement, this includes: the double variable cross-section vertical beam structure with the 1:10 conical surface section in front of the inner shaft which were used to measure five-component force and attach the model, the inner shaft and the outer tube attached at the double saddle-type beams of the middle part, which to make inner shaft become the 'force-floating' section of outer tube. A horizontal mullion structure placed at rear of inner shaft which can be combined with the eccentric cam in wind tunnel support sting to form a Scotch yoke mechanism.

The double variable cross-section vertical beam structure is regarded as strain-gauged section design. It overcomes the defects of sheet-beam structure which is good at anti-pitching torque but weak in normal stiffness, and solves the problem of column-beam structure which has large stiffness and high sensitivity but difficulty to match longitudinal and lateral loads. It is great in longitudinal load bearing capacity and lateral sensitivity output level for the new strain-gauged section.

Secondly, the outer tube is mainly used to achieve motion modulation and measurement. The front section of the outer tube attached with inner shaft is constructed of "oil film bearing+ elastic pivot" II-type beams structure. The front "I"-type structure is a fixed oil film bearing, through which the inner and outer shaft structure can be welded by electron beam, so that it can allow the rotation of inner shaft around the axis of oil film bearing. The rear "I" -type structure is a detachable elastic pivot, on which the strain gauges are attached. The elastic pivot can be replaced according to the aerodynamic trim torque. The rear of the outer tube is 1:10 conical surface section which is applied to connect the wind tunnel sting.

The role of the double-beam structure is similar to elastic bearing of existed dynamic balance [3,9], the oil film bearing adopted by the front "I"-type structure could enhance the carrying capacity and effectively reduce the mechanical damping. The rear "I"-type structure can output the pitch motion modulation signal in real time for data reducing. Besides, the double-beam structure acted as a connection structure reduces the distance to 35mm among the vibration center of developed dynamic balance, model center of mass and the calibration center. Compared with the existing dynamic balance [3,9], the developed dynamic balance effectively shortens the distance between the vibration center and the calibration center, greatly reduces the axial length of the size and improves the measurement accuracy.

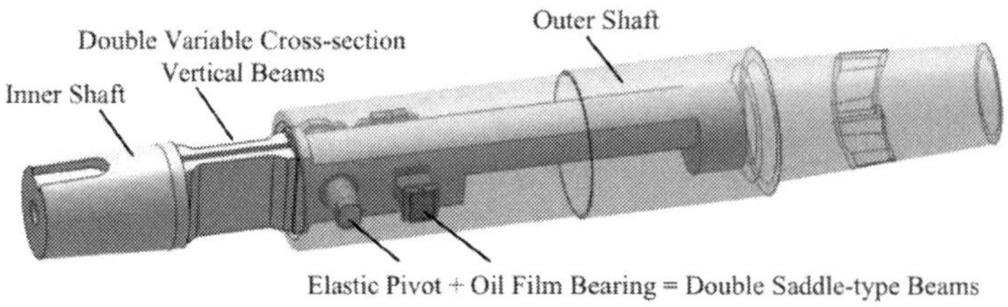

Figure 6. Structure of dynamic balance.

In summary, the saddle-type dynamic balance based on the double variable cross-section vertical beam / double saddle-type beams structure makes the high-load saddle-type dynamic balance compact and miniaturized. The high stiffness of the double-variable cross-section vertical beams structure not only can withstand the high load and model gravity, even can reduce the impact of noise, but also to ensure the lateral output magnitude. The double saddle-type beams structure which could solve the bearing problem of the motion modulation mechanism, but also effectively trim the aerodynamic torque and record motion modulated signal. The whole device owns higher measurement accuracy than before.

3.3 Miniaturization optimization design of the whole structure

The problems of high longitudinal load and extreme mismatch of longitudinal/lateral aerodynamic forces are solved by the double-variable cross-section vertical beam / double saddle-type beams structure. Due to the limitation of the wind tunnel test section and the inner space of aircraft test model, the entire dynamic balance needs to be constrained in the limited space.

The conical surface and the mullion structure of the inner shaft play the connection and motion transmission function. These parts are adopted by the high stiffness structure which will not affect the motion modulation and aerodynamic measurement function for its small deformation. The double variable cross-section vertical beam / double saddle-type beams structure are the main bearing and measuring structure, which should be optimized in necessary. In addition, the little influence of the high-stiffness structure leads to its omission in the optimization model. The optimization model should be preciseness and effective to avoid the redundant calculation.

In summary, The optimization design of the dynamic balance can be simplified as a double variable cross-section vertical beam / double saddle-type beams structure shown in figure 7.

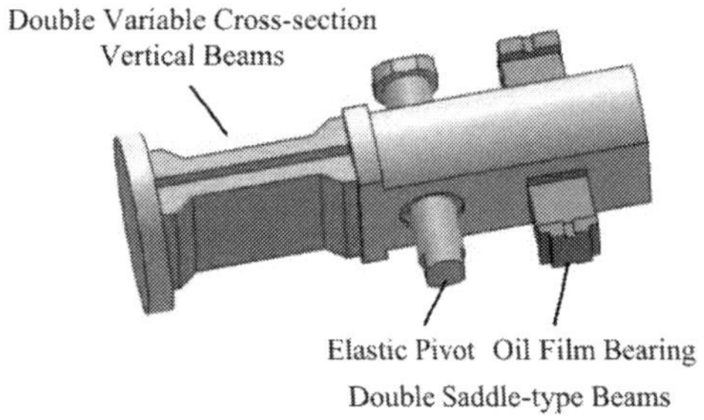

Figure 7. Double variable cross-section vertical beam / double saddle-type beams structure.

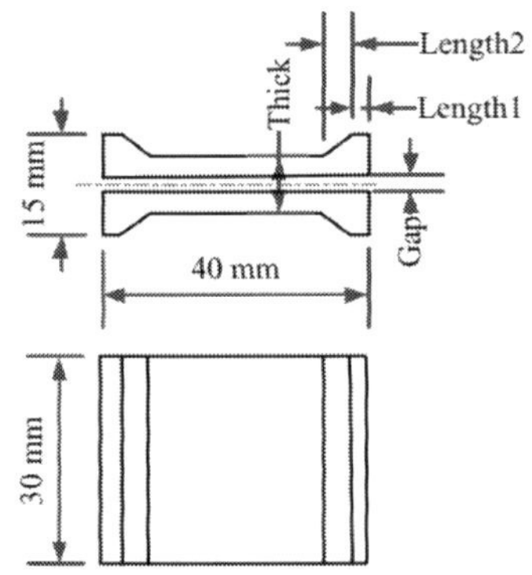

Figure 8. Dimension parameters of double variable cross-section vertical beam.

According to experiment requirements and design experience, the dynamic balance can be further optimized on the dimension of the cross section vertical beam as shown in figure 8.

Four typical results of ten outputs are chosen for processing after the five components load simulation, which include: the stress under the longitudinal force and torque applied alone, the strain under the lateral force and torque applied alone. The optimization target requires to lower stress under the longitudinal force and to increase the lateral load strain. Thus, the mathematical model of dynamic balance optimization is:

$$\begin{aligned} &\min \quad \mathrm{Y10000S}(\boldsymbol{x}), \mathrm{Mz600S}(\boldsymbol{x}) \\ &\max \quad \mathrm{Z800E}(\boldsymbol{x}), \mathrm{My50E}(\boldsymbol{x}) \\ &s.t.\begin{cases} 0.6 \le \mathrm{Gap} \le 3.0 \\ 1.0 \le \mathrm{Length2} \le 5.0 \\ 1.0 \le \mathrm{Length1} \le 5.0 \\ 8.0 \le \mathrm{Thick} \le 12.0 \end{cases} \end{aligned} \tag{10}$$

Where:

Y10000S (x) means the relationship between value x and stress under normal force Y = 10000N;
Mz600S (x) means the relationship between value x and stress under normal torque Mz = 600Nm;
Z800E(x) means the relationship between value x and strain under lateral force Z = 800N;
My500E (x) means the relationship between value x and strain under lateral torque My = 50Nm;
Value x could be written as a matrix: $\boldsymbol{x}$=[x(1) x(2) x(3) x(4)]=[Gap Length2 Length1 Thick].

The optimization method is based on the multi-objective optimization algorithm, and Design of Experiment (DOE) method is used to find the most suitable parameters as the initial values. Finally, the optimized Pareto set would be output after implementing an optimization algorithm for a plurality of sub-targets. The parameters of optimized dynamic balance also have been determined by considering the processing capacity. The length of the dynamic balance is 335mm, of which the outer diameter is 50mm and the inner diameter is 36mm. Among the parameters of double variable cross-section vertical beam, the parameter named Length1 is 3mm, Length2 is 4mm, the Thick is 9mm and Gap is 2mm.

4 Calibration results

The general static/dynamic calibration on dynamic balance could adopt two kinds of methods: the ground test and the numerical simulation test. The purpose of the static calibration is similar to the conventional static balance calibration, while dynamic calibration aims to detect the frequency modulation and amplitude modulation capability of the motion modulation mechanism as well as the dynamic response characteristics of the input / output of the dynamic balance system. In this paper, numerical method is used to calibrate the developed dynamic balance.

4.1 Finite element model of whole structure

The appropriate simplification has been done on the whole structure to facilitate the meshing and numerical calculation by omitting keyway, wedge hole and other irrelevant parts. The global meshing uses a quadratic tetrahedron mesh. The local grid refinement method is applied on the strain-gauged section and the double-variable cross-section vertical beam / double saddle-type beams structure. The total number of grid cells is 422167 and the total number of nodes is 641882, as shown in figure 9.

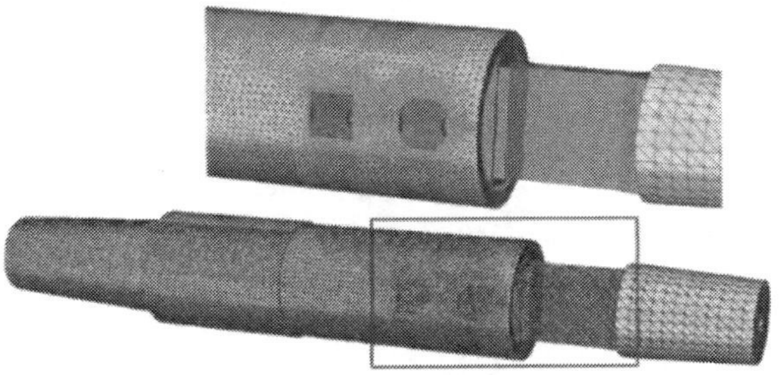

Figure 9. Finite element model of whole structure.

 DOI: 10.1051/matecconf/201711401001

F141 (00Ni18Mo5Co9TiAl) is chosen as the material of dynamic balance, whose Young modulus is 1870MPa and the yield limit is 1754MPa. The safety factor is taken as $n = 1.5$ so that the available stress [σ] is equal to 1170MPa. The details about the F141 are shown in table 2.

Table 2. Mechanical properties of material (20°C).

Material	ρ (kgm^{-3})	E (Nm^2)	μ	σ_s (MPa)	σ_b (MPa)
F141	7800	1.87e11	0.29	1754	1862

The fixed-boundary condition is adopted on the taper section of outer tube of the dynamic balance, and the binding constraints are made between the remaining parts. Load condition is considered as two situations: static calibration and dynamic calibration.

4.2 The static calibration

The whole finite element model is established as a linear model which is in accordance with the test requirements. In static analysis, linear change in the static load will produce a linear change in the results during the static analysis. The load case on the dynamic balance would only be considered in a single direction. The single-component loading results are shown in table 3.

Table 3. Single-component finite element simulation results on dynamic balance.

Load Component	Value (N,Nm)	σ_{max} (Pa)	ε_{max}	U (m)
Y	10000	1.944e8	1.057e-3	1.219e-4
Z	300	2.631e7	1.479e-4	2.699e-5
M_x	30	1.505e8	8.270e-4	1.358e-4
M_y	18	4.653e7	2.585e-4	1.551e-4
M_z	80	8.817e7	4.752e-4	1.094e-4

The strain-gauged section of dynamic balance could meet the design requirements with a bit lower about lateral output. However, the distance between the aerodynamic rotation center and the geometric center is equivalent to applying an extra moment on the dynamic balance after transferring the load from geometric center to rotation center and resulting to enhance the effect of the lateral load.

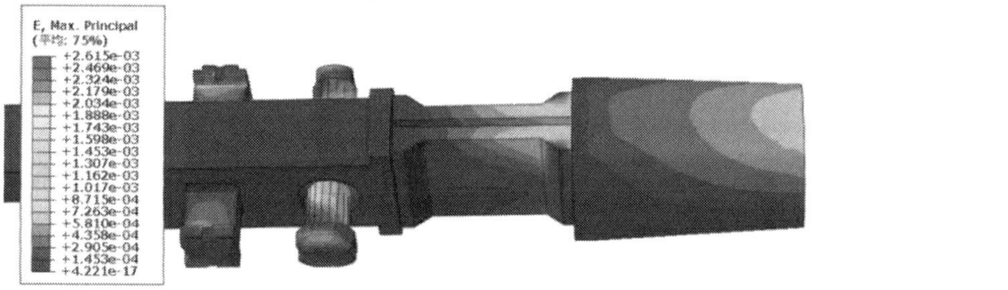

Figure10. Local strain cloud diagram of dynamic balance at Ma = 0.8 and α = 12 °.

The local strain cloud diagram in figure 10 visually shows that the root on strain-gauged section is the location where the stress concentration occurs, and the stress gradient is large with the strain increases sharply. It is necessary to ensure the stress concentration region is within allowable stress [σ]. In addition, the elastic pivot fixed together on the outer tube of dynamic balance, whose junction part is also a weak point, but the strain value and gradient are smaller to the root on strain-gauged section.

The stress and strain distributions under the wind tunnel load are plotted in figure 11 by selecting a characteristic edge on the top surface of the strain-gauged section. The effect of wind tunnel load is more prominent than the design load to meet the strain signal output requirements. Besides, the stress gradient on both sides of the curve increases rapidly, which indicates the stress concentration range at the root of the strain-gauged section and becomes more prominent at the high angle of attack.

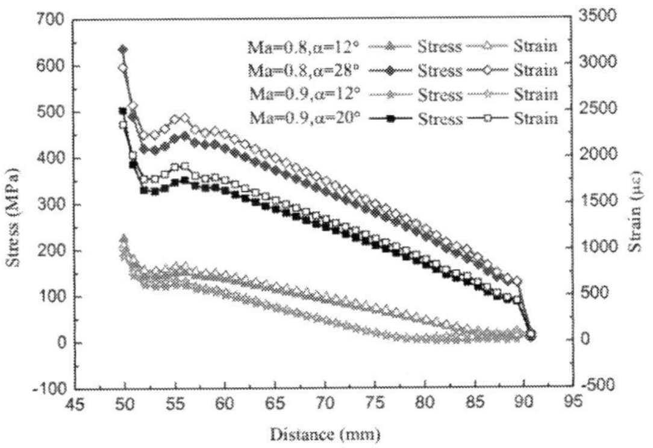

Figure 11. Stress/strain distribution on the upper surface of the strain-gauged section.

The main reason for the high σ_{max} is the stress concentration at the root of strain-gauged section and the strain gauge area can guarantee the effective output without any damage. The high σ_{max} exceeded the allowable stress [σ] at Ma = 0.8 and α = 28 ° does not reach to the yield limit of the material. Moreover, the impact load to the dynamic balance is small which will not beyond the material limitation from α = 0 °to α = 28 °after the wind tunnel uniform flow field has been established.

4.3 The dynamic calibration

The dynamic analysis of the dynamic balance contains many aspects: the modal analysis, harmonic response analysis, spectrum analysis and other dynamic analysis.

Finite element modal analysis shows the natural frequency of dynamic balance is very high which would not reach the mechanical system resonance to cause the structural damage.

The displacement function $x = 2 \times 10^{-3} \sin(10\pi t)$ is imported into the dynamic balance to simulate the forced vibration under finite element harmonic response analysis. The displacement-time curves shown in figure 12 are drawn from the points on the strain-gauged section and the oil film bearing.

A nice agreement is obtained between the output and given-frequency of the dynamic balance structure system. The response frequency is approximately 5 Hz which is close to the input frequency. The structure can effectively reproduce the forced vibration of ± 1° by the response results. A few attenuated occurs at the angular amplitude during the transmission while the effect is very small so that the forced vibration response can receive high precision. The voltage signal output of strain gauges on the oil film bearing could reach 500mV/V which can make strain gauge work in normal. The voltage signal output of stain gauges on the strain-gauged section is very low due to lack of aerodynamic load. Actually, the output signal on the strain-gauged section can meet the working requirement under wind tunnel loads, which has been verified in the static calibration.

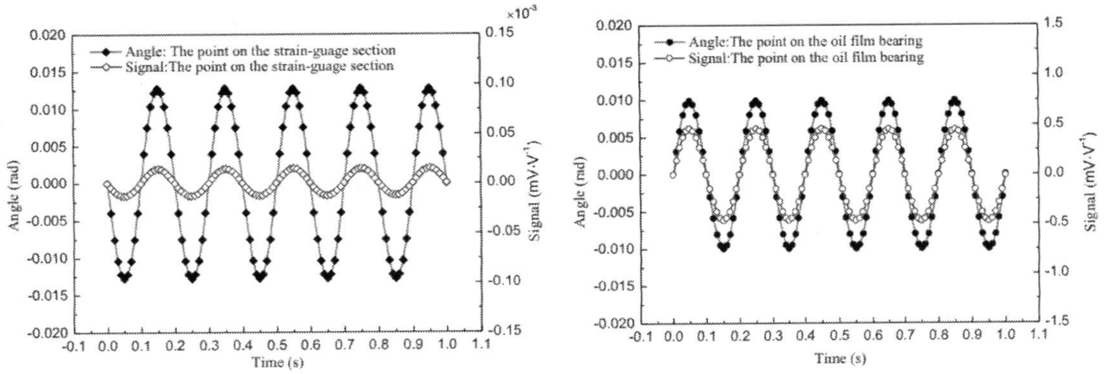

Figure 12. Displacement and output signals over time (Left: strain-gauged section, Right: oil film bearing).

Therefore, the dynamic balance of static/dynamic calibration results has reached the design requirements. It could be considered that the dynamic balance would work normally to meet the security requirements and to obtain the high test accuracy.

5 Conclusions

The wind tunnel dynamic test measurement problems brought from the dynamic characteristics of unconventional configuration aircraft result to the developed dynamic balance integrated with the double-variable cross-section vertical beam / double saddle-type beams structure. The working principle, mechanical structure principle and whole structural design are expounded to the dynamic balance, which is regarded as a special kind of electromechanical motion modulation sensor. The miniaturization about the optimization design has been done based on the whole structural characteristics of the finite element model analysis, which reduces the radial dimension of the dynamic balance effectively. A complete design method could be established from the structure, optimization and finite element analysis.

According to the numerical simulation about the calibration on static/dynamic characterizes of the dynamic balance, the bearing capacity of two key substructures: motion modulation and aerodynamic response measurement are enhanced, whose maximum longitudinal load bearing capacity is extended more than twice. The sensitivity of lateral load is improved and a great agreement could be obtained between the measurement accuracy and design requirements. Such dynamic balance could be widely applied to high-load wind tunnel test, which owns good practicability and popularization value.

References

1. K.J. Orlikruckemann, NASA-CR-114583 (1973)
2. D.M.Tomek, W.G.Sewall, S.E.Mason, AIAA 2006-3148 (2006)
3. J Liu. Master Degree Thesis (2014)
4. J.M. Luckring, O.J. Boelens, AIAA 2015-0061 (2005)
5. C.J. Schueler, L.K. Ward, Hodapp. Jr. A.E, AD-669227 (1968)
6. C Wang, Acta Aerodynamic Sinica 5(2) (1987)
7. D He, X Gu, First International Symposium on Strain Gauge Balances (1999)
8. Makoto Ueno, Hitoshi Miwa, AIAA 2001-0406 (2001)
9. J Lin, N Chen, Y Song, J Hu, K Xie, APISAT-2014 (2014)
10. R.A. Kilgore, B.T. Averett, Journal of Aircraft 1(5) (1964)
11. K.J. Orlikruckemann, J.G. Laberge, E.S. Hanff, AIAA 74-611 (1974)
12. J.C. Uselton, B.L. Uselton, AIAA 75-211 (1975)
13. S.M. Coulter, T.D. Buchanan, AIAA 80-0451 (1980)
14. D.R. Haberman, AIAA 82-0595 (1982)
15. M. Samardzic, Z. Anastasijevic , D. Marinkovski, Proc. IMechE Part G: J. Aerospace Engineering, 226(11) (2011)
16. Y Li, H Liu, Y Huang, Acta Aerodynamic Sinica 34(1) (2016)
17. D.D. Vicroy, T.D. Loeser, Schuette Andreas, AIAA 2010-4394 (2010)

Improved MPSP Method-based Cooperative Re-entry Guidance for Hypersonic Gliding Vehicles

Haiyan Chu[1], Jian Li , Yi Dong, Xiaoqian Yang and Ye Tao

College of Aeronautic Science and Engineering, National University of Defense Technology, 410073, Changsha, P.R. China

Abstract: A computationally sufficient technique is used to solve the 3-D cooperative re-entry guidance problem for hypersonic gliding vehicles. Due to the poor surrounding adaptive ability of the traditional cooperative guidance methods, a novel methodology, named as model predictive static programming (MPSP), is used to solve a class of finite-horizon optimal control problems with hard terminal constraints. The main feature of this guidance law is that it is capable of hitting the target with high accuracy for each one of the cooperative vehicles at the same time. In addition, it accurately satisfies variable constraints. Performance of the proposed MPSP-based guidance is demonstrated in 3-D nonlinear dynamics scenario. The numerical simulation results show that the proposed cooperative re-entry guidance methodology has the advantage of computational efficiency and better robustness against the perturbations.

1 Introduction

Hypersonic vehicles have been drawing a lot of attention in recent decades. In order to improve the survivability and fighting strength, it is significant to consider the multiple vehicles as a cooperative group and design an optimal trajectory for each of them. It can be foreseen that cooperative hypersonic vehicles will replace single ones for more complicated flight missions. What's more, the core task of the multi-vehicle cooperative guidance is to avoid the threat and finally reach the target to complete the task. It can be achieved in two ways. One is to make the vehicle select more curved trajectory to increase flight time, waiting for the later arrival vehicle. Another is to require the later arrival vehicle to accelerate, and thus catch up. But for the hypersonic gliding vehicles, the re-entry process has no power, so it is difficult to achieve the normal acceleration process. Therefore, the first way is desirable.

Vehicles cooperative operations can complete many tasks that a single vehicle cannot complete, such as tactical stealth, the ability of electronic countermeasures and so on [1]. Vehicles not only need to achieve their precise guidance, but also need to cooperate with other vehicles to complete a variety of operational tasks, so the overall effectiveness of one's own operations can be maximized.

Many researchers have done a lot of works on the cooperative guidance. A flight controllable guidance law, Impact-Time-Control Guidance (ITCG) is given in [2]. And in [3], a two-level cooperative guidance architecture for multi-missile attack is proposed. The architecture is composed

[1] Corresponding author: xiaoyanzi905@qq.com

with the guidance law for each missile on the lower level, and the upper level guidance law used the coordination strategy. Zhao Shiyu proposed a cooperative guidance scheme based on distributed coordination [4].

In this paper, a nonlinear suboptimal guidance scheme is developed for the re-entry phase of the reusable launch vehicles here. It is based on a quite straightforward methodology, named as MPSP [5-6], which combines the philosophies of nonlinear model predictive control (MPC) [7] and approximate dynamic programming (ADP) [8]. The method provides a finite time nonlinear suboptimal guidance law which leads to a rapid solution of the update. To the best of authors' knowledge, this method have not been applied to the cooperative guidance for hypersonic gliding vehicles.

The rest of this paper is organized as follows. The problem formulation and preliminaries are presented in the next section. The improved MPSP theory used in the cooperative guidance is given in Sec 3. The simulation results and the analysis are then presented in Sec 4. Finally, the appropriate conclusions are drawn in Sec 5.

2 Mathematic model for hypersonic gliding vehicles

2.1 System dynamics model description of gliding vehicles

The nonlinear equations assuming a spherical, rotating Earth can be described as:

$$\begin{aligned}
&\dot{r} = V\sin\gamma \\
&\dot{\theta} = -\frac{V\cos\gamma\sin\psi}{r\cos\phi} \\
&\dot{\phi} = \frac{V\cos\gamma\cos\psi}{r} \\
&\dot{V} = -D - g\sin\gamma + \omega^2 r\cos\phi(\sin\gamma\cos\phi - \cos\gamma\sin\phi\cos\psi) \\
&\dot{\gamma} = \frac{1}{V}[\mathrm{L}\cos\sigma + (\frac{V^2}{r} - g)\cos\gamma] + 2\omega\cos\phi\sin\psi + \frac{\omega^2 r}{V}\cos\phi(\cos\gamma\cos\phi - \sin\gamma\sin\phi\cos\psi) \\
&\dot{\psi} = -\frac{\mathrm{L}\sin\sigma}{V\cos\gamma} + \frac{V}{r}\cos\gamma\sin\psi\tan\phi + 2\omega(\sin\phi - \tan\gamma\cos\phi\cos\psi) + \frac{\omega^2 r}{V}\cos\gamma\cos\psi\sin\phi\cos\phi
\end{aligned} \tag{1}$$

2.2 Standard predictive equations

Consider that the cooperative strategy is deceleration and waiting. The terminal speed is unconstraint in the process of cooperative control. The state variables are $X = [\mathrm{r}, \theta, \phi, V, \gamma, \psi]^T$, and the control variables are $U = [\alpha, \sigma]^T$. The nonlinear function can be described as $\dot{X} = F(\mathrm{X}, \mathrm{U})$ and discretization handled by Euler method as the discretization systems model. The state and output are given by

$$X_{k+1} = F_k(\mathrm{X}_k + \mathrm{U}_k) \qquad \mathrm{Y}_k = X_k X_{k+1} = F_k(\mathrm{X}_k + \mathrm{U}_k) \tag{2}$$

3 Improved MPSP design: mathematical details

3.1 Solution of the sensitivity matrix

The primary principle of MPSP guidance is to update the control variable and obtain a suitable control history U_k, $k=1,2,\cdots,N-1$, so that the output at the final time step Y_N goes to a desired value Y_N^d, i.e. $Y_N \to Y_N^d$. Expanding Y_N about Y_N^d using Taylor series expansion and then we can write the error in the output as

$$\Delta Y_N \cong dY_N = \left[\frac{\partial Y_N}{\partial X_N}\right] dX_N \tag{3}$$

From (1), the error in state at time step (k+1) can be expressed as

$$dX_{k+1} = \left[\frac{\partial F_k}{\partial X_k}\right] dX_k + \left[\frac{\partial F_k}{\partial U_k}\right] dU_k \tag{4}$$

where dX_k is the error of state at time step (k+1), and dU_k is the error of state at time step (k+1). According to the function (5), we can write the error in output at time step (N-1) as

$$dY_N = \left[\frac{\partial Y_N}{\partial X_N}\right]\left(\left[\frac{\partial F_{N-1}}{\partial X_{N-1}}\right] dX_{N-1} + \left[\frac{\partial F_{N-1}}{\partial U_{N-1}}\right] dU_{N-1}\right) \tag{5}$$

And since the initial condition is specified, there is no error in state at the first step. It means

$$dY_N = B_1 dU_1 + B_2 dU_2 + \cdots + B_{N-1} dU_{N-1} = \sum_{k=1}^{N-1} B_k dU_k \tag{6}$$

Note that it is quite difficult to evaluate each of the B_k, $k=1,2,\cdots,N-1$. Especially when N is high, it will be the main impediment to computational efficiency.

3.2 Update the control

If the terminal error cannot meet the requirement, the optimal control theory will be used in the MPSP method to correct the control variable. For equation (6), each control variable is independent from others. And the cost function is described as

$$J = \frac{1}{2}\sum_{k=1}^{N-1} (dU_k)^T R_k (dU_k) \tag{7}$$

After deducing, we can get

$$dU_k = -R_k^{-1} B_k^T A_\lambda^{-1} dY_N \tag{8}$$

In each guidance cycle, the control at the k th time step is updated to $U_k = U_k^0 - dU_k$, where U_k^0 is the control history of the previous update

4 Numerical results and Discussions

4.1 Time and accuracy

To start the MPSP algorithm, the guess value of the guidance command history makes full use of prior knowledge and improves the accuracy of prediction-correction. The trajectory along with lower and upper bounds of constraints can be seen in Figure 1. All the path constraints and terminal constraints are satisfied.

The guidance cycle (for which the guidance command is updated) is assumed to be 1.8 s. The total time taken by the Vehicle A and Vehicle B to reach the desired end point conditions are around 2830s and 2780s respectively. This means that vehicle B takes off 50s more lately than the vehicle A.

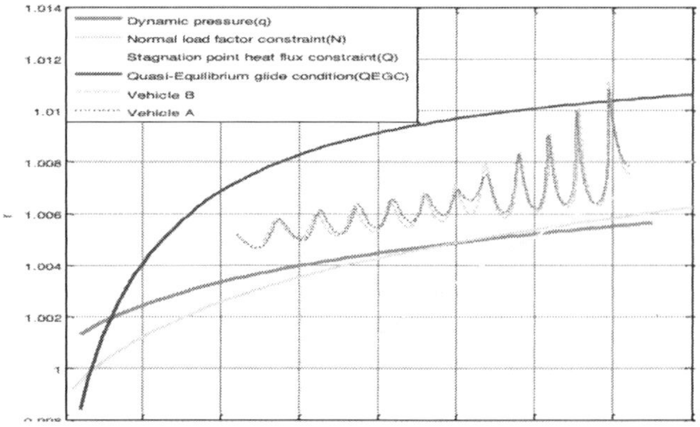

Figure 1. Velocity vs. geocentric distance

The improved MPSP prediction and correction algorithm requires only 7 iterations per iteration cycle for Vehicle B and only 3 iterations per iteration cycle for Vehicle A. It does not consume the computational time. Thus, the MPSP method-based cooperative re-entry guidance has the possibility of online implementation.

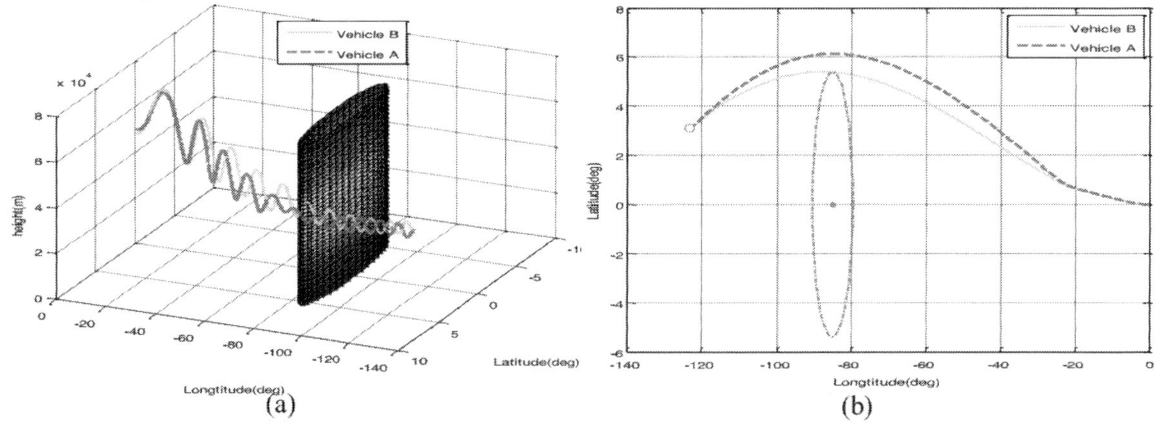

Figure 2. (a) 3-D figure of reentry trajectory (b) Flat trajectory

The no-fly zones can be described as the cylinder. Figure 2(a) and 2(b) shows that the 3-D trajectory and the flat trajectory of MPSP method-based cooperative re-entry guidance for hypersonic gliding vehicles respectively. In order to arrive at the target at the same time, Vehicle A chose to glide in a more curved trajectory. Furthermore, both Vehicle A and Vehicle B avoid the no-fly zone successfully.

4.2 The analysis of MPSP method-based cooperative re-entry guidance process

Figure 3 and 4 show the latitude and longitude, the heading angle and flight path angle. Without the MPSP method-based cooperative guidance, the trajectories of the two vehicles are very close. This means that they cannot hit the same target.

 DOI: 10.1051/matecconf/201711401002

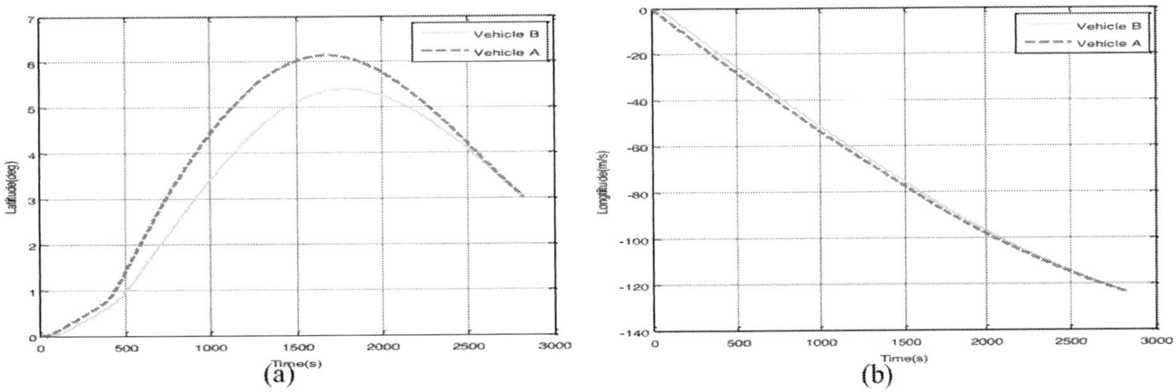

Figure 3. (a) Latitude of the cooperative vehicles (b) Longitude of the cooperative vehicles

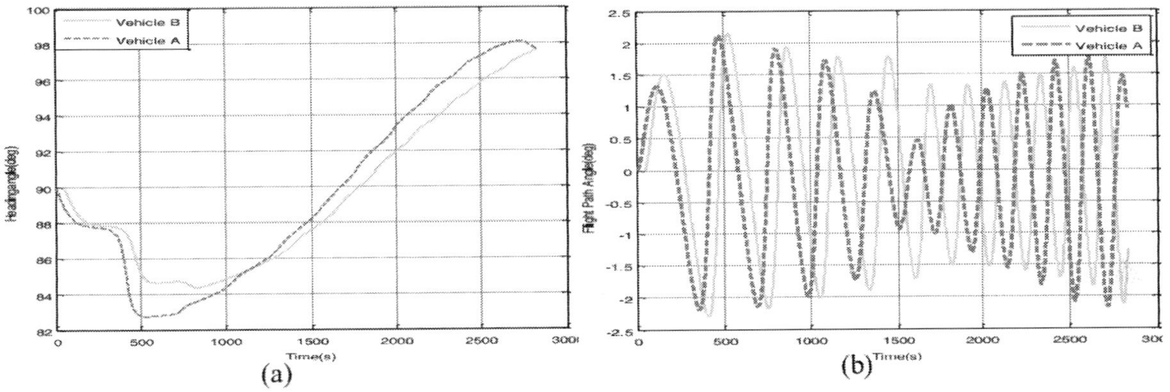

Figure 4. (a) Heading angle (b) Flight path angle

4.3 Aerodynamic coefficient perturbation

Considering the aerodynamic coefficient and the perturbation effect, take both the deviation of 20%, and get the simulation results as shown below.

Table 1. The Terminal states of different case

	Vehicle	C_L C_D	$L\,/\,km$	E_Y	$r\,/\,km$	$dL\,/\,km$
Case 1	*Vehicle A*	0	*13771.7154814*	*0.0000258*	*6411.4*	*0.4223065*
	Vehicle B			*0.0001689*	*6408.96*	*2.9496947*
Case 2	*Vehicle A*	20%	*13771.7154814*	*0.0001547*	*6411.5*	*2.2496153*
	Vehicle B			*0.0001085*	*6409.9*	*1.4978738*

The entry guidance algorithm is proved to be successful. In table 1, the aerodynamic perturbation is simulated with 20% aerodynamic perturbation in Case 2. As expected, it is seen that the MPSP method-based cooperative re-entry guidance can enhance the flexibility of the cooperative guidance law. Obviously, the 3-D cooperative guidance law has better robustness and improves the guidance accuracy of the hypersonic gliding vehicles.

5 Conclusions

There were few researches in developing the algorithms and methods for hypersonic gliding vehicles in the cooperative re-entry guidance. In this paper, a computationally sufficient MPSP technique has been used to solve the 3-D cooperative re-entry guidance problem for hypersonic gliding vehicles. The direct adaptive technique undermines the need to update the control quickly and uses for the cooperative re-entry vehicles online. Numerical simulation results validate the high performance about computational efficiency and accuracy and also show great robustness against perturbations in the aerodynamic conditions.

Acknowledgments

Authors would like to express their sincere thanks to the reviewers and the associate editor of this paper for their many constructive comments and suggestions of both technical and editorial nature. They have really helped improve the preciseness and readability of the paper.

References

1. McLain T W, Beard R W. Coordination variables, coordination functions, and cooperative timing missions[J]. Journal of Guidance, Control, and Dynamics, 2005, 28(1) : 150-161.
2. Jeon I S, Lee J I, Tahk M J. Impact-time-control guidance law for anti-ship vehicles[J]. IEEE Transactions on Control Systems Technology, 2006, 14(2) : 260 — 266
3. Zhao Shiyu. Multi-vehicle cooperative guidance law study[D]. Beijing: Beijing University of Aeronautics and Astronautics, 2009(in Chinese)
4. Zhao Shiyu, Zhou Rui. Cooperative guidance for multivehicle salvo attack[J]. Chinese Journal of Aeronautics, 2008, 21 (6) :533 — 539
5. Kothari, M. and Padhi, R., "A Hybrid Energy Insensitive Explicit Guidance Scheme for Long Range Flight Vehicle Using Solid Motors," 17th IFAC Symposium on Automatic Control in Aerosapce, Toulouse, France, June 25-29, 2007.
6. Padhi, R., "Model Predictive Static Programming: A Promising Technique for Optimal Vehicle Guidance ," To appear in Annals of Indian National Academy of Engineering (INAE).
7. Rossiter, J. A., " Model based predictive control: A Practical Approach," CRC, New York, 2003.
8. Werbos, P. J.,"Approximate Dynamic Programming for Real-time Control and Neural Modeling," In White D. A., and Sofge D.A (Eds.), Handbook of Intelligent Control, Multiscience Press, 1992.

MATEC Web of Conferences **114**, 01003 (2017) DOI: 10.1051/matecconf/201711401003
2MAE 2017

Design of Experimental Platform for the Static Stiffness of Guide Roller

Yihua Li[1,a], Jianjian Xue[1],Xin Zhang[1],Jianwei Song[1],Huawei Jing[1],Anhang Zhu and Lei Lin[1]
[1]*School of Mechanical Engineering,Anhui University of Science and Technology,Huainan Anhui 232001,china*

Abstract. In mine shaft hoisting system, The hoisting container generally adopts the guide roller which is used as the movement guiding device. Stiffness parameters and characteristic curves are very important technical indexes of the guide roller. For guide roller series, Proposing The method of stiffness performance test, Designing the experimental platform for the stiffness of the guide roller, The main functions of the experimental platform and the structure of the main units are analyzed. Results show that equipment can meet the stiffness test of the guide roller series, so as to determine the optimum stiffness value of the guide roller. Achieving the purpose of improving the life of guide roller.

1 Introduction

The guide roller is installed in shaft lifting container along the guide device running under stable tank on the road and it is one of the main components of the lifting container.On the one hand, it is used as a guide wheel running along the tank, and on the other hand, it can be connected with the lifting container and the tank, and transfer the horizontal force between them. The roller can also bear the lateral impact and vibration caused by the deviation of the shaft, the lack of the straight channel, the step of the joint and the vibration of the wire rope.Therefore, the guide roller is an important device which directly affects the safety and efficiency of the lifting container, and is interaction of media between shaft lifting container and shaft equipment, The stiffness parameter has a very important effect on the quality of the rigid equipment[1].

2 Structure and stiffness of elastic guide roller

The geometrical structure of guide roller is closely related to its mechanical properties, which directly affects the load distribution and stiffness of wheel ear. Therefore, the guide roller for the mechanical properties of the study, first of all need to master its structural features.

2.1 Elastic guide roller

The high speed shaft system generally adopts elastic guide roller, It has good cushioning and damping performance, and rigid collision between container and tank can be avoided. Elastic guide roller only

[a] Yihua Li,JiangxiNanchang,associate professor.email:825857694@qq.com

multiplies to prolong the service life of the container and tank, but also greatly improved the state of shaft hoist. So that the container running smoothly and no noise.For large scale hoisting containers for high speed operation, especially for high quality guide roller.The development of elastic roller ear is closely related to the use and development of rectangular cross section[2].At present, commonly used elastic guide roller are mainly L type guide roller, As shown in Figure 1.

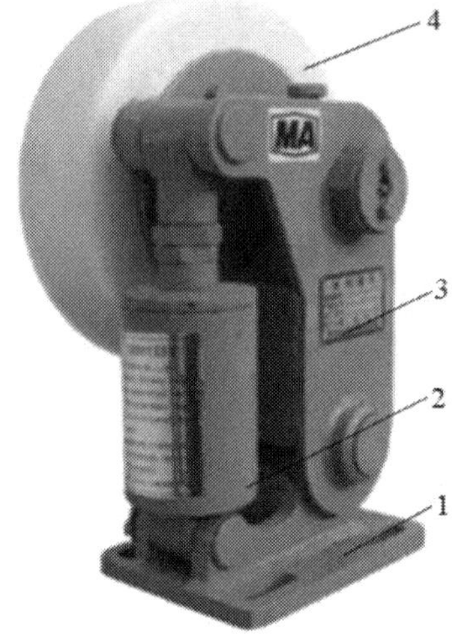

1-base 2-buffer 3-Bracket 4-roller

Figure 1. The structure of guide roller

2.2 Stiffness performance of guide roller

The guide roller is mainly composed of four parts: the base, the buffer, the bracket and the roller. The roller is made of polyurethane material, the disk spring element is adopted in the buffer, most of the impact force can be transmitted to the buffer device when the tank ear is subjected to lateral impact, and the vibration time delay will be transmitted to the damping device[3].

Mine shaft lifting container system generally adopts the guide roller as motion guiding device, its stiffness and stiffness characteristics is a very important index of the guide roller. If it is just too large, the ability of absorbing vibration is weakened, The vibration of the lifting container is intensified in the process of running along the tank is guided by the guide roller, lifting containers, tanks and roller cage body is subject to dynamic load increases, the probability of early failure or damage is increased, If the stiffness is too small the deformation of the roller ear is too large, the probability of contact between the container body and the tank is increased so as to increase the probability of failure or damage of the container and the tank ahead of time, lift system security is reduced[4].

In order to increase the service life of the guide roller and improve its stiffness characteristics , it is necessary to carry out the stiffness test of the guide roller, and get the stiffness parameter and the characteristic curve. Therefore, the design of the experiment platform of the guide roller stiffness is imminent, and the reasonable structure design and the selection of the components are the key to the design of the experimental platform[5].

3 Structural design of experimental platform for the stiffness of the guide roller

3.1 The design hydraulic system

Hydraulic loading system includes oil cylinder, manual pump, pipe joint hydraulic accessories, etc.The hydraulic loading system provides a key driving force for the design of the experimental platform.When the handle of the upper and the lower manual pumps makes the oil cylinder enter the oil, the pressure of the oil inlet cavity of the oil cylinder is larger, and the guide rod moves back and forth,the pressure head of the guide rod tightly presses the roller, the handle of the manual pump is continuously pressed, and the pressure of the roller is also increasing.Hydraulic power system provides different pressure, it is the core of the whole experimental platform[6].

Hydraulic cylinder is a hydraulic actuator can be converted into mechanical energy and reciprocating linear motion of the implementation of components.It has the advantages of simple structure, reliable operation, stable movement, flexible arrangement and high efficiency,and is widely used in various hydraulic systems.In the design of the experimental platform, the selection of the oil cylinder is very important.The basic parameters of the guide rollers are shown in Table 1.

Table 1. The basic parameters of the guide roller

Model	roller diameter /mm	Max horizontal force /KN	Center height /mm	Cushioning device stiffness KN/mm	Buffer trip/mm
L20	200	8	240	0.29~0.32	15~18
L25	250	12	260	0.43~0.48	15~18
L30	300	16	320	0.57~0.64	15~18
L35	350	24	320	0.86~0.96	15~18
L42.5	420	32	320	1.0~1.12	15~18

L represents the type, and the following number represents the diameter of the roller.Table 1 shows that the maximum horizontal forces of different guide rollers are different.Therefore, the maximum output power of the designed cylinder is more than 32KN, and the working stroke of the buffer device is more than 18mm.The main parameters of the cylinder are calculated as shown in Table 2.

One way manual oil pump is a kind of small hydraulic pump station, which can convert manual mechanical energy into hydraulic energy.The manual oil pump is composed of six parts: the pump body, the handle part, the hose part, the storage tank valve and the pressure gauge. This experiment platform uses the SYB-2 series manual oil pump.

Table 2. The main parameters of the cylinder

Pressure /MPa	Internal diameter /mm	Piston rod diameter/mm	wall thickness /mm	Cylinder bottom thickness/mm	Working stroke/mm
3.5	100	50	2.18	12.49	40

3.2 The design of high modulation system

In order to adjust the cylinder head to the center position of the guide roller, this experiment table uses the worm gear screw rod elevator, the worm gear screw lifter has the advantages of large transmission ratio, compact structure, stable operation, low noise and can not be used for other auxiliary mechanisms that can be obtained through the transfer of anti stroke self-locking. As can be seen from table 1, the center of the cylinder is different from each other, and the worm gear is used to transfer the oil cylinder to the right position in order to apply pressure to the cylinder ear.It not only improves the accuracy of the experimental platform, but also improves the efficiency of the experiment. The center height of the guide roller is shown in Figure 2.

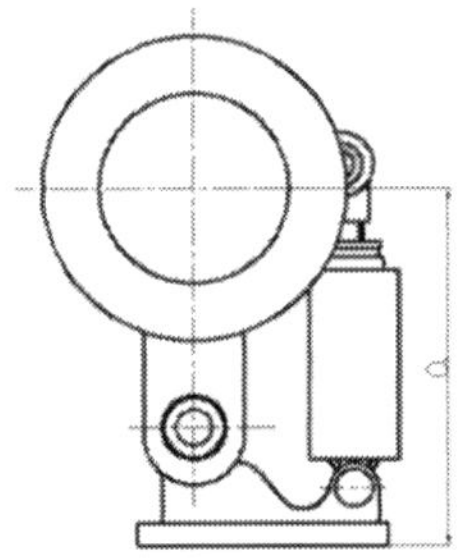

Figure 2. The center height of the guide roller

In order to achieve the cylinder head can move up and down the oil cylinder to be fixed on the flange, flange and linear guide rail are connected by bolt.By adjusting the worm wheel screw rod by hand wheel,by adjusting the worm gear screw rod elevator by hand wheel,the worm wheel elevator controls the pressure head to move up and down in the guide rail to ensure that the pressure head is aligned with the center of the guide roller,the manual pump is used to pressurize the oil cylinder to push the pressure head to exert pressure on the guide roller.

3.3 The design of measurement system

The measuring system comprises a digital display meter, a pressure sensor and a displacement sensor, A pressure sensor is a sensor that converts pressure into electrical signal output. The pressure sensor is generally composed of elastic sensitive element and displacement sensing element. The function of the elastic sensing element is to make the measured pressure acting on a certain area and convert it into a displacement or strain, and then convert the displacement sensitive element or strain gauge into an electrical signal which is related to the pressure. The pressure sensor is made of stainless steel components, imported elastomer original, with high sensitivity, stable performance, good impact resistance.

Displacement sensor, also known as linear sensor, is a kind of linear device belonging to metal induction, the function of the sensor is to convert the measured physical quantities into electricity. The experimental platform uses BWL pull rod displacement sensor. BWL pull rod displacement sensor effective stroke is 100mm, both ends have 3.5mm buffer stroke, the accuracy of 0.08% ~ 0.04%FS.BWL displacement sensor using the world's leading nano chemical conductive materials and precious metal wear-resistant new materials, It features high precision, high linearity, long life. The pressure sensor is arranged at the end of the oil cylinder, and the pressure of the pressure head can be displayed when the oil cylinder is put into the oil. The BWL displacement sensor is installed on the flange of the oil cylinder, and the pull rod ball head of the displacement sensor is consistent with the position of the head of the oil cylinder. The measured displacement precision is high, and the numerical value is displayed on the displacement digital display table. Electrical control system does not solve the pressure and displacement measurement, but also improves the accuracy and efficiency of the entire experimental platform.

4 Composition and structure of experiment platform

4.1 Composition of the experimental platform

The experiment platform mainly consists of the frame body, the mounting seat, the hydraulic loading system, the height adjusting system, the testing system and so on. As shown in Figure 3.The structure of the experimental platform is consistent with the experiment of the stiffness of the guide roller. It not only has the reasonable structure design, but also has the optimal structure arrangement of the experimental platform, which saves a lot of materials and reduces the cost. Four wheels with brakes

 DOI: 10.1051/matecconf/201711401003

are installed on the base of the experimental platform, so that the handling of the experimental platform[7].

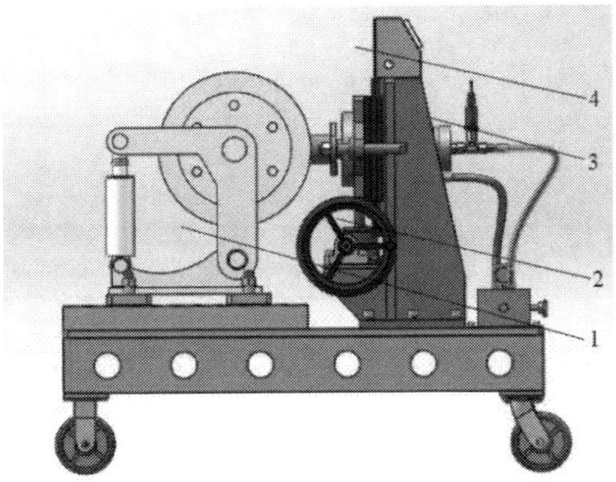

1-Roller cage ear 2-Handwheel 3-Pressure transducer 4-Digital display meter

Positive view

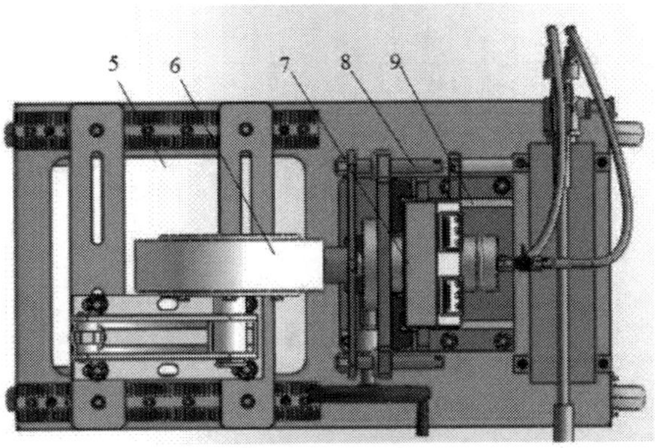

5-Mounting seat 6-Pressure head 7-Hydraulic cylinder
8-Displacement transducer 9-Manual pump

Top view

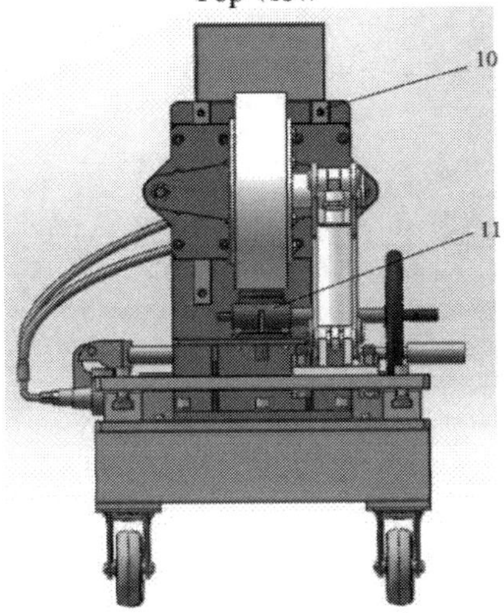

10-Linear guide rail 11-Worm screw

Figure 3. Composition and structure of experiment platform

4.2 Composition of the experimental platform

The main technical parameters are shown in Table 3.

Table 3: The main parameters of the cylinder

The size of the experimental bench is long, wide and high /mm	1100X650X750
T type groove track adjustable range/mm	560
Hydraulic cylinder stroke /mm	80
Maximum thrust of hydraulic cylinder/KN	100
Worm screw rod height adjustment range/mm	400
Manual pump rated pressure /MP_a	0-60
Hand pump the most big shake force/N	450
Adjustable range of U type groove/mm	200

Guide roller test bench is pushed to the right position,through the heavy fixed belt brake with brake casters and heavy duty casters will experiment fixed guide roller.The selection of the guide roller is hoisted on the mounting seat on the machine frame, and the position of the mounting seat is adjusted, and the mounting seat is fixed on the T groove rail by screws.There is a tooth surface on the T type slot rail, which increases the stability of the guide roller.Adjust the position of the guide roller and fix it with the bolt on the mounting seat.The mounting seat is provided with a waist shaped hole, which is convenient for the movement of the position of the guide roller.When the preparatory work is completed,horizontal force is applied to the guide roller .

Acknowledgements

This research is supported by the key project of natural science foundation for Anhui University (No.KJ2015ZD019).

References

1. XUE Jian-jian,LI Yi-hua,WANG Jing-wei.Shaft lifting application and development direction of the container roller cage ear[J].NEI JIANG KE JI,2016(8)39-40.
2. LIU Chun-feng,HAN Yan-wei.Structure analysis and development of roll guide shoes as vertical well hoisting installation[J].Coal Mine Machinery,2003(9):84-85.
3. FUN Ming-yuan,SUN Han-jing.Polyurethane elastomer and its application,Third Edition[M].BEIJING:Chemical Industry Press,1994.
4. FAN YING,Mine hoisting machinery design[M].China University of Mining and Technology
5. press,2000,12.
6. QI Xiao-nan,QIN Qiang,ZHU Jia-cheng.Analysis of Guider Roller Stiffness Based On ANSYS[J].Coal Mine Machinery,2015,36(6):130-132.
7. QIN QIANG,ZHANG An-ning,YAN Hong-xin.Static stiffness test device of guide roller:China, ZL201505641209[P].2015-09-07

MATEC Web of Conferences **114**, 01004 (2017) DOI: 10.1051/matecconf/201711401004
2MAE 2017

Dynamic Simulation of Rigid Guide Structure Based on ANSYS

ZHANG Xin[1,a] and WANG Zhe[2]

[1] *Anhui University of Science and Technology, Huainan Anhui, China*
[2] *Anhui University of Science and Technology, Huainan Anhui, China*

Abstract. In order to reflect the varying law of the deflection of the rigid guide when the relative motion occur between the rigid guide and the cage roller, transient dynamic simulation is carried out for the commonly used calculation model of rigid guide and bunton by ANSYS. Simulation of the horizontal force through a section of the guide evenly, and the deflection curves of each model are obtained. It is found that the deflection of the simply supported beam model is the largest, and the three-span continuous beam model have similar peak spans in each span with the spatial grid model, but the spatial grid model has obvious fluctuation with the horizontal force.

1 Introduction

Rigid guide is an important part of shaft equipment. At present, the simple beam model and the three-span continuous beam model is still widely used in the calculation of bending resistance of the guide, and the influence of the space structure composed of guide and bunton is not taken into account. The bending moment of the force point is too large, so that the design value tends to be conservative [1]. But the research on the difference of deflection of each model under the action of the moving load is still blank.

In this paper, the transient dynamic simulation of these three models under the action of the moving load is carried out by ANSYS to get the deflection changes in the midpoint. Then export the data to MATLAB, so the curves of the variation of deflection in each span about the simple beam model, three-span equal-span continuous beam model and spatial grid model under the moving load are presented. It has a certain guiding significance for the structural design calculation of rigid guide and bunton.

2 Calculation Model of Rigid Guide and Bunton

The dynamic load transmitted by the hoisting conveyance be resisted by rigid guide and bunton in the hoisting process, the guide structure has an important impact on the safety and reliability of the hoisting conveyance[2]. The design and calculation of guide now, the majority consider static function at one point. However, in most of the hoisting process the role of the horizontal force is not stage but continuous, so considering the load through the rigid guide at a uniform speed, a variety of rigid guide

[a] ZHANG Xin : xzhang@aust.edu.cn

deflection curve calculation model is necessary. In the stage of uniform hoisting of the hoisting conveyance, it is assumed that the transverse horizontal force acting on the guide by the cage roller does not change, so the simply supported beam model, three-span continuous beam model and spatial grid model can be built in Figure 1. The force of the cage roller on the guide is simplified as a force perpendicular to the beam, the relative motion of the cage roller and the guide into a uniform translation is simplified as a force along the axis of the beam.

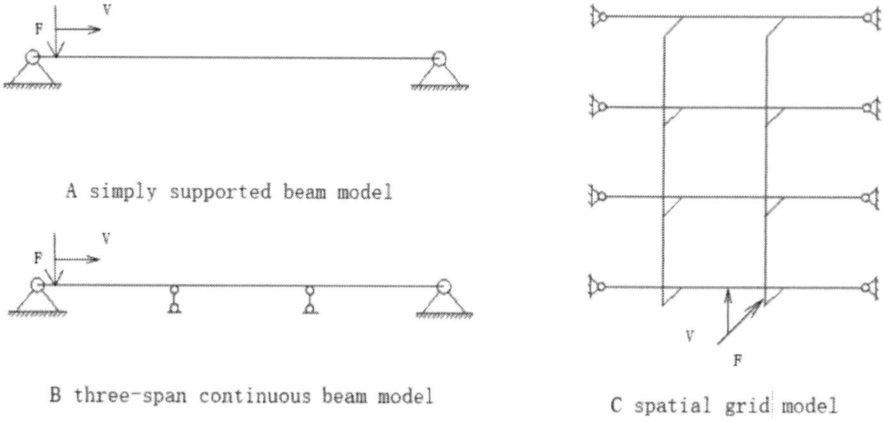

Figure 1. Three Simulation Models.

In the simple supported beam model, both ends of the guide are considered to be hinged; both ends of the three-span continuous beam model are considered to be hinged, and the middle two nodes are regarded as rigid contact

3 Calculation of Mid-Span Deflection of Model

For the maximum deflection of the simple beam model and the three-span continuous beam model, according to the mechanics of materials and the use of building structure statics to get its formula. In the simply supported beam model, the maximum deflection occurs when the load is applied to the mid-span nodes; in the three-span continuous beam model, the maximum deflection occurs when loads are applied to the three mid-span joints respectively[3].

Equations of the maximum deflection of the simply supported beam model:

$$w_{max} = -\frac{lFl^3}{48EI} \tag{1}$$

Equations of the maximum deflection of the three-span continuous beam model:

$$w_{max} = -K \times \frac{Fl^3}{100EI} \tag{2}$$

The meaning of each parameter:

- F is the force;
- l is the length of the single-layer canal (ie, the distance between the canopy beam layer);
- E is the elastic modulus of the material;
- I is the section moment of inertia;
- K is the maximum deflection calculation factor obtained by the look-up table.

According to the actual data:

$$F = 8000N$$
$$l = 4m$$
$$E = 2.06 \times 10 MP \quad (3)$$
$$I = \frac{BH^3 - bh^3}{12}$$

In the mid-span of the first span and the third span in the three-span continuous beam:

$$K = 1.458 \quad (4)$$

In the second span.:

$$K = 1.146 \quad (5)$$

So the following data can be calculated:

The maximum deflection of the simply supported beam model is:

$$w_{max} = -1.57mm \quad (6)$$

The maximum deflection of the three-span continuous beam is:

The mid-span of the first and third span:

$$w_{max} = -1.12mm \quad (7)$$

The mid-span of the second span:

$$w_{max} = -0.86mm \quad (8)$$

However, it is not easy to calculate the deflection of each span in the spatial grid structure. In order to obtain the deflection of the spatial grid structure under the moving load, it is necessary to carry on the finite element simulation analysis.

4 Finite Element Simulation Analysis

4.1 Define Materials

According to the design data of a mine, both sides of the guide are arranged in parallel. The length of the guide is 12 m, two guides are located 1.8 m apart, the length of the bunton is 4.86 m, the span of bunton is 4 m, the short beam length of the connecting guide and bunton in the spatial grid model is 0.1 m. Materials are Q235 cold-formed square steel, cross-sectional size of 180 × 180 × 10.

According to the above data beam, the analysis model is established in ANSYS, set the modulus of elasticity for $2.06 \times 10^5 MP$, poisson's ratio is 0.3, the density is 7800kg/m^3, choose the BEAM188 three-dimensional beam element.

4.2 Meshing, Applying Constraints and Loads

The size of the global control element is 0.1m, and the guide length 12m used to apply the load is divided into 120 elements with 121 nodes. For both ends of the hinge point, constrain the X, Y, Z direction of displacement; for the model of spatial grid structure, constrain the UX, UZ, ROTX, ROTY, ROTZ degrees of freedom of the short beam at the junction of the guide and the bunton; constrain the UX, UY, UZ degrees of freedom at both ends of the bunton. The spatial grid structure is shown in Figure 2. In this paper, we use the * DO loop to apply the load to the 121 nodes of the guide in order to simulate the relative motion between the guide and the cage roller by using the method of moving load in the vehicle-bridge analysis model[4-5].

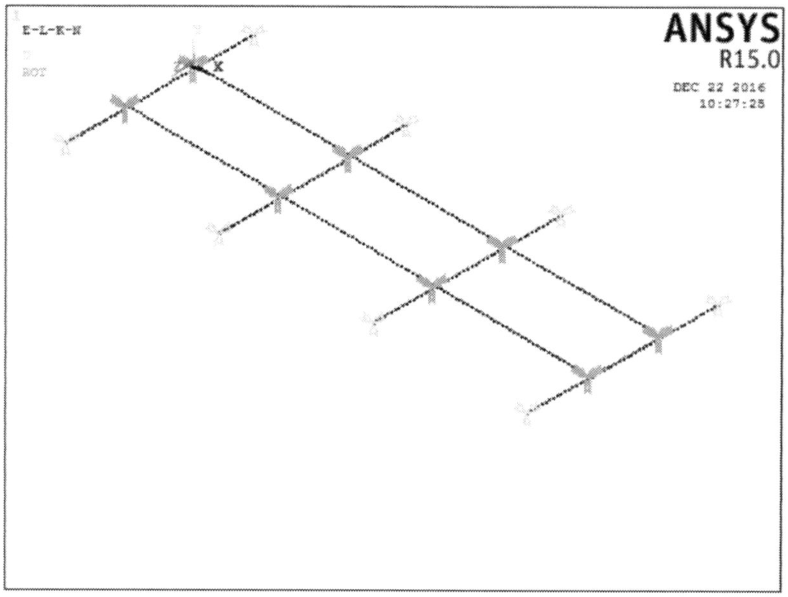

Figure 2. Spatial Grid Structure Model Meshing and Constraints.

According to the data, the pre-tightening force of the Y direction of the cage roller is 8000N, the speed of the uniform speed is 8m/s. Assuming the uniform speed does not impact on the guide, defining the moving load P is 8000N, along the Y axis negative direction, and the moving speed is 8m / s, along the X axis positive direction.

4.3 Simulation Results

The NSOL function is used to extract the displacement value of the Y direction of the simple supported beam, the displacement values of the three mid-span nodes in the Y direction of the three-span continuous beam model and the spatial grid model. Then export the data to MATLAB, take the time as the X axis, the displacement as the Y axis, draws the displacement - time curve, as shown in Figure 3

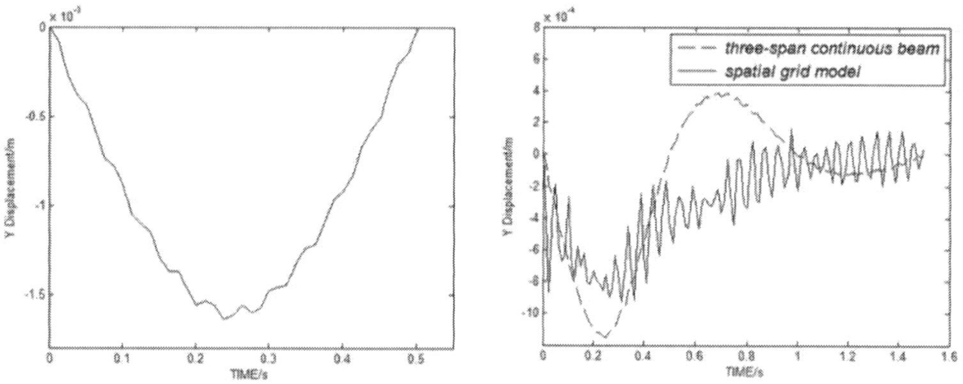

A The Mid-span of simple supported beam

B Displacement of the first mid-span

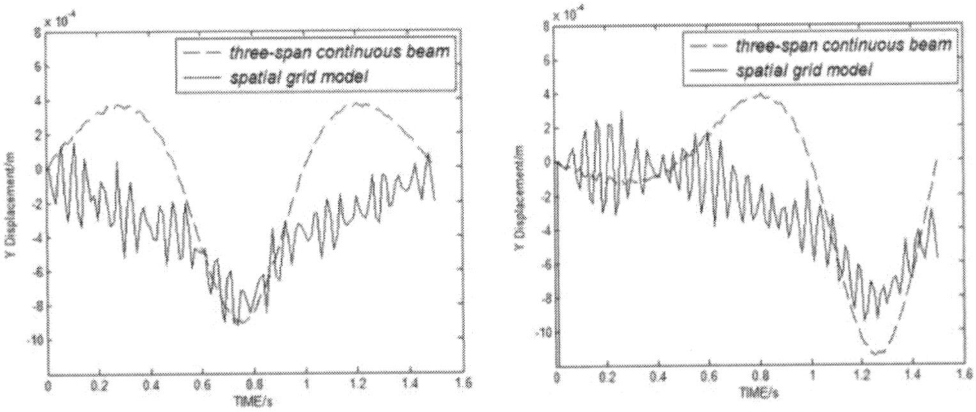

C Displacement of the second mid-span

D Displacement of the third mid-span

Figure 3. Comparison of Mid-Span Deflection of Each Model.

5 Result Analysis

According to the simulation results, the maximum deflection values in each span of the simply supported beam model, three-span continuous beam model and spatial grid model are shown in Table 1.

Table 1. Maximum deflection values in each span(mm).

	First Mid-span			Second Mid-span			Third Mid-span		
	Simulat-ion	Theoret-ical	Deviati-on	Simulat-ion	Theoret-ical	Deviati-on	Simulat-ion	Theoret-ical	Deviati-on
Simply Supported Beam				-1.64	-1.57	0.07			
Three-span Continuous Beam	-1.15	-1.12	0.03	-0.90	-0.86	0.04	-1.15	-1.12	0.03
Spatial Grid Model	-0.93			-0.92			-0.93		

It can be seen from Table 1 that when the calculation model of guide and bunton is calculated according to the simple supported beam, its mid-span deflection will be much larger than that of three-span continuous span beam and spatial grid model. So it is necessary to increase the cross-sectional area of the guide or the use of higher bending stiffness of the material, thereby increasing costs[6]. At the same time, it can be seen from the table that the maximum deflection of the mid-span of the simple beam model and the three span continuous beam model are basically the same as the theoretical calculation result.

The deflections of the three-span continuous beam model and the spatial grid model are similar at the maximum. However, according to the comparison of mid-span deflection curves, the mid-span deflection of the three-span continuous beam is relatively gentle, but the mid-span deflection of the spatial grid model is more violent, which is bound to produce vibration, impact on the stability of the guide structure

6 Conclusion

• Finite element simulation deviation within the allowable range. So the finite element method can be used to simulate the spatial structure of the guide structure;
• The deflection of the simply supported beam model is much greater than that of the three-span continuous beam and spatial grid model. The maximum deflection in each span is similar to the three-span continuous beam model when calculated by the space grid model;
• Compared with the three-span continuous beam model, when the space grid structure is used, the mid-span deflection of each span is more violent under the moving load. In the design calculation of the tank, to take into account this situation, to avoid fatigue damage.
• Compared with the theoretical calculation, the finite element method is more consistent with the actual working conditions, especially for the analysis of complex moving loads, the use of finite element simulation can provide the necessary reference for the design and calculation.

Acknowledgements

This research was supported by the Key Project of Natural Science Foundation for Anhui University (No.KJ2015ZD019).

References

1. Wang Dongquan,Shi Tiansheng,Liu Zhiqiang and Guo Jinpu. Method for Calculating the Structure of Rigid Shaft Equipment of Deep Mine. *Journal of China University of Mining& Technology*. 26(1):5～8,(1997)
2. Qin Qiang,QI Xiao-nan,Ding Jian-guo and Jian Hui. Analysis on Shaft Guide Equipment Quality. *Coal Mine Machinery* .36(6):168-170,(2015).
3. Guo Zhenxi,Zhang Shuyi. *Static Caculation Handbook for Practical Strucyure*. China Machine Press. 4:320-322,(2009)
4. Ge Junying. *Analysis of Bridge Structure Based on ANSYS*. China Railway Publishing House. 8:150-156,(2007).
5. SANMANI F S,PELLICANO F. Vibration reduction on beams subiected to moving loads using linear and nonlinear dynamic aborbers[J]. *Journal of Sound and Vibration*. 325:742～754,(2009)
6. Jiang Yuqiang. Research on Nonlinar Coupling Characteristcs and Condition Assessment of Vertical Steel Guide System. China University of Mining and Technology.(2011).

MATEC Web of Conferences **114**, 01005 (2017) DOI: 10.1051/matecconf/201711401005
2MAE 2017

Parametric Optimization Design of Brake Block Based on Reverse Engineering

JIN Hua-wei[1,a], WANG Zhe[1], ZHANG Xin[1]and WANG Sai[1]

[1]*Mechanical Engineering School, Anhui University of Science and Technology, Huainan, 232001, China*

Abstract. As one of the key part of automotive brake, the performance of brake block has a direct impact on the safety and comfort of cars. Modeling the brake block of disc brake in reverse parameterization by reverse engineering software, analyzing and optimizing the reconstructed model by CAE software. Processing the scanned point cloud by Geomagic Studio and reconstructing the CAD model of the brake block with the parametric surface function of the software, then analyzing and optimizing it by Wrokbench. The example shows that it is quick to reconstruct the CAD model of parts by using reverse parameterization method and reduce part re-design development cycle significantly.

0 Introduction

Reverse engineering is technology ,which choosing advanced production equipment, sample piece, software (including drawings, procedures, technical documents or images and so on) as the research object applies the modern design method, production engineering, materials science and relevant professional knowledge to System analysis and research, and to explore and master the key technology in order to develop more advanced products of the same kind, developed in 1990s. Geomagic Studio is an automated reverse engineering software, which can automatically generate accurate digital models through scanning point cloud[1]. Geomagic Studio can output industry standard formats, including STL, IGES, STEP and CAD and many other file formats. Parametric reconstruction model can fast analysis of product design intent and redesign the product.

Brake block is one of the carrier of automobile brake, which directly affects the braking performance and comfort level. The brake block is divided into the friction block and the brake back plate and the friction force between the friction block and the brake disc is used to eliminate the kinetic energy of the vehicle, so as to achieve the purpose of braking. The structure of the brake block has a critical influence on the stress distribution and contact state, so it indicate that analyze the brake block by finite element method would be meaningful. After working for a period time,it is inevitably that cause brake block wear and other missing. In this case, it is difficult to realize the optimal design of the brake block only by the forward modeling software. In this article,through three dimensional scanning of the wear disc brake block can get the point cloud data.By Geomagic Studio make a parametric curved surface structure, and reestablish the three dimensional CAD model. The resulting model is analyzed by ANSYS Workbench for stress and contact analysis and optimized.

[a] JIN Hua-wei (1986-),E-mail： anlijinhuawei@126.com.

1 Parameterized reconstruction of model based on Geomagic Studio

1.1 Point cloud data acquisition and processing

In this paper, the point cloud acquisition device is Tianjin Weishen Technology Vtop 130 three-dimensional scanner, which scanning method is non-contact and single scanning accuracy is 0.03mm. the point cloud data of scanning is shown in Figure 1,then input the resulting point cloud into the Geomagic Studio software to reduce the noise of the point cloud data and then packaged into the polygon processing phase. In order to facilitate the parameterized after reconstruction,at this time the polygon data are divided into two parts and handle separately.After filling holes,adjusting and repairing mesh to be finished will get the triangular surface shown in Figure 2.

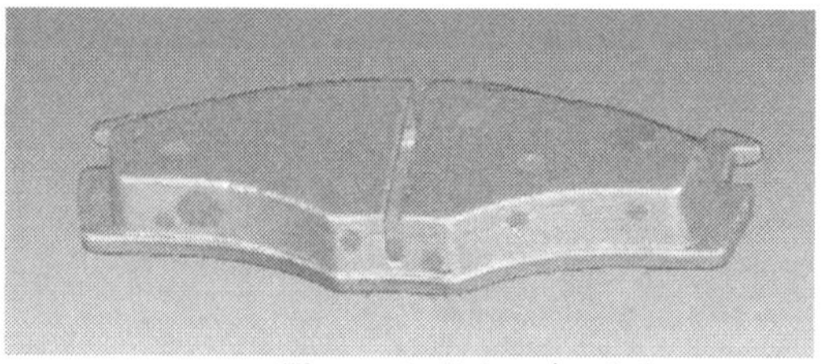

Figure 1. Point cloud data

Figure 2. Triangular surface

1.2 Constructing parametric surface and reconstructing model

After the processing of the polygon phase,it accesses Geomagic Studio stage of parametric surfaces. Surface reconstruction is an important stage of reverse engineering. After the formation of the curved surface, the model will form a uniform size of the surface patches.Then, through the software function of structural grid, we can form a uniform grid network on the surface of the model, the more uniform the grid, the more smooth the surface.Finally, the Geomagic Studio will automatically synthesize the NURBS surface on the basis of the grid network and form the required model [2]. After cutting and stitching,The CAD model shown in Figure 3 , and output STEP or other common format file

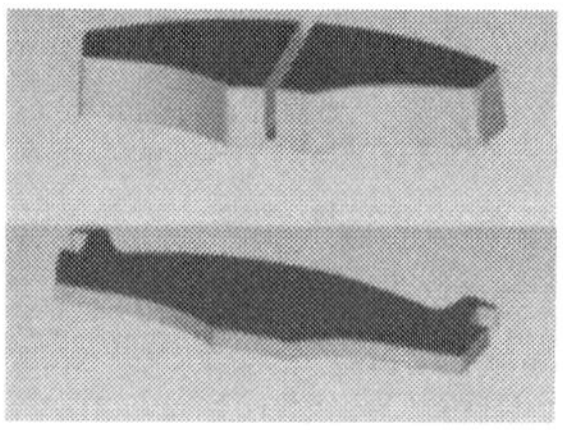

Figure 3. CAD model

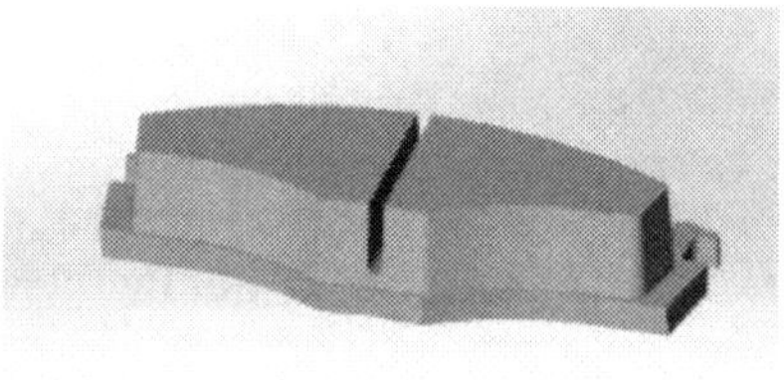

Figure 4. Reconstruction model of brake block

2 Finite element analysis and optimization of brake block

2.1 Coordination of upper and lower parts and model processing

Generally speaking, parts obtained by CAD software need a common data format to import professional CAE software, but through this conversion may make the original file data missing, resulting in broken surface and other issues, and direct affect the subsequent CAE analysis[3]. The IGS format files from Geomagic Studio need to be imported into Solidworks for input diagnosis repair and Recognition feature,mean while Solidworks can directly exchange data with ANSYS and Workbench, to avoid data loss. After Solidworks processing to get the final reconstruction of the CAD model shown in Figure 4.

2.2 Stress analysis and model definition of brake block

In the process of braking, due to the limit of the brake guide groove on both sides of the back brake block, the brake block plate can only slide along the axial direction, while the piston exerts an axial force on the brake backing under the action of hydraulic pressure, which causes the brake block and brake disk contact, and generate a braking force. In this paper, the contact between brake pads and brake discs is studied, Ignoring the effect of friction on contact analysis. Compared with the brake disc, the stiffness of the friction block is relatively small, and for simplicity of analysis, the brake disc is reduced to a rigid surface.

After importing the rebuilt model into ANSYS Workbench, it is necessary to define the material and connection. The density of the friction block is 2615kg/m3, the elastic modulus is 8600MPa, the Poisson's ratio is 0.3;the density of the brake pad is 7800kg/m3, the elastic modulus is 182000MPa, the Poisson's ratio is 0.3.The connection between the friction block and the brake backplane is defined as "bonded" and the connection between the friction block and the rigid face is defined as "rough" and the rigid face is the target face.

2.3 Mesh generation

In the process of finite element analysis, meshing is the key step in finite element analysis. The accuracy and efficiency of finite element analysis are directly affected by meshing. Hexahedral element is more accurate and has better mechanical properties than the tetrahedral element[4-5]. According to the brake backplane choose the method of multi body partition, mapping division selection of hexahedron. The friction block adopts hexahedral partition method, the overall accuracy is 1.5mm.After the completion of the mesh shown in Figure 5.

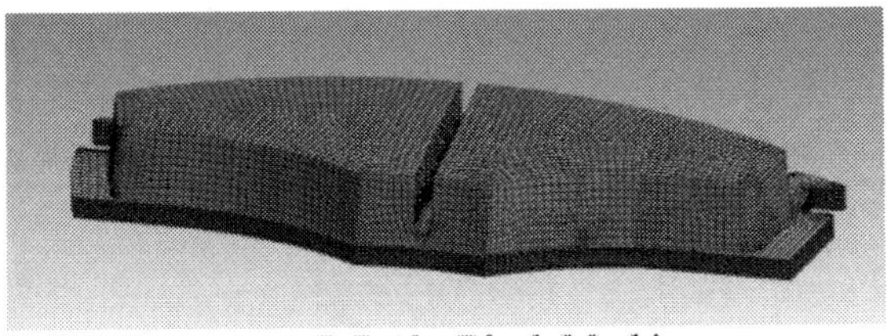

Figure 5. Brake Block Meshing

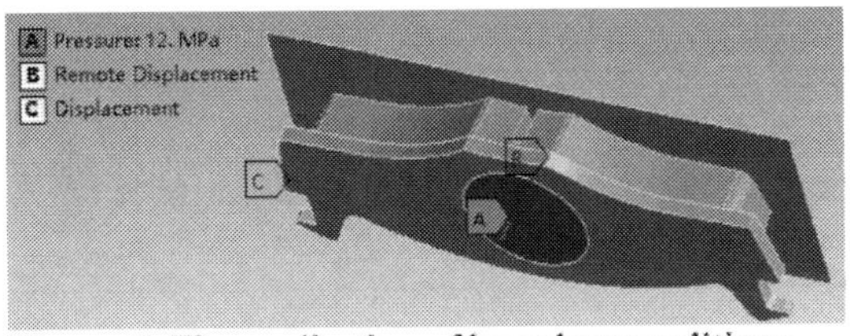

Figure 6. The application of boundary conditions

2.4 Applied loads and boundary conditions

According to the actual working condition, the 12MPa pressure is applied in the brake backplane center diameter of 35mm circular area;In the axial direction of the Z axis, in the back of the brake guide groove to apply X direction and the Y direction of the fixed displacement constraints, Z direction of freedom;The rigid surface is applied a remote displacement constraint, the X direction, the Y direction, the Z direction, and the rotation around the X axis, the Y axis is rotated, and the Z axis is set to be fixed. Open the large deformation option "Large Deflection" in the analysis settings. The result of the application is shown in figure 6.

2.5 Analyze options and results

The maximum stress point can be seen through the equivalent stress cloud as shown in Figure 7,and the maximum is 55.231MPa;The maximum stress point between the friction block and the brake disc is shown in Fig. 8, and the maximum value is 11.345MPa;The contact state between the friction block and the brake disc is shown in figure 9.

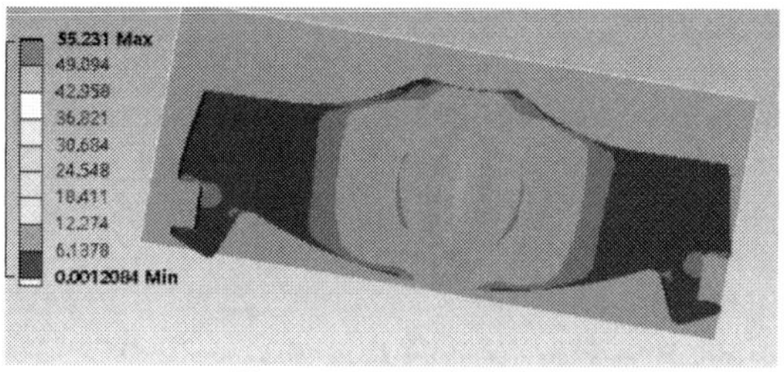

Figure 7. Equivalent stress

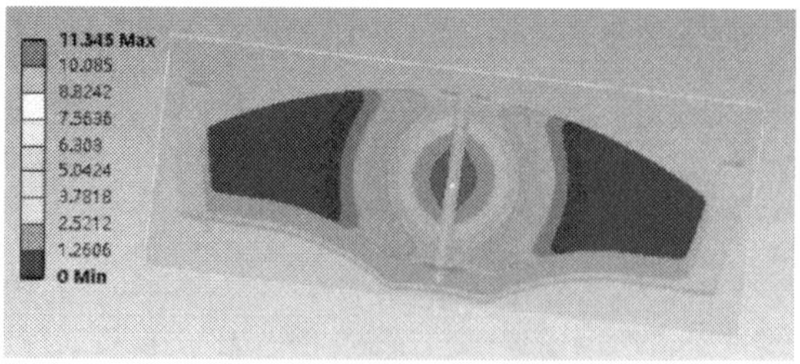

Figure 8. Contact stress

Figure 9. Contact state

3 Optimization of brake block structure

3.1 The scheme determination of optimization

The 3D model reconstructed in Geomagic Studio can not be optimized directly using ANSYS Workbench, which needs to be imported into Solidworks,and add the ds_ prefix to the parameter name that needs to be optimized[6].The above analysis shows that there are more non contact surfaces on both sides of the friction block with the brake disc, and the brake block equivalent stress, will make the brake block wear uneven and affect the service life.The shape of the friction block, the shape and thickness of the brake back plate will cause the change of contact state and contact pressure.In this paper, the influence of the thickness of the brake pad on the contact state is studied.The input parameters are the thickness of the brake back plate, and the output parameters are the equivalent stress and contact stress of the brake block.Set the original thickness of the brake back plate is 4mm, the optimization range is set to 3.4mm to 4.6mm, divided into five design points.

3.2 Optimization results and analysis

Through analysis and calculation, the input and output parameters of the 5 design points are shown in table 1.It can be seen that when the thickness of the brake plate is less than the original size, the contact stress is relatively increased, but the maximum equivalent stress is relatively reduced; when the thickness of the brake plate is larger than the original size, the contact stress is relatively reduced and the maximum equivalent stress is also reduced slightly.

Table 1. Design point input and output parameters

Serial number	Brake backing thickness (mm)	Maximum contact stress (MPa)	Maximum equivalent stress (MPa)
1	3.4	11.963	55.399
2	3.7	11.653	50.515
3	4.0	11.345	55.231

4	4.3	11.035	53.020
5	4.6	10.768	52.255

To ensure that the contact state, should make the contact stress to optimize the highest priority, and for the convenience of processing should select thickness of 4.5mm. Choosing the optimal size set to the current design point, the optimized brake block equivalent stress nephogram after re-calculated shown in Figure 10, the maximum equivalent stress of 51.592MPa;The contact stress cloud is shown in Figure 11, the maximum contact stress is 10.866MPa; contact state diagram shown in Figure 12.

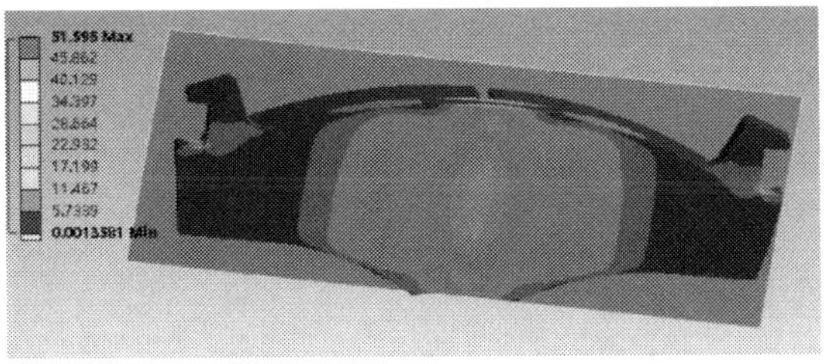

Figure 10. equivalent stress after optimization

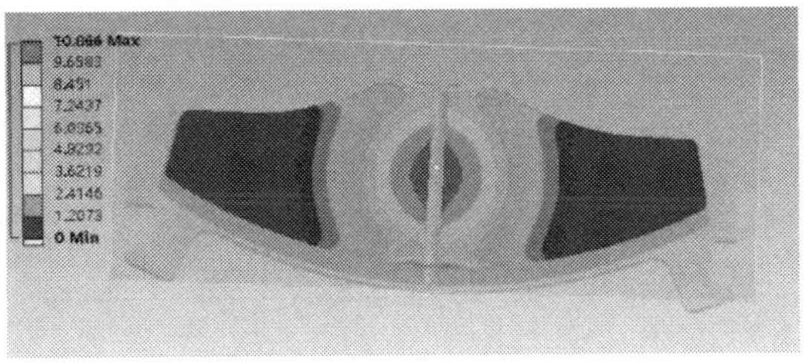

Figure 11. contact stress after optimization

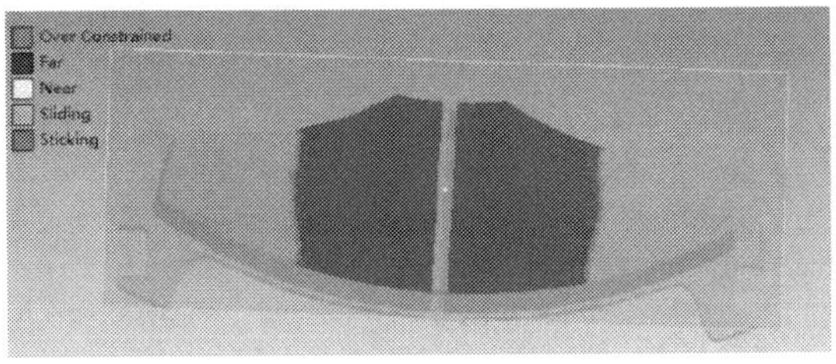

Figure 12. Optimized contact state

The optimized maximum contact stress, maximum equivalent stress is compared with that before optimization, as shown in Table 2. It can be seen that the maximum contact stress after optimization is reduced by 0.479 MPa and the maximum equivalent stress is reduced by 3.639 MPa.

Table 2 . Comparison before and after optimization

	Brake backing thickness (mm)	Maximum contact stress (MPa)	Maximum equivalent stress (MPa)
Pre-optimization values	4.0	11.345	55.231
Optimized value	4.5	10.866	51.592
Change value		-0.479	-3.639
Optimization rate		4.2%	6.6%

4 Conclusion

The results show that the real model and parameters can be quickly restored by the reverse engineering software, and the parameters of the model can be reconstructed quickly, especially in the case of parts with wear or other defects, compared with single forward modeling, and it can be redesigned with most of the forward modeling software. Through this method, the development process of parts re-design can be greatly reduced and the market competitiveness of the products can be improved.

Acknowledgements

This research is supported by the key project of Anhui Province Natural Science Foundation(Working mechanism and structure research of automotive micro electrical and mechanical brake device), the project of Anhui Province Science and Technology Research Project (1301022074,1301022075), the key project of Anhui Province Natural Science Foundation (No.KJ2015ZD019) and the key project of Young Teachers' Scientific Research Fund of Anhui University of Science and Technology (QN201416).

References

1. JinTao,Tong Yong-guang. Reverse Engineering Technology[M].BeiJing.China Machine Press.(8):1-5 (2003).
2. JuLin-guang,Bian Xianhu,Heavy truck bumper Geomagic Catia software and the drag hook gate number mode reconstruction[J]. machinery. based on the.39 (4):6 10 (2012).
3. ChouCan-hua,ChengSi-yuan,ZhangXiang-wei,and so on. Finite element analysis of parts based on reverse engineering technology[J]. Machinery Design and Manufacture.(9):35-36 (2009).
4. Che X,Liang X,Li Q.G1 continuity conditions of adja—cent NURBS surfaces[J].Computer Aided Geometric De.sign.22(4):285—298 (2005).
5. Shi X.Wang T,Wang T,et a1.Reconstruction of convergent G1 smooth B—spline surface[J]. Computer Aided Geometric Design.21(9):893—913 (2004).
6. HanJiang,MengChao,YaoYin-ge,and so on.Parametric optimization design based on Reverse Engineering[J]. Combined machine tool and automatic machining technology.(3):38-39 (2016).

MATEC Web of Conferences **114**, 01006 (2017) DOI: 10.1051/matecconf/201711401006
2MAE 2017

Influences of Lubricant Viscosity Ratio on Static Characteristics of Hydrostatic Thrust Bearing

Chuan-Chieh Liao, Sheng-Yen Hu, Chao-Ping Huang and Yuan Kang[a]
Chung Yuan Christian University, Department of Mechanical Engineering, 32023 Chung Li, Taiwan, R.O.C

Abstract. This paper uses numerical simulation to study the influences of viscosity ratio on the static characteristics of hydrostatic thrust bearing compensated by various restrictors. The design parameters of open type hydrostatic bearing include: effective area, supply pressure, restriction constants of restrictor and bearing pad, and specific value of average viscosity of the lubricant passing through restrictor and bearing pad film. The optimal design requirements of restrictors and hydrostatic bearing and the corresponding range of design parameters for excellent bearing characteristics can be obtained from the relationships between these design parameters together with their composed parameters and ratios, and the static load capacity and film stiffness and worktable displacement of hydrostatic bearing. Fixing the other design parameters and changing viscosity ratio, the simulation results depict that the lubricant viscosity during restrictor being greater than that in bearing pad film will make the load capacity decrease, but will improve the film static stiffness.

1 Introduction

Hydrostatic bearings according to their applications can be divided into thrust, radial and biaxial three kind bearing systems. Lubricating oil was pumped by an external pump through a restrictor into recess to form restriction pressure, then flew into the bearing clearance and provided bearing pad pressure to form a lubricating film supporting the load. The restrictors can apply constant compensations such as capillary, orifice and slot, or variable compensations such as pressure feedback driving membrane-type and spool-type restrictors, or use one recess connected to one constant flow pump or valve to adjust supply pressure into working pressure by constant flow compensation.

Malanoski and Loeb [1] studied hydrostatic bearings using capillary, orifice or flow control valve as the pressure compensation elements. They investigated the influences of these three compensated types on the film stiffness of hydrostatic bearings, and found the stiffness using flow control valve will be the greatest, followed by using orifice, then using capillary. Wang [2] investigated the load capacity and film stiffness of hydrostatic slide, and exploded the influences of film thickness and slider clearance on worktable characteristics. Kang et al. [3] investigated the constant compensations. They utilized the equations of flow equilibrium to determine the worktable displacement with respect to the recess pressure. The influences of bearing pad and restrictor geometric design parameters on its static characteristics: load capacities and stiffness coefficients were investigated.

[a] Corresponding author: yk@cycu.edu.tw

Nikolakopoulos and Papadopoulos [4] analytically and experimentally studied ER fluid lubricated journal bearings, and indicated ER fluid can be used to create "smart" journal bearings and vibration controllers can be constructed to control the dynamics and stability of the ER fluid lubricated bearings. Bouzidane and Thomas [5-6] studied the vibration control of an ER hydrostatic journal bearing. They found that the viscosity of a negative ER (NER) fluid will decrease with an electric field applied. They concluded that using NER fluid lubrication is very efficient when the force transmitted to the base must be reduced. Yu et al [7] analyzed the influence of rotational speed and viscosity for 3D temperature distribution. But they didn't discussed the influence of viscosity caused by temperature for the load capacity and static stiffness.

These previous studies of hydrostatic bearing applied electromagnetic rheological fluid as lubricant. When the local external magnetic field affected lubricant viscosity as it pass through at the same time, the characteristics of bearing have been changed. This is mainly caused by the difference between the viscosity of lubricant flowing through the restrictor and supporting film within the system. Therefore, this study investigated the influences of viscosity ratio of lubricant passing through restrictor to supporting film on the static characteristics and performances of hydrostatic bearing.

2 Numerical formulations

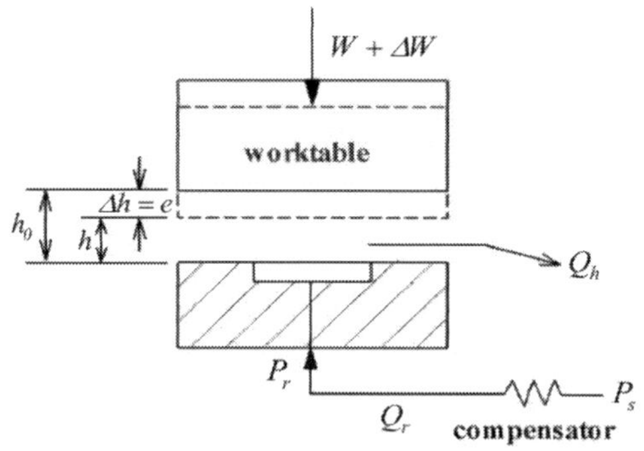

Figure 1. Open-type hydrostatic planar bearing

For an open-type hydrostatic bearing as shown in Fig. 1, the volumetric flow rate Q_h out of bearing pad will be

$$Q_h = \left(P_r h^3 \gamma\right) / \left(12\mu_h\right) \tag{1}$$

where P_r is the work pressure, h is the film thickness, μ_h is the lubricant viscosity of oil film, and γ denoted for dimensionless parameter of pad restriction that is a proportional constant of film flow due to pad geometry. Furthermore, dividing $(P_s h_0^3)/(12\mu_r)$ can obtain the dimensionless flow is:

$$\bar{Q}_h = \frac{12\mu_r Q_h}{P_s h_0^3} = \frac{P_s}{P_r}\frac{\mu_r}{\mu_h}\left(\frac{h}{h_0}\right)^3 \gamma = \bar{P}_r \bar{\mu} \bar{h}^{-3} \gamma \tag{2}$$

where μ_r denotes the viscous coefficient of oil during compensator, P_s is the system supply pressure, is the characteristic clearance of open-type hydrostatic bearing as the bearing pad and restrictor been designed, $\bar{\mu}$ is the viscosity ratio of lubricant in restrictor to recess (μ_r/μ_h), and $\bar{h} = h/h_0$ is the film thickness ratio. The load capacity of an open-type hydrostatic bearing can be determined by $W = P_r A_{eff}$, and its non-dimensional form is:

$$\bar{W} = W / \left(P_s A_{eff}\right) = \bar{P}_r \tag{3}$$

where A_{eff} is the effective area of bearing pad, which is smaller than the pad area, thus, a greater A_{eff} is indicated for a larger recess proportion of bearing pad. The film static stiffness is defined by the load variation with respect to the displacement variation of worktable and expressed as

$$K = \frac{dW}{de} = \frac{-dW}{dh} = \frac{-P_s A_{eff}}{h_0}\frac{d\bar{P}_r}{dh} \tag{4}$$

where $e = h_0 - h$ is the displacement of worktable. Dividing Eq. (4) by $P_s A_{eff}/h_0$ gives the dimensionless film stiffness:

$$\bar{K} = \frac{Kh_0}{P_s A_{eff}} = \frac{dW}{de}\frac{h_0}{P_s A_{eff}} = \frac{-d\bar{P}_r}{d\bar{h}} \tag{5}$$

In order to make the relationship between the bearing design parameters and characteristic parameters generalized, the equations established with dimensionless parameters are adopted. Hence, according to Eqs. (3) and (5), the dimensionless load capacity $\bar{W}$ and static stiffness coefficient $\bar{K}$ can be obtained from design conditions respectively, and show that load capacity is proportional to $P_s A_{eff}$ and static stiffness is proportional to $(P_s A_{eff})/h_0$.

2.1 Constant flow compensation

The constant flow compensation can be realized by giving constant flow for recess pressure varying to bear the load fluctuation, and the continuity equation can be expressed as:

$$Q_p = (P_r h^3 \gamma)/(12\mu_h) = \text{constant} \tag{6}$$

Assuming the design film thickness for constant flow compensation to be $h_0^3 = (12Q_p \mu_r)/P_s$, which gives a dimensionless form of Eq. (6) as below:

$$\gamma \bar{\mu} \bar{P}_r \bar{h}^3 = 1 \tag{7}$$

Therefore, the unit dimensionless load capacity for an open-type hydrostatic bearing compensated by constant flow compensation will be

$$\bar{W} = \bar{P}_r = (\gamma \bar{\mu} \bar{h}^3)^{-1} \tag{8}$$

Differentiating equation (8) with respect to $\bar{h}$, then, substituting it into equation (5) gives the non-dimensional film stiffness as:

$$\bar{K} = \frac{-d\bar{P}_r}{d\bar{h}} = 3(\gamma \bar{\mu} \bar{h}^4)^{-1} = 3(\gamma \bar{\mu})^{1/3} (\bar{P}_r)^{4/3} \tag{9}$$

where $\bar{h} = (\gamma \bar{\mu} \bar{P}_r)^{-1/3}$.

2.2 Capillary compensation

The flow rate through a capillary restrictor can be determined by:

$$Q_c = \frac{\pi d_c^4}{128\mu_r \ell_c}(P_s - P_r) = \frac{\kappa_c}{\mu_r}(P_s - P_r) \tag{10}$$

where d_c and ℓ_c are diameter and length of capillary, and κ_c is the restriction constant of capillary. Therefore, the continuity equation of flow through capillary restrictor to bearing pad film is

$$\frac{P_r h^3}{12\mu_h}\gamma = \frac{\kappa_c}{\mu_r}(P_s - P_r) \tag{11}$$

Letting the design film thickness of open-type bearing compensated by capillary to be $h_0^3 = 3\pi d_c^4 / 32\ell_c = 12\kappa_c$, which gives a dimensionless form of equation (11) as below:

$$\bar{P}_r \bar{h}^3 \gamma \bar{\mu} = 1 - \bar{P}_r \tag{12}$$

The dimensionless load capacity for capillary compensated bearing can be observed by rearranging equation (12) as following:

$$\overline{W} = \overline{P}_r = (\gamma \overline{\mu} \overline{h}^{-3} + 1)^{-1} \tag{13}$$

To observe the formulation of non-dimensional film stiffness, we first differentiate equation (13) with respect to $\overline{h}$, then, substituting it into equation (5) gives the final formulation as:

$$\overline{K} = \frac{-d\overline{P}_r}{d\overline{h}} = \frac{3\gamma \overline{\mu} \overline{h}^{-2}}{(\gamma \overline{\mu} \overline{h}^{-3} + 1)^2} = 3\sqrt[3]{\gamma \overline{\mu} \overline{P}_r^{-1} (1 - \overline{P}_r)^2} \tag{14}$$

where $\overline{h} = \sqrt[3]{(\overline{P}_r^{-1} - 1)(\gamma \overline{\mu})^{-1}}$.

2.3 Orifice compensation

The flow through an orifice can be expressed by

$$Q_o = c_o \frac{\pi d_o^2}{4} \sqrt{\frac{2(P_s - P_r)}{\rho}} = \kappa_o \sqrt{P_s - P_r} \tag{15}$$

where c_o is geometric factor of orifice, d_o is orifice diameter, ρ is oil density, and $\kappa_o = \left(c_o \pi d_o^2\right) \times \left(\sqrt{8\rho}\right)^{-1}$ is restriction constant of orifice. Here, we assume the applied lubricant is an incompressible fluid and κ_o is constant. Then, the flow rate through the orifice is proportional to κ_o . Therefore, the continuity equation of flow through orifice restrictor to bearing pad film is

$$\left(P_r h^3\right) / \left(12\mu_h\right) = \kappa_o \sqrt{P_s - P_r} \tag{16}$$

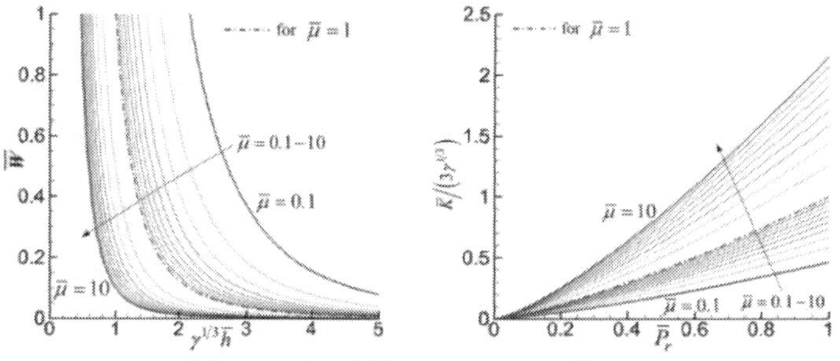

(a) constant flow compensation

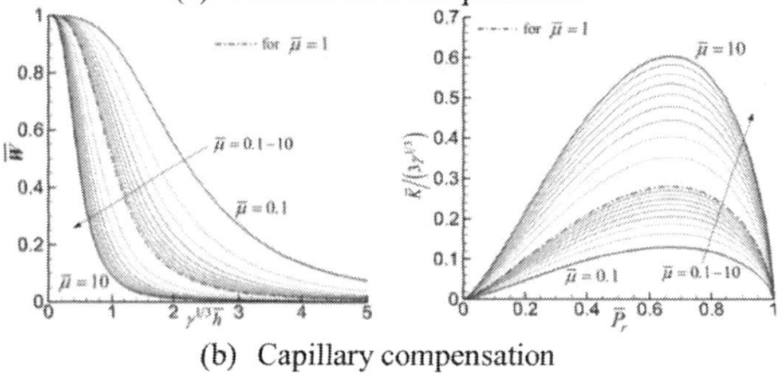

(b) Capillary compensation

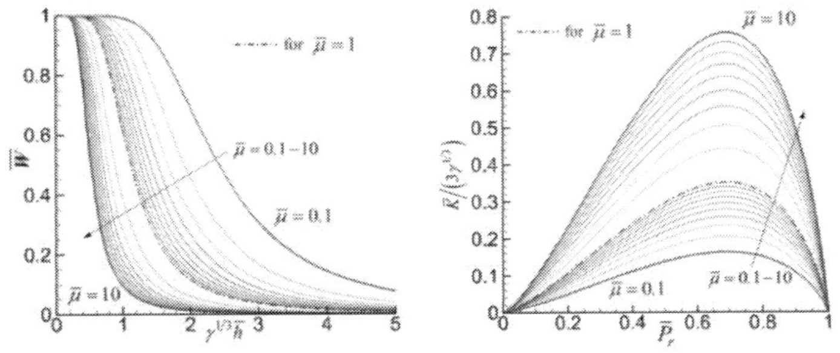

(c) Orifice compensation

Figure 2. $\bar{W}$ versus $\gamma^{1/3}\bar{h}$ (see left column) and $\bar{K}/(3\gamma^{1/3})$ versus $\bar{P}_r$ (see right column) for different $\bar{\mu}$ using different compensations of open-type hydrostatic bearing.

Setting the design film thickness of open-type bearing compensated by orifice to be $h_0^3 = \sqrt{8}\pi d_o^2 c_o \mu_r / \sqrt{P_s \rho} = 12\mu_r \kappa_o / \sqrt{P_s}$, which gives a dimensionless form of equation (16) as:

$$\bar{P}_r \bar{h}^{-3} \gamma \bar{\mu} = \sqrt{1-\bar{P}_r} \tag{17}$$

The dimensionless load capacity for orifice compensated bearing can be observed by rearranging equation (18) as following:

$$\bar{W} = \bar{P}_r = \frac{-1+\sqrt{1+4\gamma^2\bar{\mu}^2\bar{h}^{-6}}}{2\gamma^2\bar{\mu}^2\bar{h}^{-6}} \tag{18}$$

To observe the formulation of non-dimensional film stiffness, we first differentiate equation (18) with respect to $\bar{h}$, then, substituting it into equation (5) gives the final formulation as:

$$\bar{K} = \frac{-d\bar{P}_r}{d\bar{h}} = \frac{6\gamma^{1/3}\bar{\mu}^{1/3}\left(1-\bar{P}_r\right)^{5/6}\bar{P}_r^{4/3}}{2-\bar{P}_r} \tag{19}$$

where $\bar{h} = \left[\sqrt{1-\bar{P}_r}\left(\gamma\bar{\mu}\bar{P}_r\right)^{-1}\right]^{1/3}$.

3 Results

According to Eq. (8) for constant flow compensation, Eq. (13) for capillary restriction, and Eq. (18) for orifice restriction, the load capacity $\bar{W}$ is equal to the pressure ratio $\bar{P}_r$. The film thickness $\bar{h}$ is inversely proportional to the $\bar{\mu}^{1/3}$ under the same load, in other words, the $\bar{W}$ is lower as $\bar{\mu}$ being greater with the same film thickness. However, according to Eqs. (9), (14) and (19) for the relationship between static stiffness $\bar{K}$ and $\bar{P}_r$, the stiffness is proportional to $\bar{\mu}^{1/3}$. Plot $\bar{W}$ versus $\gamma^{1/3}\bar{h}$ and $\bar{K}/(3\gamma^{1/3})$ versus $\bar{P}_r$ for various $\bar{\mu}$ gives Fig. 2. $\bar{W}$ will be smaller as $\bar{\mu}$ being greater with the same $\bar{h}$ and $\bar{h}$ will be smaller as $\bar{\mu}$ being greater with the same $\bar{W}$. We can get the relationship on $\bar{W}$ and $\bar{K}$. Thickness ratio of oil film $\bar{h}$ versus pressure ratio $\bar{P}_r$ for different pad restriction constant using three kinds pressure compensation methods as shown in Fig. 3.

In Fig. 2, the film thickness increases as viscosity ratio decreasing with the same load. This means that cooling lubricant at bearing pad makes film viscosity greater than supply oil viscosity will increase the film thickness also increase the load capacity, but the film stiffness increases as the viscosity ratio increasing. On the contrary, because of the friction force of lubricant making the film temperature rising makes the film viscosity decreasing, while decreases the film thickness but increases the film static stiffness. In Fig. 3, we can find the restriction constant is an important design parameter for the film thickness and load capacity when viscosity is constant. The film thickness and the load capacity both are inversely proportional to the restriction constant.

4 Conclusions

This study analyses the influences of thickness ratio of oil film, lubricant viscosity ratio, and restriction constant on its performances: load capacity and film static stiffness coefficient, and concludes as below: (1) When the viscosity ratio of lubrication fluid in restrictor stage to bearing recess stage is increasing, the load capacity of hydrostatic bearing will cause a certain amount of decrease, but its amplitude will reduce as the restriction constant increasing. (2) The viscosity ratio of lubrication fluid in restrictor stage to bearing recess stage can enhance the film static stiffness of hydrostatic bearing effectively, especially when greater than 1. (3) When the viscosity ratio of lubrication fluid in restrictor stage to bearing recess stage and restriction constant are increasing, the variation of thickness ratio of oil film will be stable more which mean having the good static stiffness. (4) Comparing three types of constant compensations in hydrostatic bearing, the constant flow has the best static performances, followed by the orifice restrictor, then the capillary restrictor.

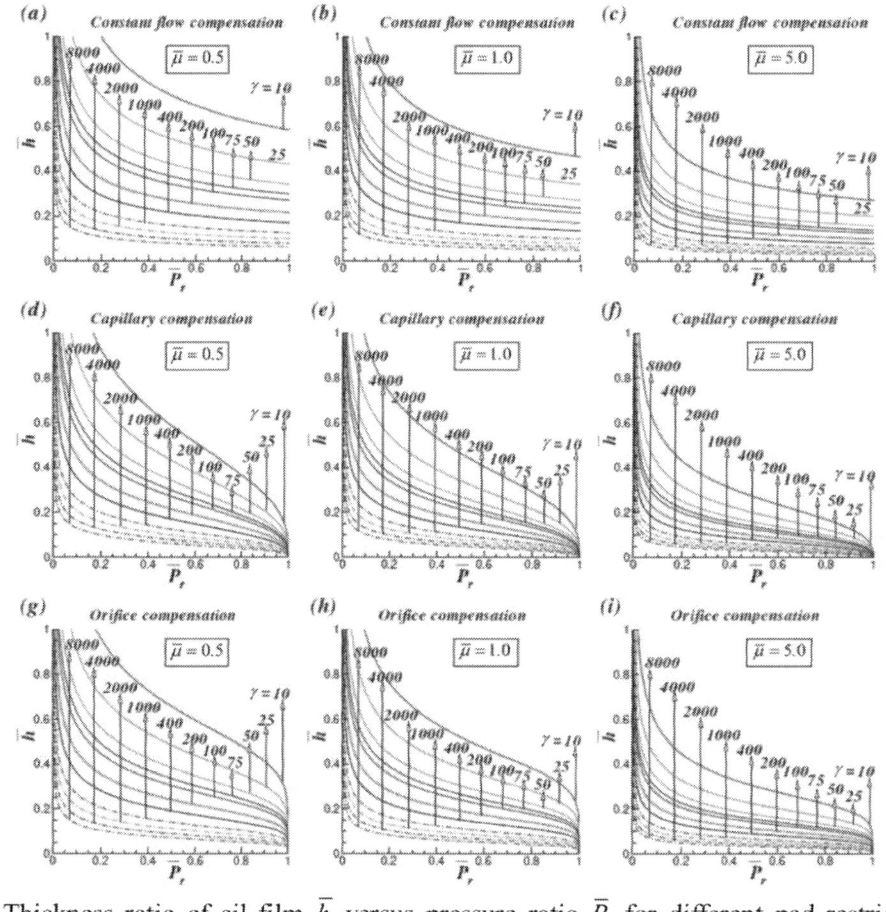

Figure 3. Thickness ratio of oil film $\bar{h}$ versus pressure ratio $\bar{P}_r$ for different pad restriction using constant flow compensation [(a)-(c)], capillary compensation [(d)-(f)] and orifice compensation [(g)-(i)].

Acknowledgement

The authors greatly appreciate the support of this work by Grants NSC 103-2221-E-003-016-MY3 from the National Science Council Taiwan R. O. C.

References

1. Malanoski S.B., Loeb A.M., The effect of the method compensation on hydrostatic bearing stiffness, J Basic Eng-T ASME, **83(2)**, 179-187 (1961)
2. Wang D., Hydrostatic slideway and its design research, LUBR ENG, **4**, 114-118 (2004)
3. Kang Y., Lee J.L., Huang H.C., Lin C.Y., Lee H.H., Peng D.X., Design for static stiffness of hydrostatic plain bearings: constant compensations, Ind Lubr Tribol, **63(3)**, 178-191 (2011)
4. Nikolakopoulos P.G., Papadopoulos C.A., Controllable high speed journal bearings, lubricated with electro-rheological fluids. An analytical and experimental approach, Tribol Int, **31**, 225-234 (1998)
5. Bouzidane A., Thomas M., An electrorheological hydrostatic journal bearing for controlling rotor vibration, Comput Struct, **86**, 463-472 (2008)
6. Bouzidane A., Thomas M., Nonlinear dynamic behavior of a flexible shaft supported by smart hydrostatic squeeze film dampers, J Tribol, **135**, 031701-1~9 (2013)
7. Yu X.D., Fu X., Meng X.L., Liu D., Zhang Y.Q., Wang W.S., Experimental and numerical study on the temperature performance of high-speed circular hydrostatic thrust bearing, J Comput Theor Nanosci, **12(8)**, 1540-1545(6) (2015).

MATEC Web of Conferences **114**, 01007 (2017) DOI: 10.1051/matecconf/201711401007
2MAE 2017

Dynamic Modeling & Stability Analysis of a Generic UAV in Glide Phase

Imran Mir[a], Adnan Maqsood[1] and Suhail Akhtar[2]

[1]*Research Center for Modeling and Simulation, National University of Sciences and Technology, H-12 Campus, Islamabad, Pakistan*
[2]*College of Aeronautical Engineering (CAE), National University of Sciences & Technology, Risalpur, Pakistan*

Abstract. In this paper, we present dynamic modelling and stability analysis of a generic UAV in the glide phase under engine failure condition. When such extreme phenomena occurs, the most desirable requirement is to survive that stage by keeping the vehicle controllable by maintaining its orientation and to glide the vehicle towards the intended direction with maximum extended range. This study investigates the stability aspects of one such aerial vehicle under engine failure condition. In the proposed architecture, a six degree of freedom vehicle dynamic simulation model is implemented through a set of coupled non-linear differential equations. The aerodynamic forces and moments encountered by the UAV during various phases of the flight are ascertained through empirical / non-empirical techniques. Non-linear constrained optimization technique is employed to evaluate the steady state values of the optimized trajectory for the complete flight regime. Results from dynamical systems theory are applied to investigate local stability characteristics of UAV around the steady state. Complete set of dynamic modes of UAV throughout the glide phase are evaluated and mode content in each of the motion variable is determined using modal decomposition technique. The dynamic characteristics of the open-loop configuration are assessed to generate adequate benchmark performance for closed-loop controller design

1 Introduction

Unmanned Air Vehicles (UAV) have poor safety records than manned counterparts [1, 2]. By bringing human out of the cockpit, both mission and cost effectiveness are enhanced by multiple folds. Medium Altitude Long Endurance (MALE) [3] or High Altitude Long Endurance (HALE) [4] configurations are worthy of performing Dull, Dangerous and Dirty (DDD) missions. However, since the safety factor during design considerations of UAVs is relatively low, therefore, system and sub-system level failures are a common occurrence [5]. This research deals with one undesirable phenomenon of engine failure during flight. Generally, UAVs fly with one engine on-board and typically are equipped with pusher type propeller. The engine failures can be caused by mechanical or electrical malfunctioning or as a result of enemy fire. Under such scenarios, successful retrieval of the platform in integrated state back at the operating base becomes of vital importance. In this paper, dynamic modeling and stability analysis is carried out for one such generic Unmanned Air Vehicle with two all moving control surfaces, which can stabilize and accurately navigate to a reasonably

[a] Corresponding author : imranmir56@yahoomail.com

distant base station in case of engine failure. In this proposed architecture, a 6-DOF vehicle modeling is performed through a set of highly coupled non-linear differential equations [6]. In order to evaluate aerodynamic forces and moments during various phases of the flight, empirical and non-empirical techniques such as Computational Fluid Dynamics (CFD) [7], USAF Stability and Control Digital Datcom [8] and so on [9] were employed. Non-linear constrained optimization technique is employed to evaluate the steady state values of the optimized trajectory for the complete flight regime to ensure vehicle movement at an optimum path. Trim point analysis and model linearization was performed in Matlab® environment. Eigen value analysis of the linearized system was than performed to determine the response of the open loop system under stated problem. System stability aspects were than evaluated to determine the possibility of designing a suitable control architecture which can keep the system stabilize and ensure that the UAV follows the optimized trajectory under engine failure condition.

2 UAV Flight Dynamics Model

Vehicle simulations are commonly designed to work with three-degree of freedom (3 DOF) [10, 11] or six-degree of freedom (6 DOF) [12] equations of motion. Each system has its own pros and cons, which make them more suitable for specific applications. In this study, 6-DOF model because of its capability to cater for both the translational and rotational dynamics is utilized to model UAV dynamics.

2.1 UAV Equation of Motion

The rigid aircraft motion equations are presented in the form of system of first order nonlinear differential equations describing translational/ angular accelerations and kinematics of angular motion The governing equations (refer equation(1)-(12)) based upon two valid assumptions of flat and non-rotating Earth, are given below:-

$$\dot{U} = RV - QW - g_D \sin\theta + (X_A + X_T)/m) \tag{1}$$

$$\dot{V} = -RU + PW + g_D \sin\phi\cos\theta + (Y_A + Y_T)/m \tag{2}$$

$$\dot{W} = QU - PV + g_D \cos\phi\cos\theta + (Z_A + Z_T)/m \tag{3}$$

$$\dot{\phi} = P + \tan\theta(Q\sin\phi + R\cos\phi) \tag{4}$$

$$\dot{\theta} = Q\cos\phi - R\sin\phi \tag{5}$$

$$\dot{\psi} = (Q\sin\phi + R\cos\phi)/\cos\theta \tag{6}$$

$$\Gamma\dot{P} = J_{xz}[J_x - J_y + J_z]PQ - [J_z(J_z - J_y) + J^2{}_{xz}]QR + J_z l + J_{xz} n \tag{7}$$

$$J_y\dot{Q} = (J_z - J_x)PR - J_{xz}(P^2 - R^2) + m \tag{8}$$

$$\Gamma\dot{R} = [(J_x - J_y)J_x + J^2{}_{xz}]PQ - J_{xz}[J_x - J_y + J_z]QR + J_{xz} l + J_x n \tag{9}$$

$$\dot{x} = U c\theta c\psi + +V(-c\phi s\psi + s\phi s\theta c\psi) + W(s\phi s\psi + c\phi s\theta c\psi) \tag{10}$$

$$\dot{y} = U c\theta s\psi + V(c\phi c\psi + s\phi s\theta s\psi) + W(-s\phi c\psi + c\phi s\theta s\psi) \tag{11}$$

$$\dot{h} = U s\theta - V s\phi c\theta - W c\phi c\theta \tag{12}$$

where
$$U = V_T \cos\alpha \cos\beta,\ V = V_T \sin\beta,\ W = V_T \sin\alpha \cos\beta$$
$$X_A = \bar{q}SC_x,\ Y_A = \bar{q}SC_y,\ Z_A = \bar{q}SC_z,\ \Gamma = J_x J_z - J^2_{xz}$$

also ϕ,θ,ψ ,P,Q,R are the roll, pitch yaw angles and rates respectively , U,V,W and $J_{\otimes}$ are the velocity components and moments along the three axis, γ is the flight path angle , X_A,X_T are the axial and tangential forces

2.2 Aerodynamic Parameters Estimation

Multi-dimensional look-up tables based high fidelity aerodynamic model is developed from experimental data obtained by employing various empirical and non-empirical techniques. Broadly each aerodynamic coefficient in aircraft body axes system is expressed as a summation of aerodynamic effects due to static affects and dynamic affects (refer equation (13)).

$$C_i = C_{i,static}(\alpha,\beta,\delta,M) + C_{i,dynamic}(\dot{\alpha},\dot{\beta},p,q,r) \tag{13}$$

where $C_i = C_L, C_D, C_Y, C_l, C_m, C_n$

Different methods were subsequently adopted to ascertain the static and dynamic components of these aerodynamic coefficients. Evaluation of static (basic) coefficient data (refer equation (14)) was performed employing Computational Fluid Dynamics (CFD) [7, 13] technique. All the parameters were obtained as a function of control effort, angle of attack (α), side slip (β) and mach number (M).

$$C_{i,static} = C_{Db}, C_{Lb},, C_{Yb}, C_{lb}, C_{mb}, C_{nb} \tag{14}$$

Dynamic coefficient data (refer equation (15)) consisting of damping/rate derivatives and acceleration derivatives were calculated utilizing 'USAF stability and Control DATCOM' [8] and other empirical techniques [9].

$$C_{i,dynamic} = dampingderivatives + Accelerationderivatives \tag{15}$$

where the damping and acceleration derivatives are defined in equation (16) and equation (17) respectively

$$dampingderivatives = (C_{Yr}, C_{lr}, C_{nr}) + (C_{Yp}, C_{lp}, C_{np}) + (C_{Yr}, C_{lr}, C_{nr}) \tag{16}$$
$$accelerationderivatives = (C_{L\dot{\alpha}}, C_{D\dot{\alpha}}, C_{m\dot{\alpha}}) + (C_{Y\dot{\beta}}, C_{l\dot{\beta}}, C_{n\dot{\beta}}) \tag{17}$$

Through this accurate estimation of the aerodynamic affects (refer equation (18)) were ascertained for the complete flight regime.

DOI: 10.1051/matecconf/201711401007

$$C_Y = C_{Yb} + C_{Y\dot{\beta}} * \dot{\beta} * (b/2V_T) + C_{Yp} * p * (b/2V_T) + C_{Yr} * r * (b/2V_T)\,,\; C_l = C_{lb} + C_{l\dot{\beta}} * \dot{\beta} * (b/2V_T) + C_{lp} * p * (b/2V_T) + C_{lr} * r * (b/2V_T)$$

$$C_L = C_{Lb} + C_{L\dot{\alpha}} * \dot{\alpha} * (\bar{c}/2V_T) + C_{Lq} * q * (\bar{c}/2V_T), \quad C_D = C_{Db} + C_{D\dot{\alpha}} * \dot{\alpha} * (\bar{c}/2V_T) + C_{Dq} * q * (\bar{c}/2V_T), \tag{18}$$

$$C_m = C_{mb} + C_{m\dot{\alpha}} * \dot{\alpha} * (\bar{c}/2V_T) + C_{mq} * q * (\bar{c}/2V_T), \; C_n = C_{nb} + C_{n\dot{\beta}} * \dot{\beta} * (b/2V_T) + C_{np} * p * (b/2V_T) + C_{nr} * r * (b/2V_T)$$

3 Trajectory Optimization

3.1 Generation of Optimized Velocity Profile

Generation of optimized velocity profile (ascertaining velocity, dynamic pressure and Mach no for the complete flight regime) that yields the greatest performance plays a substantial role in the effective design of flight vehicles. This is important since the use of ad-hoc profile or control policies to evaluate competing configurations may inappropriately penalize the performance of one configuration over another. Thus, to guarantee optimized trajectory, it is important to optimize the velocity profile and control policy for each configuration early in the design process. Subsequently, two different types of optimized velocity profiles were created to acquire ranges of 100 km and 120 km respectively once vehicle engine failure condition occurs (refer 'figure 1')

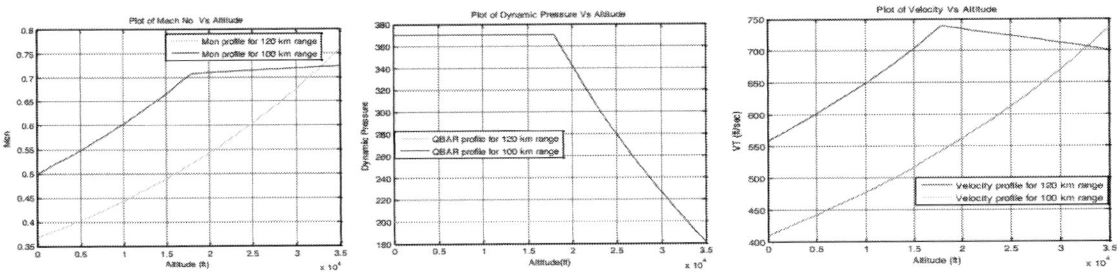

Figure 1. Generation of optimized velocity

In regards to the velocity profile for 328000ft (100 km) range, dynamic pressure of 370.5 is maintained for the altitude ranging from 0 to 18000 ft. Using the standard atmospheric model, Mach number (M) and Velocity (VT) to have the desired altitude and dynamic pressure are subsequently ascertained. For high altitudes (19000-35000ft) dynamic pressure is slowly varied to maintain desired mach number (refer Table 1). Similarly velocity profile for 120 km range was generated by maintaining a constant dynamic pressure of 200 at all the altitudes (0-35000ft). Using the standard atmospheric model, corresponding Mach number (M) and Velocity (VT) to have the desired altitude and dynamic pressure were calculated for all the individual 36 flight conditions (0-35000ft with an increment of 1000).

Table 1. Velocity profile for achieving 100km range

No	Altitude (ft)	Velocity (ft/sec)	Mach No	Dynamic Pressure	No	Altitude (ft)	Velocity (ft/sec)	Mach No	Dynamic Pressure
1	0	558.334	0.500	370.5	19	18000	738.79	0.707	370.5
2	1000	566.54	0.50	370.5	20	19000	736.74	0.708	356.32
3	2000	574.94	0.51	370.5	21	20000	734.67	0.709	342.57
4	3000	583.521	0.52	370.5	22	21000	732.56	0.710	329.24
5	4000	592.292	0.53	370.5	23	22000	730.47	0.711	316.31
6	5000	601.260	0.54	370.5	24	23000	728.35	0.712	303.78
7	6000	610.431	0.55	370.5	25	24000	726.26	0.713	291.65
8	7000	619.811	0.57	370.5	26	25000	724.03	0.714	279.89
9	8000	629.406	0.58	370.5	27	26000	721.84	0.715	268.51
10	9000	639.223	0.59	370.5	28	27000	719.63	0.715	257.49
11	10000	649.269	0.60	370.5	29	28000	717.40	0.716	246.83
12	11000	659.55	0.61	370.5	30	29000	715.14	0.717	236.51
13	12000	670.076	0.627	370.5	31	30000	712.8703	0.718	226.541
14	13000	680.852	0.639	370.5	32	31000	710.5723	0.719	216.8965
15	14000	691.88	0.652	370.5	33	32000	708.252	0.720	207.5758
16	15000	703.190	0.665	370.5	34	33000	705.9092	0.721	198.571
17	16000	714.769	0.679	370.5	35	34000	703.5435	0.722	189.8744
18	17000	726.632	0.693	370.5	36	35000	699.9891	0.721	180.8756

3.2 Trajectory Optimization

In order to ascertain steady states for the optimized trajectory, the problem was formulated as non-linear dynamic optimization problem, with an objective to determine, in open loop control, a set of time dependent decision variables that optimizes a specified performance index subject to certain constrains [14]. Matlab optimization toolbox was used to solve the problem utilizing *fminsearch* solver which is based upon 'Nelder-Mead simplex direct search' algorithm [15]. Steady state for the optimized trajectory were subsequently obtained for a coordinated turn flight as per the following constraints (refer equation (19)-(24))

$$\text{Performance function} = [\dot{V}_T, \dot{\alpha}, \dot{\beta}, \dot{p}, \dot{q}, \dot{r}] * \begin{bmatrix} 1 & 0 & 0 & 0 & 0 & 0 \\ 0 & 1 & 0 & 0 & 0 & 0 \\ 0 & 0 & 1 & 0 & 0 & 0 \\ 0 & 0 & 0 & 1 & 0 & 0 \\ 0 & 0 & 0 & 0 & 1 & 0 \\ 0 & 0 & 0 & 0 & 0 & 1 \end{bmatrix} * [\dot{V}_T, \dot{\alpha}, \dot{\beta}, \dot{p}, \dot{q}, \dot{r}]' \tag{19}$$

$$\text{Fixed States} = [V_T; h; \gamma; \psi; \dot{\phi}; \dot{\theta}; \dot{\psi}] \tag{20}$$

$$\text{Variable States} = [\alpha; \beta; LCF; RCF] \tag{21}$$

$$\text{Constrained States} = [\phi; \theta; p; q; r] \tag{22}$$

$$\text{Turn rate } (\dot{\psi}) = [-10\ -7\ -5 - 2\ -1\ 0\ 1\ 2\ \ 5\ 7\ 10]\ (\text{deg/sec}) \tag{23}$$

$$\text{Altitude} = [1000{:}1000{:}35000]) \tag{24}$$

Where [LCF,RCF] is the control vector for left and right control fins and all other variables already defined.

 DOI: 10.1051/matecconf/201711401007

This makes a total of 35x11 flight conditions, about which the trim points (for both states and controls) were ascertained for the generation of optimized vehicle trajectory. This enhanced flight envelope frame work was developed to ensure controlled vehicle movement under a wide range of operating conditions.

4 Open Loop Analysis

4.1 Model Linearization & Eigen Value Analysis

In order to linearize the non-linear system about the steady states trim points, simulink model was developed in matlab environment [16], in which the 6 DOF non-linear model along with all dependencies were incorporated. Matlab linearization tool linmod was used for model linearization, as a result of which, linearized state space representation of the model about the trim point was obtained for the optimized trajectory. Open loop analysis of the linearized system, revealed that system is unstable as it contained eigen values in open right half plane at various flight conditions.

As an example linearized system response at two different flight conditions corresponding to altitude 10,000ft , $\dot{\psi}$ =0 deg and altitude 30,000ft , $\dot{\psi}$ =0 deg is depicted at 'figure 2', which is clearly unstable as it contains eigen value in the open right half plane. The open loop unstable eigen values at various flight conditions was the cause of vehicle instability and movement in an uncontrolled manner under engine failure condition. System uncontrolled behavior was subsequently analyzed to ascertain system state response and to determine the feasibility of design of a control architecture which can stabilize the system and ensure system stability under such stated problem.

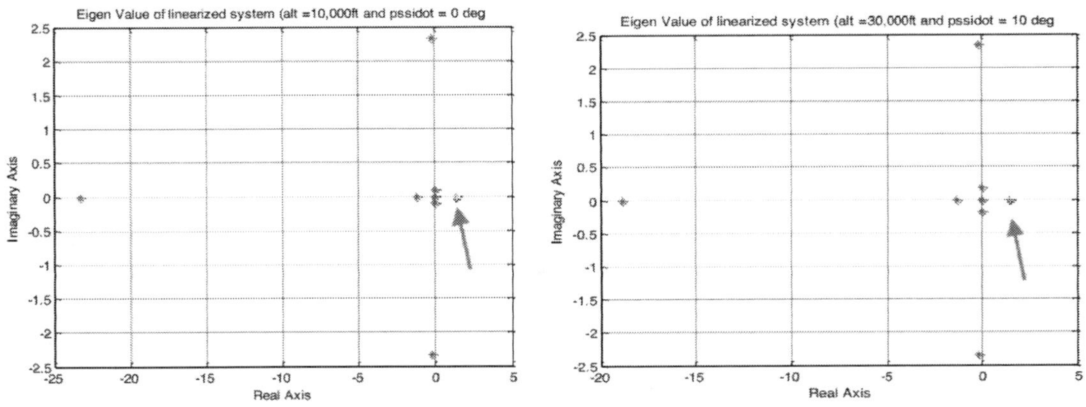

Figure 2. Eigen value analysis

4.2 Open Loop System Response

In order to analyze system response under engine failure condition, a simulink model was developed in the matlab environment. In this, the vehicle once subjected to the stated problem, showed a complete unstable response. The UAV instead of following the optimized trajectory refer 'figure 3 (a)', exhibited an unstable response with a glide range of just 15 Km (refer 'figure 3 (b)'). This problem of low range was encountered as the vehicle after a short time became totally unstable and exhibited a steep vertical decent instead of following the optimized trajectory.

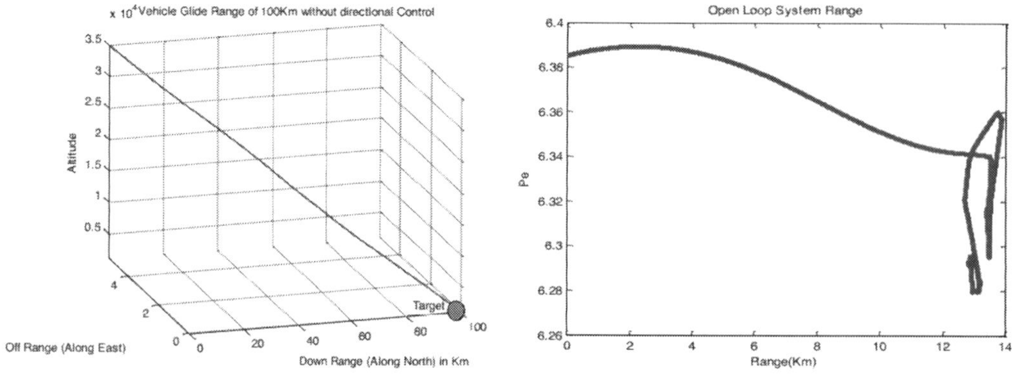

Figure 3. (a) Desired response (b) Actual system response

Apart from low range, system longitudinal and lateral dynamics also exhibited unstable responses (refer 'figure 4').

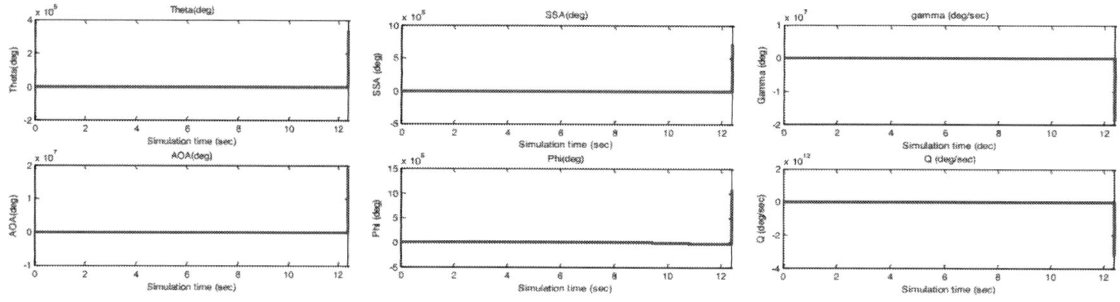

Figure 4. Open loop system unstable response

4.3 Stability Analysis

In order to keep the linearized system stable about the operating points and to make the system follow desired trajectory, requirement existed for the presence of a closed loop system with dedicated control architecture. Linearized system when evaluated for controllability and observability was found to be both controllable and observable. This ensured that control architecture can always be designed for close loop system stability and with the incorporation of appropriate stability controller, system states can be forced to follow the optimized trajectory. To design a closed loop system which can follow the optimized trajectory, response of various types of controllers such as full state feedback LQR controller [17], PID controller [18] (both SISO [19] and MIMO [20]) can be investigated. A promising choice in this regard could be the utilization of robust state feedback LQR controller in case the states are measurable or else an associated Kalman filter [21] can be designed for the state estimation. Implementation of such robust architecture will not only ensure close loop system stability but will also dramatically reduce the sensitivity of the system to modeling uncertainties and provides disturbance rejection to those quantities that detract from the desired aircraft motion. Apart from this, much enhanced vehicle glide phase range will also be achieved in comparison to the open loop range of just 15km. This range enhancement is a consequence of the system states forced to follow the optimized trajectory. In case controller is designed utilizing 6 DOF equations of motion, problem of system tracking the intended direction might be experienced. This is so because the estimates for directional orientation provided by navigational equation (refer equation (10-12)) are not used by other state equations. Due this reason, heading angle (ψ) is not controlled as per the actual direction of flight. Accurate navigation to the target, therefore requires incorporation of an additional tracking controller which utilizes the orientation information provided by navigational equation to compute the desired heading and then compares it with current heading to compute the difference in heading for necessary compensation. This computed value may be then subsequently utilized for steering the aircraft in the desired direction. With the incorporation of suitable close loop stabilization controller and a tracking controller, control system architecture can be developed to ensure UAV survivability under stated problem, and that can keep the vehicle stable and glide towards intended direction.

5 Conclusion

In this study, dynamic modelling and stability analysis of UAV was performed under engine failure condition. Vehicle dynamic simulation model was implemented through a set of highly coupled non-linear 6 DOF differential equations. Established empirical/non-empirical techniques were employed to accurately determine, the aerodynamic affects during various phases of the flight. Steady state analysis of the linearized system revealed various unstable modes that caused vehicle instability and short range of just few km. System was however ascertained to be fully controller and observable that guaranteed design of a close loop system for system stability. Vehicle response under stated problem was analyzed and a framework was presented to ensure vehicle survivability. Simulations results depicted considerable enhancement in the UAV range to 100km from free fall unguided range of just 15km. The potential benefit provided by this study remains two folds in which not only considerable range enhancement is achieved to attain high percentage of vehicle survivability but the intrinsic simplicity of the optimization algorithm guarantee its on board implementation. At the end a suitable control architecture comprising of stabilization and tracking controllers was also briefly discussed (scope of next work) that could provide steering commands to follow the optimized trajectory.

References

1. Dalamagkidis, K., K. Valavanis, and L. Piegl, *On unmanned aircraft systems issues, challenges and operational restrictions preventing integration into the National Airspace System.* Progress in Aerospace Sciences, 2008. **44**(7): p. 503-519.
2. Wild, G., J. Murray, and G. Baxter, *Exploring drone accidents and incidents to help prevent potential air disasters.* Aerospace, 2016. **3**(3): p. 1-11.
3. Scarselli, G., L. Lecce, and F. Nicolosi, *DEVELOPMENT OF A MALE TURBO-PROP UNMANNED AERIAL VEHICLE FOR CIVIL APPLICATION.* Aerotecnica Missili & Spazio, 2016. **86**(1): p. 22-35.
4. Cestino, E., *Design of solar high altitude long endurance aircraft for multi payload & operations.* Aerospace science and technology, 2006. **10**(6): p. 541-550.
5. Nickol, C.L., et al., *High altitude long endurance air vehicle analysis of alternatives and technology requirements development.* AIAA Paper, 2007. **1050**: p. 2007.
6. Roaskam, J., *Airplane Flight Dynamics and Automatic Flight Controls.* Vol. Part1. 2001: DARcoperation, USA.
7. Petterson, K., *CFD analysis of the low-speed aerodynamic characteristics of a UCAV.* AIAA Paper, 2006. **1259**: p. 2006.
8. *USAF Stability and Control Datcom.* Vol. 1. 1999: McDonnell Douglas Astronautics Company , St. Louis Division, St Louis, Missouri 63166.
9. Roskam, D.J., *Airplane Design Vol 1-7. Roskam Aviation and Engineering Corporation.* 1985
10. Tischler, M.B. and R.K. Remple, *Aircraft and rotorcraft system identification.* AIAA education series, 2006.
11. Wickenheiser, A.M. and E. Garcia, *Optimization of perching maneuvers through vehicle morphing.* Journal of Guidance, Control, and Dynamics, 2008. **31**(4): p. 815-823.
12. Jung, D. and P. Tsiotras, *Modeling and hardware-in-the-loop simulation for a small unmanned aerial vehicle.* AIAA Infotech at Aerospace, AIAA, 2007: p. 07-2763.
13. Buning, P.G., R.J. Gomez, and W.I. Scallion, *CFD approaches for simulation of wing-body stage*

separation. AIAA Paper, 2004. **4838**: p. 2004.
14. Maqsood, A. and T. Hiong Go, *Longitudinal flight dynamic analysis of an agile UAV.* Aircraft Engineering and Aerospace Technology, 2010. **82**(5): p. 288-295.
15. Lagarias, J.C., et al., *Convergence properties of the Nelder--Mead simplex method in low dimensions.* SIAM Journal on optimization, 1998. **9**(1): p. 112-147.
16. Li, H., X.-G. Jing, and L.-M. Xu, *The Simulation of 6-DOF Motion of Launch Vehicle Based on Matlab/Simulink [J].* Journal of Astronautics, 2005. **5**: p. 016.
17. Hespanha, J.P., *LQG/LQR controller design.* Undergraduate Lecture Notes, University of California, Santa Barbara, California, USA, 2007.
18. Rivera, D.E., M. Morari, and S. Skogestad, *Internal model control: PID controller design.* Industrial & engineering chemistry process design and development, 1986. **25**(1): p. 252-265.
19. Ramasamy, M. and S. Sundaramoorthy, *PID controller tuning for desired closed-loop responses for SISO systems using impulse response.* Computers & Chemical Engineering, 2008. **32**(8): p. 1773-1788.
20. Vilanova, R. and A. Visioli, *PID control in the third millennium.* 2012: Springer.
21. Hajiyev, C. and S.Y. Vural, *LQR controller with Kalman estimator applied to UAV longitudinal dynamics.* Positioning, 2013. **4**(01): p. 36.

Starting Fault Experiment of a Certain Turbo Shaft Engine

Xie Xiaodong[1,a], Li Benwei[1], Wu Heng[1] and Yu Fulei[2]

[1]Department of Airborne Vehicle Engineering, Naval Aeronautical and Astronautical University, Yantai Shandong 264001,China
[2]Aircraft Repair Station, Naval Aviation Institute, Changzhi Shanxi 046000, China

Summary: In order to diagnose the starting fail fault of the certain turbo shaft engine which often occurs in daily use, the experiments for the micro pump and the fuel filter were carried out by the method of contrast test. Through the comparison and analysis, the differences between domestic components and French-made ones were found. The results showed that: The outlet pressure of domestic micro pump is smaller under the engine's working condition; the flux of it is significantly lower under high outlet pressure; the pressure-flux characteristic line of domestic micro pump can close to the French-made pump when the French-made fuel valve and regulating valve were installed on it; the structure size of domestic filter frame is very different from the French-made one, which leads to its greater flow resistance. According to the performance indicators of French-made components, the improvement measures were put forward in order to improve the success rate of engine ground starting.

1 Introduction

As the heart of the helicopter, the turbo shaft engine must guarantee the stability of rapid starting and health work in the whole process of tasks [1]. After the localization of the certain turbo shaft engine's fuel system components, the fault of unsuccessful starting gradually appeared with the increase of installed sets and frequency of usage, especially in the first starting after a long time placement. Over the years, the problem has not been thoroughly solved, and it increased the helicopter mission preparation time and affected the normal flight.

Aiming at the faults of turbo shaft engine, X.X. Cai [2] developed the turbo shaft engine health management system for bench test and flight test phase; S.L. Hou [3] did the preliminary analysis for the basic composition and design characteristic of a certain turbo shaft engine fuel control system; T. Wang [4] studied the surge fault in the starting process by analysis the starting control process and fuel supply rate. But the study about the starting failure fault of this certain turbo shaft engine at home and abroad is less. Most of its components were developed by surveying and mapping the French-made ones, some manufacturing imperfections which affected the engine's working performance occurred because of the technical limits the lack of accurate and reliable components manufacturing standards. In this paper, according to the starting failure fault of this type of engine, contrast experiment of the domestic and French-made components were carried out by establishing the fault tree and analyzing the main bottom events. The experimental results show that, there are some big differences between some domestic components and French-made ones in the starting fuel system and

[a] Corresponding author : xxdxxd163@163.com

main fuel system, and they have great influences to the starting performance of the engine. Considering the vulnerabilities of domestic components, some corresponding improvement measures were put forward in order to improve the engine starting performance.

2 Fault Analysis

When the engine is starting, the starting fuel system is used to produce enough ignition torch to light the main combustion chamber and the main fuel system is used to guarantee the combustion stability of the main chamber. [5]In order to find out the cause of the starting failure fault accurately, the fault tree which is shown in Figure 1 were built according to the phenomenon and characteristics of it and combining with the structure and principle of the starting fuel system and main fuel system.

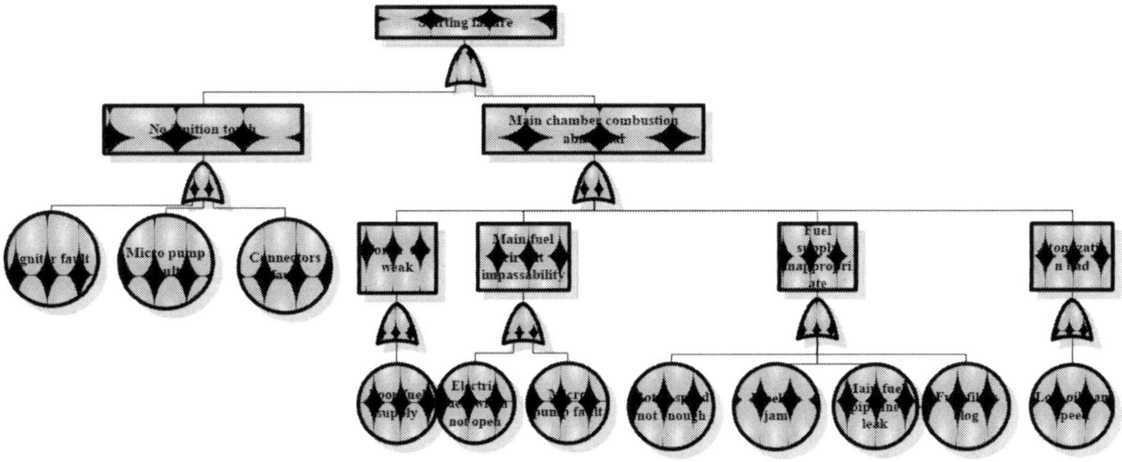

Figure 1. Starting failure fault tree of this type of turbo shaft engine.

According to the former troubleshooting conditions of this kind of engine, faults such as electric fuel switch not open, pipeline jam and main fuel pipeline leak are very rare, they are not the main events of starting fault, so we don’t take them as the research object. [6] But the components such as micro pump, igniter, starting fuel filter are close to the bottom events and they require in-depth research and analysis. In this paper, the experiments for the micro pump and the fuel filter were carried out by the method of contrast test. Through the comparison and analysis, the differences between domestic components and French-made ones were found, and the improvement measures were put forward in order to improve the success rate of engine ground start.

3 Experiments and analysis

3.1 Micro pump experiments

RCB-4 micro pump was produced by surveying and mapping the French-made pump. The purpose of it is to supply fuel to the igniter and switch on the fuel cock in the main fuel system. Its main components are one-way valve, regulating valve, fuel valve, etc. The structure of it is shown in Figure 2.

DOI: 10.1051/matecconf/201711401008

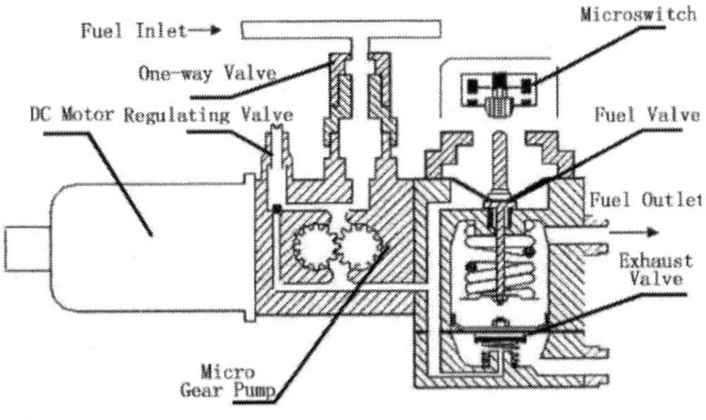

Figure 2. Structure of RCB-4 micro pump.

3.1.1 Micro pump-igniter experiment

Select French-made micro pump (NO.05AT) and domestic micro pump (NO.1011017), and connect them to the domestic and French-made igniter respectively. Connect them with other starting fuel system components through cross valve and fuel pipe to simulate the engine working condition. Set the inlet pressure of micro pump to 40kPa and carry out the atomization experiment under the voltage of 28V, measure the outlet pressure of it. The Table 1 is the experiment data.

Table 1. Micro pump outlet pressure.

Micro pump	Igniter	Outlet pressure（kPa）
05AT	Domestic	555
05AT	French-made	555
1011017	Domestic	523
1011017	French-made	530

Through the table we can find that when the same micro pump is connected to the different igniter, the outlet pressure of it are almost the same, it is obvious that the fuel holes in French-made and domestic pump have the similar throttling ability. When connected to the same igniter, the outlet pressure of French-made pump is about 550kPa, it is obviously higher than the domestic one, which is about 525kPa.

3.1.2 Pressure-flux experiment

Set the inlet pressure of the two micro pump above to 40kPa, perform the pressure-flux experiment under the voltage of 28V, and measure the flow rate of them under different outlet pressure.

Figure 3 shows us the pressure-flux curve of these two micro pumps. It can be seen that: when the outlet pressure is less than 400kPa, the domestic micro pump has the similar flux with the French-made one; when it is greater than or equal to 450kPa, the flux of French pump is greater than the domestic pump.

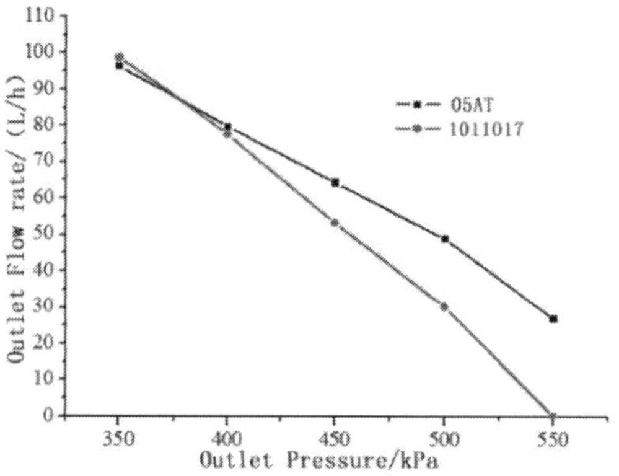

Figure 3. Flow rate of these two kinds of pump under different outlet pressure (L/h).

3.1.3 Pressure-flux experiment after adjusting regulating valve

Adjust the regulating valve of domestic micro pump, set the flux to 267L/h when the outlet pressure is 550kPa and the voltage id 28V, which is similar to the French-made pump. Then perform the pressure-flux experiment for these to pumps. The outlet flow rate of each pump is shown in Figure 4.

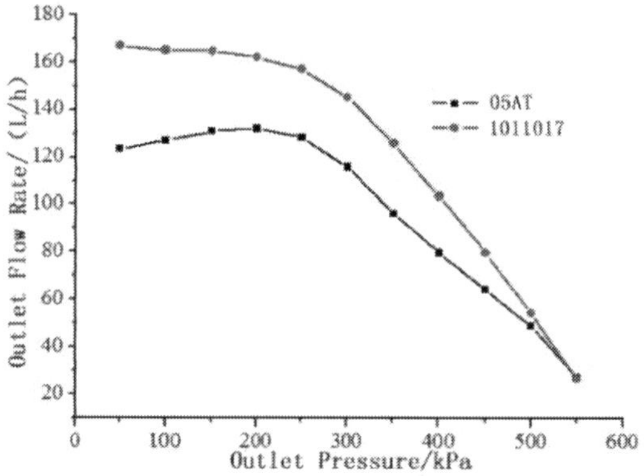

Figure 4. Flow rate under different outlet pressure after adjusting regulating valve (L/h).

After adjusting the regulating valve, the flow rate of domestic pump is promoted under the corresponding pressure, which can meet with the level of French-made pump at working point basically. But in this condition, the flow rate of domestic pump is clearly greater when the outlet pressure is under 500kPa.

3.1.4 Pressure-flux experiment when install French-made valves on domestic pump

Install the French-made regulating valve and fuel valve on the domestic micro pump and adjust them. Set the flow rate to 26.7L/h when the outlet pressure is 550kPa and the voltage is 28V, then set the flow rate to 98L/h when the outlet pressure is 25kPa. The results of this experiment are shown in Figure 5.

 DOI: 10.1051/matecconf/201711401008

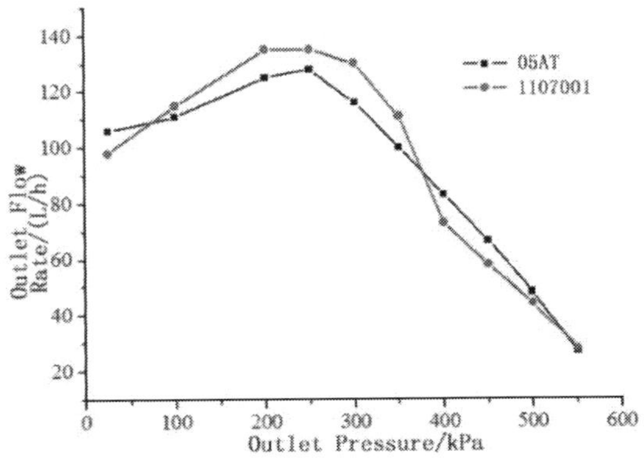

Figure 5. Flow rate under different outlet pressure after installing French-made valves on domestic pump.

The results show that after installing the French-made regulating valve and fuel valve and adjusting them correspondingly, the pressure-flux characteristic of domestic micro pump can be close to the French-made one.

3.1.5 Structure comparison

Decompose, measure and compare the domestic and French-made micro pump, then some differences can be found in the structure.
1) Regulating valve differences
There are four main diferences in the structure of regulating valve, which are listed in Table 2.

Table 2. Components structural differences in regulating valve.

Regulating valve	05AT	1107001
Discharge slot size（mm）	0.8-0.9	1.2
Steel ball diameter（mm）	Φ4	Φ4
Spring elasticity coefficient（N/mm）	0.49/0.44/0.46	0.7/0.67/0.69
Valve seat shape	With convergence	Without convergence

2) Fuel valve differences
After measuring, some differences in shaft and spring was found between domestic and French-made fuel valve. These differences are very obvious and can be seen in Figure 6.

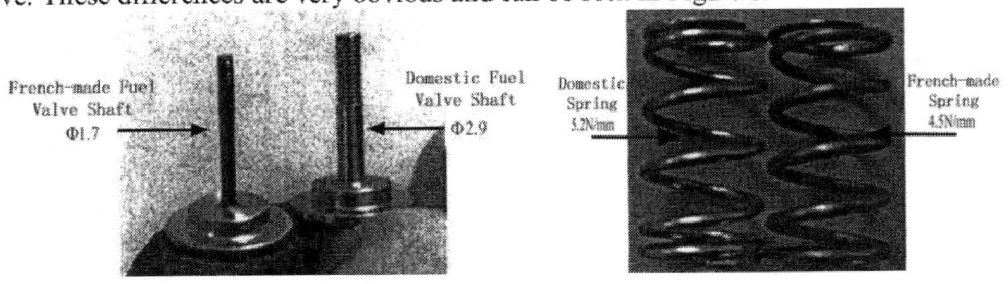

Figure 6. Structure differences in fuel valve.

3.1.6 Influence analysis

The main role of micro pump in the process of starting is to supply fuel to the igniter by pressure. The outlet pressure of domestic micro pump at the working point is lower than the French-made pump, which affects the fuel's atomization performance. This will reduce the flame intensity around the igniter and reduce the rate of lighting the main fuel successfully. The pressure-flux characteristic of micro pump is mainly determined by the motor and gear pump, and can be adjusted through the pressure regulating valve. The results confirm that, by adjusting the regulating valve and fuel valve the pressure-flux characteristic of domestic micro pump can be consistent with the French-made one at the working point. The structural differences of these two valves have great influences on the pressure-flux characteristic, so we can make the domestic pump have the performance which is as good as the French-made one by adjusting the structure of it. The specific improvement measures can be: Improve the elasticity coefficient of the spring in regulating valve and fuel valve to make them consistent with the French-made ones as far as possible; change the fuel valve shaft diameter from $\Phi 2.9$ to $\Phi 1.7$; remeasure and determine the installation compression of the fuel valve spring.

3.2 Starting fuel filter experiments

The starting fuel filter is installed between the fuel regulator and starting adjusting device. The fuel from the pump flows to the regulating device after the filtration, and the flow resistance of filter will directly affect the smooth degree of starting fuel. The fuel filter is mainly composed of shell, filter element, aprons, etc.

3.2.1 Flow resistance experiment

Select two domestic fuel filters (NO.Rl1-011-1, NO.RL-011-2) and two French-made filters (NO.J202B, NO.J135B). Take the RP-3 fuel as the test medium, measure the flow resistance of these filters under the flow rate of 1.7L/min, 2.2L/min, 2.6L/min and 3L/min at room temperature. The results of experiment is shown in Figure 7.

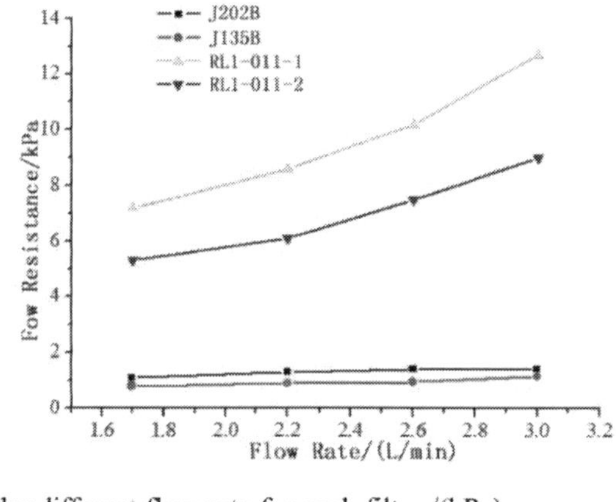

Figure 7. Flow resistance under different flow rate for each filter /(kPa).

The result confirm that, the flow resistance of French-made filter is much lower than the domestic filter in the same conditions.

3.2.2 Filtration precision experiment

Select these two kinds of filter, and measure the filtration ratio of them for the particle from 20 to 50 μm diameter. Figure 8 shows the results of filtration.

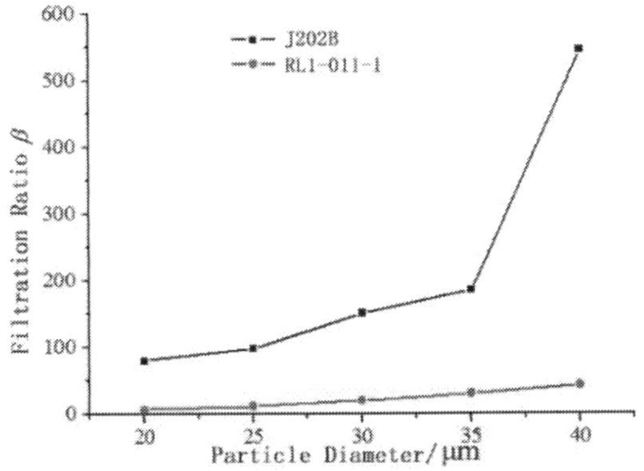

Figure 8. Filtration ratio for each diameter.

The original filtration precision requirement is $\beta(30\ \mu m) \geq 20$, it is obvious that the filtration precision of domestic filter is much higher than this, but the corresponding data of French-made is lower.

3.2.3 Structure comparison

By surveying and mapping, the diameter of French-made filter wire is 0.093~0.1mm, it is about the same size with the RL-1-011-1, which is 0098~0.101mm. But there are great differences in the frame structure and wire spacing. By looking under microscope (100 times magnified), these differences can be found. The microscope images are shown in Figure 9.

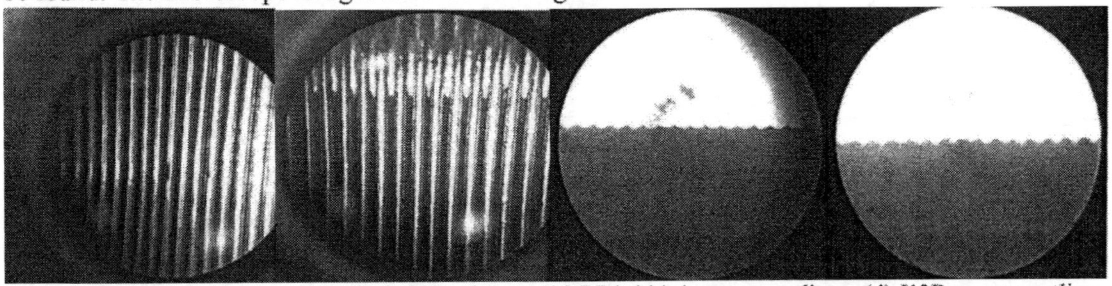

(a) RL1-011-1 wire (b) J135B wire (c) RL1-011-1 groove outline (d) J13B groove outline

Figure 9. Filter structure comparison under microscope(100 times magnified).

And there are also some structural differences of the filters which are listed in Table 3.

Table 3. Differences in filter frame structure.

Parameter	RL-1-011-1	J202B
Groove spacing	0.105~0.11mm	0.12~0.125mm
Number of holes	8×24	6×24
Hole spacing 1	5mm	7mm
Hole spacing 2	1.5mm	2.6mm
Wire slot depth	0.024~0.03mm	0.045~0.064mm
Slot angle	80°	90°

3.2.4 Influence analysis

The starting fuel filter is used for fuel filtration, mainly to ensure the cleaning of the fuel in air pressure regulator, starting regulator and fuel regulator. Its flow resistance will directly affect the flow rate of main fuel pipe in the process of starting, because it is installed between the starting regulator and fuel regulator. Confirmed by the manufacturer, the French-made filter is able to meet the filtration precision. The flow resistance of domestic filter is greater than the French-made one, which reduces the flow rate of fuel in the process of starting. The low flow rate of the main fuel will cause the low ignite possibility of it, which influence the starting performance. Therefore, the filtration precision can be reduced appropriately to reduce its impact on the main fuel flow rate.

4 Conclusion

In this paper, according to the starting failure fault of this type of engine, contrast experiment of the domestic and French-made micro pump and fuel filter were carried out by establishing the fault tree and analysing the main bottom events. By comparing and analysing the experiment data we can get that:

1) There are great differences in manufacturing process between domestic and French-made fuel valve and regulating valve in micro pump, which impact their performance distinctly. The outlet pressure of domestic micro pump is lower at the engine's work condition, and its flow rate is also lower than the French-made pump under high outlet pressure. This situation improves by installing the French-made fuel valve and regulation valve on it.

2) The flow rate of domestic fuel filter is much greater than the French-made filter. There are plenty of differences in filter frame structure between them, which make the filtration of domestic filter higher than the French-made one, and it's much higher than the requirement of $\beta(30\mu m) \geq 20$.

3) The structure of domestic micro pump and starting fuel filter is unfavourable to the fuel flow, and reduce the fuel supply to igniter and fuel pan in the starting process. The starting success rate of this engine can be increased by improving the accuracy of surveying and mapping and perfecting the components performance requirements.

References

1. Arita M M, Syrmos V L, Vian J L, et al. Multi source data integration for aircraft health management.J. Montana: IEEE Aerospace Conference, (2008).
2. X.X. Cai, X.J. Guo, C.L. Wu, Investigation of EHM technique for turbo shaft engines.J. Aeroengine, **39**, 39(2013).
3. S.H. Hou, Fault analysis of fuel control system for WZ-8A engine.J. Heli. Tech. **4**, 30(1998).
4. T. Wang, H.G. Zhao, J. Jian, Study on starting control law and trouble shooting of turboshaft engine.J. Aero. Sci. & Tech. **27**, 20(2016).
5. X.H. Peng, Y.B. Hu, K.R. Zhang, A study on the failure of engine start parameter, J. Gas Tur. Exp. and Res.J.18, 38(2005).
6. H.W. Li, J. Hong, Integrated fault detection and diagnosis system for a turboshaft engine.J. Aeroengine, **30**, 56(2004).

 DOI: 10.1051/matecconf/201711401009

Virtual Design and Dynamic Simulation of Valve Trains

Guantao Xuan[1,2] and Yuanyuan Shao[1,a]

[1]*College of Mechanical and Electronic Engineering, Shandong Agricultural University, 271018 Tai'an, China*
[2]*Shandong Province Key Laboratory of Horticultural Machinery and Equipment, 271018 Tai'an, China*

Abstract. The valve train plays an important role in the engine, which has a great influence on engine performance and reliability. In this paper, valve trains are studied based on multi-body dynamics theory and virtual prototype technology. Multi-rigid-body dynamic model of the valve train is established with Adams software, and then simulation and calculation are carried out. Dynamic characteristics of the valve train are obtained including force and motion of each component. The research results have important theoretical significance and practical engineering value in the design, dynamic analysis and optimization of valve trains.

1 Introduction

The valve train in an internal combustion engine is responsible for controlling the air/fuel mixture into the cylinder during intake and to allow the exhaust gases to exit after combustion [1].It carries out complicated and heavy working, and suffers intense vibration and shock [2]. Operating these engines in high speed usually causes some dynamic problems such as spring surge, friction and fatigue damage, valve bounce, lifter/cam pair separation, etc. So it is important for engines to improve the valve train's design quality. Traditionally, valve trains are designed by project comparison and tests, which are waste of time and delay the research period. The valve train is also designed as an elastic dynamic system, but it is calculated complicatedly and not demonstrated clearly [3-6]. With the development of computer, CAE simulation technology is more and more applied to design the valve train and predict its characteristics, such as finite element analysis, CFD and Adams software [7-11].

In this paper, applying multi-body dynamics theory and virtual prototype technology, the multi-body dynamic model of the valve train is established with Adams software, and simulation is carried out by using the virtual prototype technology. The dynamics of components are obtained such as the lift, velocity, acceleration of valve, the contact stress between cam and lifter. The results show that the simulation of the valve train is feasible during the initial developing stage without physical prototype.

2 Multi-body dynamic equation of valve trains

Valve trains typically are composed of camshafts, lifters, pushrods, rocker arms, valves and valve springs. Therefore, the valve train can be considered as a multi-body system composed of many parts, and its analysis is carried out by using the multi-body dynamics theory. Using the Lagrange equation as the multi-body dynamic analysis method, dynamic equation of the valve train is established. Every

[a] Corresponding author : xoyo111@163.com

rigid-body i of the valve train is described with six generalized coordinates as shown in equation (1), which are Cartesian coordinates of mass center in the inertial reference frame and Euler angles reflecting azimuth of the rigid-body.

$$q_i = [x, y, z, \phi, \theta, \varphi]_i^T, \quad q = [q_1^T, \ldots\ldots q_n^T]^T \tag{1}$$

And differential equation of the valve train motion is also established as in (2) with lagrangian multiplier method.

$$\frac{d}{dt}\left(\frac{\partial T}{\partial \dot{q}}\right)^T - \left(\frac{\partial T}{\partial q}\right)^T + \varphi_q^T \rho + \theta_{\dot{q}}^T \mu - Q = 0 \tag{2}$$

Integrity constraint equation is defined as follows.

$$\varphi(q, t) = 0 \tag{3}$$

Non-holonomic constraint equation is generated as follows.

$$\theta(q, \dot{q}, t) = 0 \tag{4}$$

Where, T is system energy, q is generalized coordinates array, Q is generalized force array, ρ is lagrangian multiplier array for integrity constraint equations, and μ is lagrangian multiplier array for non-holonomic constraint.

3 Multi-body dynamic model

The geometric model of the valve train is established by using multi-body dynamics software Adams/engine, and then the constraint, drive and load are defined as shown in Table 1 and Table 2. Finally, the multi-rigid body dynamic model of the valve train is established as shown in Figure 1.

Table 1. Component parameters

Symbol	component	Quality (Kg)	Rotary inertia(Kg·mm2)		
			Ixx	Iyy	Izz
A1	Cam	0.1688	22.6021	30.9096	49.4604
A2	Cam Shaft	0.221	50.640	46.560	28.812
A3	Valve	4.2708	43.5432	43.5432	1.8958
A4	Spring Seat	1.8821	1.2187	1.2187	2.4091

Table 2. Main constraints

Symbol	Object	Constraint
B1	Cam and Ground	Revolute joints
B2	Cam Shaft and Cam	Fixed Joints
B3	Cam and Lifter	Contact Joints
B4	Valve and Ground	Cylinder Joints
B5	Valve and Ground	Collision Constraint
B6	Valve and Spring Seat	Fixed Joints

DOI: 10.1051/matecconf/201711401009

Figure 1. Multi-body dynamic model

4 Dynamic simulation and analysis

After establishing the virtual prototype model of the valve train, a rotary drive is applied to the camshaft to simulate the actual motion. When the engine works at the rated speed 3000r/min, multi-body dynamic simulation of the valve train is carried out, and the camshaft, cam, valve, valve springs, and valve performance are obtained and analyzed.

4.1 Force between cam and lifter

In valve trains, the contact force between cam and lifter is very lager, and is prone to premature wear or scratch, split and other faults. From Figure 2, it can be seen that the maximum force between cam and lifter is 3740N. This impact force will do harm to the valve train and engine if it can't be controlled properly.

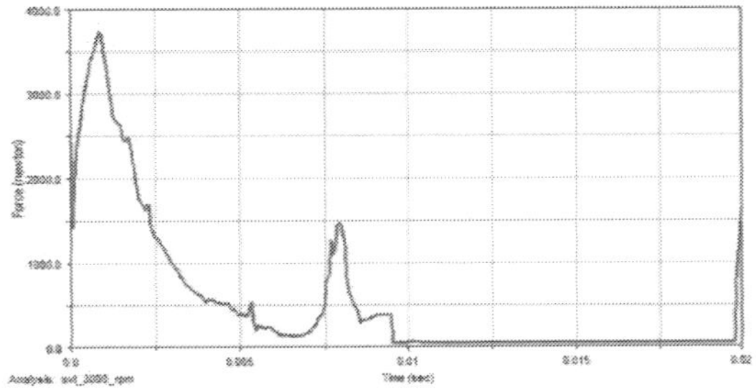

Figure 2. Force between cam and lifter

4.2 The valve seating force

In valve trains, the valve is suffering continuous impact from the valve seating force under no lubrication. The friction and abrasion caused by the valve seating force are important reasons for the failure of the valve train to work. Figure 3 shows the maximum valve seating force is 2600N, which is caused by burst pressure of valve as shown in Figure 4.

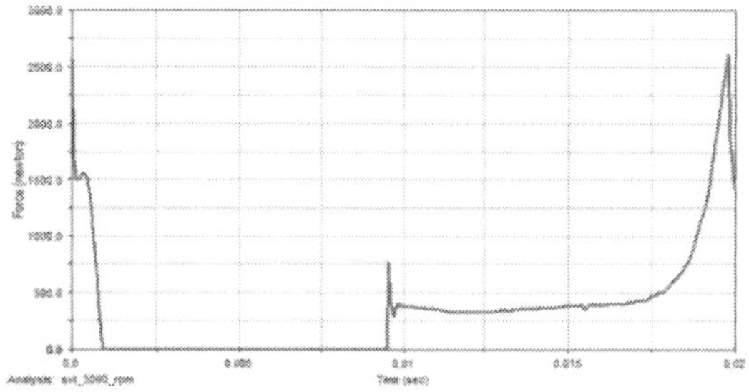

Figure 3. The valve seating force

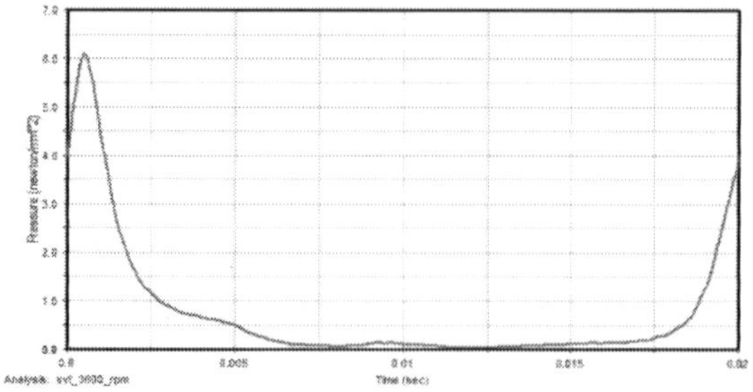

Figure 4. Pressure of valve

4.3 Motion of valve

Motion of valve is generated as shown in Figure 5 and Figure 6, and some conclusions can be drawn: at rated camshaft speed 3000 r/min, valve lift is about 8.5mm, and the maximum velocity of valve is 3540mm/s. Through Adams simulation, a variety of force and kinematic parameters are provided for the further finite element analysis. These data can also be compared with experimental data in order to further optimize the model.

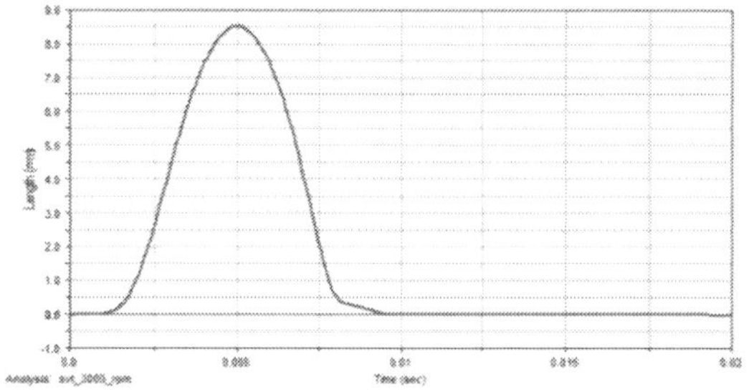

Figure 5. Valve lift

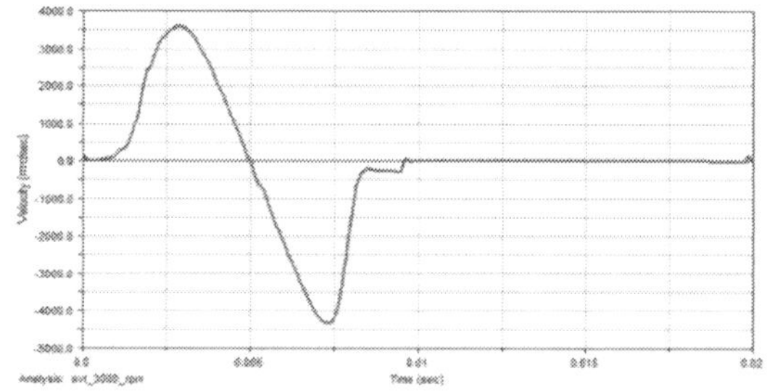

Figure 6. Velocity of valve

4.4 Acceleration analysis of valve

Valve acceleration is an important dynamic parameter of valve trains. There is a great acceleration impact at the moment of valve opening and seat, and the impact should be reduced as much as possible. In the cam profile design process, the positive peak acceleration and reverse peak acceleration are often used as constraints to control the cam contact stress and prevent the valve movement of valve trains. In this section, the valve acceleration is taken as the analysis object, and the following are the acceleration curves of the valve under different camshaft speeds.

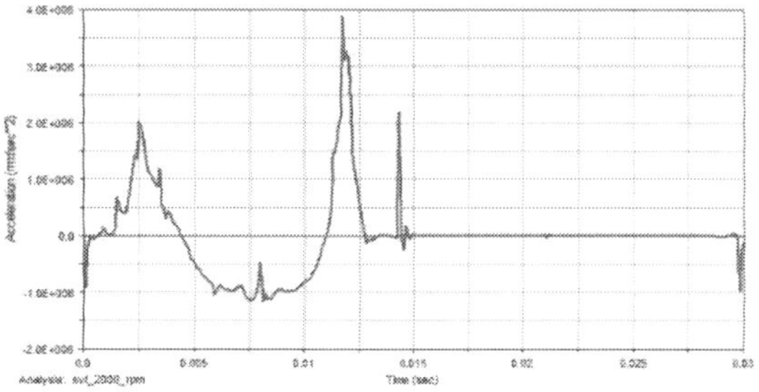

(a) at the rated camshaft speed 2000r/min

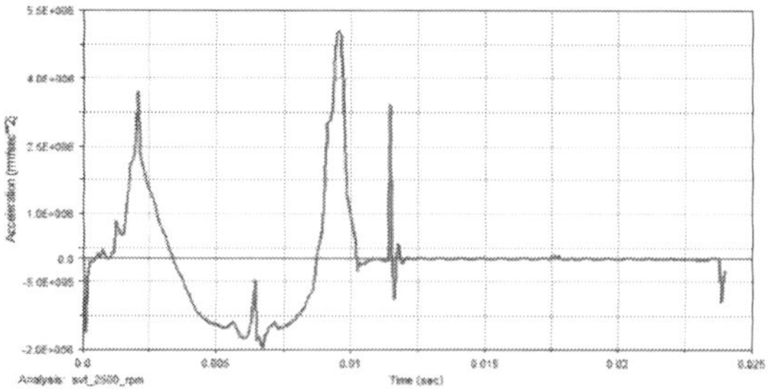

(b) at the rated camshaft speed 2500r/min

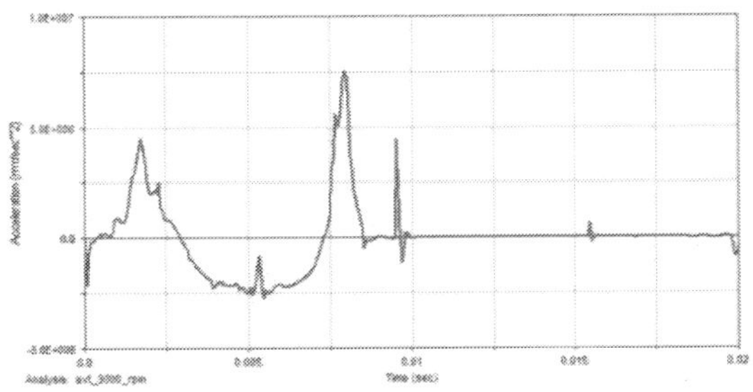

(c) at the rated camshaft speed 3000r/min

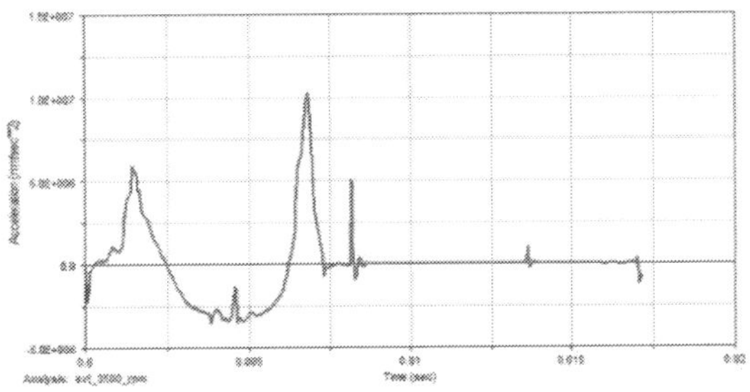

(d) at the rated camshaft speed 3500r/min

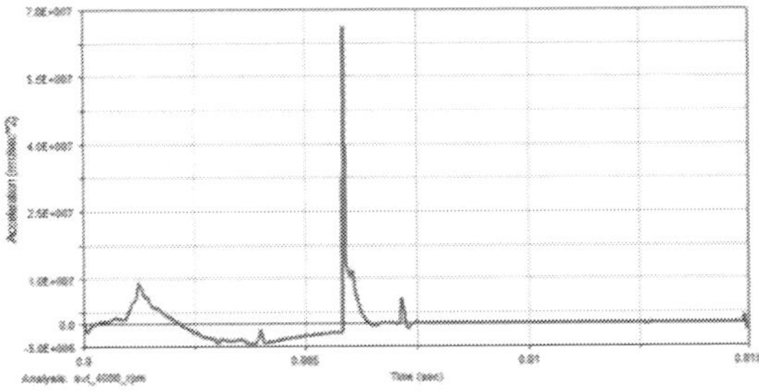

(e) at the rated camshaft speed 4000r/min

Figure 7. Acceleration of valve under different camshaft speeds

Some analysis conclusions are drawn from Figure 7 (a) - (e). With the increase of camshaft speed, valve acceleration is more and more high, which brings greater impact force to valve and has a great influence on engine operating characteristics. There is acceleration vibration of the valve at different camshaft speeds. But the valve vibration is weak at low speed, and there is obvious vibration at high speed. The negative acceleration width is one of the important indexes in the design of the cam profile. The greater the width is, the longer the valve opening time is and the larger time-area value is. As can be seen from the Figure 7, negative acceleration section of valve is wide at low speed. With increasing speed, negative valve acceleration section becomes gradually narrow. It shows engine can obtain higher time-area value in the low speed stage, and the time-area value is relatively low in the high speed stage due to the work cycle time shortened.

5 Conclusions

In this study, it is concluded that the valve train model is established well using multi-body dynamics theory, and dynamic response of the valve train can be analysed truly by simulation with Adams software. It is considered that the dynamic analysis of the valve train by using the virtual prototype technology is an important method for the modern design of valve trains.

Acknowledgments

We would like to thank "National Natural Science Foundation of China (31671632)", "Natural Science Foundation of Shandong Province (BS2015NY012)" and "Key Research and Development Program of Shandong Province (2016GNC112007)" for financial supports.

References

1. C. F. Taylor, *The Internal-Combustion Engine in Theory and Practice* (The M.I.T.Press, 1985)

2. Simon K. Chen, Advanced Cylinder Head, Valve Train, Combustion and Fuel Systems for High Power-Density Truck Diesel Engines, Alternative Fuels and Heavy Duty Engines, **1247**(1997)
3. Rohde S M, Whicker D, and Browne A L, Dynamic Analysis of Elasto-Hydrodynamic Squeeze Films, ASME Journal of Lubrication Technology **98**, 3(1976)
4. H. Sakai And K. Tsuda, Analysis of valve motion in overhead valve linkages, Bulletin of the JSME **13**, 55(1970)
5. A. P. Pisano, and F. Freudenstein, An experimental and analysis investigation of the dynamic response of a high-speed cam-follower system, Part 1:Experimental Investigation, ASME J. of Mechanisms, Transactions, and Automation in Design **105**, 12(1983)
6. Zuomin Liu, and Childs THC, Material dissipative in automatic engine exhaust valve-seat wear dissipative processes in tribology, Tribology Series, **27**(1994)
7. Liu H, and Kim D, Effects of end coils on the natural frequency of automotive engine valve springs, International Journal of Automotive Technology **10**, 4(2009)
8. Cardona A, and Geradin M, Kinematical and dynamic analysis of mechanical with Cams, Computer Methods in Applied Mechanics and Engineering **103**, 2(1993)
9. Qin Wen-jie, Liao Ri-dong, Zuo Zheng-xing, and Liang Li-wei, Applying multibody system dynamics to the improvement of valve train's cam profiles, Journal of Beijing Institute of Technolog **11**, 2(2002)
10. E. Sher, and T. Bar-Kohany, Optimization of variable valve timing for maximizing performance of an unthrottled SI engine-a theoretical study, Energy, **27**(2002)
11. J.Lampinen, Cam shape optimization by genetic algorithm, Computer-Aided Design, **35** (2003)

Modelling and Simulation of Vessel Alongside Replenishment Motion Response Based on Sea Conditions Analysis

Zhengnan JI[1,a], Limin TAO[1] and Wei LV[1]

[1]Science and Technology on Integrated Logistics Support Laboratory, National University of Defense Technology, 410073 Changsha, P.R. China

Abstract. Offshore accurate replenishment technology is a vital basement of conducting ocean research and deep ocean exploitation for every advanced country. However, It's necessary to compensate waving for ensuring the security and accuracy of offshore project due to the fact that wave makes vessel generate 6-DOF motions. This paper set offshore accurate supplying for vessels as the background, basing on Froude-Krylov assumption, to do research on 6-DOF motion responses of different sea conditions. This paper builds the vessel motion response formulas on the basis of the 3-dimensional potential theory and then resolve it by AQWA. The simulation results indicate the regularity of vessel motions, which can be considered as requirement indicators during designing AHC (Active Heaving Compensation) system.

1 Introduction

The vessel sailing in the blue ocean is not only an important platform for conducting ocean research for every institutes but also an elementary unit of every country's navy for executing the escort task [1]. Moreover, with the increasingly complexity and high-consuming during the offshore working, the lifting of target is becoming an essential project gradually.

Offshore replenishment refers to the process of vessel offering supplies (energy, oil etc.) to other vessels in the ocean [2], which is a significant way to achieve deep ocean sailing for vessels. Generally speaking, offshore replenishment can be divided into four styles according to the double vessels' relative positions, including horizontal replenishment, longitudinal replenishment, vertical replenishment and alongside replenishment. Their specifications have been shown in the Table 1 [2-5].

Table 1. The specifications of four kinds of replenishment.

Types	Theorical Accuracy	Loads per Time	Theorical Security	Financial Cost	Technology Maturity
Horizontal	Fair	Good	Fair	Lowest	Good
Longitudinal	Poor	Fair	Poor	Low	Fair
Vertical	Good	Poor	Good	High	Fair
Alongside	Very good	Very good	Good	Low	Poor

[a] Corresponding author: banzhu365@163.com

The vessel will generate 6-DOF motions under the restoring force and inertia force of the seawater due to the influence of wind force [3]. There are translation vectors paralleled to X, Y, Z axis (surging x, swaying y, heaving z) and rotation vectors using X, Y, Z axis as pivot respectively (rolling φ, pitching θ and yawing ψ). In order to compensate for the wave motions in the alongside replenishment, AHC technology is always adopted to ensure the security of hoisting. It is necessary to gain the knowledge of vessels time response and its amplitude for acquiring the range of detection and control in the development of AHC mechanism. Therefore, the project concerning alongside replenishment is vital to comprehensive analysis of double vessel relative motions in the sea.

So far, a variety of institutes has conducted some research in different fields. Yan built the differential formulas of vessel motions and solved it with Runge-Kutta algorithm [4]. Shi conducted the frequency domain's research based on the three-dimensional potential theory and then built the simulation model of alongside replenishment [5]. Lu designed the infinitely variable speeds mechanism to conduct simulative experiment of alongside replenishment for missile supplying between two vessels [6]. Ebbesen studied on the FPSO's structure whose anchor lines and production risers are linked to the floater and then simulated the vessel motions based on SIMO and RIFLEX [7].

2 Geometry Modelling and Relative Theory

The vessel models are derived from comprehensive replenishment vessel and warship served in the navy. According to the actual condition of replenishment in the sea, the model scheme has considered mooring and protection in the simulation process.

2.1 Vessel Models Selection

Take secrecy into consideration, the paper adopts approximate data to do research. We adopt non-uniform scaling methodology to model vessel. Details about modeling are shown in the Table 2.

Table 2. The parameters of two vessels.

Type	Length	Length Ratio	Beam	Beam Ratio	Gravity Height	Water Line
A	171.4 m	0.840	24.6 m	0.683	3.0 m	6.0 m
B	156.8 m	0.768	18.9 m	0.525	2.0 m	4.0 m

The paper designs a prototype vessel whose scale is 204m×36m×16m. Two vessels were generated by scaling the vessel and their center of gravity are both located in the 0.5m along the Z axis positive direction. Then we can acquire rotation rigidness value from these three empirical equations.

$$K_{xx} = 0.34 \times Beam\,,\; K_{yy} = 0.25 \times Length\,,\; K_{zz} = 0.26 \times Length \quad (1)$$

In order to describe the wave, vessel motion and flow field, we introduce these three coordinates showed in the Figure 1. The 6-DOF motion of single vessel are showed in the Figure 1 as well.

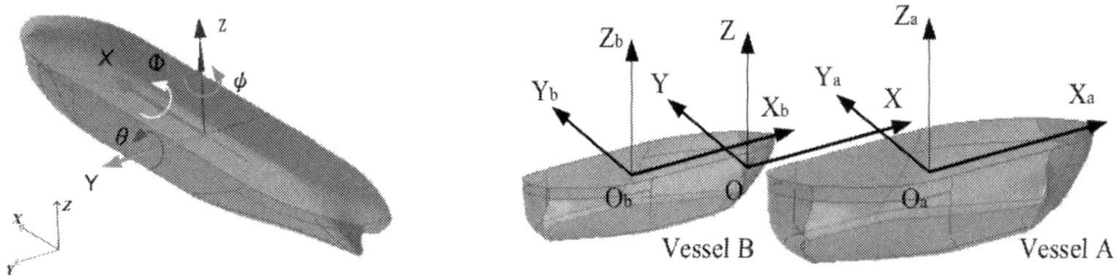

Figure 1. Coordinates setting.

In the Figure 1, there are world fixed coordinates O-XYZ (O-XY plane is located in the undisturbed plane. O-Z axis is vertical to upright.) and vessel motion coordinates O_i-$X_iY_iZ_i$ (i=a, b) fixed in the vessel (coordinates O_i is selected in the gravity center. X_i axis is paralleled to vessel baseline pointing at vessel bow. Z_i axis is vertical to vessel waterline.). The second coordinates will not sway with the motion of the vessel and its Z axis will keep vertical to upright forever. Wave angle is represented by m, Wave is called top wave when m=180° and wave is called starboard wave when m= 90°.

2.2 Three-Dimensional Potential Theory

Potential theory is the prevalent solution to calculate the wave force of structural surface which sets velocity potential, Laplace formula and four boundary conditions. The velocity potential can be uniquely determined according to Laplace formula and the boundary conditions. Then, the force of the wet surface can be calculated by Bernoulli Formula [8]. There is some difference between single vessel and double vessel working in the sea. It is that the velocity potential can be divided into wave incident potential Φ_I, self-diffraction potential, adjacent diffraction Φ_D and radiation potential Φ_R [5].

$$\Phi = \Phi_I + \Phi_D + \Phi_R = \mathrm{Re}\{[\phi_I(x,y,z) + \phi_D(x,y,z) + \phi_R(x,y,z)]e^{-i\omega t}\} \tag{2}$$

It should be noted that there is an assupmtion (Small Amplitude Wave Theory) that the fluid is ideal for incompressible fluid, non-sticky, without spin, and infinite depth.

3 Loads of Alongside Replenishment

Sea loads refer to the loads derived from environmental impact indirectly or immediately, including all of external forces (mooring force, motion inertial force and sloshing force etc.) [9]. In order to describe the model clearly, the paper considered and calculated the wind, wave, current sea loads, mooring loads and protection loads based on Froude-Krylov assumption, as shown in the Figure 2.

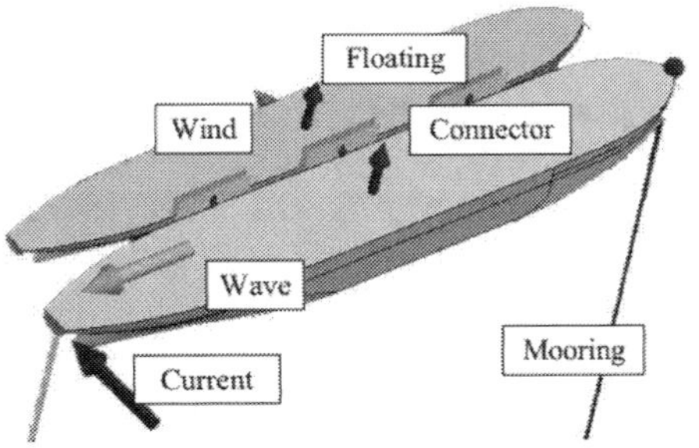

Figure 2. The sketch charts of sea loads.

Meanwhile, the paper ignores unimportant factors like deck loads and vibration derived from Karman Vortex. The 6-DOF motion formula of the structure can be listed based on Newton's second law.

$$\sum_{k=1}^{6} M_{jk}\ddot{x}_k = F_{HS} + F_{EX} + F_R + F_M \tag{3}$$

3.1 Environmental Wind Load

As for the infinite working platform, the minimum wind velocity should be 51.5m/s and 36.0m/s in different working mode respectively [10]. Environmental wind load can be calculated by the formula according to the selected wind velocity.

$$P = 0.613\times10^{-3} C_h C_S S V^2 \tag{4}$$

Among that, P is the wind force. S is the positive projection area. C_h is height coefficient. C_s is shape coefficient. V is designed wind velocity. The C_h and C_s should be selected by height and shape.

3.2 Wave Effect Load

Wave load p is able to be divided into three parts corresponding to incident potential, diffraction potential and radiation potential. The first part is the force dominating the distorted force. The second part is diffraction force. The third part is liquid impact inverse force induced by radiation potential. The detailed equation is shown in the Formula 5. ρ is density of seawater and z is depth of it.

$$p = -\rho g z - \rho\frac{\partial \Phi}{\partial t} = -\rho g z - \rho(\frac{\partial \Phi_I}{\partial t} + \frac{\partial \Phi_D}{\partial t} + \frac{\partial \Phi_R}{\partial t}) \tag{5}$$

Finally, the wave force of the structure F and M can be acquired with these following Formulas 6.

$$F = \iint_S -pnds \quad M = \iint_S -p(\mathrm{r}\times n)ds \tag{6}$$

3.3 Current Effect Load

The designed flow velocity should be selected to the maximum velocity, which includes tidal current, velocity storm, surge wind velocity. In addition, the vertical flow velocity distribution should be considered. The flow effect load can be calculated by Formula 7 if we only consider current.

$$F = \frac{1}{2}C_D \rho_W V^2 A \tag{7}$$

Among that, C_d is drag force coefficient. P_w is seawater density. V is designed current velocity. A is the projected area of the member on the plane perpendicular to the flow velocity v_c.

$$V_c = \frac{1}{T}\int_0^T (v_c)^2 ds \tag{8}$$

Designed current velocity V_c can be calculated by Formula 8 when we apply diffraction theory.

3.4 Mooring and Protection Mechanism

It is difficult to control the vessels relative motions in the high level sea conditions. Thus we have to let the vessel moored to the bottom of the ocean. Meanwhile, for the sake of security, replenishment vessel and warship are supposed to be connected by cables and fenders [5].

The cables between two riggings should be flexible and its length is supposed to be suitable. It will generate trailing reaction if the riggings are too long. However, the flexible riggings will become a rigid body if the riggings are too short. Moreover, it is necessary to design protections mechanism (fenders) between two vessels because the vessels will generate clash in the process of replenishment. The fender is floating structure which cannot only keep safety but also decrease financial cost.

4 Motion Response Simulation Based on AQWA

With the development of finite element theory, more and more institutes try to apply the AQWA to solve water hydrodynamic and hydrostatic problem [11]. The paper adopts the AQWA to build the analysis model in the simulative condition and explains the hydrodynamic problem. The red curves refer to replenishment vessel and blue curves refer to the warship in all of the following figures.

4.1 RAOs Analysis of Still Water

AQWA-LINE module can calculate the wave force derived from diffraction and radiation around arbitrary shape structure. This module will use typical Green Function to resolve the wave force of floating structure. Each element in the gridding will be given a simulating source.

As can be illustrated in the Figure 3, there is a peak frequency in every DOF motion in the simulative curves except for the Y axis translation. The reason is that the fenders and cables can offer the comfortable restraint. The motions heaving z, rolling φ and pitching θ are important factors in the process of hydrostatic analysis. They are about 0.81 rad/s, 0.57 rad/s and 0.82 rad/s for z, φ and θ respectively. The phenomenon indicates that the replenishment process is able to avoid the resonance frequency easily. And it can draw a conclusion that the model can be applied to do research in terms of hydrodynamic analysis in the next section. By the way, the resonance frequency can be adjusted by the rigidness of the fender in the actual application.

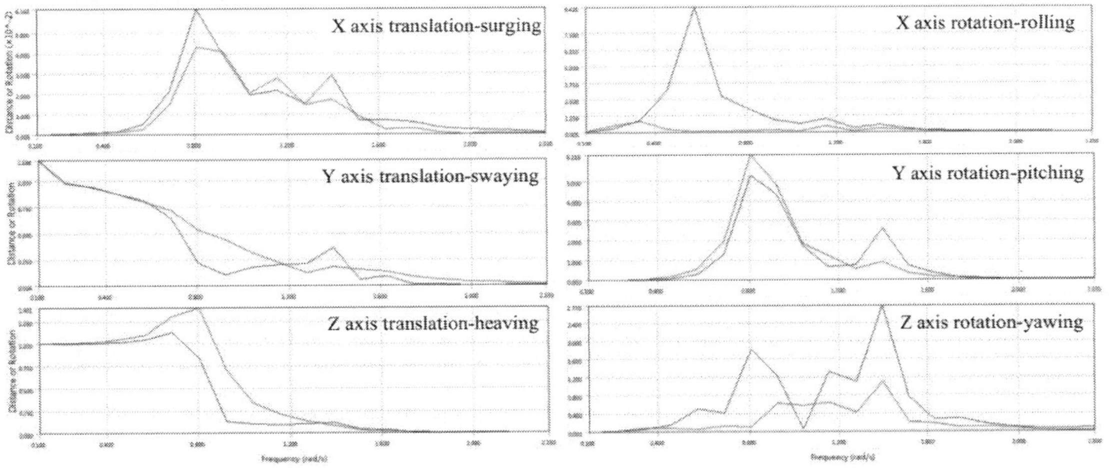

Figure 3. RAOs curves in the still water analysis.

4.2 Time Response Simulation

The AQWA-DRIFT module can be applied to calculate the time response and the corresponding loads in the total process under the random wave condition.

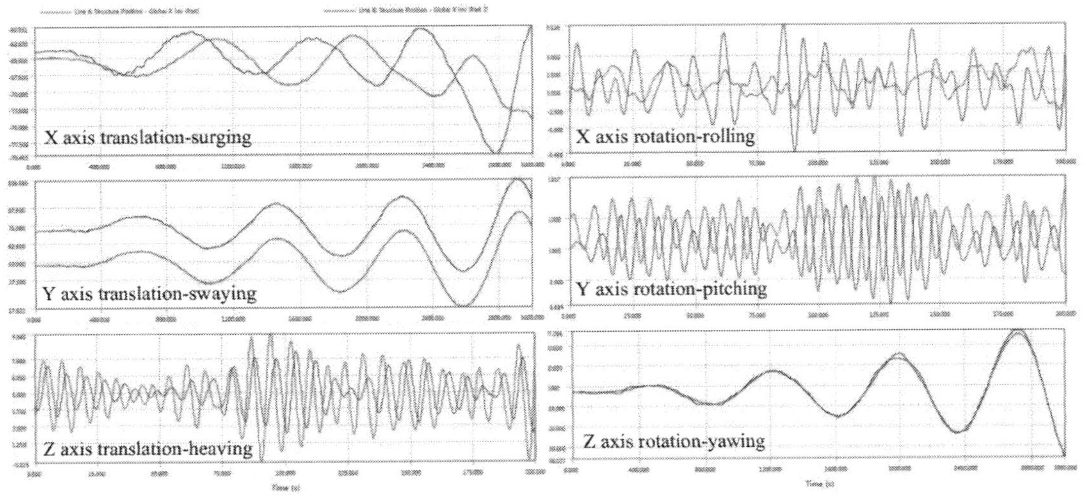

Figure 4. Time response curves of double vessel.

As can be illustrated in the Figure 4, surging x, swaying y, yawing ψ are low frequency motions due to the existence of cables and fenders again. We are only interested in their amplitude maximum values. They are 4.413m, 32.86m and 30.5° (absolute values of single vessel motion data) for x, y and ψ respectively. High frequency motion heaving z, rolling φ and pitching θ are significant factors in the process of hydrostatic analysis. In order to display the result clearly, these three curve's horizontal axes have been shortened to 200 seconds so that we can acquire vessel motion response frequency values easily. They are about 1.8 Hz, 2.0 Hz and 2.1 Hz for z, φ and θ respectively.

4.2.1 Comparison of Motion Response Varied from Different Wave Models.

In the analysis of the Hydrodynamic response, the AQWA module has offered three kinds of irregular wave spectrum. They are JONSWAP wave, Pierson-Moskowitz wave and Gaussian wave. This section shows the comparison of the rolling in the 6-DOF motions under level 2 sea condition.

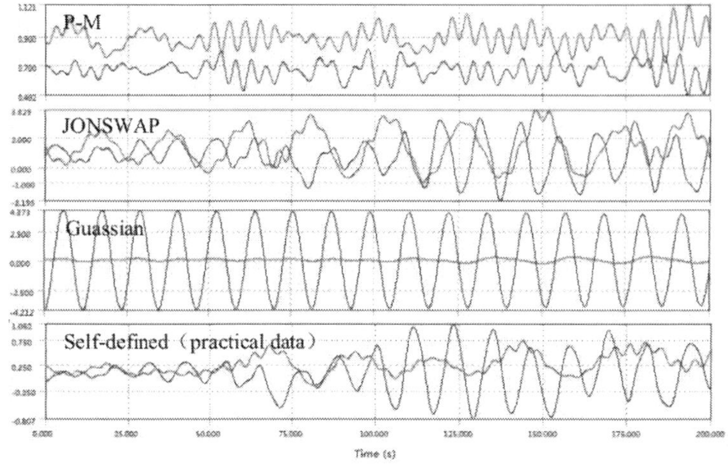

Figure 5. Comparison of motion response varied from different wave models (rolling).

As can be illustrated in the Figure 5, there is obviously much difference between four results based on different wave model. The rolling angle are about 1.121°, 3.829°, 4.373° and 1.062° for Pierson-Moseowitz specturm, JONSWAP specturm, Gaussian specturm and self-defined spectrum. Obviously, the data from P-M specturm is closer to self-prepared data. P-M spectrum represents the observed quantity of growing waving that is special for ideal sea condition. Whereas, JONSWAP is more suitable and compatible in the actual application in terms of the waving's relative motion status.

4.2.2 Comparison of Motion Response Varied from Different Sea Levels

This section selects four sea levels from 2 to 5 to simulate the motion of two vessels in the sea. The parameter of simulation condition is ranged from 0.5978m to 3.612m in terms of significant average wave height. The angular frequency is ranged from 0.4973Hz to 3.225Hz [10].

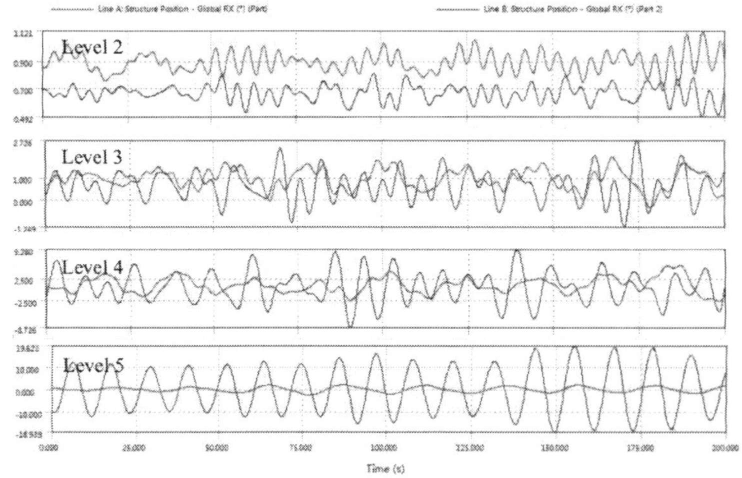

Figure 6. Comparison of motion response varied from different sea levels (rolling).

As can be illustrated in the Figure 6, the motion frequency and phase position is various because of the difference of the time response amplitude under different sea condition levels. In the Figure 6, there are four levels from level 2 to level 5. The motion trails in level 2 is almost same for two vessels due to gentle sea condition. However, in level 3 and 4, the warship motions are more intensive. Specially, the motions of two vessels will become steady and regular in level 5. Detailed results about relative poses' maximum amplitude between two vessels have been settled in the Table 3.

Table 3. Comparison of motion response varied from different sea levels (relative value).

Sea Level	Average Wave Height	Angular Limit	x(m)	y(m)	z(m)	φ(°)	θ(°)	ψ(°)
2	0.5978m	1.222~3.225 rad/s	0.888	0.641	0.041	0.111	0.375	0.018
3	1.1440m	0.8836~2.331 rad/s	2.221	0.646	0.191	0.575	0.459	0.176
4	1.8410m	0.6966~1.838 rad/s	3.199	0.789	0.333	1.274	0.734	0.351
5	3.6120m	0.4973~1.312 rad/s	5.876	1.238	0.520	8.701	0.796	1.181

Learned from the above analysis, 6-DOF motion response maximum value will increase with the increase of sea level. Among that, the average wave height under level 2 is relatively low. Specifically, the level 4 and 5 sea condition are our key researching target in this project. The heaving (0.333m), rolling (1.274°) and pitching (0.734°) under level 4 sea condition will offer reference data to develop AHC detection system. Moreover, it is easy to know that surging and rolling have relatively high response amplitude from the response curves because there is no restraint in this direction.

5 SUMMARY

This research has successfully accomplished the 6-DOF motion simulation of vessel waving in the ocean under various sea conditions, taking hoisting in the sea as the background and ANSYS AQWA software as the experimental platform. Firstly, the paper simplifies the vessel modeling process under the reasonable hypothesis. Then, the velocity trend function was researched combined with three - dimensional potential theory. Next, the paper analyses and calculates the sea loads respectively based on step-by-step modelling methodology. Lastly, we verify the load design and insert the vessel model to resolve the static solution and time response with AQWA module in the ANSYS workbench platform. We can draw conclusions that the wave motion is non-linear transformation relationship. And the simulative data can guide the design of AHC detection system.

References

1. T. M. Williams, Heuristic Scheduling of Ship Replenishment at Sea, Journal of the Operational Research Society **43.1**, 11-18(1992)
2. G. Thomas, Ship motions during replenishment at sea operations in head seas, International Journal of Maritime Engineering **152.4**, A181-A196(2010)
3. X. J. Xu, X. Chen, J. Z. Shang, The Principle and Mathematical Modeling of a New Active Heave Compensation System, Journal of National Univ. of Defense Technology. **29(3)**, 118-122 (2007)
4. L. Yan, The research on key technology of vessel motion simulation based on sea conditions. Shanghai Jiao Tong University (2009)
5. P. A. Shi, Research on Wave Motion Compensation of Two Side-by-Side Positioned Ships, Jiao Tong University for Master Degree of Engineering (2013)
6. W. J. Lu, Study on system of missile sealifts with wave compensation for navy ships connected paralledly, Nanjing University of Science and Technology (2006)
7. C. Ebbesen, *Analysis of Motions and Anchor Line Forces for Floating Production Units* (Institute for Marin Teknikk, Norway, 2013)

8. Telukunta, Srinivas, S. Mukherjee, The extended boundary node method for three-dimensional potential theory, Computers & Structures,**17–18**, 1503-1514 (2005)
9. M. O. Faltinsen, *Sea loads on ships and offshore structures* (Cambridge University Press, Cambridge, New York & Melbourne, 1990)
10. China classification society: *Specification for Marine Mobile Platforms* (China classification society, Beijing, 2016)
11. G. Yan, J. Ou, Dynamic response analysis of TLP based on AQWA, Journal of Southeast University, **47(S1)**, S130–S137 (2009)

MATEC Web of Conferences **114**, 01011 (2017) DOI: 10.1051/matecconf/201711401011
2MAE 2017

A Design of LPDA Antenna with Patch Dipoles for Airbone Communication

Ying Liu[1,a], Yuwen Wang[2,b], Yongwei Li[2,c] and Ting Yin[2, d]

School of Aeronautics and Astronautics , University of Electronic Science and Technology of China, Chengdu611731, China

Abstract. This article designs a Log-periodic Dipole Antenna(LPDA) for airborne communica--tion. This antenna replaces traditional dipoles with patch dipoles, the feeder adopt gradual changing structure, its working frequency is from 200MHz to 450MHz.Use HFSS which is the special software for high-frequency electromagnetic to calculate, optimize and simulate. The simulation rsults are in good agreement with the test results. The designed antenna improves its flexibility and reduces profile, it provide a good method for the design of the airborne antenna.

1 Introduction

LPDA(Log Periodic Antenna) is proposed in 1957,it is a kind of directional antenna[1-6],LPDA is a FI(frequency independent)antenna, FI is that the antenna is transformed by a proportional factor τ,the performance of antenna operating requency is equal to antenna operating frequency.LPDA has wider bandwith, which can be designed as 10:1(the ratio of the highest frequency to the lowest frequency),obviously,LPDA belongs to UWB(Ultra Wide Band)antenna[7].The profile of traditional LPDA[8]is relatively high,and its flexibility is not very ideal,but we can replace traditional dipoles with patch dipoles to research airborne antenna.Patch-dipoles LPDA is better to keep correspondence with the wing of aircraft,what's more,its electrical properties is less affected by the deformation of the wing.so we design a patch-dipoles LPDA applied to airborne antenna.

2 The Design of Patch-dipoles LPDA

A patched-dipoles LPDA operating at 200MHz-450MHz was designed for a gain of 10dB.the structure of LPDA[9] is affected by the combination of the scaling factor Tau(τ) and spacing factor Sigma(σ)[10],which can be calculated from carrel curves[11], τ corresponding to the best design is 0.95, σ is 0.17. α is 8.411 degree,N(the numbers of dipoles) is 17.From the operating principle of LPDA,we can know that for obtainning higher gain,we must add the number of dipoles,so we add four dipoles,the number of dipoles of LPDA is 21.

The input impedence of LPDA is about 80ohm,it is easy to match with coaxial cable which characteristic impedance is 75ohm. In the field of military radio communication,the characteristic impedance of the mutual of each communication module is 50 ohm,obviously,LPDAwhich input impedance is 80ohm connecting with coaxial cable which characteristic impedance is 50ohm will

[a]1535354598@qq.com, [b]83211668@qq.com, [c]1014972000@qq.com,[d]996328534@qq.com

cause impedance adaptation and cause VSWR[12] too large,so we need to connect an impedance matching device between the feeder terminal of LPDA and coaxial cable,however,the antenna selected operate at 200MHz to 450MHz,its belong to UWB antenna,it is difficult to design impedance matching device of UWB antenna system,besides,The introduction of he impedance matching device will not only cause the complex structure of the antenna,increase the size of antenna,but aslo bring a certain insertion loss,which affect the efficiency.for solving the difficult problem,the feeder of LPDA antenna we select adopt gradual changing structure,the width of LPDA in high frequency is 2 millimeter,the width of LPDA in low frequency is 20 millimeter,in this way,the input impedance of patch-dipoles LPDA is very close to 50ohm,it match with coaxial cable well.

3 The Structure and Results of LPDA In HFSS

The structure of simulation of LPDA is showed in Fig.1.

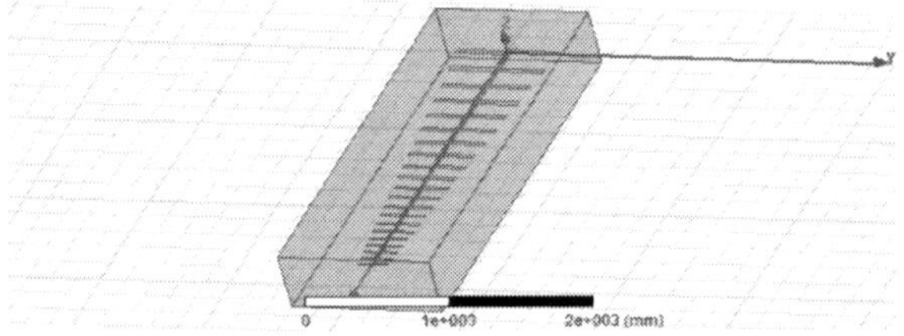

Figure 1. The structure of LPDA

By optimize,we can get the radiation patterns of LPDA antenna which dipoles are 21.The E-plane and H-plane radiation patterns of LPDA operating at 200MHz are showed in Fig.2 and Fig.3. The E-plane and H-plane radiation patterns of LPDA operating at 250MHz are showed in Fig.4 and Fig.5. The E-plane and H-plane radiation patterns of LPDA operating at 300MHz are showed in Fig.6 and Fig.7. The E-plane and H-plane radiation patterns of LPDA operating at 350MHz are showed in Fig.8 and Fig.9. The E-plane and H-plane radiation patterns of LPDA operating at 400MHz are showed in Fig.10 and Fig.11. The E-plane and H-plane radiation patterns of LPDA operating at 450MHz are showed in Fig.12 and Fig.13.From the results of simulation,we can see that the biggest gain at 200MHz is11.34dB,the biggest gain at 450MHz is 10.8dB,in the whole frequency band,LPDA can aslo get high gains.

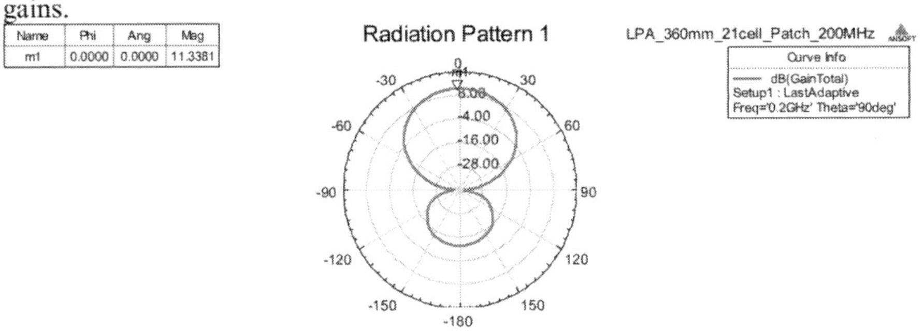

Figure 2. E-plane radiation patterns of LPDA (200MHz)

DOI: 10.1051/matecconf/201711401011

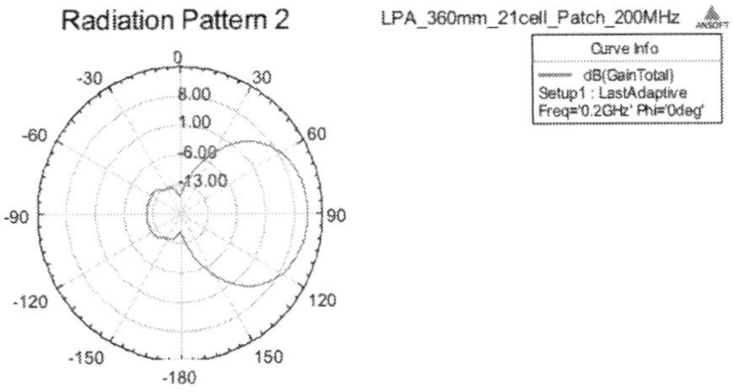

Figure 3. H-plane radiation patterns of LPDA(200MHz)

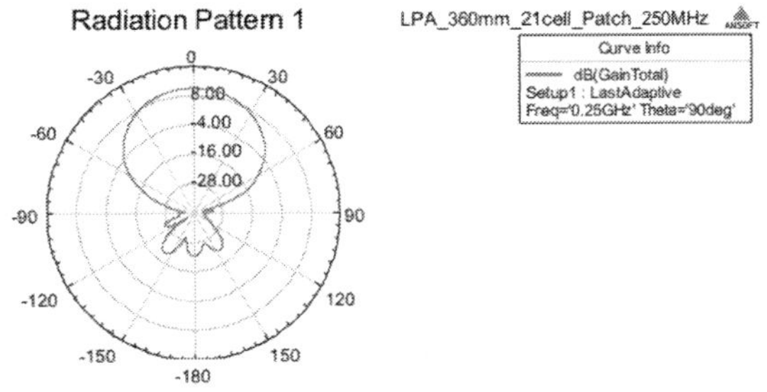

Figure 4. E-plane radiation patterns of LPDA (250MHz)

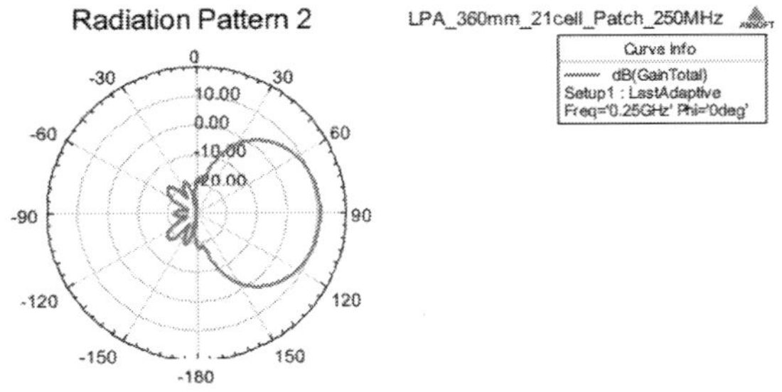

Figure 5. H-plane radiation patterns of LPDA(250MHz)

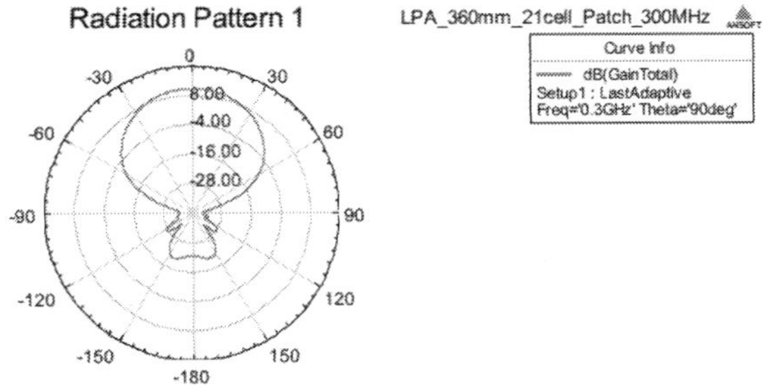

Figure 6. E-plane radiation patterns of LPDA(300MHz)

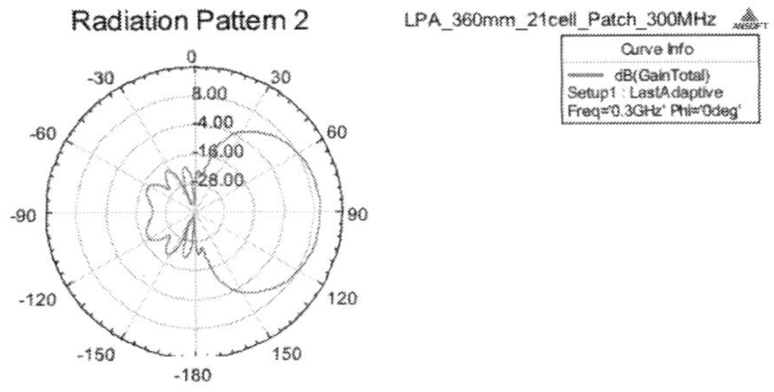

Figure 7. H-plane radiation patterns of LPDA(300MHz)

DOI: 10.1051/matecconf/201711401011

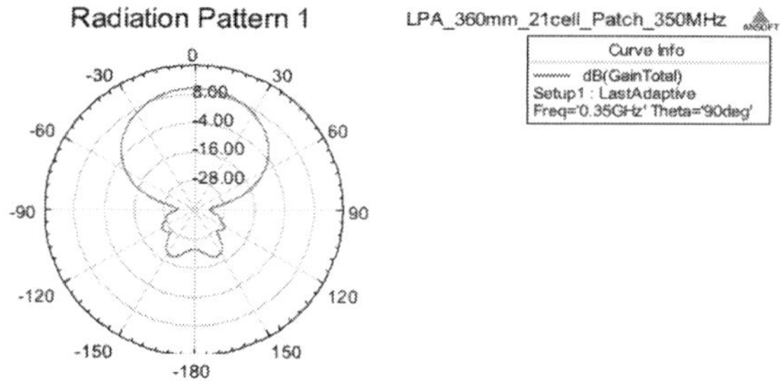

Figure 8. E-plane radiation patterns of LPDA(350MHz)

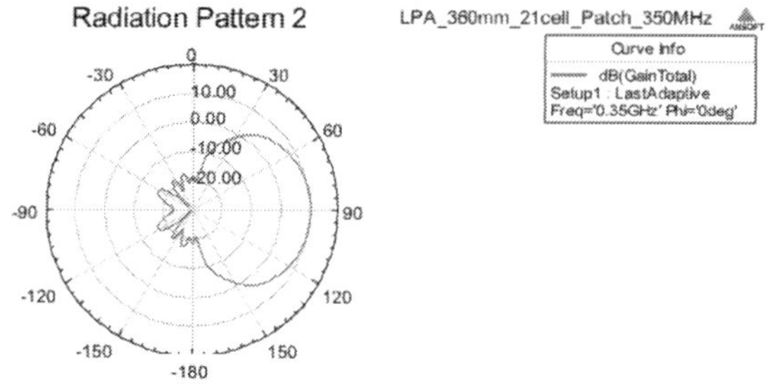

Figure 9. H-plane radiation patterns of LPDA(350MHz)

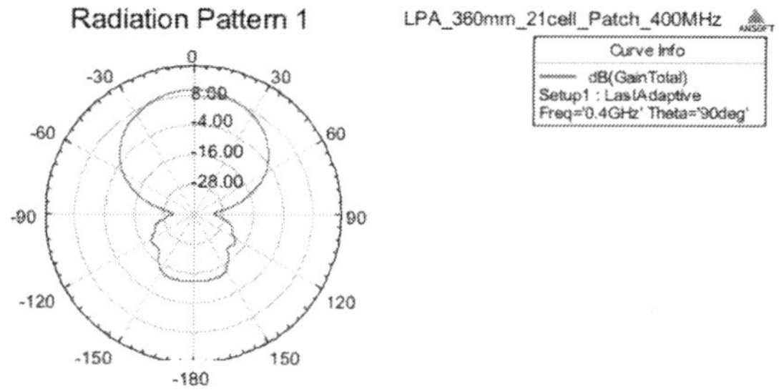

Figure 10. E-plane radiation patterns of LPDA(400MHz)

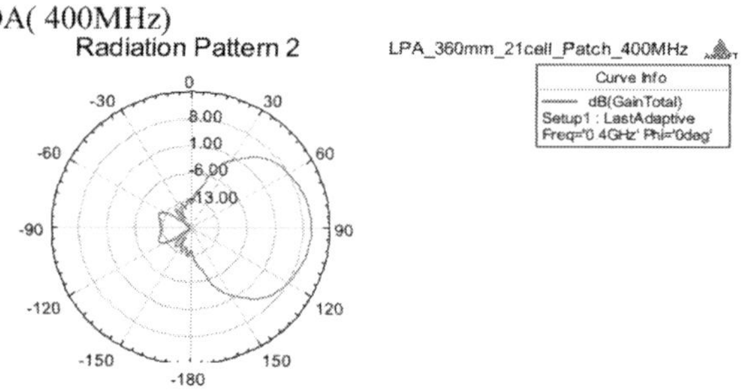

Figure 11. H-plane radiation patterns of LPDA(400MHz)

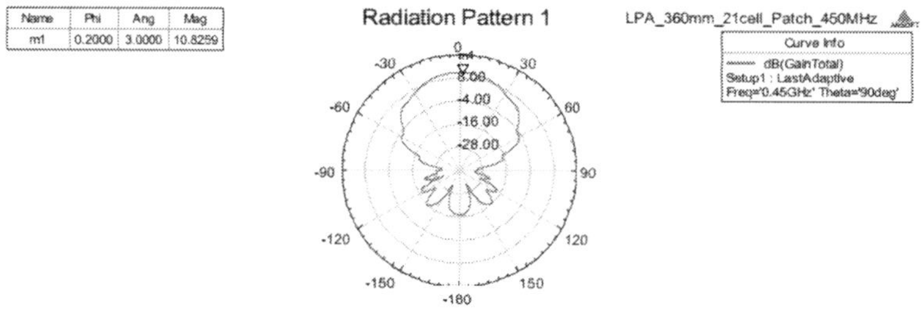

Figure 12. E-plane radiation patterns of LPDA(450MHz)

DOI: 10.1051/matecconf/201711401011

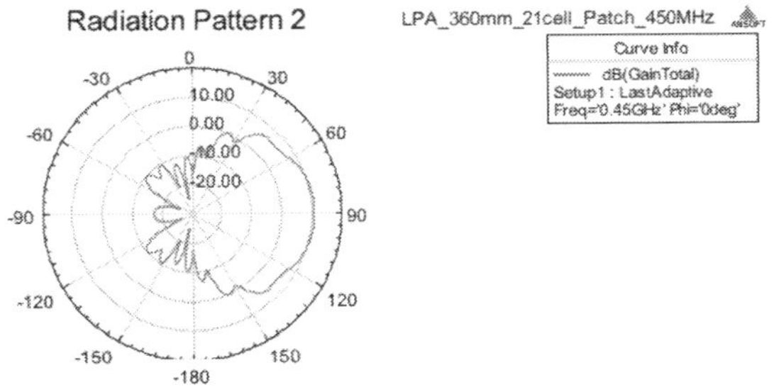

Figure 13. H-plane radiation patterns of LPDA(450MHz)

4 The Physical Design of LPDA

Since the size of LPDA operating at 200MHz to 450MHz is too large,it is not easy to test its electrical properties,for solving this problem,we adopt scaling shrunk simulation measuring method,scaling factor is 4,the size of antenna for the test is 1/4 of the original LPDA antenna,in this way,we can solve many problems caused by the too large size of antenna when we measure antenna.when the scaling factor is 4,the operating frequency of LPDA will be changed into 800MHz to 1800MHz.The physical LPDA is showed in Fig.14.In the laboratory of College of Aeronautics and Astronautics of UESTC,we use Vector Network Analyzer (Type:N5242A) designed by Alilent to measure the scaling LPDA.The reflective characters of LPDA measured is showed in Fig.15.The measurment of antenna is showned inFig.16.From Fig.15,we can see that within the frequency band of 600MHz to 2000MHz,the reflective characters of antenna(scaling factor is 4) is below -10dB,this shows that the antenna(actual size) we designed operating frequency at 150MHz to 500MHz,its reflective characters is below -10dB too.The actual bandwidth of LPDA we designed(150MHz-500MHz) is better than the original bandwidth(200MHz-450MHz).

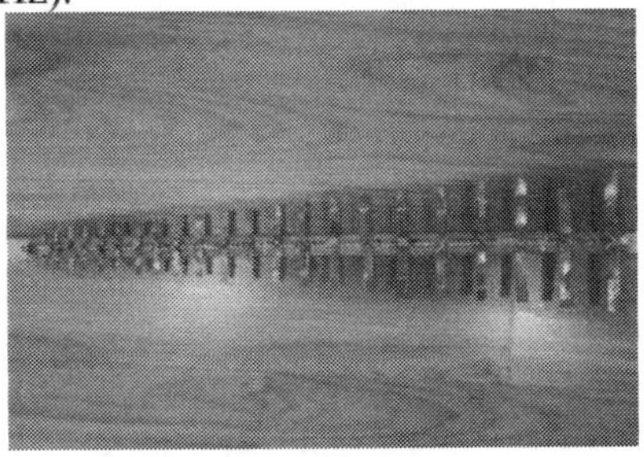

Figure 14. The physical LPDA

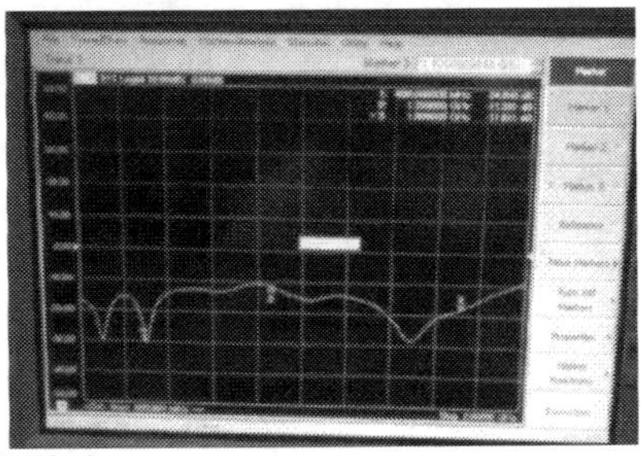

Figure 15. The reflective characters of LPDA

Figure 16. The measurment of antenna

5 Summary

The patch-dipoles LPDA we designed operating at 200MHz to 450MHz has wide bandwidth, high gain, low profile, good flexibility and light weight etc. It provides a reliable choice for airborne antenna, particularly in the antenna of solar powered warning drones in Near Space.

References

1. DuHamel R H, Ore F. Logarithmically Periodic Antenna Designs [C]. IRE International Coventional Record, 2001(6): 139-151.
2. Nguyen C. A Wide-band Microstrip-fed Log-periodic Antenna [C]. Antennas and Propagation Society International Symposium, 2006(1):33-36.
3. Mohammad M T. Wideband Planar Log -periodic Antenna [C]. Antenna technology: Small and Smart Antennas Metamaterials and Applications, 2007.IWAT '07. International Workshop on, 2007:331-334.
4. Orlob C,Dao Q H, Kornek D. Dual-polarized Log -periodic Antenna on a Conical MID Substrate[C]. Antennas and Propagation(EUCAP), Proceedings of the 5th European Conference on, 2011:361-364.
5. D.E.Isbell. Log-periodic dipole arrays[J]. IRE Trans. Antennas Propag., 2011, AP-8: 260-267.
6. R.L.Carrel. The design of log-periodic antennas[J]. IRE Int.Conent.. Rec., 2010, 9(1): 61-75.
7. Praveen Nartam, Abhijeet Gulawani, and Subhajit Chatterjee, "Design of an Ultra wide Band antenna using defected ground structure,"ICCPEIC, pp. 75-79, 2016.
8. Zou Aihua and Xu Changwu, "A small dual stacked log periodic dipole array antenna(LPDA) with improved gain of low frequency for immunity,"APEMC,pp.208-211,2015.
9. Wang Zhefei and Fu Jiahui, "A design of miniaturization LPDA antenna with ultra sideband,"2015 IEEE 6th International Symposium on Microwave,Antenna,Propagation,and EMC Technologies,pp819-821,2015.
10. Jerin Jose,Tessa Mathew,Annu Thomas,Haripriya N,Mahima Cherian,and D.D.Krishna, "A Cost Effective Hybrid-Log Periodic Dipole Antenna,"2015 Fifth International Conference on Advances in Computing and Communications,pp263-265,2015.
11. Carrel,R.l, "Analysis and design of the log-periodic dipole antenna,"Technical Report,No.52,Antenna Laboratory,University of Illinois 1961.
12. T,Taniguchi and T.Kobayashi, "An omnidirectional and low-VSWR antenna for ultra-wideband wireless systems",Radio andWireless and Conference,pp.145-148,2002.

Design of LEO Constellations Providing Internet Services Based on SOC Method

Quan Chen[1,a], Yuzhu Bai[1], Lihu Chen[1] and Ziyan Pang[2]

[1]*College of Aerospace Science and Engineering, National University of Defense Technology, Changsha 410072, China*
[2]*College of Basic Education for Commanding Officers, National University of Defense Technology, Changsha 410072, China*

Abstract. Narrow-beamwidth antenna is applied to LEO satellite constellations to provide high-speed internet services for the globe. Low-altitude, limited coverage region of single satellite and large scale are the main features of them. In this paper, under the constraint of specified half-beam angle of antenna, both inclined and polar orbit constellations are analysed based on Streets-of-Coverage method to provide continuous coverage. The results indicate that the coverage performance of single satellite is the key factor which can affect the configuration and number of constellations. When the orbit altitude is 1200 km and half-beam angle is 32 degrees, up to thousands of satellites are required for constellations providing internet services for zonal region 60 degrees from the equator. Non-symmetrical polar constellations show the best performance of continuous coverage and require approximately half of the satellites of inclined orbit constellations. The required satellites with different single coverage performance are also investigated as a reference for future research.

1 Introduction

A major trend of satellite development is a decrease of satellite size and cost. Instead of traditional single larger satellite, distributed micro-satellite systems are widely deployed to implement complex missions [1]. This has led to the rapid development of constellations with a large number of satellites. OneWeb [2] has announced a constellation consisting of up to 648 satellites to provide high-speed internet services. Meanwhile, SpaceX is planning to launch more than 4000 satellites to build a gigantic space-based internet communication system [3].

Satellite with a higher altitude usually has a better performance in coverage, but the corresponding attenuation of space link is greater, which is not suitable for high-speed broadband signal transmission. Generally, the optimal orbital altitude for LEO satellites ranges from 500 to 1500km. Elliptical-orbit satellites are able to stay for a longer time above certain hemisphere. However, the variation of altitude brings about a higher requirement for the on-board components when satellites fly across the Van Allen radiation belt. Circular low earth orbit consequently becomes the most popularly used in communication constellations [4].

For satellite systems providing internet services, a large zonal or the global region should be continuously covered to ensure the availability at any time and any place. Traditional methods and

[a] Corresponding author : chenquan11@foxmail.com

results are based on the constraint of a lower minimum elevation angle because the coverage region of single satellite is large enough [5]. However, to transmit high-speed and broadband signals, satellites are equipped with narrow-beamwidth antennas which limit the coverage of single satellite [6]. Then the satellite amount can be extremely large such as OneWeb and SpaceX satellite constellations.

In this paper, the half-beam angle of antenna is given as the primary constraint, and Streets-of-Coverage method [7] is applied to design the optimal constellation providing continuous coverage of the mid-and low-latitude zonal regions. The influence of inclination and the number of orbital planes is discussed. Both inclined and polar satellite constellations are analyzed and compared so as to find out the optimal results. Finally the results of required satellite number at different altitudes are listed.

2 Single coverage

The coverage region of single satellite, illustrated by Figure 1, is defined as the spherical surface formed by the intersection of the satellite-centered-cone and the earth. Minimum elevation angle is the common constraint in traditional analysis, and a point is covered by a satellite if the local elevation angle is greater than the specified minimum value. However, in terms of LEO satellites offering high-speed data communication service, Ku, Ka or even higher frequency is used which leads to greater attenuation of space link [6, 8]. Then high-gain antenna is on demand to compensate for the attenuation. The relation of the antenna gain and beamwidth is given by

$$G = \eta \left(\frac{\pi \times 70}{\alpha_{0.5}} \right)^2 \tag{1}$$

where G is the antenna gain, $\alpha_{0.5}$ is the half power beamwidth in degree and η is the antenna efficiency. It is obvious that reducing beamwidth is an effective approach to improve the gain. Thus narrow-beamwidth antenna is designed to transmit high-speed signals in satellite-ground links. Although multi-beam antenna is adopted to enlarge the whole beamwidth, the coverage region is much smaller than traditional communication satellites or observation satellites. In this situation, beam-angle becomes the key factor in the design of satellite constellations providing internet services.

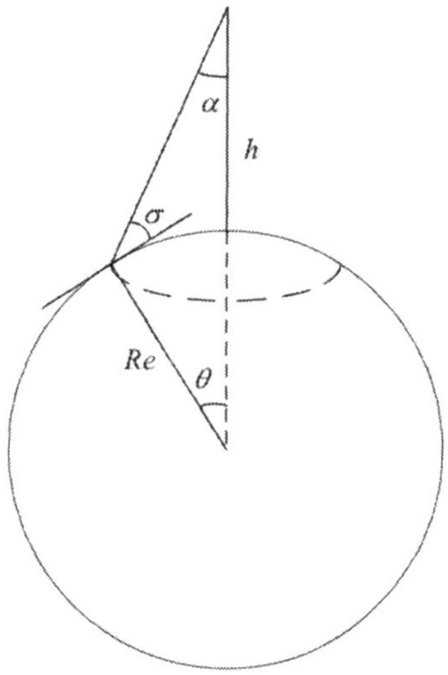

Figure 1. Coverage geometry.

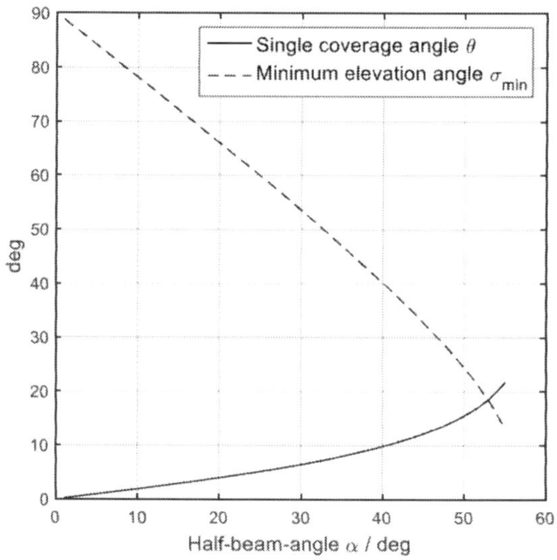

Figure 2. Coverage angle and minimum elevation angle versus half-beam angle.

A satellite with altitude h and half-beam angle α is able to cover a spherical segment of the earth's surface. The geocentric half-cone-angle of the coverage region is defined as coverage angle θ in this paper. From the geometry, it follows that

$$\theta = \arcsin\left(\frac{h+\mathrm{Re}}{\mathrm{Re}}\sin\alpha - \alpha\right) \quad (2)$$

where Re is the radius of the earth. When the satellite altitude is fixed, following equation shows the relation between half-beam angle α and minimum elevation angle $\sigma_{\min}$ of the coverage region.

$$\sigma_{\min} = \arccos\left(\frac{h+\mathrm{Re}}{\mathrm{Re}}\sin\alpha\right) \quad (3)$$

Figure 2 shows the relation between θ, $\sigma_{\min}$ and α, when the satellite altitude is fixed(h =1200km). According to the curves, with the increase of half-beam angle, the single coverage region reduces while the minimum elevation angle of the coverage region increases gradually. In the case of α =30°, corresponding angles are θ=6.45° and $\sigma_{\min}$ =53.55°.

3 Constellation design method based on SOC

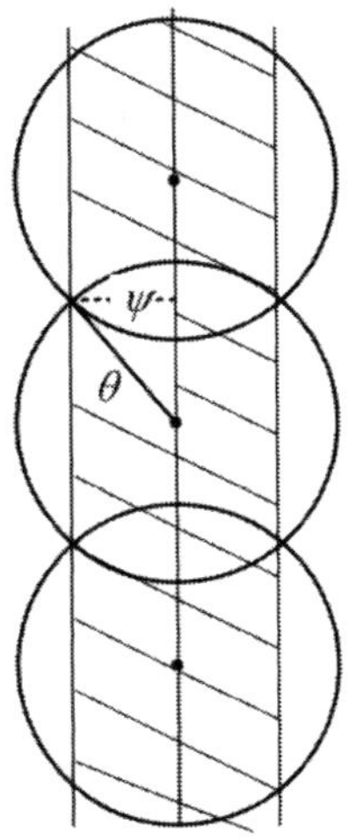

Figure 3. Streets of coverage.

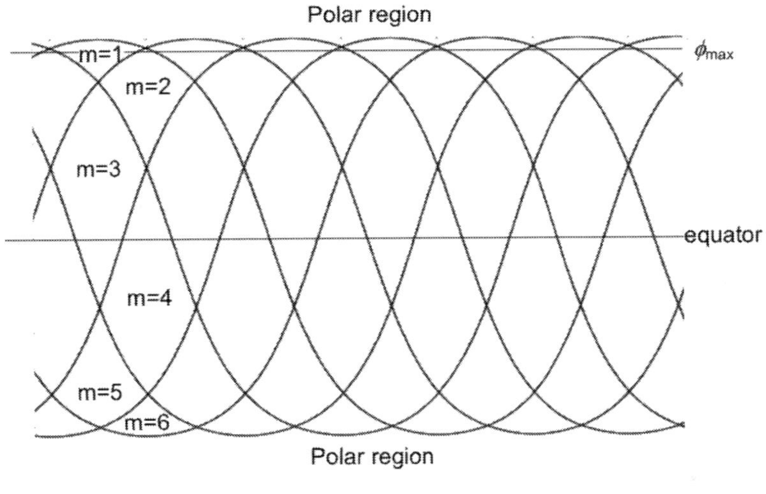

Figure 4. Geometry of the mesh (n_1=7).

3.1 Streets-of-Coverage methods

Streets-of-Coverage (SOC) method solves the problem of continuous coverage for a zonal or global region using the overlapped coverage regions of adjacent intra-plane satellites which can be continuously viewed. In SOC method, constellation configuration shows following features: 1) All of the orbits are circular with the same altitude and inclination. 2) The constellation consisting of n_1 orbit planes and each contains n_2 satellites. 3) Satellites in each plane are evenly distributed.

It can be seen from Figure 3, when single coverage angle θ satisfies $\theta > \pi / n_2$, streets viewed continuously are formed by overlapping coverage regions of adjacent intra-plane satellites. By means of spherical trigonometry, the street width follows

$$\cos\theta = \cos\psi\cos(\pi/n_2) \quad (4)$$

In SOC method, only streets covered continuously are considered while the coverage beyond the streets is neglected. Consequently, the phase difference between adjacent plane satellites can be eliminated and the circular coverage regions are simplified as strips corresponding to each plane. Then the goal of the constellation design is to calculate the required street width of each constellation configuration so that the specified zonal or global region can be completely covered [7].

3.2 Method for inclined orbit constellation

Considering the zonal region to be continuously covered that ranges from the equator to latitude ϕ_{max}, the orbit inclination is i, and n_1 orbital planes are evenly distributed along the equator. By projecting the orbit trace at certain moment, the earth's surface can be divided into n_1-1 layer meshes and two polar regions, as Figure 4 shows. The latitude of the intersections from north to south is

$$\phi_j = \arctan\left[\tan i \cos\left(j\pi / n_1\right)\right] \tag{5}$$

where $j = 1,\ldots,n_1-1$ and $-i < \phi_j < +i$. Then define $\phi_0 = i$ and $\phi_n = -i$. Note that ϕ_1 is the latitude of the northernmost grid node. If $\phi_{max} \le \phi_1$, the only coverage of inside mesh regions needs to be considered. Figure 5 gives a general geometry of mth-layer mesh. Point P is the center of the inscribed circle that is tangent to all the four boundaries of the mesh. X_U is the angle from P to the upper vertex. Let $\psi_{\min,m}$ be the required minimum width of streets to cover the mth-layer mesh and it is also the angle between P and either of the boundaries. $\psi_{\min,m}$ can be calculated by

$$\sin\psi_{\min,m} = \sin X_U \cos i / \cos\phi_{m-1} \tag{6}$$

where

$$X_U = \arctan\left\{\sin\left(\frac{m\pi}{n_1}\right)\sin\left(\frac{\pi}{n_1}\right)\tan i\left\{1+\cos\left(\frac{m\pi}{n_1}\right)\cos\left(\frac{\pi}{n_1}\right)\cos\left[(m-1)\frac{\pi}{n_1}\right]\tan^2 i\right\}^{-1}\right\} \tag{7}$$

and the overall required width of streets $\psi_{\min}$ is the maximum value of $\psi_{\min,m}$ for meshes needed to be covered.

If $\phi_{max} > \phi_1$, additional coverage outside the n_1-1 layer meshes should be considered as well, as Figure 6 illustrates. The required street width to cover the external region is given by

$$\psi_{ext} = \arcsin\left[\sin\phi_{max}\cos i - \cos\phi_{max}\sin i\cos\left(\pi / n_1\right)\right] \tag{8}$$

In this situation, the required minimum width is

$$\psi_{\min} = \max\left\{\psi_{\min,1},\psi_{\min,2},\ldots,\psi_{\min,n_1-1},\psi_{ext}\right\} \tag{9}$$

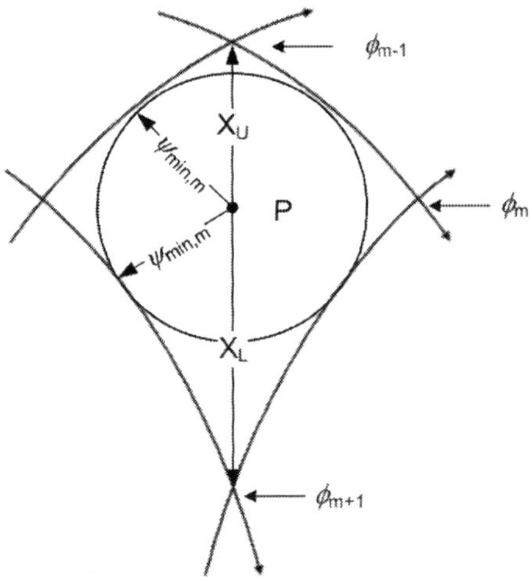

Figure 5. General geometry of *m*th-layer mesh.

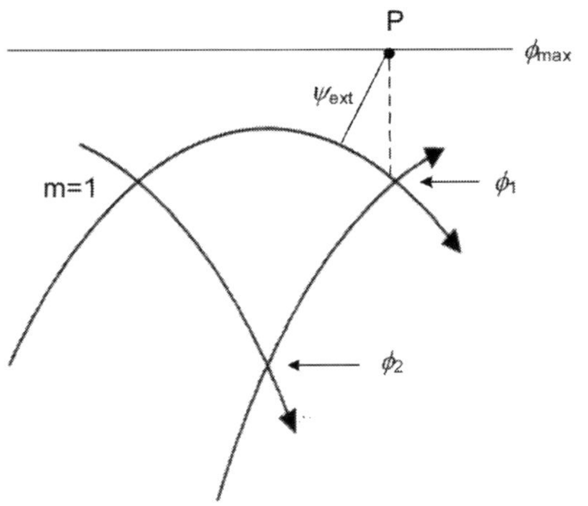

Figure 6. Coverage for regions beyond the mesh.

When the orbit altitude h and half-beam angle α are given, the coverage angle $\theta = \theta(h,\alpha)$ can be solved from equation (2). If $\theta \le \psi_{\min}$, i.e. the coverage angle of single satellite is no greater than the needed minimum street width, then it is impossible to form the streets with required width. When $\theta > \psi_{min}$, combining $\psi > \psi_{min}$ and equation (4) yields the required number of satellite per plane.

$$n_2 \geq \frac{\pi}{\arccos(\cos\theta/\cos\psi_{min})} \tag{10}$$

Then the required minimum number of the satellite N can be obtained by $N = n_1 n_2$

$$N = n_1 \left\lceil \frac{\pi}{\arccos(\cos\theta/\cos\psi_{min})} \right\rceil, \qquad \theta > \psi_{min} \tag{11}$$

Note that if the inclination is much smaller than ϕ_{max}, the value of ψ_{ext} will be the crucial factor with a large value. If the performance of single coverage is poor, then the constraint $\theta > \psi_{min}$ will not be satisfied at any n_1. In addition, ψ_{min} will be great at a relatively small n_1, which makes it impossible to form streets with required width. Consequently, both i and n_1 should be greater than certain values to ensure the existence of a reasonable solution. Limitation of i and n_1 can be obtained by iteratively calculating N through different i and n_1. The solution procedures are listed in Table 1.

Table 1. Solution procedures of inclined orbit constellations.

***Step* 1.**	Calculate the single coverage angle θ.
***Step* 2.**	Set the number of orbital planes n_1 and inclination i.
***Step* 3.**	Calculate required minimum street width of each candidate mesh region $\psi_{min,m}$ and ψ_{ext}.
***Step* 4.**	Solve the overall ψ_{min}.
***Step* 5.**	Compare ψ_{min} with θ. If $\theta \leq \psi_{min}$, no solution for given (n_1, i) set, go to ***Step* 2**; if $\theta > \psi_{min}$, go to ***Step* 6**.
***Step* 6.**	Calculate the required number of satellites per plane n_2.
***Step* 7.**	Obtain the minimum total number of satellite N of given (n_1, i) set.

3.3 Method for polar constellation

3.3.1 Symmetrical polar constellation

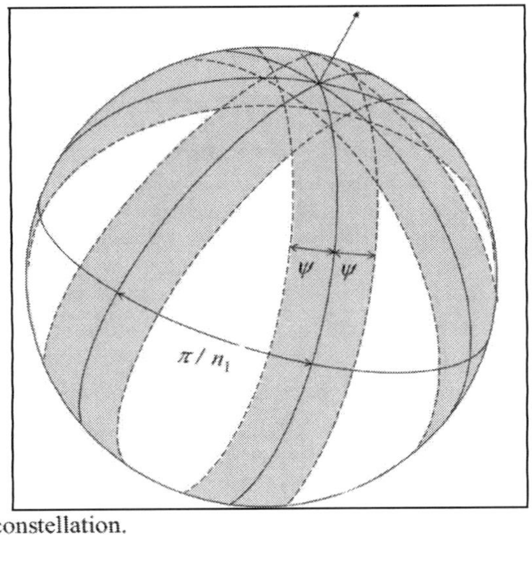

Figure 7. Symmetrical polar constellation.

The inclination of polar satellites is 90° and generally the ascending nodes spread within $(0,\pi)$ to fully utilize the coverage of ascending and descending orbits. For symmetrical polar constellation, the n_1 planes are uniformly separated by Δ, where $\Delta=\pi/n_1$, as is illustrated in Figure 7.

Since all the orbits of polar constellation intersect at two poles, polar regions are the most likely ones to be covered while lower-latitude regions are more difficult. Continuous global coverage can be achieved once the equator is continuously covered. Consequently, to cover a zonal region including the equator, the required minimum street width should be

$$\psi_{\min}=\frac{\pi}{2n_1} \tag{12}$$

The following steps are similar to those of inclined orbit constellation design. After given the orbit altitude h and half-beam angle α, one can obtain the coverage angle $\theta=\theta(h,\alpha)$. The combination of equations (4), (10), (11) and (12) yields the required number of satellites N in a symmetrical polar constellation to cover the entire globe.

$$N=n_1\left\lceil\frac{\pi}{\arccos\left(\cos\theta/\cos(\pi/2n_1)\right)}\right\rceil,\qquad \theta>\frac{\pi}{2n_1} \tag{13}$$

3.3.2 Non-symmetrical polar constellation

The coverage beyond the street in the above-mentioned constellations is not considered, which wastes the extra coverage between two planes. For polar constellations, there are two kinds of interface of two adjacent orbits, i.e. co-rotating interface and counter-rotating interface. In co-rotating interface, the relative positions of adjacent satellites are fixed and the phase separation between satellites in adjacent orbits can be arranged so that the largest continuously covered regions can be generated through the optimal phase shift. However, since the rapid variation of relative positions in counter-rotating interfaces, adjacent orbits should be close enough to form a continuous Streets-of-Coverage. Therefore, the orbital planes can be asymmetrically arranged to make full use of the coverage regions.

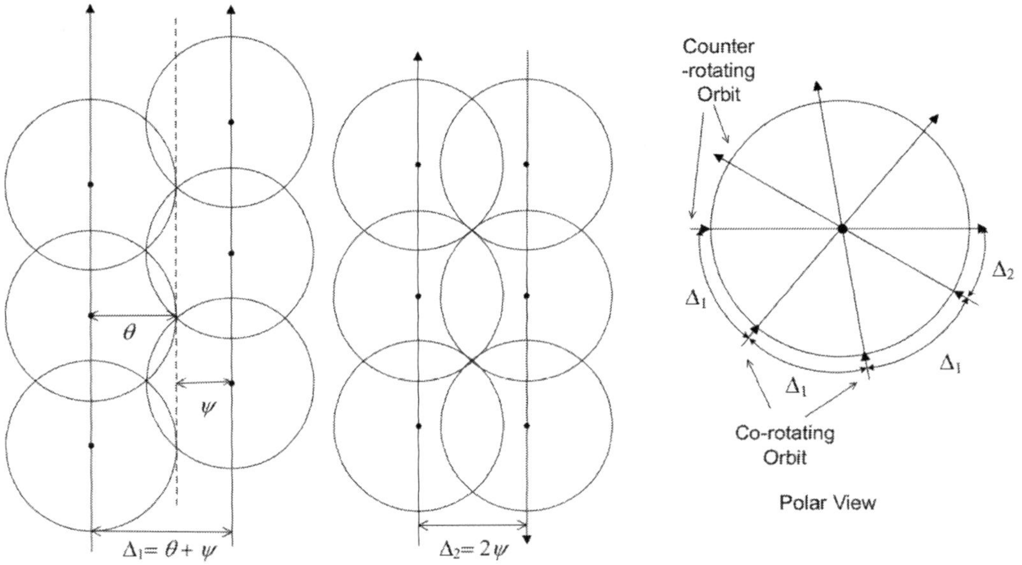

Figure 8. Co- and counter-rotating orbits and the seam of non-symmetrical polar constellation.

According to Figure 8, the angular separations of co-rotating orbits are

$$\Delta_1=\theta+\psi \tag{14}$$

and those of counter-rotating orbits are

$$\Delta_2=2\psi \tag{15}$$

For the polar constellation with n_1 planes, it contains $2(n_1\text{-}1)$ co-rotating interfaces and two counter-rotating interfaces. The relation between them is given by

$$(n_1-1)\Delta_1+\Delta_2=\pi \tag{16}$$

When the single coverage angle and number of orbit planes are given, the required minimum street width can be calculated by combining equation (14), (15),(16) and $\theta>\psi_{\min}$

$$\psi_{\min}=\frac{\pi\text{-}(n_1-1)\theta}{n_1+1},\qquad n_1>\frac{\pi}{2\theta} \tag{17}$$

Note that in this step, the single coverage angle θ should be known. Then the required satellites can be obtained through equation (11).

4 Case study

4.1 Inclined orbit constellation case

Based on the constraint of the on-board payload, the satellite altitude and half-beam angle are given (h=1200km, α=32°). Considering the distribution characteristics of global population and economic activities, zonal region 60 degrees from the equator is required to be continuously covered (i.e. $\phi_{\max}$=60°). The required minimum street width $\psi_{\min}$ and number of satellites N at different inclinations are analyzed.

Figure 9 illustrates that the required street width gradually decreases with the increase of the number of orbital planes n_1. Comparison between curves of 60° and 70° inclination shows that constellations with larger inclination require more satellites. The curve of 50° inclination decreases rapidly when the number of planes is less than 15 and then remains stable.

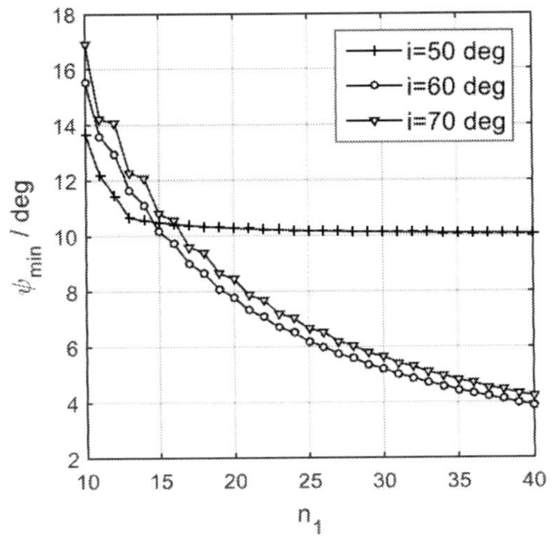

Figure 9. Required street width at different inclinations.

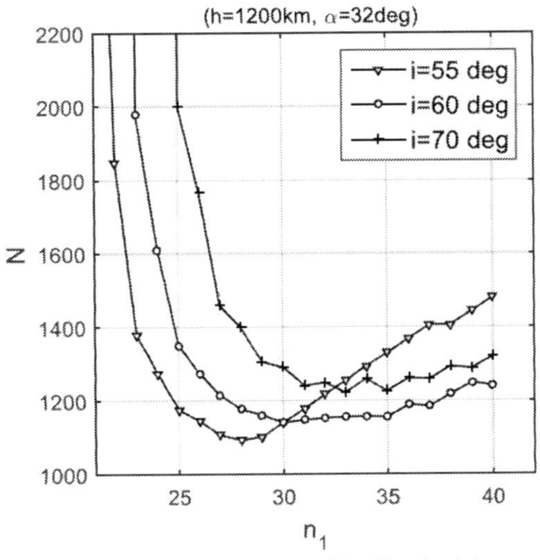

Figure 10. Satellite number of inclined orbit constellations.

When h=1200km, α=32°, the coverage angle θ=7.03° can be solved through equation(2). By iterating through the available combinations of n_1 and i, it is found that the orbit inclination should be greater than 53.1° for $\theta>\psi_{\min}$. By means of the foregoing method in Table 1, the required satellite amount N affected by n_1 can be obtained, as is displayed in Figure 10. When n_1 is much small, the given single coverage is unable to form the required large ψ_{min}. In this case, when i =55°, it is found that n_1 should be greater than 21 through iterative calculations. N is directly determined by n_1 and n_2. And the increase of n_1 will induce smaller ψ_{min} and then lower the following n_2. Thus there is an optimal n_1 for constellations with minimum satellites. By calculating required number of satellites N at each set of n_1 and i repeatedly, one can obtain the relations among them. Figure 10 shows that the required number of satellites has a rapid decrease firstly and then grows with the increase of planes. The least number of satellites exists when the optimal value of n_1 is close to 30. The optimal constellations at different inclinations are investigated and the optimal results are listed in Table 2. It can be seen that when the inclination is 55.1°, it requires the least number of satellites 1073 in 29 planes.

Table 2. Optimal inclined orbit constellations with different inclinations for continuously coverage (at h =1200km, α =32deg, ϕ_{max} =60deg).

i /deg	N	n_1	n_2	$_{min}$/deg
55	1092	28	39	5.26
60	1140	30	38	5.19
70	1221	33	37	5.08
80	1254	33	38	5.18
55.1	1073	29	37	5.06

4.2 Polar constellation case

To continuously cover the same zonal region as in previous case, polar constellations should be able to continuously cover the equator, i.e. the entire globe. The influence of the number of orbital planes n_1 on satellites number N is obtained when the coverage performance of single satellite is the same as that in inclined constellation case. In Figure 11, similar to inclined orbit constellation, N decreases with the growth of n_1 and then increases in both symmetrical and non-symmetrical cases. The optimal result of symmetrical polar constellation is 663 satellites in 17 orbital planes while 522 satellites in 18 planes in non-symmetrical case.

Table 3 lists the optimal results of different constellation configurations. It shows that the required satellites are substantially cut down compared to inclined cases under the same single coverage performance. The first two results indicate that polar constellation requires less orbital planes while offering a wider Streets-of-Coverage than inclined orbit constellation. Therefore, in polar constellation, the overlapped regions between adjacent planes are smaller, which manifests a higher efficiency of single coverage utilization. In addition, non-symmetrical polar configuration is better than symmetrical configuration since phase arrangement enlarges the continuous coverage region and thus reduces the number of satellites per orbital plane. Note that ψ_{min} of non-symmetrical case in the table is the required streets width in counter-rotating interface.

Table 3. Optimal constellation results of different configurations (at h =1200km, α =32deg, ϕ_{max} =60deg).

	N	i /deg	n_1	n_2	$_{min}$/deg
Inclined orbit constellation	1073	55.1	29	37	5.06
Symmetrical polar constellation	663	90	17	39	5.29
Non-symmetrical polar constellation	522	90	18	29	3.18

All the analyses above are based on specified single coverage performance. Then the effect of single coverage angle θ is investigated by changing the orbit altitude. The required satellite numbers of the three kinds of constellations are calculated at different orbit altitudes. As Figure 12 shows, because of the growth of θ by increasing orbit altitude, both the required n_1 and n_2 decrease, which results in the decrease of corresponding total number of satellites. Comparing the three kinds of constellations, one can find that non-symmetrical constellation requires the least satellites to achieve the same coverage. The detailed parameters of the optimal constellations at different altitudes are listed in Table 4. It can be found that required satellites in inclined orbit constellations are approximately twice the number of non-symmetrical polar constellations.

Note that only the coverage of street region is considered in SOC method, thus the extra coverage beyond the streets is neglected in the design of inclined orbit constellations. Then the number of satellite can be further reduced when the extra coverage is exploited. However, the analysis will be much more complicated. The sufficiently covered results in this paper will provide theoretical guidance and preliminary reference for constellation design.

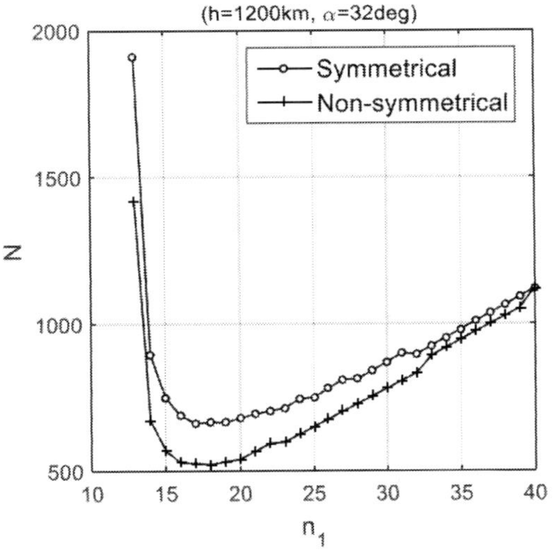

Figure 11. Number of satellites of polar constellations.

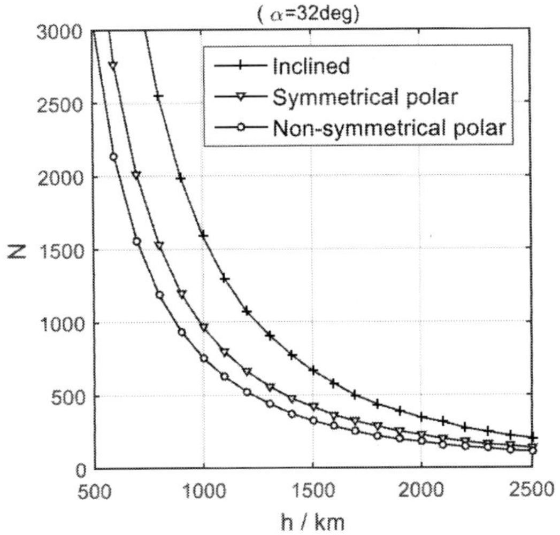

Figure 12. Satellite number at different altitudes.

Table 4. Optimal constellation configurations at different orbit altitudes.

		Inclined orbit constellation				Non-symmetrical Polar constellation			
h /km	θ /deg	N	n_1	n_2	i /deg	N	n_1	n_2	$N_{polar}/N_{inclined}$
500	2.86	6860	70	98	57.90	3096	43	72	0.451
600	3.44	4636	61	76	57.60	2135	35	61	0.461
700	4.03	3366	51	66	57.10	1560	30	52	0.463
800	4.62	2552	44	58	56.70	1188	27	44	0.466
900	5.21	1989	39	51	56.30	936	24	39	0.471
1000	5.81	1591	37	43	56.10	756	21	36	0.475
1200	7.03	1073	29	37	55.10	522	18	29	0.486
1500	8.90	667	23	29	53.90	325	13	25	0.487
2000	12.13	345	15	23	51.20	180	10	18	0.522
2500	15.55	204	12	17	49.40	112	8	14	0.549

5 Conclusion

In this paper, Streets-of-Coverage method is applied to design inclined and polar orbit constellations providing continuous coverage of mid-and low-latitude zonal regions under the constraint of prescribed half-beam angle. The results indicate that single coverage angle is the core factor that influences the constellation configuration and corresponding number of satellite. When narrow-beamwidth antenna is adopted to provide internet services, up to thousands of LEO satellites are required to continuously cover mid-and low-latitude regions due to the small single coverage.

When the single coverage performance is specified, inclined orbit constellation is mainly affected by inclination and the number of orbital planes. The inclination of the optimal result is close to the highest latitude to be covered while slightly lower than it. If the inclination is fixed, the required satellite firstly decreases then increases with the increment of orbit planes. Compared to inclined orbit constellations, polar constellations are able to save half of the satellites to provide the same coverage because of the less overlapped regions between two adjacent planes. When the phase difference and coverage beyond streets are considered, non-symmetrical configuration is better than symmetrical case. In the case of h =1200km, α =32°, the required number of satellites of inclined orbit, symmetrical polar and non-symmetrical polar cases are 1073, 663 and 522 respectively. Increase of orbital altitude and satellite scale are the two effective methods to improve continuous coverage performance.

Acknowledgment

This work was supported by the National Natural Science Foundation of China (61302092) and Pre-research Project of National University of Defense Technology (ZDYYJCYJ20140701).

References

1. D.J. Barnhart, T. Vladimirova, M.N. Sweeting, Very-small-satellite design for distributed space missions, J Spacecr Rockets, **44,** 1294-1306 (2007)
2. Home – OneWeb, http://oneweb.world/#solution (Accessed 14 November 2016)
3. SpaceX satellite constellation, http://en.wikipedia.org/wiki/SpaceX_satellite_constellation (Accessed 14 November 2016)
4. G. Maral, J.J.d Ridder, B.G. Evans, M. Richharia, Low earth orbit satellite systems for communications, Int J Satell Commun Network, **9**, 209-225 (1991)
5. G. Maral, M. Bousquet, Satellite Communications Systems: Systems, Techniques and Technology, (John Wiley & Sons, Chichester, 2010)
6. M. Ibnkahla, Q.M. Rahman, A.I. Sulyman, H.A. Al-Asady, Y. Jun, A. Safwat, High-speed satellite mobile communications: technologies and challenges, Proc IEEE, **92**, 312-339 (2004)
7. R.D. Luders, L.J. Ginsberg, Continuous zonal coverage: a generalized analysis, AIAA Mechanics and Control of Flight Conference, NO 74-842 (1974)
8. D. Palombini, S. Arena, T. Cavanna, E. Limiti, Compact sub-harmonic mixer for Q-band satellite communications, 2014 9th European Microwave Integrated Circuit Conference, 61-64 (2014)

MATEC Web of Conferences **114**, 01013 (2017) DOI: 10.1051/matecconf/201711401013
2MAE 2017

3D-Calibration for IMU of the Strapdown Inertial Navigation Systems

V.V. Avrutov[1,a], M.D. Geraimchuk[1] and Xing Xiangming[2]

[1]*National Technical University of Ukraine, Kiev Polytechnic Institute, Kiev, Ukraine*
[2]*Beijing Institute of Control Devices, Beijing, China*

Abstract: A new calibration method for Inertial Measurement Unit (IMU) of Strapdown Inertial Navigation Systems was presented. IMU has been composed of accelerometers, gyroscopes and a circuit of signal processing. Normally, a rate transfer test and multi-position tests are using for IMU calibration. The new calibration method is based on whole angle rotation or finite rotation. In fact it's suggested to turn over IMU around three axes simultaneously. In order to solve the equation of calibration, it is necessary to provide an equality of a rank of basic matrix into degree of basic matrix. The results of simulated IMU data presented to demonstrate the performance of the new calibration method.

1 Introduction

Inertial Measurement Unit (IMU) is a base for developing of Strapdown Inertial Navigation Systems. Each one IMU consist at least three accelerometers and three gyroscopes.

Normally, accelerometers and gyroscopes should go via autonomous testing before they will be assembled to IMU. But it is very important to determinate their features and parameters in whole IMU, because output signals of accelerometers and gyroscopes would be tied to IMU's frame. Therefore IMU calibration is an integral stage of his preparing to work or alignment of the Inertial Navigation System (INS). The calibration is a determination of IMU parameters or his errors for further using or compensation during INS's working.

Usually the IMU calibration is going on by multi-position tests [1-3] for sets of accelerometers, using a precision dividing head. This equipment has a setting accuracy near one second of arc, enables the sensitive axis of an accelerometer rotate with respect to the gravity vector. For calibration of a set of accelerometers it is need to measure output signals, at first to turn the them around ox axis, then to reinstall the set of accelerometers for to measure of output signals by turning around oy axis, and finally to repeat the procedure to measure of output signals by turning around oz axis. Frankly speaking such technology is an artificial case. Actually the vehicle or body turns for whole angle rotation or finite rotation, which is a result of turnings around two or three axes.

For the gyroscopes calibration in IMU by rate transfer tests it is using rate table (precision turntable). During the transfer test the IMU is mounted on the turntable at first with one sensitive axis. For

[a] Corresponding author: vyshgorod@gmail.com

example, axis *ox is* parallel to the axis of rotation of the rate table and rate table is stepped through a series of angular rates ω_{xi} starting at zero by recording data at each stage. Such test repeated for rotation around axis *oy* with angular rates ω_{yi} and, finally, for rotation around axis *oz* with angular rates ω_{zi}, i – the number of tests. Need to mark that such method is required a lot of time and at real life the vehicle or body is rotates around the each one of rotation axes.

There are some methods for estimation and compensation of sensors output signal noises. Most known is Kalman filter, used in [4-7]. For estimation of noise features were used Allan variance methods and some time – Wavelet transform [8]. Also, it is well known of using of fuzzy logic techniques [9] and neural networks [10, 11].

There is the scalar method of IMU calibration used for gyroscopes and accelerometers [12, 13], based on a scalar reference input motion. For the gyroscopes in the Earth's gravitational field such scalar value is the rotation rate Ω, and for accelerometers – the value of gravity vector g.

In this paper was suggested to use a new method of 3D-calibration. The new calibration method based on whole angle rotation. In fact it is suggests to turn over IMU around three axes simultaneously.

2 Calibration of IMU's set of axial accelerometers

Output signals of the IMU's axial accelerometers may be expressed according to source [1]:

$$\begin{bmatrix} U_{ax} \\ U_{ay} \\ U_{az} \end{bmatrix} = \begin{bmatrix} B_{ax} \\ B_{ay} \\ B_{az} \end{bmatrix} + \begin{bmatrix} k_{11} & k_{12} & k_{13} \\ k_{21} & k_{22} & k_{23} \\ k_{31} & k_{32} & k_{33} \end{bmatrix} \cdot \begin{bmatrix} a_x \\ a_y \\ a_z \end{bmatrix} + \begin{bmatrix} n_{ax} \\ n_{ay} \\ n_{az} \end{bmatrix} \tag{1}$$

where B_{ax}, B_{ay}, B_{az} – g – independent bias, a_x, a_y, a_z – accelerations acting along the x, y and z axes of the sensor respectively, n_{ax}, n_{ay}, n_{az} – zero-mean random biases or output measurements noises, *Oxyz* – body frame, 3x3 matrix representing the g – dependent biases induced by accelerations a_x, a_y, a_z, k_{11}, k_{22}, k_{33} – scale-factor coefficients of accelerometers, another one coefficients of the matrix are cross-coupling coefficients.

For reducing noises of output measurements we will average output signals during 100 seconds. Therefore in the future we will ignore of n_{ax}, n_{ay}, n_{az} value. Let's write the expression (1) for each one accelerometer and for i – position of IMU testing:

$$U_{axi} = B_{ax} + k_{11}a_{xi} + k_{12}a_{yi} + k_{13}a_{zi};$$

$$U_{ayi} = B_{ay} + k_{21}a_{xi} + k_{22}a_{yi} + k_{23}a_{zi};$$

$$U_{azi} = B_{az} + k_{31}a_{xi} + k_{32}a_{yi} + k_{33}a_{zi}.$$

We will do a set of testing measurements, which are undertaken multi-position tests for turn angles α, β, γ around three axes simultaneously with exact step of turn (Fig.1).

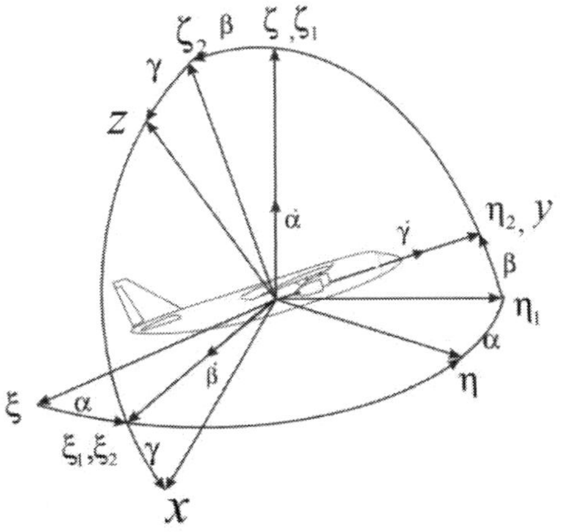

Figure 1. Turn angles α, β, γ of axial accelerometers

Normally the precision dividing head is using for calibration of accelerometers. But in this case, when IMU is turning around three axes simultaneously or whole angle rotation or finite rotation, the three-axes turntable is used instead a typical precision dividing head.

Let's represent output signal of first accelerometer:

1-st measurement:

$$U_{ax1} = B_{ax} + k_{11}a_{x1} + k_{12}a_{y1} + k_{13}a_{z1};$$

2-nd measurement:

$$U_{ax2} = B_{ax} + k_{11}a_{x2} + k_{12}a_{y2} + k_{13}a_{z2};$$

....

n-st measurement:

$$U_{axn} = B_{ax} + k_{11}a_{xn} + k_{12}a_{yn} + k_{13}a_{zn}.$$

Above set of equations we can represent in matrix form for first, second and third accelerometers:

$$\begin{bmatrix} U_{ax1} \\ U_{ax2} \\ \vdots \\ U_{axn} \end{bmatrix} = \begin{bmatrix} 1 & a_{x1} & a_{y1} & a_{z1} \\ 1 & a_{x2} & a_{y2} & a_{z2} \\ . & . & . & . \\ 1 & a_{xn} & a_{yn} & a_{zn} \end{bmatrix} \begin{bmatrix} B_{ax} \\ k_{11} \\ k_{12} \\ k_{13} \end{bmatrix}; \begin{bmatrix} U_{ay1} \\ U_{ay2} \\ \vdots \\ U_{ayn} \end{bmatrix} = \begin{bmatrix} 1 & a_{x1} & a_{y1} & a_{z1} \\ 1 & a_{x2} & a_{y2} & a_{z2} \\ . & . & . & . \\ 1 & a_{xn} & a_{yn} & a_{zn} \end{bmatrix} \begin{bmatrix} B_{ay} \\ k_{21} \\ k_{22} \\ k_{23} \end{bmatrix};$$

$$\begin{bmatrix} U_{az1} \\ U_{az2} \\ \vdots \\ U_{azn} \end{bmatrix} = \begin{bmatrix} 1 & a_{x1} & a_{y1} & a_{z1} \\ 1 & a_{x2} & a_{y2} & a_{z2} \\ . & . & . & . \\ 1 & a_{xn} & a_{yn} & a_{zn} \end{bmatrix} \begin{bmatrix} B_{az} \\ k_{31} \\ k_{32} \\ k_{33} \end{bmatrix}$$

After that let's to combine the received matrix equations to single 'equation of calibration' for IMU's set of accelerometers:

$$\mathbf{U_{a1}} = \mathbf{G_{n\times4}} \cdot \mathbf{X_1}, \qquad (2)$$

where

$$\mathbf{U_{a1}} = \begin{bmatrix} U_{ax1} & U_{ay1} & U_{az1} \\ U_{ax2} & U_{ay2} & U_{az2} \\ . & . & . \\ U_{axn} & U_{ayn} & U_{azn} \end{bmatrix}, \quad \mathbf{G_{n\times4}} = \begin{bmatrix} 1 & a_{x1} & a_{y1} & a_{z1} \\ 1 & a_{x2} & a_{y2} & a_{z2} \\ . & . & . & . \\ 1 & a_{xn} & a_{yn} & a_{zn} \end{bmatrix}, \quad \mathbf{X_1} = \begin{bmatrix} B_{ax} & B_{ay} & B_{ax} \\ k_{11} & k_{21} & k_{31} \\ k_{12} & k_{22} & k_{32} \\ k_{13} & k_{23} & k_{33} \end{bmatrix}.$$

We will solve the last matrix equation by least-squares method:

$$\hat{\mathbf{X}}_1 = \left(\mathbf{G}^{\mathbf{T}}_{\mathbf{n\times4}}\mathbf{G}_{\mathbf{n\times4}}\right)^{-1}\mathbf{G}^{\mathbf{T}}_{\mathbf{n\times4}}\mathbf{U}_{\mathbf{a1}}. \qquad (3)$$

Here symbol 'T' is mean a transposed matrix.

Sample 1. We will consider the IMU's accelerometers with the below followings nominal parameters:

$B_{ax} = B_{ay} = B_{az} = 2.5\ V/g;\ k_{11} = k_{22} = k_{33} = 1.0\ V/g;$
$k_{12} = 0.01\ V/g,\ k_{13} = -0.01\ V/g,\ k_{21} = -0.01\ V/g,$
$k_{23} = 0.01\ V/g,\ k_{31} = 0.01\ V/g,\ k_{32} = -0.01\ V/g,$

Body's turns with angles α, β, γ wich will change from 0 to 400 degree with the same step of 10 degree (the number of positions is 40).

The calculated values of output signals (2) for matrix $\mathbf{G}_{n\times 4}$ are shown on Fig. 2.

After calculations according to (3), we will have:

$$\hat{\mathbf{X}}_1 = \begin{bmatrix} 2.5 & 2.5 & 2.5 \\ 1.0 & -0.01 & 0.009999 \\ 0.01 & 1.0 & -0.009999 \\ -0{,}009999 & 0.01 & 1.0 \end{bmatrix}.$$

Thus, as a result of measuring (2) and calculations (3) it is succeeded to get estimations of the biases $\hat{\mathbf{B}}_{\mathbf{a}}$, scale factors and cross-coupling coefficients – elements of matrix 3×3.

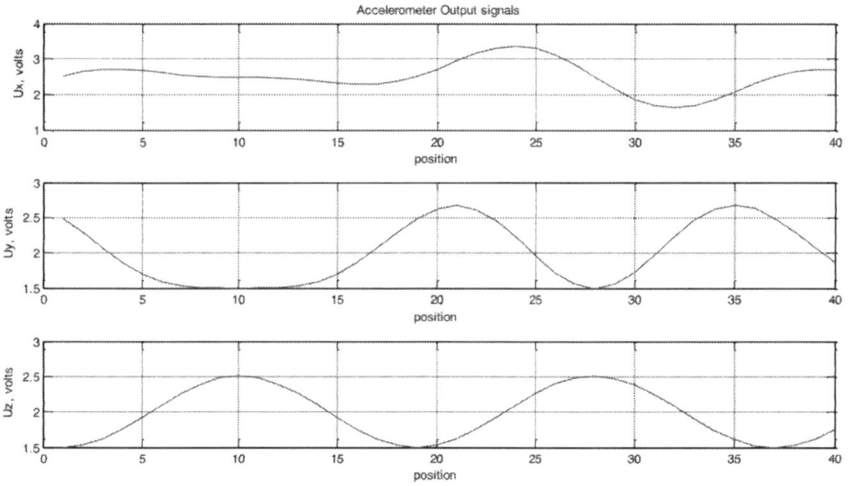

Figure 2. Averaging output signals of axial accelerometers

3 Calibration of IMU's set of gyroscopes

Output signals of the IMU's gyroscopes may be expressed according to source [1]:

$$\begin{bmatrix} U_{\omega x} \\ U_{\omega y} \\ U_{\omega z} \end{bmatrix} = \begin{bmatrix} B^*_{\omega x} \\ B^*_{\omega y} \\ B^*_{\omega z} \end{bmatrix} + \begin{bmatrix} n_{11} & n_{12} & n_{13} \\ n_{21} & n_{22} & n_{23} \\ n_{31} & n_{32} & n_{33} \end{bmatrix} \cdot \begin{bmatrix} \omega_x \\ \omega_y \\ \omega_z \end{bmatrix} + \begin{bmatrix} n_{\omega x} \\ n_{\omega y} \\ n_{\omega z} \end{bmatrix}. \quad (4)$$

Here $B^*_{\omega x}, B^*_{\omega y}, B^*_{\omega z}$ - biases of each one gyroscopes, which may be depends from g and g^2 – drifts, ω_x, ω_y, ω_z – turn rates, acting along the *x*, *y* and *z* axes of the sensor respectively, $n_{\omega x}$, $n_{\omega y}$, $n_{\omega z}$ - zero-mean random biases or output measurements noises, 3x3 matrix representing the biases induced by ω_x, ω_y, ω_z, n_{11}, n_{22}, n_{33} – scale-factor coefficients of gyroscopes, another one coefficients of the matrix are cross-coupling coefficients.

It's known [1], that g and g^2 dependent biases are usually insignificant in optical sensors such as fiber optic and ring laser gyroscopes.

Consider a calibration such optical sensors when IMU turns around three axes simultaneously. When the set of n-measurements will be done with turn rates ω_{xi}, ω_{yi}, ω_{zi} and after averaging of output signals during 100 seconds we will have a new matrix 'equation of calibration':

$$\mathbf{U}_{\omega 3}=\boldsymbol{\omega}_{\mathbf{n}\times 4}\cdot\mathbf{X}_3, \qquad (5)$$

where

$$\mathbf{U}_{\omega 3}=\begin{bmatrix} U_{\omega x1} & U_{\omega y1} & U_{\omega z1} \\ U_{\omega x2} & U_{\omega y2} & U_{\omega z2} \\ . & . & . \\ U_{\omega xn} & U_{\omega yn} & U_{\omega zn} \end{bmatrix}, \mathbf{X}_3=\begin{bmatrix} B^*_{\omega x} & B^*_{\omega y} & B^*_{\omega z} \\ n_{11} & n_{21} & n_{31} \\ n_{12} & n_{22} & n_{32} \\ n_{13} & n_{23} & n_{33} \end{bmatrix}, \boldsymbol{\omega}_{\mathbf{n}\times 4}=\begin{bmatrix} 1 & \omega_{x1} & \omega_{y1} & \omega_{z1} \\ 1 & \omega_{x2} & \omega_{y2} & \omega_{z2} \\ . & . & . & . \\ 1 & \omega_{xn} & \omega_{yn} & \omega_{zn} \end{bmatrix}.$$

On the face of it the structure of the equation (5) is very similar to equation (2). The difference is that matrix $\boldsymbol{\omega}_{\mathbf{n}\times 4}$ consists from turn rates ω_{xi}, ω_{yi}, ω_{zi} and matrix $\mathbf{G}_{\mathbf{n}\times 4}$ - from accelerations a_{xi}, a_{yi}, a_{zi}. However if we have ordered turn series without any limitations for calibration of IMU's set of accelerometers, otherwise we should provide a below following rule:

An equation of calibration will have a decision if the rank of coefficient matrix $\boldsymbol{\omega}_{\mathbf{n}\times\mathbf{m}}$ *is equal to number of columns of the same matrix or by other words* $rank\,\boldsymbol{\omega}_{\mathbf{n}\times\mathbf{m}}=\mathbf{m}$.

The above rule is based on Rouche-Capelli or Kronecker-Capelli theorem [14]. A system of linear equations with n variables has a solution if and only if the rank of its coefficient matrix $\boldsymbol{\omega}_{\mathbf{n}\times 4}$ is equal to the rank of its augmented matrix $\left[\boldsymbol{\omega}_{\mathbf{n}\times 4},\mathbf{U}_{\omega 3}\right]$:

$$rank\,\boldsymbol{\omega}_{\mathbf{n}\times 4}=rank\left[\boldsymbol{\omega}_{\mathbf{n}\times 4},\mathbf{U}_{\omega 3}\right].$$

This condition could be provided, for example due nonlinear depending of last columns of matrix $\boldsymbol{\omega}_{\mathbf{n}\times 4}$.

Solving the last matrix equation (5) by least-squares method:

$$\hat{\mathbf{X}}_3=\left(\boldsymbol{\omega}^{\mathbf{T}}_{\mathbf{n}\times 4}\boldsymbol{\omega}_{\mathbf{n}\times 4}\right)^{-1}\boldsymbol{\omega}^{\mathbf{T}}_{\mathbf{n}\times 4}\mathbf{U}_{\omega 3}. \qquad (6)$$

Sample 3. Consider the IMU's gyroscopes with the below followings nominal parameters:
$B^*_{\omega x}=B^*_{\omega y}=B^*_{\omega z}=2.0\ V$;
$n_{11}=0.01\ V/(d/s)$; $n_{12}=0.01\ V/(d/s)$; $n_{13}=-0.02\ V/(d/s)$;
$n_{21}=-0.01\ V/(d/s)$; $n_{22}=0.1\ V/(d/s)$; $n_{23}=0.01\ V/(d/s)$;
$n_{31}=0.03\ V/(d/s)$; $n_{32}=-0.01\ V/(d/s)$; $n_{33}=0.1\ V/(d/s)$;

To provide the rank of matrix $\boldsymbol{\omega}_{\mathbf{n}\times 4}$ to order of matrix:

$$rank\,\boldsymbol{\omega}_{\mathbf{n}\times 4}=4,$$

we will arrange the projections of angular rate of turntable in below following series:

$$\omega_{xi}=\omega_i,\quad \omega_{yi}=\omega_i^{1/2},\quad \omega_{zi}=\omega_i^{1/3}.$$

We will change the angular rate ω_i from 0 to 100 degree/sec with the same step of 10 degree/sec. The number of tests is 10.

Solving the equation of calibration according to (6) by least-squares method we will have

$$\hat{\mathbf{X}}_3 = \begin{bmatrix} 2.0 & 2.0 & 2.0 \\ 1.0 & -0.01 & 0.03 \\ 0.01 & 1.0 & -0.01 \\ -0.02 & 0.01 & 1.0 \end{bmatrix}.$$

Thus, as a result of measuring (5) and calculations (6) it is succeeded to get estimations of the gyro's biases and elements of the matrix 3×3.

4 Conclusions

A new calibration method for IMU of Strapdown Inertial Navigation Systems was suggested. The new calibration method is based on whole angle rotation or finite rotation. In fact it is suggested to turn over Inertial Measurement Unit around three axes simultaneously. In order to solve the equation of calibration, it is necessary to provide an equality of a rank of basic matrix to degree of basic matrix. The results of simulated IMU data are presented to demonstrate the performance of the new calibration method.

References

1. Titterton D. H. and Weston J. L. Strapdown Inertial Navigation Technology. - IEE Radar, Sonar, Navigation and Avionics Series 17, 2004. – pp. 558.
2. Nebot E., Durrant-Whyte H., Initial Calibration and Alignment of Low Cost Inertial Navigation Units for Land Vehicle Applications. – *Journal of Robotics Systems*, Vol. **16**, No. 2, February 1999, pp. 81-92.
3. Artese G., Trecroci A. Calibration of a low cost MEMS INS sensor for an integrated navigation system. – *The International Archives of the Photogrammetry, Remote Sensing and Spatial Information Sciences. Vol. XXXVII. Part B5. Beijing 2008,* pp. 877-882.
4. Grewal M. S., Henderson, V. D., Miysako R. S. Application of Kalman filtering to the calibration and alignment of inertial navigation systems. *IEEE Transactions on Automatic Control*, 1991, vol. **36**, pp. 3-13.
5. Hide C., Moore T., Smith M. Adaptive Kalman filtering for low-cost INS/GPS. *The Journal of Navigation*, 2003, vol. **56,** pp. 143-152.
6. Nikbakht N., Mazlom M., and Khayatian A. Evaluation of solid-state accelerometer for positioning of vehicle. *Proceeding of the IEEE International Conference on Industrial Technology*, Hong Kong, 2005, pp. 729-733.
7. Pang G., Liu, H. Evaluation of a low-cost MEMS accelerometer for distance measurement. *The Journal of Intelligent and Robotic Systems*, 2001, vol. **30**, pp. 249-265.
8. Sheng-Chih Shen, Chia-Jung Chen, Hsin-Jung Huang. A new calibration method for low cost MEMS inertial sensor module. *Journal of Marine Science and Technology,* 2010, vol. **18**, No. 6, pp. 819-824.
9. Gaysse, J. A low cost absolute position calculation system. *Proceeding of SICE-ICASE International Joint Conference*, Busan, Korea, 2006, pp. 5658-5661.
10. El-Rabbany A., El-Diasty, M. An efficient neural modal for denoising of MEMS-based inertial data. *The Journal of Navigation*, 2004, vol. **57**, pp. 407-415.
11. Wang H., Tian W. Modeling the random drift of micro-machined gyroscope with neural network. *Neural Processing Letters*, 2005, vol. **22**, pp. 235-247.
12. Avrutov V. On Scalar Calibration of an Inertial Measurement Unit / V. Avrutov, S. Golovach, T. Mazepa // 19-th St. Petersburg International Conference on Integrated Navigation Systems. St. Petersburg, Russia, 2012. - State Research Center (CSRI) Elektropribor, 2012, pp.117-121.

13. Avrutov V. Scalar Diagnostics of the Inertial Measurement Unit// *I.J. Intelligent Systems and Applications,* 2015, vol.**11**, pp.1-9.
14. Bugrov Y.S., Nikolsky S.M. Higher mathematics. Elements of linear algebra and analytic geometry. – Moscow: 'Nauka', 1984. – pp.192 (*In Russian*).

Gas Holding mechanism and Impeller Structure Optimization of High Efficiency Gas-Liquid Mixer

Jing Rui Li[1], Ting Jun Yan[1,a] and Bai Ru Zhang[1]

[1]*Mechanical and Electrical Engineering Institute, Beijing University of Chemical Technology, Beijing 100029, China*

Abstract. This research aims at researching one high efficiency gas-liquid mixer which can be applied under high gas content condition and can overcome the limited application of sedimentation and rotary oil-gas separator for electric submersible pump. Standard k-ε model and population balance model of the Fluent are applied to simulate the gas-liquid two-phase flow in high efficiency gas-liquid mixer. The internal flow field of impeller opening with different radius and amounts are simulated. The gas holding mechanism of high efficiency gas-liquid mixer is also analyzed. Then according to the gas particle size at the output of impeller, the optimized structure parameter can be obtained. The simulation results show that part of the high pressure liquid at impeller outlet would form cross flow with fluid in normal flowing inside the impeller passage to produce strong shear friction inside the area and burst a large amount of bubbles accumulated in the area. Also, it is proved to be more beneficial for breaking bubble and dealing with gas under the circumstances that gas content at the input is higher than 35%, opening diameter of impeller is Φ4.8mm and opening amount of impeller is 2.

1 Introduction

Generally, the output liquid of the oilfield contains natural gas, water and the other substance. If the content of the gas in the well fluid exceeds designed allowable value of the ESP, it can result in the significant reduction of displacement, head and efficiency of electric submersible pump, so that the working performance can be impacted [1-3]. Aiming at improving applicability of the ESP in live wellbore, it is general to add a gas-processing equipment (oil-gas separator) at its input. However, the normal ESP can not work effectively if gas content is higher than 35% [4-5]. In this case, one high efficiency gas-liquid mixer (e.g. from Schlumberger) can solve the problem and improve the performance of the ESP [6]. In China, several ESP production companies such as Chittagong Machinery Manufacturing Company, Victory Pump Company, Daqing oilfield power of God have produced gas-processing equipment, which is effectively applicable for high gas content considering the related productions in other countries. Henceforth, researching this lack plays an important role for extracting oil in high gas content oil well.

[a] Corresponding author: Ting Jun Yan: yantj555@163.com.

2 Gas holding mechanism of high efficiency gas-liquid mixer

2.1 The impeller structure of high efficiency gas-liquid mixer

High efficiency gas-liquid mixer is installed in the entrance of the ESP and its structure is similar to multistage centrifugal pump, which is composed of multistage impeller. The impeller structure is showed in figure 1. The impeller has 6 blades, and the low pressure area of the impeller near the root of each blade is provided with two equal diameter holes. In figure 1, the holes are labeled as A and B, and the two holes are arranged along the flow direction of the blade, the total amount of holes is 12.

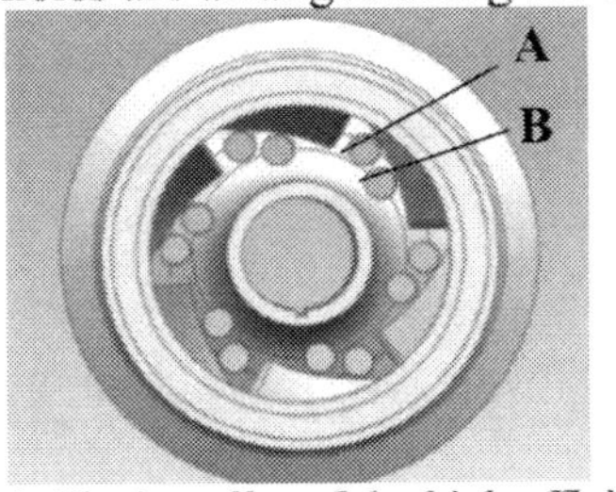

Figure 1. The impeller of the high efficiency gas-liquid mixer

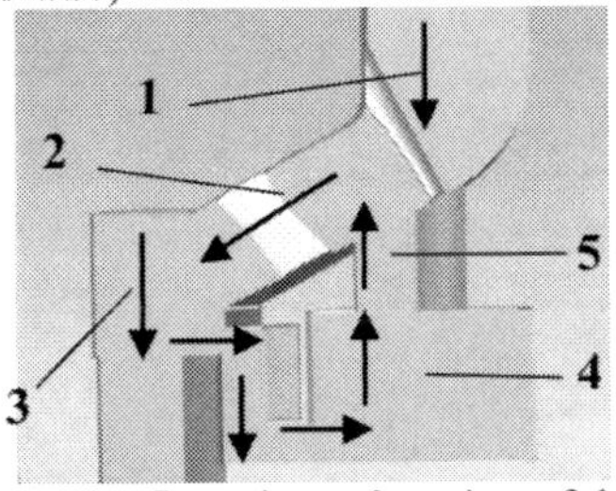

Figure 2. The flow channel section of the high efficiency gas-liquid mixer cross hole A

Figure 2 shows the flow passage section of the high efficiency gas-liquid mixer cross hole A. The flow passage section is composed of the flow passage 1 of impeller entrance, the rotating impeller passage 2, impeller outlet 3, the reflux cavity 4 and the flow passage 5 of holes A. The hole A is connected with the reflux cavity and the impeller entrance. The direction of the arrow in the figure shows the direction of the fluid which flows in the impeller of the high efficiency gas-liquid mixer. The fluid enters into the impeller passage 2 through the impeller entrance 1, then outflows from the outlet 3 of the impeller, into the roulette. Among them, there is a small amount of fluid flow along the clearance from the impeller outlet 3 to the back of the impeller, flow through the reflux cavity 4 and roulette hole A and B, return to the impeller inlet port 1.

2.2 Liquid phase velocity and pressure distribution of high efficiency gas-liquid mixer

According to the impeller structure of high efficiency gas-liquid mixer, standard k-ε model and population balance model in Fluent software can be used to simulate the internal flow field of the impeller. Figure 3 shows the liquid phase velocity vector in the high efficiency gas-liquid mixer through hole A and hole B. The discussion of high efficiency gas-liquid mixer mixing mechanism is the preliminary studies of our group which have been published [7]. As shown in Figure 3a, the flow velocity in the reflux cavity and hole A is large and flowing uniformity. However, the flow velocity in the reflux cavity and hole B is small and flowing weakly as shown in figure 3b. Thus, due to the hole A approaches the impeller outlet and the hole B is located at the impeller inlet, when the fluid flows through the hole A and hole B, the resistance of flow passage of hole A is relatively small, so that most of the return fluid flows to the impeller inlet through the flow passage of hole A, while a small amount of fluid flows through the flow passage of hole B.

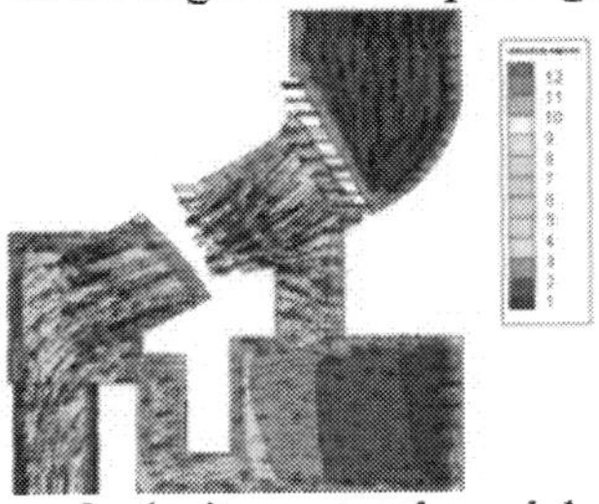

a. The picture of velocity vector through hole A

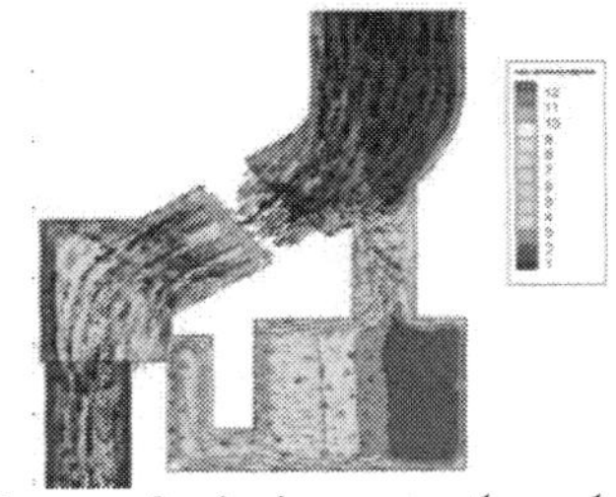

b. The picture of velocity vector through hole B

Figure 3. Liquid phase velocity vector in the high efficiency gas-liquid mixer

2.3 Gas-liquid mixing principle of high efficiency gas-liquid mixer

Figure 4 and Figure 5 show the fluid velocity vector and the turbulent kinetic energy contours of the high efficiency gas-liquid mixer impeller near the openings. A part of the high pressure fluid flows through the high efficiency gas-liquid mixer roulette hole A and B, and then flows to the impeller inlet port back. The arrows direction of I and II regional showed in FIG. 4 is the fluid velocity direction within roulette holes (A, B). The flow direction and movement direction of the normal flow of fluid in the impeller passage formed cross flow because there is a certain Angle. So there is relative strong shearing action between two parts of fluid. When shear force is greater than the surface tension of the air bubble in the fluid, the large diameter of bubble gathered here is torn and broken into smaller bubbles. Figure 5 shows the high velocity of the fluid at the openings so that the turbulent kinetic energy near the openings is large, so small bubbles mixed with the surrounding liquid by strong turbulent action. Through cutting and mixing after flowing out of the roulette holes (A, B), the gas particle size will be decreased and the distribution of gas and liquid is uniform, as similar as single-phase flow, thus can improve the electric submersible pump suction condition, provides guarantee for stable operation of the electric submersible pump.

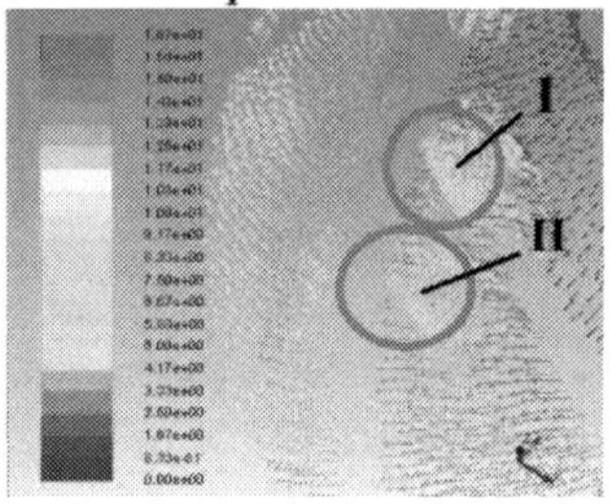

Figure 4. The fluid velocity vector of the high efficiency gas-liquid mixer impeller near the openings

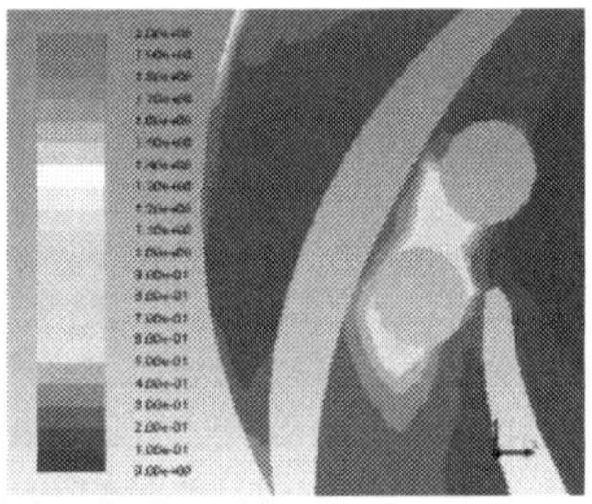

Figure 5. The turbulent kinetic energy contours of the high efficiency gas-liquid mixer impeller near the openings

3 The mixing effect impacted by the impeller parameters of high efficiency gas-liquid mixer

Simulate the flow field of gas-liquid two-phase mixing process within the high efficiency gas-liquid mixer by using Fluent. The unstructured mesh arrangement with hexahedral elements is used in calculating watershed. The periodic boundary is applied to the circumferential of impeller. Velocity inlet is set for entry of computation region and free flow condition is inducted for outlet. Standard k-ε model is used to serve as a turbulent flow model with the object of simulate being the gas-liquid two-phase flow. The population balance model is implemented in the two-fluid model that accounts for the coalescence and breakup of bubble. According to the oil field, choose to use 98 pumps which is used widely, the other calculation parameters are: high efficiency gas-liquid mixer operating pressure is 6 MPa (submergence depth is 600 m), the air as the gas phase, the water as the liquid phase, inlet velocity 1.5 m/s (Q is 210 m^3 / d), the impeller speed 2850 r/min, the initial bubble diameter is set to Φ1 mm.

3.1 The influence of high efficiency gas-liquid mixer impeller opening diameter

The size of the opening diameter of the impeller would affect the performance of the high efficiency gas-liquid mixer. High efficiency gas-liquid mixers with the opening diameter respectively of Φ2 mm, Φ3 mm, Φ4 mm, Φ4.8 mm, Φ5 mm and Φ6 mm were compared. And as the inlet gas rates are respectively 5%, 10%, 15%, 20%, 25%, 30%, 35%, 40%, 45%, we found that under different inlet gas rates, the internal flow field distribution of high efficiency gas-liquid mixer with 6 different kinds of opening diameter is similar. Here inlet gas rate is selected as 15% and the gas particle size distribution on the flow channel section of 6 kinds of high efficiency gas-liquid mixers is analyzed.

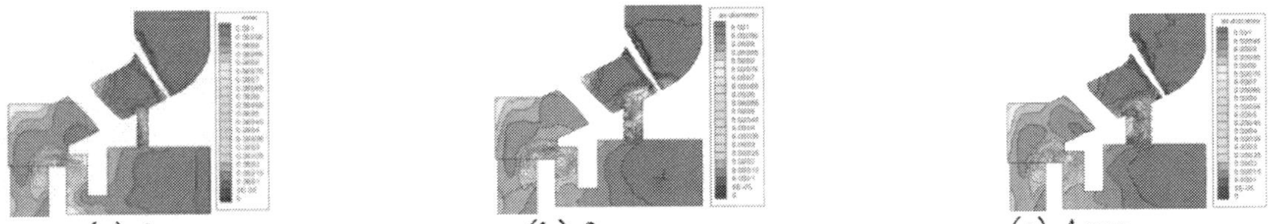

(a) 2 mm (b) 3 mm (c) 4 mm

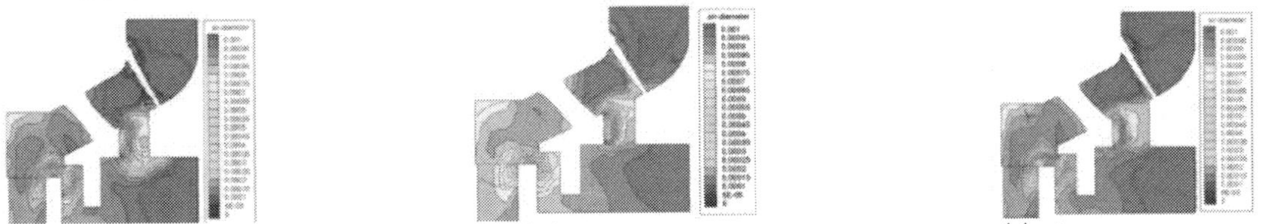

(e) 5 mm (d) 4.8 mm (f) 6 mm

Figure 6. The gas diameter contours of the high efficiency gas-liquid mixer with different aperture openings

Figure 6 shows the gas diameter contours on the cross hole A section of the high efficiency gas-liquid mixer with different aperture openings. Each of the flow channel's cross-sections is corresponding to Figure 2. From the gas diameter contours it can be seen that when the high efficiency gas-liquid mixer aperture opening is respectively Φ2 mm, Φ3 mm, Φ4 mm, Φ4.8 mm, Φ5 mm and Φ6 mm, the average gas particle diameter of the impeller's outlet was respectively: Φ0.806 mm, Φ0 784 mm, Φ0 780 mm, Φ0.740 mm, Φ0.790 mm, Φ0.786 mm. It's found that the high efficiency gas-liquid mixer with Φ 4.8 mm opening diameter has the least average gas particle diameter of the impeller's outlet compared to the other 5 kinds, which illustrates that the opening diameter of Φ4.8mm is more beneficial to bubble's broken.

3.2 The influence of efficient gas-liquid mixer impeller opening number

As for the effect of the number of openings on the performance of high efficiency gas-liquid mixer, according to the results of the discussion, through exploring the size of the gas particle of the impeller outlet with single hole and double holes (each opening hole area is same). Therefore, the optimal number of openings can be determined.

Figure 7. The gas diameter contours of the high efficiency gas-liquid mixer with two holes in axial and radial section of the flow path cross hole A

Figure 8. The gas diameter contours of the high efficiency gas-liquid mixer with one holes in axial and radial section of the flow path cross hole A

Figure 7 shows the gas diameter contours of the high efficiency gas-liquid mixer with two holes in axial and radial section of the flow path cross hole A. Figure 8 shows the gas diameter contours of the high efficiency gas-liquid mixer with one hole in axial and radial section of the flow path cross hole A. According to the gas particle size distribution of the high efficiency gas-liquid mixer with double and single hole, the average gas particle size of the impeller outlet is Φ0.746 mm and Φ0.811 mm. By contrasting, it's found that the high efficiency gas-liquid mixer with double holes had smaller gas average particle size at impeller outlet, thus the two-hole high efficiency gas-liquid mixer is more conducive to bubble broken.

4 Conclusion

1. The high pressure liquid at impeller outlet would return to the zone on the blade non-working surface with the minimum pressure through the opening on the high efficiency gas-liquid mixer, and form cross flow with fluid in normal flowing inside the impeller passage, thereby dispersing, chopping the bubbles, and mixing the bubbles with the liquid.
2. The opening radius is 4.8 mm, the opening number is 2, more conducive to the bubble breaking, the gas treatment effects in better.

References

1. H. Pineda, J. Biazussi, F. Lopez, et al. Phase distribution analysis in an Electrical Submersible Pump (ESP) inlet handling water-air two-phase flow using Computational Fluid Dynamics (CFD)[J]. Journal of Petroleum Science and Engineering. 2016; **139**: 49-61.
2. H Stel, T Sirino, F J Ponce, et al. Numerical investigation of the flow in a multistage electric submersible pump[J]. Journal of Petroleum Science and Engineering. 2015; **136**: 41-54.
3. H Q Zhao, Z L Li, H Y Sun. Numerical Simulation of Flow Field in Downhole Hydrocyclone Oil-gas Separator[J]. Journal of China University of Petroleum (Edition of Natural Sciences), 2007; **31**(4): 94-97.
4. H P Tang, Q B Wang, G Y Yang, et al. A Honeycomb-Structured Ti-6Al-4V Oil-Gas Separation Rotor Additively Manufactured by Selective Electron Beam Melting for Aero-engine Applications[J]. JOM. 2016; **68**(3): 799-805.
5. Z L Li, H Y Sun. Performance Simulation Test of Downhole Centrifugal Cyclone and High Efficiency Oil-gas Separator[J]. China Petroleum Machinery. 2007; **35**(12): 5-8.
6. Y.L Woon, Bartlesville, OKla. Downhole Pumping System for Recovering Liquids and Gas: United States, 5628616[P]. 1997-05-13.
7. B R Zhang, T J Yan, J Li, et al. Theoretical Research on Gas Holding of a New Kind Mixer with High Gas-liquid Ratio[J]. China Petroleum Machinery, 2016, **44**(9): 52-57.

Effect of System Pressure on the Surface Microstructure of Spherical Porous Basic Magnesium Carbonate

Yu Wei [1], Hong Lei [1], Weikang Peng [1] and Zhiliang Huang [1, a]

[1] *School of Materials Science and Engineering, Wuhan Institute of Technology, Wuhan 430074, People's Republic of China*

Abstract: spherical porous basic magnesium carbonate ($Mg_5(CO_3)_4(OH)_2 \cdot 4H_2O$) has been prepared by using ammonium bicarbonate and magnesium chloride hexahydrate as raw materials. A novel CO_2 bubble template was constructed to explain the formation of pore structure of porous $Mg_5(CO_3)_4(OH)_2 \cdot 4H_2O$. System pressure was found to be important in controlling the final pore size of spherical porous $Mg_5(CO_3)_4(OH)_2 \cdot 4H_2O$. By the investigation of the particles from pressure-dependent experiments, scanning electron microscopy observations, X-ray diffraction, and Fourier transform infrared spectra were used to follow the reactions and identify the products. The results show that the pore size of spherical porous $Mg_5(CO_3)_4(OH)_2 \cdot 4H_2O$ gradually decreased via increasing system pressure.

1 Introduction

For magnesium resource utilization, the new development direction is that the abundant magnesium chloride is used to synthesize the functional magnesium carbonate with excellent performance. Basic magnesium carbonate (BMC) is a kind of important magnesium inorganic material. With good fluidity, filling and dispersion [1], BMC has played an more and more important part in many industries and fields, such as pharmaceuticals, rubber industry, insulation materials, dental preparations, senior glass, food industry and ingredients of various kinds of chemical products [2].At present, many studies about preparation of BMC have been reported and most crystal microstructure are spherical[3], rod-like [4] and microtubes [5]. Because of the larger specific surface area and higher decomposition temperature, spherical structure attract extensive attention, many methods were used to synthesis BMC. For example, *Zaki* et al have synthesized spherical BMC by using Cr solution as promoter [6]. However, the study on the surface microstructure of spherical porous basic magnesium carbonate is rare. It is important to control the surface microstructure with simple condition.

In this paper, the spherical porous BMC was synthesized without using surfactants. It is supposed the assembling of spherical porous BMC with CO_2 bubble template according to phase transformation. We try to use several equation to describe relationship between pore size and pressure. Further more, Effect of system pressure on the surface microstructure of spherical porous basic magnesium

[a] Corresponding author: hzl6455@126.com

carbonate was studied.

2 Experimental

2.1 Materials and Synthesis

Magnesium chloride hexahydrate and ammonium bicarbonate were used as analytical grade in a typical experiment. Then, aqueous solution were transferred into special autoclave. The pressure of the system can be accurately adjust at any time, and the system temperature remains steady at 373 k. After reaction finished, the special autoclave was taken out to age for 5 h, and the mixture was repeatedly washed by distilled water and ethanol. After that, vacuum oven was used to dry mixture. Finally, the solid products were obtained at different system pressure as follows: S_1, 4 atm; S_2, 6 atm; S_3, 8 atm; S_4, 10 atm.

2.2 Characterization

Fourier transform infrared and powder X-ray diffraction were employed to analyze the chemical composition of BMC. A scanning electron microscope was used to observe The microstructure and the surface changes of spherical BMC.

3 Results and discussion

Spectra for the obtained products at different system pressure are shown in Fig.1, the spectrum of the S_1 synthesized at 4 atm shows typical bands at 3648.76, 3514.54, 3448.70, 797.74 and 593.77 cm^{-1} which are due to structural hydroxyl (O-H), with those at 1483.55, 1421.56, 1120.24, 885.31, 852.99 and 713.11 cm^{-1} are due to carbonate anion (CO_3^{2-}). Apparently, spectra of S_2, S_3, S_4 are similar to S_1, XRD patterns of S_1, S_2, S_3 and S_4 were shown in Fig. 2, According to the standard JCPDS card NO. 25-0513, it is obviously that all those samples are considered to be BMC with high purity crystalline. In addition, An interesting phenomenon was found from the 4 patterns was that diffraction peaks become more acute with the increase of pressure. Thus, from the analysis of the XRD and FT-IR, all of the products are BMC.

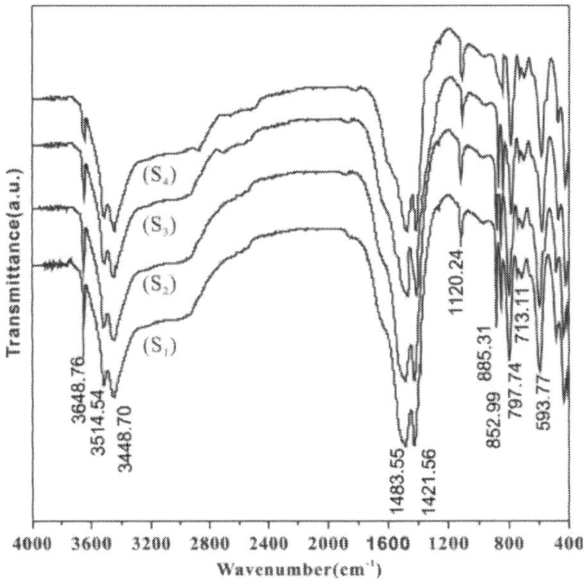

Figure 1. Infrared spectra of solid products synthesized at different system pressures (S_1: 4 atm; S_2: 6 atm; S_3: 8 atm; S_4: 10 atm)

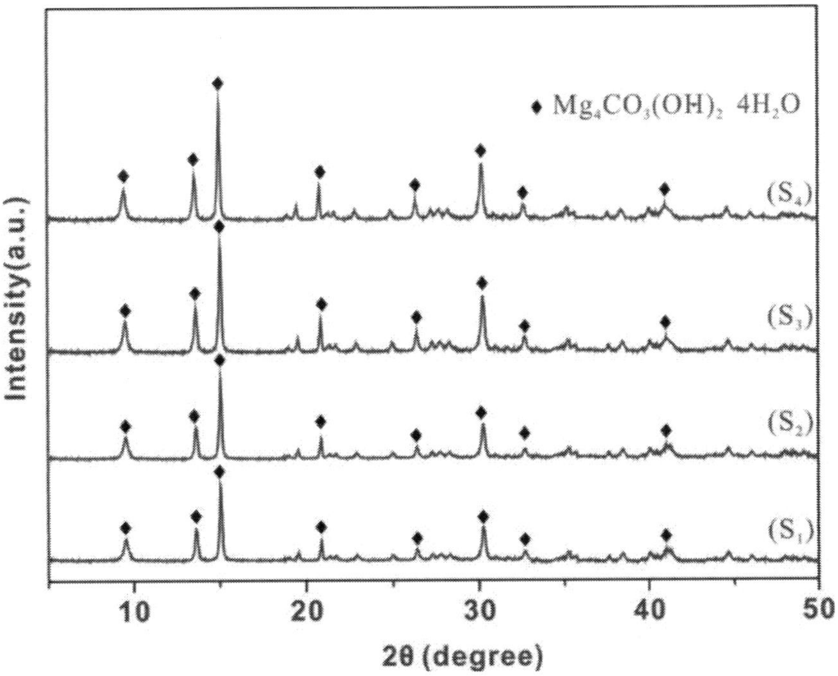

Figure 2. XRD pattern of samples synthesized at different system pressures (S_1: 4 atm; S_2: 6 atm; S_3: 8 atm; S_4: 10 atm)

SEM images of S_1 is shown in Fig. 3, the surface of BMC is assembled by the flakes, most of which interconnect and form an porous sphere. *Cheng* et al have discovered that gas bubbles can be seen as a type of template for crystal growth, In this paper, CO_2 bubbles is considered to be a temple to form such porous sphere of BMC. Growth process of $Mg_5(CO_3)_4(OH)_2{\cdot}4H_2O$ is shown Figure 4. The existence and aggregation of many bubbles lead to form a foam system. In this system. the plateaus emerge with highest energy density is in the center of at least three CO_2 bubbles where BMC begin to nucleate steadily. After that, BMC grow up via the the passage between the CO_2 bubbles by self-assembling. Lastly, when reaction is over, CO_2 bubbles was washed and spherical porous basic magnesium carbonate was obtained.

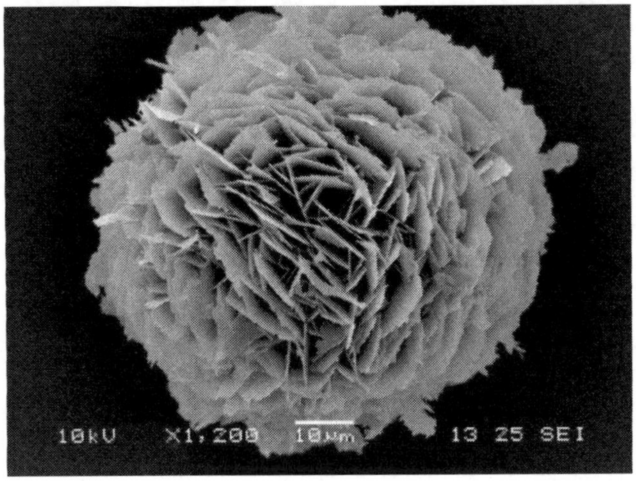

Figure 3. SEM images of S_1.

DOI: 10.1051/matecconf/201711401015

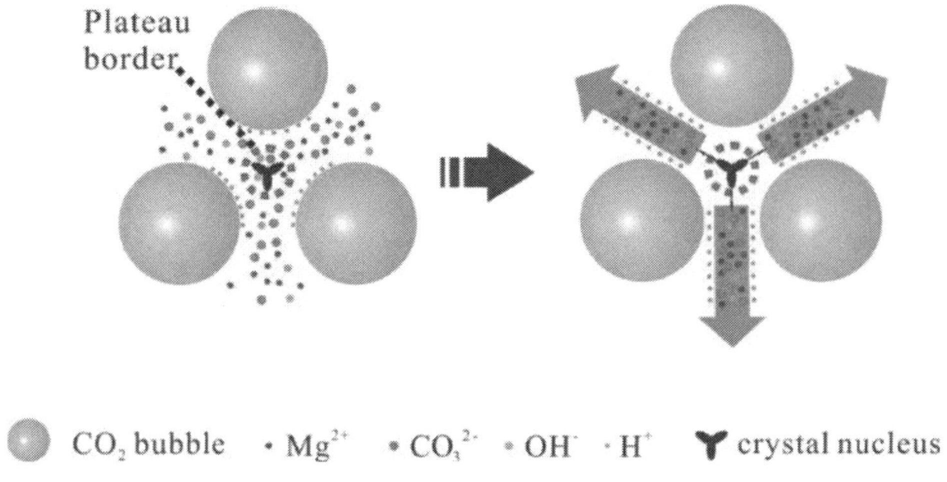

Figure 4. Illustration of the growth process of the flakes of BMC.

The effect of the system pressures on the Pore Size of the spherical porous BMC was studied. Fig.5 shows SEM images of the BMC obtained under different system pressures. BMC microspheres were found different under different given pressures. It is obvious that the arrangement of lamellae is denser with the system pressures increasing. Due to the increased system, volume of CO_2 bubbles will become smaller. Then, the channel between the CO_2 bubbles become narrower and the distance between the adjacent channels decrease, which results in smaller lamellae and denser arrangement of lamellae. This phenomena indicated that with increasing system pressure, the pore sizes of BMC microspheres decreased.

Figure 5. SEM images of spherical porous BMC synthesized at different system pressures (S_1: 4 atm; S_2: 6 atm; S_3: 8 atm; S_4: 10 atm)

According to the viewpoint of thermodynamics, CO_2 bubble is constant in solution, which can be described by Laplace, Kelvin equation and Clapeyron equations as follows [7] :

$$\Delta P = P_{in} - P = \frac{2\delta}{r} \quad (1)$$

$$P_{in} = \frac{nRT}{\frac{3}{4}r^3} + P_r^* \quad (2)$$

$$\ln\frac{P_r^*}{P^*} = -\frac{2dM}{\rho RTr} \quad (3)$$

According to the three equations 1, 2 and 3 above, combination equation can be obtained :

$$P = \frac{nRT}{\frac{4}{3}r^3} + P^* \exp(-\frac{2\delta M}{\rho RTr}) - \frac{2\delta}{r} \quad (4)$$

The final Equation 4 reveals the increasing pressure P has a major impact on the radius of CO_2 bubble, which results in smaller pore size of $Mg_5(CO_3)_4(OH)_2 \cdot 4H_2O$ with increasing pressure mentioned above. Thus, it proves that the formation of pore structure is based on CO_2 bubbles as template.

4 Conclusions

(1) spherical porous $Mg_5(CO_3)_4(OH)_2 \cdot 4H_2O$ have been prepared by using ammonium bicarbonate and magnesium chloride hexahydrate as raw materials.
(2) System pressure was important in controlling the final pore size of spherical porous BMC. In addition, with increasing pressure, the pore sizes of BMC microspheres decreased.
(3) A novel CO_2 bubble template was constructed to explain the formation of pore structure of porous BMC. And the theoretic equation is described by Laplace, Kelvin equation and Clapeyron equations.

References

1. D. Bhattacharjya, T. Selvamani, I. Mukhopadhyay, J. Therm. Anal. Calorim. **107**, 439-445 (2011)
2. H. Pang, P. Tian, J. Wang, X. Wang, G. Ning, Y. Lin, Mater. Lett. **131**, 206-209 (2014)
3. X. Ma, Q. Zhang, Z. Luo, X. Lin, G. Wu, Mater. Design. **89**, 71-77 (2016)
4. C. Yan, D. Xue, J. Phy. Chem. B. **109**, 12358-12361 (2005)
5. Y.P. Jiang, T.J. Peng, H.J. Sun, Adv. Mater. Res. **178**, 230-235 (2010)
6. M.I. Zaki, S.A.A. Mansour, J.Mater. Sci. Let. **13** 505-507 (1994)
7. X.H. Zhang, A. Khan, W.A. Ducker, Phys. Rev. Lett. **98**, 136101 (2007)

Preparation of Aragonite Whiskers by Dolomite

W.R Xu[a], J Chen and Z.L Huang

School of Materials and Engineering, Wuhan Institute of Technology, Wuhan 430074, China

Abstract. Aragonite $CaCO_3$ with complicated morphologies needle-like structrues have been successfully synthesized through dolomite as raw materials by the decomposition of urea. The phase of the aragonite whiskers were characterized by X-ray diffraction and Fourier transform infrared, the structrue of the samples was characterized by scanning electron microscope and the effects of different ratio of materials were studied. The result show that the concentration of urea was found to have an effect on the morphology of aragonite crystals. All the prepared crystals were almost aragonite and there was no phase under varied experiment conditions.

1 Introduction

Aragonite whisker is a new kinds of calcium carbonate material under development recently. It is a mineral of common occurrence with very high mechanical strength[1]. The aragonite crystals were initially separated from the hot solution saturated with carbon dioxide, or by addition of calcium chloride to a hot solution of ammonium or sodium carbonate, when calcium carbonate crystallized. Water-soluble additives, such as, polymers[2], inorganic metal ions[3], and surfactants[4], have been widely used to modify the sharps and control the nucleation and growth of aragonite crystals. In order to synthesize aragonite whiskers, generally, higher temperature, low concentration of salt carbonate solution were to the benefit of the formation of aragonite whiskers[5].

Typically, the uniform needle-like aragonite whiskers were obtained by aging solution of calcium salts with the presence of urea at 90°C by L Wang. While Yu and co-workers synthesized pure aragonite crystals with rod-like shape in aqueous solution of 1:3 ratio of urea to water, and needle-like crystals made of aragonite nanorods with 1:1 ratio of urea to water without any organic additives[6].

In numerous studies, calcium carbonate whiskers are a new kind of inorganic single crystal with a production cost of about $236 per ton. These crystals have been used as microscopic reinforcements in cementitious composites to improve toughness and reduce shrinkage. With a length of 20-30 um and a diameter of 0.5-2 um, $CaCO_3$ whikers differ from other widely used fibers, like steel, glass, or polyvinyl alcohol fibers, in terms of the "particle-like" small size and large specific surface area. Compared with the addition of steel, glass, or PVA macrofibers, the same content of $CaCO_3$ whiskers added in cementitious composites may cause fresh pastes[7].

[a] W.R Xu: 740146159@qq.com

2 Experimental

2.1 Material

All starting materials, dolomite(provided by LingShou HengDa minerals processing factory in Hebei, China), hydrochloric acid(obtained from XiLong chemical CO,. LTD. about 37% of mass fraction), urea($(NH_2)_2CO$, obtained from XiLong chemical CO,. LTD.) were of A.R. Grade. The water used in this experiment was deionized water.

2.2 Synthesis procedure

6.0g dolomite was mixed with 25ml hydeochiloric acid, in the presence of urea in 90°C bath as described. Specifically, equal volumes of urea solution were firstly at room temperature. Then transparent solution was transferred into a 60ml reaction still. Then, the reaction still was incubated and unstirred in a 90°C water bath for 24h. All of the particle nucleated and grew inside the reaction still, not on the wall of the flask. The precipitate was finally dried in a vacuum oven overnight at 100°C. The synthesized condition of the aragonite whiskers was listed as in Table 1. The concentration of dolomite was kept at 1.0M, and the concentration of urea was varied from 1.0 to 2.0M.

2.3 Characterization of the particles

The phase composition and structure were identified by X-ray diffraction recorded on a(Bruker XD-5A) X-ray powder diffractometer with Cu Kα radiation(λ=1.5405Å) at scanning rate of 4° min^{-1}. The infrared spectra were characterizated by using Bruker Fourier transform infrared (FT-IR) spectrometer on KBr pellets. The morphology and structure of as-prepared particles were examined by JEM-2210LV scanning electron microscope operating at an accelerating voltage of 20kV after sputtered gold film. The average grain size D was determined from Scherrer`s equation:

$$D=K\lambda/\beta\cos\theta \quad (1)$$

In the equation(1), K is expressed as dimensionless constant, λ is expressed as wavelength of the X-ray radiation(1.5405Å), β is the full width at half-maximum of the diffraction peak, and θ is half of the diffraction angle.

3 Result and discussion

3.1 XRD analysis

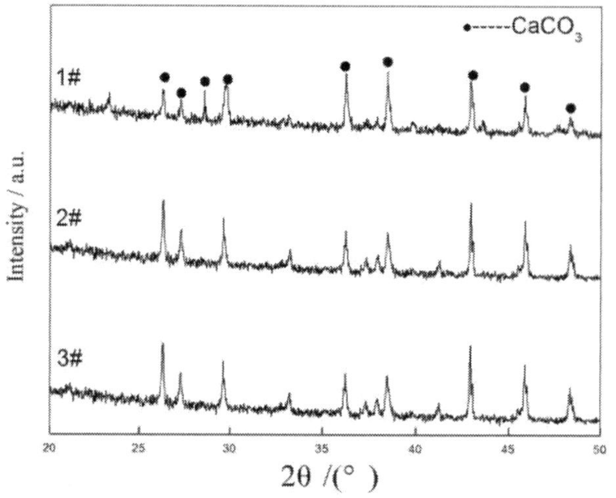

Figure.1 The XRD patterns of samples(1#, 2#, 3#)

The above structures were showed by the X-ray diffraction patterns in Figure. 1. All of the samples of the diffraction peaks, compared with the standard JCPDS files (aragonite: 41-1475; calcite: 05-0586; calcite,magnesian: 43-0697), in the XRD patterns clearly showed the peaks corresponding to h k l of 1 1 1, 0 2 1, 0 1 2, 2 0 0, 1 1 2, 1 3 0, 1 2 2 et al., which agreed with those of aragonite phase. No other polymorph of $CaCO_3$ was detected in the XRD pattern. The sharp and strong peaks indicated that $CaCO_3$ was well crystalline. Scherrer`s equation was applied to (1 1 1) (0 2 1) (2 2 1) peaks and provided a mean grain size D shows that the average crystalline size for the prepared $CaCO_3$ crystals around 35.6-38.7nm, indicating that the resulted aragonite clusters were composed of aggregates of crystals.

3.2 SEM analysis

The scanning electron microscopy images of the fabricated $CaCO_3$ crystals with 1M of dolomite, in varied concentration of urea were shown as in Fig.2. All pictures in Fig.2 showed that the prepared $CaCO_3$ particles exhibited mostly needle-like. When the concentration of urea was 1.0M, the needle-like aragonite with about 15~16 of aspect ratio, which was defined as the ratio of length to diameter, was obtained as Fig.2(1#), and from which some of the crystals showed the tendency of orientation and needle-like structure. Fig.2(2#) exhibits majority of aragonite crystals with clavate needle-like structure, made of the rods approximately with 9~10 of aspect ratio in addition to minority of needle-like crystals, the bundle-like crystals were formed. with 1.5M concentration of urea. With the concentration of urea increased to 2M, the lager needle-like crystals were observed in Fig.2(3#) with about 6~7 of aspect ratio, the bundle-like crystals and dumbbell-like crystals were observed, the aggregation phenomenon of whiskers were also observed.

DOI: 10.1051/matecconf/201711401016

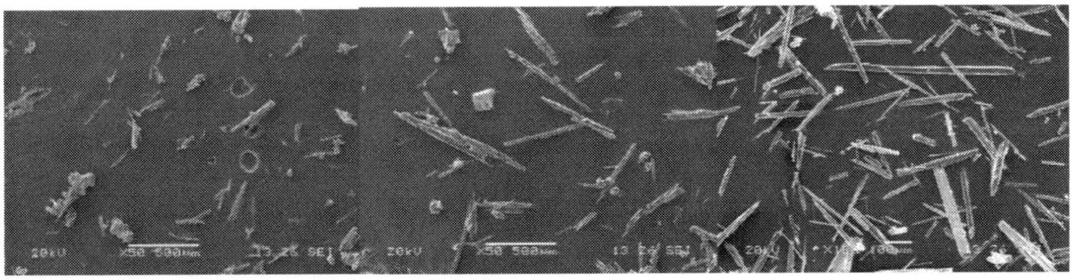

Figure.2 SEM images of aragnite crystals preparad at varid concentration of urea(1#, 2#, 3#)

3.3 FT-IR analysis

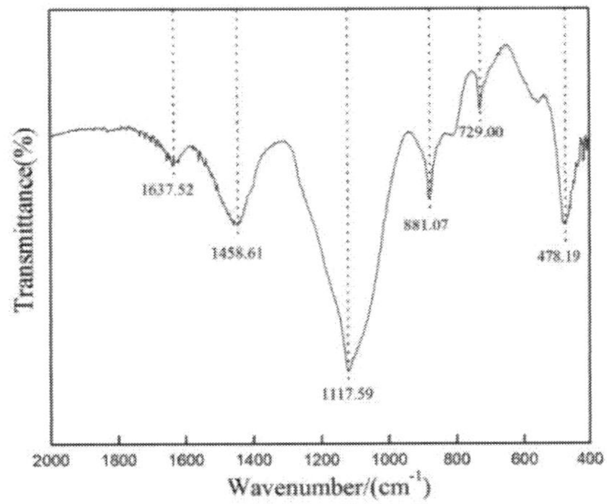

Figure.3 FT-IR spectrum of aragonite(1#)

The aragonite phase of the sample was further confirmed by Fourier transform infrared spectroscopy spectrum shown in Figure.3. All spectrum spectra in Figure.3 showed vibrational bonds at 1117.59cm^{-1} that can be attributed to the characteristic aragonite symmetric carbonate stretching (v_1 mode). The spctrum spectra sharp peak located at 729.00cm^{-1} was associated with the carbonate in-plane bending modes (v_4) of aragonite. Moreover, the bands at about 881.07cm^{-1} can also be attributed to the out-of-plane bending (v_2 mode) vibration. For aragonite, the positions of the corresponding v_2 vibrations appear at about 875cm^{-1}, the vibration appears at 881.07cm^{-1}, is very similar and confirmed as the pure phase of aragonite. The two characteristic absorption peaks labled by number, located at 1637.52cm^{-1} and 1458.61cm^{-1}, were assigned to the C=O stretching vibration and C-O bending vibration.

4 conclusion

In summary, aragonite $CaCO_3$ with complicated morphologies needle-like structrues have been successfully synthesized through dolomite as raw materials by the decomposition of urea. The experiment parameters, such as the concentration of urea was found to have an effect on the morphology of aragonite crystals. All the prepared crystals were almost aragonite and there was no phase under varied experiment conditions. The reduction of aspect ratio of aragonite whisker was focued with increasing of concentration of urea under invariant conditions of dolomite.

References

1. A.L. Litvin, S.Valiyaeettil, D.L. Kaplan, Template-directed synthesis of aragonite under supramolecular hydrogen-bonded Langmuir monolayers, Adv. Mater. **9,** 124(1997).
2. Y. Pan, X. Zhao, Y. Sheng et al, Biomimetic synthesis of dendrite-shaped aragonite particles with single-crystal feature by polyacrylic acid, Colloids Surf. A: Physicochem. Eng. Asp. **297**(2007).
3. J. Yu, M. Lei, B. Cheng, X. Zhao, Facile preparation of calcium carbonate particles with unusual morphologies by precipitation reaction, J. Cryst. Growth **261**, 566(2004).
4. Z. Nan, X. Chen, Q. Yang, Structure transition from aragonite to vaterite and calcite by the assistance of SDBS, J. Colloid Interface Sci. **325**(2008).
5. Z. Hu, Y. Deng, Synthesis of needle-like aragonite from calcium chloride and sparingly soluble magnesium carbonate, Powder Technol. **140**, 10(2004).
6. S. Chen, S. Yu, J. Jiang, Polymorph discrimination of $CaCO_3$ mineral in an ethanol/water solution: formation of complex vaterite superstructures and aragonite rods, Chen. Mater. **18**, 115(2006).
7. M. Cao, C. Zhang, Mechanical response and shrinkage performance of cementitious composites with a new fiber hybridization. Constr Build Master. **57**, 45(2014).

Single satellite beam scanning positioning based on Neural Network BP algorithm

Yongwei Li[1,a], Yuwen Wang[2,b], Hanjing Li[2,c], Qiang Shu[2,d] and Manchuan Zhao[2,e]

[1]*University of Electronic Science and Technology of China,School of Aeronautics and Astronautics, Chengdu, China;*
[2]*University of Electronic Science and Technology of China,School of Aeronautics and Astronautics, Chengdu, China;*

Abstract. In this paper, the principle of single line positioning based on beam scanning and the neural network algorithm are analysing, and the neural network BP algorithm is applying to the single satellite positioning. At the same time, this paper presents a new algorithm based on electron beam (MEO) for the single scan positioning (Middle Earth orbit). Finally, through theoretical analysis and simulation, it is proving that the neural network BP algorithm of single satellite beam scanning is feasible in fast positioning.

1 INTRODUCTION

Single-satellite beam scanning positioning technology, that beam information from a satellite, with beam scanning positioning technology to detect the target and the coordinates of the target location solution [1,2]. When the satellite is running near the apogee, the satellite can cover the target area for a long time. The handheld terminal in the target area can receive the beam scanning information of the satellite within the time period covered by the satellite pair area, and the handheld terminal collects enough beam information through the signal processing, positioning operations can calculate their own latitude and longitude coordinates [3].

[a] Yongwei Li: ywli0314@sina.cn
[b] Yuwen Wang: 83211668@qq.com
[c] Hanjing Li: 448063460@qq.com
[d] Qiang Shu: 874611955@qq.com
[e] Manchuan Zhao: zmc_zhao@163.com

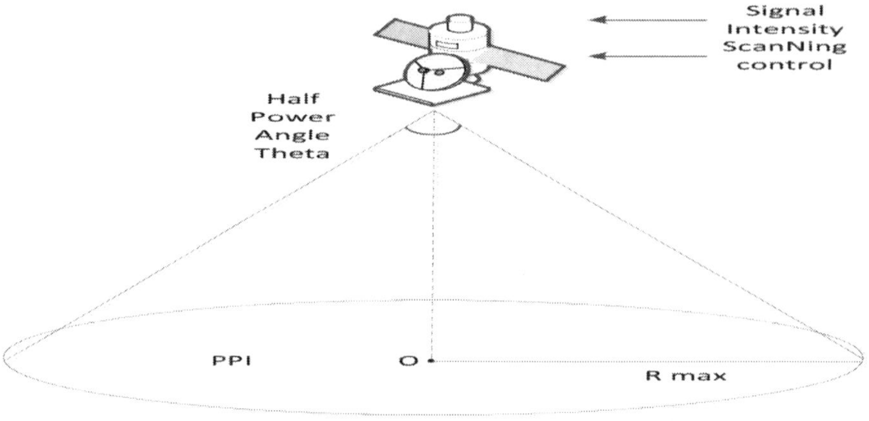

Figure.1 Schematic diagram of single-satellite beam scanning

As shown in Fig. 1, the beam information transmitted by the satellite can form an approximately circular "spot" on the ground [4], and all the receivers in this "spot" can receive the beam information. A circular map-style image of the "spot" and the centre position of the "spot" are displayed on the receiver's flat panel display (PPI). By controlling the centre position of the beam scanning, the satellite can realize the full scan of the scanning area, so that all the targets in the scanning area can be detected. The target can realize the solution of the position coordinates according to the received beam information.

For a certain target area, the position information of the satellite is the main factor to determine the positioning accuracy [5-7]. When the satellite deviates from the target area, the target area is mainly scanned with a smaller half-power angle to improve the accuracy of satellite positioning. When the satellite is moving closer to the target area, the positioning accuracy of the satellite will be improved. In order to meet the requirement of positioning accuracy, we use the neural network BP algorithm.

2 Neural Network

Neural network can be divided into biological neural network and artificial neural network. Biological neural network: generally, refers to the biological brain neurons, cells, contacts and other components of the network for the production of biological consciousness, to help biological thinking and action [8]. Artificial Neural Networks (ANNs) are networks of widely parallel interconnected networks composed of adaptive simple units, which are able to simulate the interaction of the biological nervous system with real-world objects. This network relies on the complexity of the system, by adjusting the internal connections among a large number of nodes, so as achieve the purpose of processing information [9,10].

2.1 Neural Network Introduction

Neural network is composed of a large number of simple processing units, connected by variable weight of the parallel-distributed system. Neuron is the basic processing unit of artificial neural network; it is a multi-input-single output nonlinear device [11,12]. The structure is showing in Fig.2. In Figure.2, x_i is the input signal, w_{ij} is the connection weight from the ith neuron to the jth neuron, θ_j and is the threshold value of the jth neuron. Let s_j be an external input signal and y_j be an output signal. The transformation of the jth neuron in the above model can be described as:

DOI: 10.1051/matecconf/201711401017

$$y_i = f(\sum_j w_{ij} x_i - \theta_j + s_j) \tag{1}$$

Where the non-linear function f(x) used here can be a step function, a piecewise function, and a sigmoid type function.

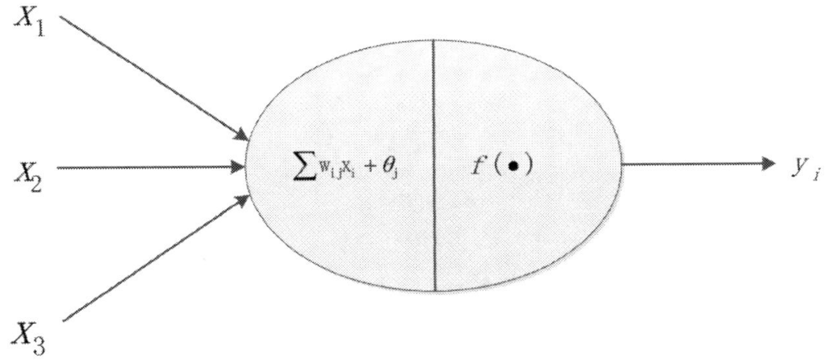

Figure.2 General Description of neuron

For each neuron, there is an output, and through the connection weights to the output to its connected processing unit, the output signal directly depends on the state of the processing unit or stimulus value. This dependency is represented by the effect of the output transform function f_j on the processing unit u_j. Suppose we use $z_j(t)$ to define the u_i output of the neuron at time t

$$z_j(t) = f_j(x_j(t)) \tag{2}$$

Or in vector form

$$Z(t) = f(X(t)) \tag{3}$$

Where, Z(t) is the output vector of the neural network, f is defining as the corresponding function of the state vector and each component, usually bounded on the interval (0, 1).

2.2 Single Satellite Location Algorithm Based on BP Neural Network

BP neural network is a multi-layer feed forward neural network, the main characteristics of the network is the signal forward transmission, error back transfer. In the forward transmission, the input signal from the input layer through the hidden layer by layer to the output layer, each layer of the neuron state only affects the next layer of neuron state. If the output layer does not get the desired output, it goes back-propagation and adjusts the network threshold and weight according to the prediction error, so that the BP neural network can predict the expected output continuously. The topological structure of BP neural network is showing in Figure.3

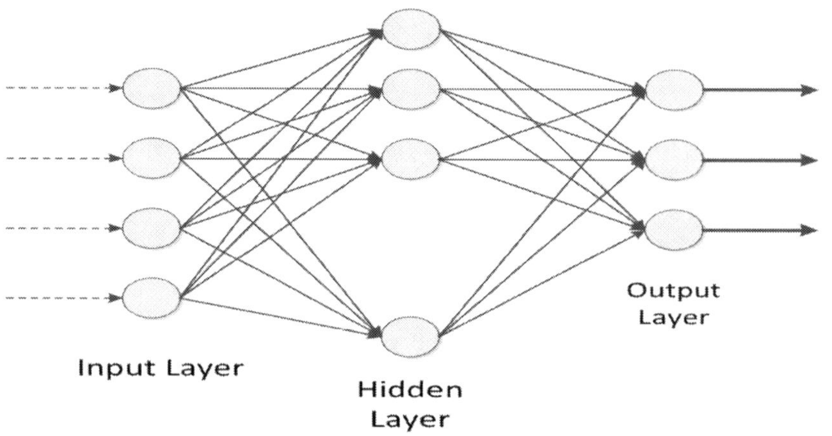

Figure.3 Topological structure of BP neural network

The main idea of the error back propagation algorithm is to divide the learning process into two phases: the first stage (forward propagation process), which gives the input information through the input layer through the hidden layer processing and calculate the actual output value of each unit; (Reverse process). If the desired output value is not obtained at the output layer, the difference (i.e., error) between the actual output and the desired output is recursively calculated on a layer-by-layer basis so as to adjust the weight value according to the difference, it is possible to calculate the product of the error value of the reception unit and the activation value of the transmission unit for each weight. BP neural network based on multi-layer feed forward network structure is showing in Figure.4.

This network not only has input layer nodes, output layer nodes, but also one or more layers of hidden nodes. For the input information, first propagate to the node of the hidden layer, and then propagate the output information of the hidden node to the output node after the activation function (also called the function function, transfer function) of each element. Finally, result. The learning process of the network consists of forward and reverse propagation.

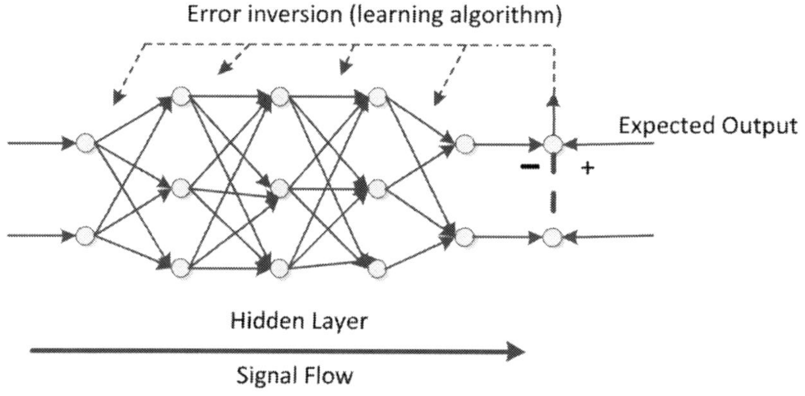

Figure.4 Multi - layer Feedforward Network Structure Based on BP Algorithm

All received beam information is substituting into the neural network to learn and predict the result. (X,Y) is set to be $X=[RX_1,RX_2,\cdots RX_m]^T, Y=[rx_1,rx_2,\cdots,rx_m]^T, RX=[x_1,x_2,\cdots x_N]^T$ is the X coordinate of all the beam information received by the receiving device at a certain location, and X is the X of all the beam information received by the m experiments. X is the X coordinate of the received beam. In addition, Y is the coordinate of the receiving device in the X direction. And the hidden layer neuron is $O=[O_1,O_2,\cdots O_l]$, the weight matrix W^1 between the input layer and the hidden layer neuron, and the weight between the hidden layer and output layer neuron W^2 respectively.

$$W^1 = \begin{bmatrix} w_{11}^1 & w_{12}^1 & \cdots & w_{1m}^1 \\ w_{21}^1 & w_{22}^1 & \cdots & w_{2m}^1 \\ \vdots & \vdots & \vdots & \vdots \\ w_{l1}^1 & w_{l2}^1 & \cdots & w_{lm}^1 \end{bmatrix}, W^2 = \begin{bmatrix} w_{11}^2 & w_{12}^2 & \cdots & w_{1l}^2 \\ w_{21}^2 & w_{22}^2 & \cdots & w_{2l}^1 \\ \vdots & \vdots & \vdots & \vdots \\ w_{n1}^2 & w_{n2}^2 & \cdots & w_{nl}^2 \end{bmatrix} \quad (4)$$

Where, m is the number of neurons in the input layer, l is the number of hidden layer neurons, n is the number of output layer neurons. The threshold θ^1 of hidden layer neurons and the threshold θ^2 of output layer neurons are respectively

$$\theta^1 = [\theta_1^1, \theta_2^1, \cdots \theta_l^1]^T, \theta^2 = [\theta_1^2, \theta_2^2, \cdots \theta_n^2]^T \quad (5)$$

The output of hidden layer neurons is

$$o_j = f\left(\sum_{i=1}^{m} w_{ji}^1 x_i - \theta_j^1\right) = f\left(net_j\right), j = 1, 2, \cdots, l \quad (6)$$

The output of neurons in the output layer is

$$z_k = g\left(\sum_{j=1}^{l} w_{kj}^2 o_j - \theta_k^2\right) = g\left(net_k\right), k = 1, 2, \cdots, n \quad (7)$$

Network output and the expected value of the error

$$\begin{aligned} E &= \frac{1}{2}\sum_{k=1}^{n}(rx_k - z_k)^2 \\ &= \frac{1}{2}\sum_{k=1}^{n}\left(rx_k - g(\sum_{j=1}^{l} w_{kj}^2 o_j - \theta_k^2)\right)^2 \\ &= \frac{1}{2}\sum_{k=1}^{n}\left\{rx_k - g\left[\sum_{j=1}^{l} w_{kj}^2 f\left(\sum_{i=1}^{m} w_{ij}^1 x_i - \theta_j^1\right) - \theta_k^2\right]\right\}^2 \end{aligned} \quad (8)$$

The partial derivative of the error E for the weight w_kj^2 between the hidden layer and the output layer neuron is

$$\frac{\partial E}{\partial w_{kj}^2} = \frac{\partial E}{\partial z_k}\frac{\partial z_k}{\partial w_{kj}^2} = -(y_k - z_k)g`(net_k)o_j = -\delta_k^2 o_j \quad (9)$$

where, $\delta_k^2 = (y_k - z_k)g`(net_k)$.

The partial derivative of error E for the weights w_{ji}^1 between input and hidden neurons is

$$\frac{\partial E}{\partial w_{ji}^1} = \sum_{k=1}^{n}\sum_{j=1}^{l}\frac{\partial E}{\partial z_k}\frac{\partial z_k}{\partial o_j}\frac{\partial o_j}{\partial w_{ji}^1} = \sum_{k=1}^{n}(y_k - z_k)g`(net_k)w_{kj}^2 f`(net_j)x_i = \delta_j^1 \quad (10)$$

By (9) and (10) available weight adjustment formula for

$$\begin{aligned} w_{ji}^1(t+1) &= w_{ji}^1(t) + \Delta w_{ji}^1 = w_{ji}^1(t) - \eta^1 \frac{\partial E}{\partial w_{kj}^1} = w_{ji}^1(t) + \eta^1 \delta_j^1 x_i \\ w_{kj}^2(t+1) &= w_{kj}^2(t) + \Delta w_{kj}^2 = w_{kj}^2(t) - \eta^2 \frac{\partial E}{\partial w_{kj}^2} = w_{kj}^2(t) + \eta^2 \delta_k^2 o_j \end{aligned} \tag{11}$$

Where η^1 and η^2 are the learning steps of the hidden layer and the output layer, respectively. Similarly, the partial derivative of the error E to the threshold θ_k^2 of the output layer neuron is

$$\frac{\partial E}{\partial \theta_k^2} = \frac{\partial E}{\partial z_k}\frac{\partial z_k}{\partial \theta_k^2} = (y_k - z_k)g`(net_k) = \delta_k^2 \tag{12}$$

The partial derivative of error E for the threshold θ_j^1 of hidden neurons is

$$\frac{\partial E}{\partial w_{ji}^1} = \sum_{k=1}^{n}\sum_{j=1}^{l}\frac{\partial E}{\partial z_k}\frac{\partial z_k}{\partial o_j}\frac{\partial o_j}{\partial w_{ji}^1} = \sum_{k=1}^{n}(y_k - z_k)g`(net_k)w_{kj}^2 f`(net_j)x_i = \delta_j^1 \tag{13}$$

From equation, (12) and (13) can be adjusted threshold for the formula

$$\begin{aligned} \theta_j^1(t+1) &= \theta_j^1(t) + \Delta\theta_j^1 = \theta_j^1(t) - \eta^1 \frac{\partial E}{\partial \theta_j^1} = \theta_j^1(t) + \eta^1 \delta_j^1 \\ \theta_k^2(t+1) &= \theta_k^2(t) + \Delta\theta_k^2 = \theta_k^2(t) - \eta^2 \frac{\partial E}{\partial \theta_k^2} = \theta_k^2(t) + \eta^2 \delta_k^2 \end{aligned} \tag{14}$$

By predicting the output of the neural network, the optimal prediction results are obtaining by constantly modifying the network thresholds and weights.

3 SIMULATION

According to our study, the positioning accuracy and the relative position of the satellite with the scanning area are highly correlated. Therefore, we selected the data of seven positioning moments of the apogee, 30 minutes apart for each positioning time point and 3 hours for the selected positioning point. The latitude and longitude corresponding to the position of the satellite at the seven positioning time points are showing in Table 1. The time points of the seven positioning points are T1, T2, T3, T4, T5, T6 and T7.

Table 1 satellite positioning time corresponding latitude and longitude

positioning time lat/lon	T1	T2	T3	T4	T5	T6	T7
latitude	30.194	14.002	1.333	-10.035	-21.401	-34.061	-50.218
longitude	112.254	106.421	100.047	93.542	87.112	81.034	76.185

We study the position of θ = 0.5 °, d = 10km, and choose the information of 2000 points as the training set of the neural network. We can set the parameter of BP network; predict the output of any 100 coordinates of the location information. The obtained prediction results are showing in Figure.5 and Figure.6.

DOI: 10.1051/matecconf/201711401017

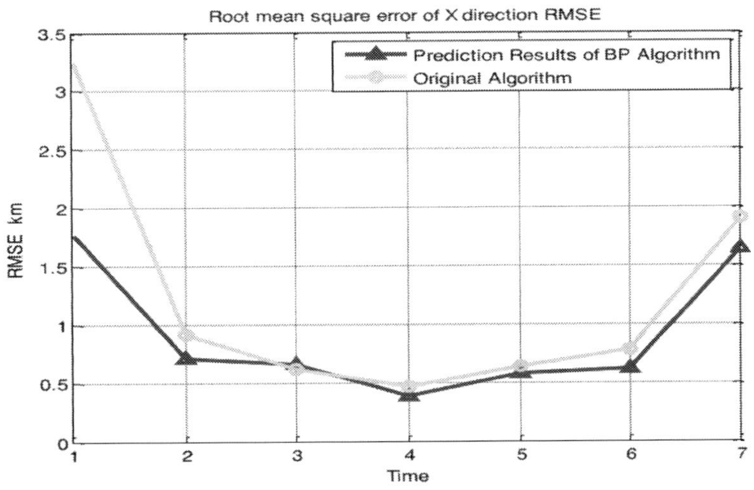

Figure.5 Root-mean-square error of BP prediction in X direction

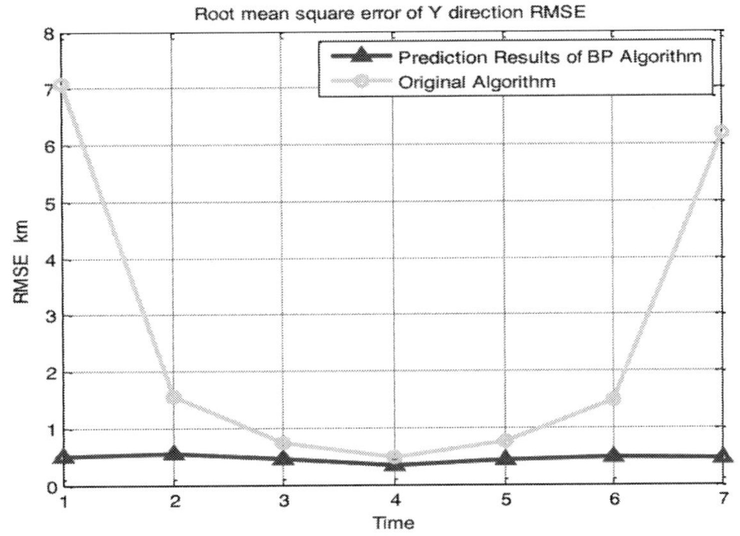

Figure.6 Root-mean-square error of BP prediction in Y direction

From the above output results, BP neural network has a good learning function, can better predict the output of the true location of the target information. Compared with the original algorithm, especially for the satellite deviation from the target area, the accuracy is more obvious the accuracy of BP prediction output is improved. We can see from the result, which BP neural network can improve the positioning accuracy of single-satellite beam scanning.

4 Conclusion

This paper, the neural network BP algorithm combined with beam scanning is using to calculate the position coordinates of the receiver. The beam scanning positioning technique is used to detect the target and to calculate the coordinates of the target position. When a satellite is in the far near the site, the satellite can cover the target area for a long time, the receiver within the target area in the period of the satellite coverage within the received beam scanning information of the satellite receiver to collect enough information through the beam signal processing, positioning calculation can calculate their longitude coordinate degree. In this paper, we proposed a new method： neural network combined with beam scanning, it is based on the scanning beam received to calculate the location information.

References

1. Bagrov A V, Leonov V A, Mitkin A S, et al. Single-satellite global positioning system [J]. Acta Astronautica, 2015, 117:332-337.
2. Zhang J, Lv J, Yu Y. Research on the Positioning by Single Satellite in Constellation Communication Systems [J]. Journal of Telemetry Tracking & Command, 2009.
3. Zhao M, Wang Y, Du S, et al. The study of rapid localization algorithm from beam scanning by single satellite based on neural network[C]// IEEE International Conference on Computer Communication & the Internet. IEEE, 2016:186-190.
4. Harchowdhury A, Sarkar B K, Bandyopadhyay K K. Beam scanning for reception performance improvement of satellite-based AIS[C]// IEEE International Conference on Aerospace Electronics and Remote Sensing Technology. IEEE, 2015:1-5.
5. An-Nan L U, Kong X Z. Passive localization from frequency measurements by single satellite [J]. Journal of China Institute of Communications, 2004.
6. Lu A N, Yang X N. Passive location from the combined set of frequency and phase difference measurements by single satellite [J]. Systems Engineering & Electronics, 2010, 32(2):244-247.
7. Chan Y T, Towers J J. Passive localization from Doppler shifted frequency measurements[C]// International Conference on Acoustics, Speech, and Signal Processing. IEEE, 1991:1465-1468 vol.2.
8. Wang Y. GENERALIZED NEURAL NETWORK SYSTEM AND BP ALGORITHM[J]. Mini-micro Systems, 2001.
9. Liang J Z, He X G, Zhou J Q. DYNAMIC ANALYSIS OF BP ALGORITHM FOR NEURAL NETWORKS[J]. 2002(5):729-735.
10. Zhang H. A New BP Algorithm of Multiplayer Neural Network[J]. Communications Technology, 2003.
11. Jing G, Du W, Guo Y. Studies on prediction of separation percent in electrodialysis process via BP neural networks and improved BP algorithms[J]. Desalination, 2012, 291(14):78-93.
12. Yu F, Xu X. A short-term load forecasting model of natural gas based on optimized genetic algorithm and improved BP neural network[J]. Applied Energy, 2014, 134(134):102–113.

Research on Ultrasonic Testing Methods for Adhesion Quality of Ceramic Coatings

Yongjie LI[1], Zandong HAN[1,a] and Yifang CHEN[1]

[1]*Key Laboratory for Advanced Materials Processing Technology, Ministry of Education; State Key Laboratory of Tribology, Tsinghua University, Beijing 100084, China*

Abstract. Ceramic coatings are widely used in aerospace field as thermal barrier materials for their high-temperature performance. The adhesion quality of ceramic coatings on metallic substrate is an important index for coating performance. The uniformity of coating thickness is also important since it determines the thermal stresses in the coatings, which means the coating performance is closely related with the uniformity. The reflection characteristic of the coating-substrate interface is an essential factor for evaluating the adhesion quality. Since the acoustic impedance of ceramic coatings is very close to that of substrate, it is not easy to distinguish their reflection wave in the received waves. In this paper, the propagation characteristics of ultrasonic wave in coatings with different thicknesses on the substrate were researched using immersion ultrasonic testing with pulse-echo method. An algorithm based on the amplitude and phase of reflection waves was proposed for the evaluation of adhesive quality. The results show that, pulse-echo method is highly reliable for the measurement of coating thickness and that the algorithm is effective for the evaluation of adhesion quality of ceramic coatings.

1 Introduction

Metallic materials, especially steels and aluminum alloys, are now the most widely-used engineering materials in the society for their excellent comprehensive mechanical properties. However, in certain extreme conditions, such as high temperature, strong corrosion and wear, etc., ceramic materials perform better than traditional metallic materials. In order to enhance the performance of metallic materials in extreme conditions mentioned above, ceramic materials are usually sprayed or bonded on the surface of metallic substrate as coatings [1, 2]. The ceramic coatings are usually very thin, in the order of several hundred micrometers, and are made by two main methods, i.e. atmosphere plasma spraying (ASP) and electron beam physical vapor deposition (EB-PVD) [3,4]. In the paper, thin ceramic layers with thickness from several hundred micrometers to several millimeters were bonded on the surface of metallic substrate as coatings.

Several bonding theories have been proposed in the past few decades, including mechanical interlocking theory, adsorption interaction theory, diffusion mechanisms theory, electrostatic forces theory and acid-base interaction theory [5, 6]. The adhesive between coating and substrate can be

[a] Corresponding author: hanzd@tsinghua.edu.cn

considered as a viscoelasticity model. There are mainly two models used to describe the adhesion properties, namely Maxwell model and Voigt model. In Maxwell model, the adhesive is considered as a spring and a damper in series and is usually used to describe stress relaxation. While in Voigt model, a spring and a damper in parallel and is usually used to describe creep. Both Maxwell model and Voigt model can be generalized by using a group of such models to better the model [7]. For the model of interface of sample, Tattersall [8] proposed a spring-based model. In his model, the interfacial forces are represented by a density of springs between the two media and the springs are light enough to transmit stress instantaneously across the boundary. L. Singher [9] regarded the interface between adhesive and adherend as a spring-mass structure, and proposed a quasi-static approach, in which a density of springs with constant stiffness were used to describe the shear mechanical behavior of the interfacial adhesive-adherend.

These models provide the theory foundation for nondestructive testing (NDT). Several NDT methods are used for the detection of adhesion quality, for example, acoustic emission testing [10], microwave testing [11], infrared testing [12], ultrasonic testing [13], etc. Ultrasonic testing is widely used for its high sensitivity, strong penetrability, wide detection-range and simple equipment. The spring model mentioned above provides an idea to use ultrasonic wave as a tool to test the adhesion quality of interface, since the ultrasonic wave is one kind of mechanical wave. A. Wegner [14] believes the an-harmonic wave is caused by weak bonds and manifests itself at high dynamic strains exerted by the ultrasonic waves, and proposed a distortion factor as a quantitative index for characterizing the adhesion quality of fusion-welded silicon wafers. L. Czarnecki [15] mainly uses the changes of a mean square value parameter (MS) and the value of variation coefficient (CV) of MS to characterize the propagation of ultrasonic waves through the polymer coating and concrete substrate system. S. I. Rokhlin [16] used high frequency pulsed angle beam ultrasonic waves modulated by low frequency vibrations to test the bonded structure, and used parametric and nonlinear mixing frequencies to characterize the adhesion degradation.

In this paper, an immersion-based pulse-echo ultrasonic testing method is used to measure the thickness of ceramic coatings and to evaluate the adhesion quality of samples.

2 Basic theory

As shown in Fig. 1, there are 4 kinds of media on the propagation route of ultrasonic waves, i.e. water, coating layer, bonding layer or adhesive layer and substrate. The thickness of coating layer is much thinner than that of substrate. And the bonding layer is thin enough comparing with coating layer and substrate. The bonding status of the bonding layer affects the propagation a lot. If an immersion ultrasonic transducer is used for ultrasonic testing, water between the transducer and coating not only serves as couplant, but overcomes the influence of the blind area of ultrasonic testing.

Suppose the complex form of the incident wave can be written as

$$s_i(t) = A_i \exp\left[i\omega(t - x/c_0)\right] \tag{1}$$

where A_i, ω, and c_0 are the amplitude, angular frequency and velocity in water of the incident wave, respectively, and $x = 0$ at the water-coating interface.

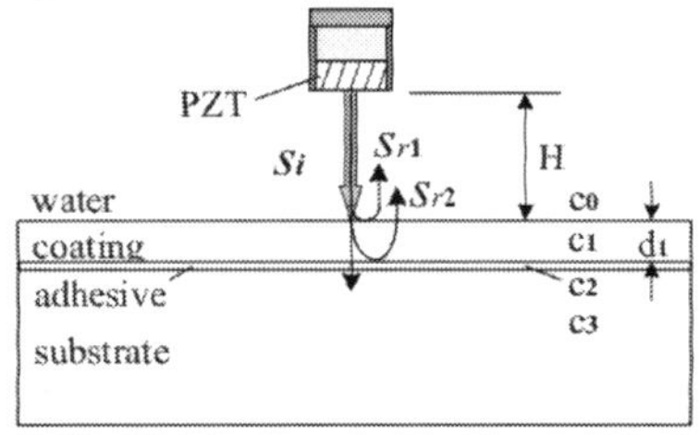

Figure1. Ultrasonic testing for adhesion quality of coating

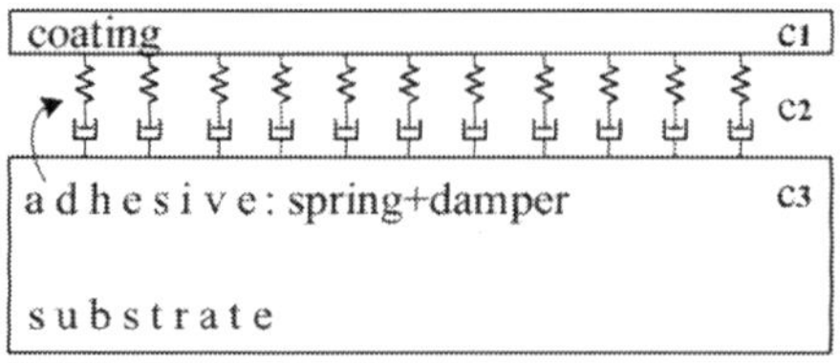

Figure2. Generalized Maxwell model for adhesive layer

Then the complex form of reflection waves from the water-coating interface and coating-substrate interface (adhesive layer ignored) can be written as

$$s_{r1}(t) = A_{r1} \exp\left[i\omega(t + x/c_0)\right] \tag{2}$$

and

$$s_{r2}(t) = A_{r2}\exp[(i\omega(t + x/c_0 - 2d_1/c_1)] \quad (3)$$

where d_1 is the thickness of the coating, and other symbols are similar to those in equation (1).

The relationship between A_{r1} and A_i can be determined by reflection coefficient and transmission coefficient of acoustic pressure of the water-coating interface, given by J. L. Rose [17]:

$$r_{01} = (Z_1 - Z_0)/(Z_0 + Z_1) \quad (4)$$

$$t_{01} = 2Z_1/(Z_0 + Z_1) \quad (5)$$

where, Z_0 and Z_1 are the acoustic impedance of water and ceramic coating.

The generalized Maxwell model is used to describe the adhesive layer, which is considered as a series of model of a spring and a damper in series, as shown in Fig. 2.

In this model, the stiffness coefficient K of the spring characterizes the bonding status of the adhesive layer. If K is small, the bonding statue is poor. Vice versa. The viscosity of the damper influences the attenuation coefficient of ultrasonic waves.

According to Q. Zhang [18], there are three types of attenuation in the propagation of ultrasonic waves, i.e. viscous attenuation, heat conduction attenuation and scattering attenuation, and the majority (about 92%) of the attenuation is caused by the last one. The coefficient of scattering attenuation α_s can be described by the equation below:

$$\alpha_s = \frac{8}{3}\pi^4 r^3 f^4/c^4 \quad (6)$$

where, r is the size of the medium particles, c and f are the velocity and frequency of the ultrasonic waves in the medium. From equation (6), we can see that, since the velocity and the particle size are basically constants in certain samples, scattering attenuation is strongly affected by frequency.

Take the scattering attenuation in the coating into consideration, then the amplitude of wave can be written as:

$$A = A_i e^{-\alpha_s x} \quad (7)$$

where x is the propagation distance in the coating. In the coating, the propagation distance is the thickness of coating d_1, which is quite thin, in the order of 10^{-3}m and the size of the ceramic coating is small, typically 3~5 μm. Then the attenuation factor α_s can then be calculated, which is in the order of 0.1m^{-1}. So the factor $e^{-2\alpha_s d_1}$ is very close to 1, which means the attenuation in the ceramic coating can be ignored.

Since the attenuation in the coating is ignored, the generalized Maxwell model can be simplified to spring model. The adhesion quality can be evaluated by the percentage of energy across the interface. This can be described by adhesion coefficient (CA).

$$CA = \frac{I_{t2}}{I_{t1}} = 1 - \frac{I_{r2}}{I_i(1 - I_{r1}/I_i)^2} = 1 - \frac{\eta_2}{(1-\eta_1)^2} \quad (8)$$

where, $\eta_1 = I_{r1}/I_i$, $\eta_2 = I_{r2}/I_i$.

In equation (8), the symbol r_{12} is directly related to the coating adhesion quality, so are I_{r2} and the adhesion coefficient CA. If the coating is well-bonded, the factor CA is relatively high. If the coating is weakly-bonded, the factor CA is relatively low. Since the parameters η_1 and η_2 can be measured by reflection method of normal incidence, then the adhesion coefficient can be determined.

3 Experimental setup

Two group of samples with different thicknesses were made for the measurement of coating thickness and the evaluation of adhesion quality. The samples consist mainly three layers, namely, coating layer, bonding layer and substrate layer. The coating layer is a relatively thin ceramic layer with thickness ranging from several hundred micrometers to a few millimeter. The bonding layer is an epoxy resin layer much thinner than that of the coating layer. The substrate layer is a stainless steel layer, usually much thicker than that of the coating layer.

The coatings of group I are relatively thick, in the order of a few millimeters, used for characterizing the propagation of ultrasonic waves in the multi-layer samples. The coatings of group II are relatively thin, in the order of several hundred micrometers, and both well-bonded and weakly-bonded samples were made in group II. These samples are used for the thickness measurement and the evaluation of adhesion quality. The mechanical properties of samples are listed in table 1. And the acoustic parameters of each layer of samples are listed in table 2.

Table 1. Coating properties of samples

Sample number	Coating thickness /mm	Bonding status
I-1	2.3	Well
I-2	1.5	Well
I-3	0.67	Well
II-1	0.35	Well
II-2	0.35	Weakly
II-3	0.35	Debonding

Table 2. Acoustic parameters of samples ($\times 10^3$)

Layers	Velocity/ $m \cdot s^{-1}$	Density/ $kg \cdot m^{-3}$	Impedance/ $g \cdot mm^{-2}s^{-1}$
Water	1.50	1.0	1.5
Coating	10.0	3.8	38.0
Adhesion	2.54	1.2	3.0
Substrate	5.72	7.9	45.2

The detection system is shown in Fig. 3. In this system, a normal-incidence pulse-echo mode is used for the measurement of coating thickness and the evaluation of adhesion quality. The immersion piezoelectric ultrasonic probe is excited by an ultrasonic frequency generator that produces pulse square waves. The ultrasonic waves generated by the transducer, which is immersed in water, propagates in the water medium above and around the samples. And the ultrasonic reflection waves received by ultrasonic probe can be transformed into electrical signals, according to the piezoelectric effect. The signals then will be sampled by an oscilloscope, transforming the analog signals into digital signals. The received digital signals then will be analyzed by spectral analysis method and time-frequency analysis method. Finally, the coating thickness and adhesion quality will be evaluated by the characteristics researched in the analyses above.

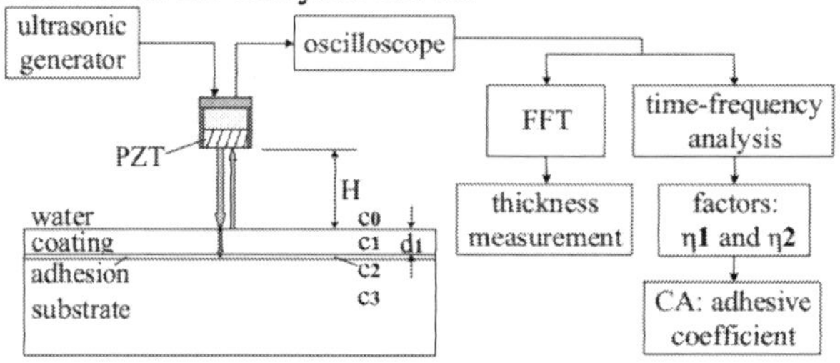

Figure 3. Schematic of detection system

4 Propagation characteristics

When the coating is quite thin, the received waves are a superposition of a series of reflected waves, which is not easy to distinguish. Therefore, thicker samples (group I) are made to research the propagation characteristics of ultrasonic waves. In order to better distinguish the wave series of each pulse excitation, a high-resolution ultrasonic transducer of 15MHz was used.

Fig. 4 shows the ultrasonic waves of samples with different thickness received by an immersion transducer, which also serve as a generator. The received waves consist of a surface echo wave, which is relatively large, and a series of waves that reflects once or more in the coating. The coating thickness of Fig. 4(a) and 4(b) are relatively thick, and every wave received can be clearly distinguished. In Fig. 4(c), the coating thickness is relatively thin, and the received waves are overlapping. It is not easy to distinguish each wave.

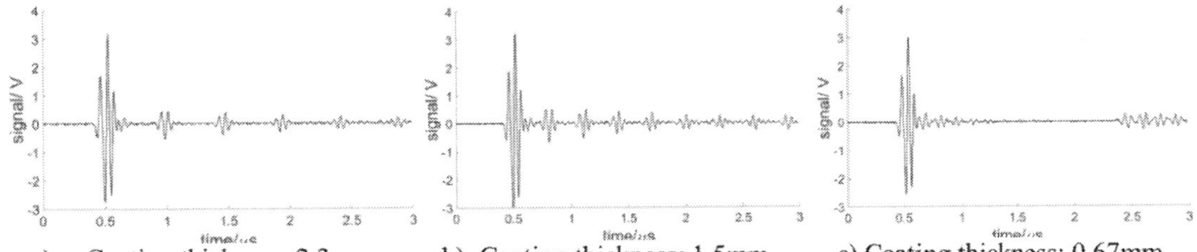

a) Coating thickness: 2.3mm   b) Coating thickness: 1.5mm   c) Coating thickness: 0.67mm

Figure 4. Ultrasonic waves received by transducer

The attenuation of peak-to-peak amplitude of received waves in the ceramic coating can be described by Fig. 5. We can see that the relative amplitude decrease nearly linearly from the second to the sixth received wave, and that the degree of attenuation is not dependent on the coating thickness. Thus, this attenuation is mainly caused by the bonding status.

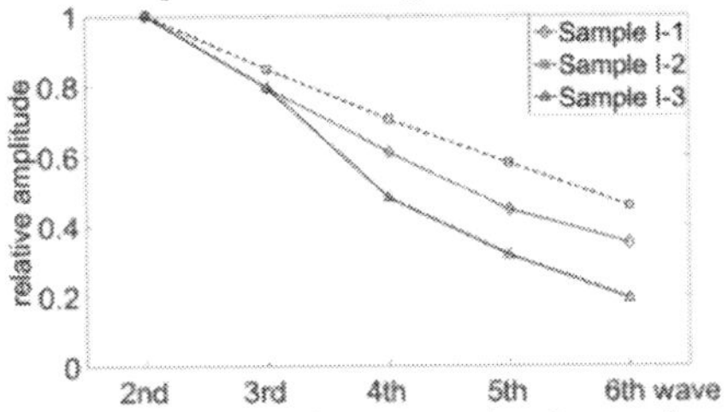

Figure 5. Ultrasonic attenuation in coatings

5 Thickness measurement

The thickness of coating can be measured by time-domain method and frequency-domain method. The time-domain method is commonly used and can be described by the follow equation

$$d = ct/2 \tag{9}$$

where d is the thickness of coating, c is the velocity of ultrasonic waves and t is the time interval of two peaks in the time-domain waves.

The frequency-domain method is based on the ultrasonic resonance in coating. According to the superposition principle of wave, when the thickness of coating is an integer multiple of half-wavelength, the reflection wave gets an extreme value. Since the ultrasonic probe covers a wide bandwidth, several peaks can be got in the frequency-domain spectrum. This method can be described by the follow equation

$$d = c/2\Delta f \tag{10}$$

where Δf is the frequency interval of two peaks in the frequency-domain waves.

When the thickness is thick, for instance 1 mm or more, both methods are available. When the thickness is very thin, the time-domain method is not available since the waves are superimposed, but the frequency-domain method is still available. The frequency-domain method is mainly based on fast Fourier transform (FFT). The first received wave is ignored in FFT, because it is not associated with the thickness of coating. Fig. 6 shows the time-domain waves and frequency-domain spectrum of 0.35mm thick coating.

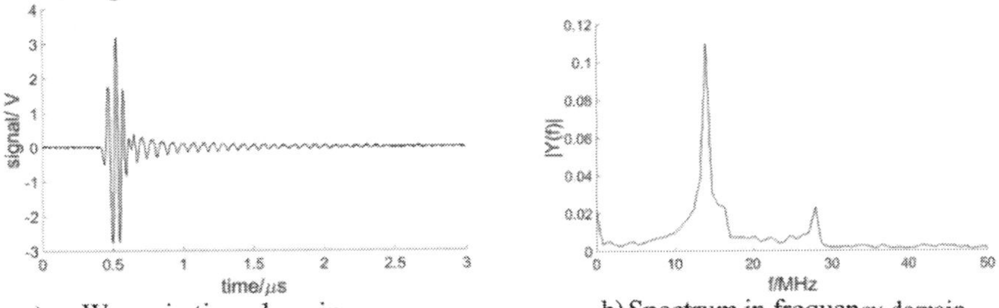

a) Waves in time-domain   b) Spectrum in frequency-domain

Figure 6. Received ultrasonic waves and spectrum

Apparently, the received waves in Fig. 6(a) are overlapped and it is not easy to calculate the time difference of a round trip in the coating. In the frequency domain, as shown in Fig. 6(b), there are clearly two peaks, the frequency interval of which is 13.96MHz. Then the thickness we get by equation (10) is 0.358mm, which is very close to the nominal value.

6 Adhesion quality evaluation

In order to distinguish the reflection wave of the coating-substrate interface (i.e. the second received wave) from the full received waves, an algorithm based on the amplitude and phase of the received waves is proposed. The received waves, or the synthesized waves mainly consist of the first received wave and the second received wave. According to the synthesis rules of waves, if the amplitudes of the synthesized waves and the first received wave are known and so is the phase difference of the first and the second received waves, the amplitude of the second received wave then can be determined.

Suppose the equations of the full received waves and the first and the second received waves can be written as:

$$s = A\sin(\omega n - \varphi).* \mathrm{Hn}(n - \varphi/\omega) \tag{11}$$

and

$$s_1 = A_1\sin(\omega n).* \mathrm{Hn}(n) \tag{12}$$

and

$$s_2 = A_2\sin(\omega n - \beta).* \mathrm{Hn}(n - \beta/\omega) \tag{13}$$

where A, A_1 and β are known, and A_2 and φ are unknown. As the thickness of the coating has been measured by the means of FFT, then the phase difference of s_1 and s_2, namely β, can be determined. The amplitudes of s and s_1, namely A and A_1, can be determined by pulse-echo method with thick ceramic plates (the same properties as the ceramic coatings) and samples to be detected. The symbol dot with star (.*) denotes the dot product of the vectors. And N is the length of wave series and Hn represents the Hann function:

$$\mathrm{Hn}(n) = \sin^2[\pi n/(N-1)] \tag{14}$$

Since we know the regular form of all the three waves, we can choose a specific feature point as reference point to determine the unknowns. In this paper, the time position of the absolute maximum of wave was chosen as the reference point. The algorithm can be described as follow:

a. Fix the position of s_1 on the *t*-axis, and the time position of the absolute maximum of s_1 is t_{max1}.
b. Mark the time position t_{max2} of the absolute maximum of s_2 on the *t*-axis.
c. Move the wave s along the *t*-axis, and calculate $\Delta s = s - s_1$. Determine the position $t_{max\Delta}$ of the absolute maximum of Δs.
d. If $|t_{max\Delta} - t_{max2}|$ is less than δ (δ is a tiny constant), then the time position of s is exactly the position it should be. Meanwhile, Δs is approximately equal to s_2.

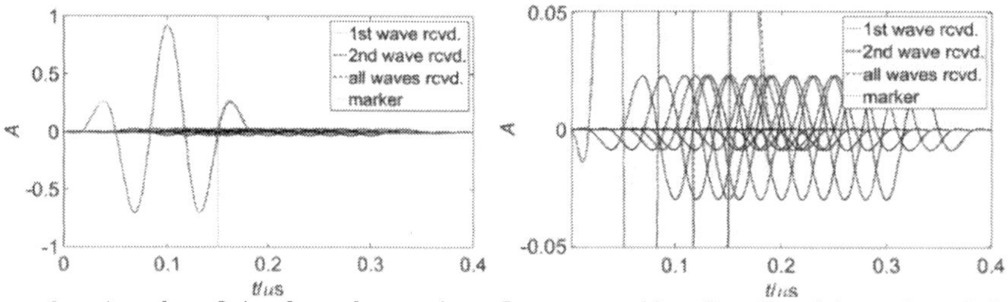

a) A series of Δs from the moving of s b) Details of the series of Δs

Figure 7. Schematic of determination of s_2

In step c, a series of Δs waves were obtained from the moving of the full waves s along the time-axis, as shown in Fig. 7. In these waves, only the one whose time position of the maximum coincides with the marker is exactly the wave s_2 we are looking for. Meanwhile, the amplitude A_2

of wave s_2 is determined. Since A, A_1 and A_2 are known now, we can calculate the adhesion coefficient CA with equation (8).

The absolute maximums of the three samples in group II measured by the algorithm are listed in table 3, as well as the adhesion coefficient CA. The parameter η_1 is about 0.8, according the reflection coefficient of acoustic pressure from water to alumina ceramic coating and the attenuation of ultrasonic waves in propagation. The results obtained by the algorithm above are consistent with expectations: the better the bonding status, the higher the adhesion coefficient CA.

Table 3. Results of evaluation of adhesion quality

Sample number	Absolute maximum	η_1	η_2	Adhesion coefficient
II-1	0.28	0.80	0.014	0.66
II-2	0.32	0.80	0.018	0.55
II-3	0.39	0.80	0.027	0.34

7 Discussion

For the thickness measurement. Generally, an ultrasonic transducer with a higher center frequency has a higher resolution. When the thickness of coating is smaller, transducers with a higher center frequency are needed. In order to obtain a higher amplitude in frequency domain, the center frequency of the transducer should be close to the resonant frequency of the coating, and bandwidth of the transducer should be wide enough for two peaks if possible.

For the evaluation of adhesion quality. If the reflection waves are distinguishable, i.e. the waves are not overlapping, we can use the reflection energy method to evaluation the adhesion quality. However, in many cases, the reflection waves are overlapping (when the coating is very thin), the reflection energy method cannot be used directly. In this paper, the algorithm is proposed to discover the reflection energy hidden in the surface echo wave (i.e. the first received wave).

8 Conclusion

In this paper, the propagation characteristics of ultrasonic wave in coatings with different thicknesses on the substrate were researched using ultrasonic testing with simulation method. An algorithm based on the amplitude and phase of reflection waves was proposed for the evaluation of adhesion quality. From the experimental results, we can conclude that pulse-echo method with FFT is reliable for the measurement of coating thickness and that the algorithm is feasible for the evaluation of adhesion quality of ceramic coatings.

References

1. J. Zhou, J. Li, W. Jiang, J. Ceram., **25**, 179-185 (2004).
2. P. Zhang, Q. Zhu, H. Qin, Mater. Rev., **28**, 27-31 (2014).
3. Vaßen R, Jarligo M O, Steinke T, et al, Surf. Coat. Technol., **205**, 938-942 (2010).
4. Padture N P, Gell M, Science, **296**, 280-284 (2002).
5. K. W. Allen. J. Adhes., **21**, 261-277 (1987).
6. Butt M A, Chughtai A, Ahmad J, et al, J. Faculty Eng. Technol., **15**, 21-45 (2007).
7. A V Pocius. *Adhesion Science and Engineering: Surfaces, Chemistry and Applications* (Elsevier, 2002).
8. Tattersall H G., J. Phys. D: Appl. Phys., **6**, 819-832 (1973).
9. Singher, L., Ultrason., **35**, 305-315 (1997).
10. Z. Zhong, *Quantitative Failure Assessment for Surface Cracking and Interface Delamination of Thermal Barrier Coatings by Acoustic Emission* (Xiangtan University, 2013).
11. Y. Yang, D. Zhao, Y. An, et al. Chin. J. Sci. Instrum., **36**, 1215-1220 (2015).

12. C. Feng, X. Hua, Appl. Sci. Technol., **42**, 15-18 (2015).
13. Achenbach J D, Parikh O K. Ultrasonic analysis of nonlinear response and strength of adhesive bonds [J]. J. Adhes. Sci. Technol., **5**, 601-618 (1991).
14. Wegner, A., et al., Ultrason., **38**, 316-21 (2000).
15. Czarnecki, L., A. Garbacz and M. Krystosiak ,Cem. Concr. Compos., **28**, 360-369 (2006).
16. Rokhlin, S.I., et al., Ultrason., **42**, 1037-1047 (2004).
17. Joseph L. Rose, *Ultrasonic Waves in Solid Media* (Cambridge University Press, 1999).
18. Q. Zhang, *Study on the Attenuation Laws of Ultrasonic Propagation in the Inhomogeneous Media* (Shenyang University of Technology, 2015).

MATEC Web of Conferences 114, 02002 (2017) DOI: 10.1051/matecconf/201711402002
2MAE 2017

Studies on Thermal Oxidation Stability of Aviation Lubricating Oils

Nan Wu[1,2], Zhi-Min Zong[1,a], Yi-Wei Fei[2] and Jun Ma[2]

[1] Key Laboratory of Coal Processing and Efficient Utilization, Ministry of Education, China University of Mining & Technology, Xuzhou 221116, China.
[2] Department of Aviation Oil and Material, Air Force Logistics Institute, Xuzhou 221000, China

Abstract. Simulating the operating condition of aviation engine via autoclave experiment of high temperature and pressure, we studied the physic-chemical property of poly-α-olefin base oil samples mixed with antioxidants of 2,6-di-tert-butyl-4-methylphenol and p,p'-diisooctyl diphenylamine at different temperature. The mechanism of degradation of PAO aviation lubricating oil was analyzed according to the oxidized products by modern analytical instruments. The results showed that the aviation lubricating oil produced a large number of low molecule compounds while increasing the temperature, and resulted in the viscosity decreasing and acid value increasing which indicated that the thermal oxidation of the oil sample underwent a radical process.

1 Introduction

Lubricating oils help lubricate, clean, and cool aircraft engines during operation [1, 2]. The thermal oxidation stability of lubricating oil is a pivotal indicator of its quality, which directly affects the normal operation of the aircraft as well as the service life of the equipment [3, 4]. In recent years, the structure of aircraft engines has become more complex, and there is a need for aviation lubricating oils that can be used in harsh working environments. High temperature and pressure conditions as well as high shear conditions have increased the demand for aviation lubricating oils with high thermal oxidation stabilities. Degradation of the lubricating oil can have extremely adverse effects on aircraft engine performance and result in major accidents, including plane crashes.

Previous studies on aviation lubricating oils have indicated that the oils undergo obvious changes inside the engines. The oils turn darker, their viscosity increases, and they form a large number of deposits. Therefore, in order to ensure flight safety, it is essential to elucidate the thermal oxidation degradation behavior of aviation lubricating oils.

The aviation lubricating oil investigated in this study consisted of a synthetic poly-α-olefin (PAO) oil (the base oil) to which 2,6-di-tert-butyl-4-methylphenol (DBPC) and p,p'-diisooctyl diphenylamine (DODPA) were added as antioxidants. Other additives such as detergent-based dispersants and friction-modifying agents were not used. The normal operational temperature of aviation lubricating oils is approximately 200°C; however, it increases to 250–300°C while the engine is slowing or shutting down, because the amount of air flowing into the engine decreases or the flow of air stops

[a] Corresponding author:515811945@qq.com

completely, and the bearings are cooled primarily by the lubricating oil [5-8]. Therefore, it is important to obtain more reliable data on the lubricating oil at high temperatures, as this will allow effective use of the oil.

As stated above, the oil samples used in this study were prepared by mixing the PAO base oil with DBPC and DODPA as antioxidants. The physicochemical properties of the oil at different temperatures and pressures were tested in an autoclave, which was used to simulate the operating conditions of an aircraft engine. Finally, based on the oxidation products formed at high temperatures, which were analyzed using a variety of modern analytical techniques, a mechanism to explain the degradation of PAO-based aviation lubricating oils by thermal oxidation is proposed.

2 Experimental

2.1 The main materials and equipment

6890/5973 GC/MS, American Agilent Company; IR-560, American Nicolet magna Company; high temperature and pressure autoclave, self-made.

2.2 Preparation of experimental oil sample

Taken 50g PAO base oil and the mixture of 50g PAO base oil, 0.25g DBPC and 0.25g DODPA into 500mL autoclave and stirred at speed of 800r/min and kept the temperatures of 170°C, 200°C, 230°C, 270°C and 300°C for 2h. After the reaction putted the reactor into an ice-water bath to cool to room temperature, then toke out the oil sample. For the convenience of description, the products of PAO crude oil at room temperature and other oil sample at 170°C, 200°C, 230°C, 270°C and 300°C reaction were marked as P20, P170, P200, P230, P270, P300; the products of mixture oil with antioxidants were labeled as P'20, P'170, P'200, P'230, P'270 and P'300.

3 Results and Discussion

3.1 Changes of Kinematic Viscosity and Acid Value of the Oil Sample

The kinematic viscosity of the oil sample oxidized at different temperature were shown in Table 1.

TABLE 1. Kinematic viscosity at 40°C and acid value of the oil samples

Item	kinematic viscosity (40°C) /($mm^2\ s^{-1}$)	Item	kinematic viscosity (40°C) /($mm^2\ s^{-1}$)
P_{20}	17.94	P'_{20}	18.39
P_{170}	17.97	P'_{170}	18.89
P_{200}	17.88	P'_{200}	18.11
P_{230}	17.23	P'_{230}	17.79
P_{270}	14.39	P'_{270}	17.56
P_{300}	8.28	P'_{300}	12.40

It can be seen from Table 1 that the viscosity of the PAO oil sample mixed with antioxidants was slightly greater than PAO base oil, which may be due to the steric structure of the antioxidant dissolved increasing the flow resistance of the lubricating oil[9-11]. Enhancing the temperature, the viscosity of the PAO base oil decreased gradually, what's more, when the temperature rose to above 230°C it declined sharply. However, there is no obviously changes at lower temperature (less than

230°C). While the viscosity of the PAO oil samples mixed with the antioxidants decreased flatly as long as the temperature increasing, but when the reaction temperature was higher than 270°C, it would change greatly. As can be seen from Table 1, the viscosity of the PAO base oil decreased from 17.97 $mm^2 \cdot s^{-1}$ to 8.28 $mm^2 \cdot s^{-1}$ with decline of 53.9%, while the decline of the viscosity of the PAO oil samples was 34.4% with the attenuation decreasing. Generally, the oxidation stability of PAO oil would be enhanced by adding antioxidants which effectively inhibited the viscosity attenuation of the lubricating oil.

3.2 Changes of oil samples' composition

3.2.1 FTIR Analysis

Fig. 1 shows the infrared spectra of the products oxidized at different temperature of PAO base oil and PAO oil samples mixed with antioxidants.

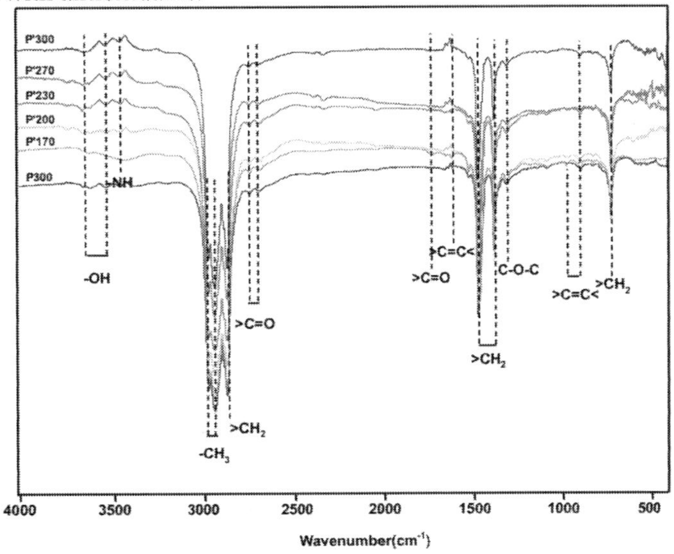

Figure 1. Infrared spectra of PAOs after oxidation test

As can be seen from Figure 1, the strong absorption peak at 2975-2850 cm^{-1} descripts a large number of functional groups of $-CH_3$ and $-CH_2$-; the sharp and strong absorption peak at 721 cm^{-1} descripts many units connected by more than four groups of $-CH_2$-; the broad absorption peak at 3650-3429 cm^{-1} descripts groups of -OH and-NH; the small and weak peaks at 2730 cm^{-1} and 1741 cm^{-1} descript groups of C=O; the peaks at 966 cm^{-1} and 888 cm^{-1} descript groups of C=C. The presence of -OH, C=O, C=C and other functional groups in the infrared spectra shows that the PAO lubricating oil was oxidized to decay under high temperature and generated olefins, alcohols, acids and other products[12-15]. But the infrared spectra can't descript the effects of antioxidants on the oil sample degradation quantitatively because of the little amount of products.

3.2.2 GC/MS analysis

Fig. 2 and 3 show the ion chromatogram of the products oxidized at different temperature of PAO base oil and PAO oil samples added with antioxidants. The GC/MS chromatogram of former 17 min retention time was mainly studied, because the compounds detected in the chromatogram of the later part were mainly polymers. The category and relative content of the products oxidized such as n-alkanes, iso-paraffins and olefins are shown in Table 2.

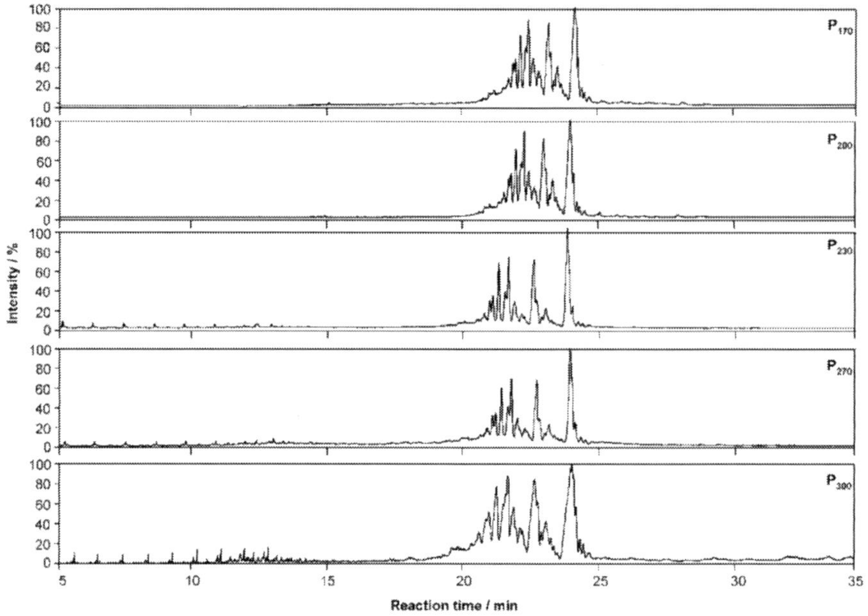

Figure 2. The ion chromatogram of PAOs after oxidation test

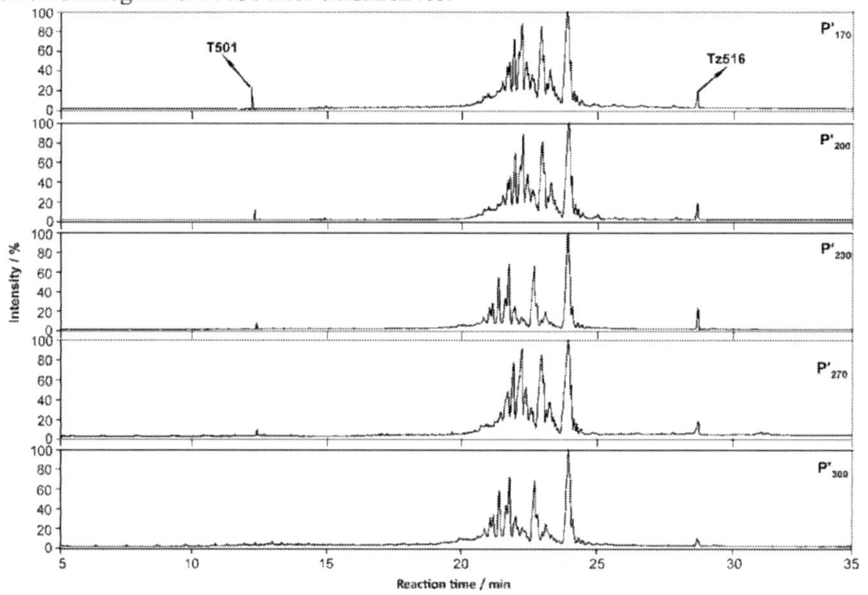

Figure 3. The ion chromatogram of PAOs added antioxidants after oxidation test

Table 2 Category and relative content of PAOs after oxidation test

Item	n-alkanes		iso-paraffin		olefin	
	category	relative content (%)	category	relative content (%)	category	relative content (%)
P_{170}	2	0.067	4	0.370	5	0.224
P_{200}	5	0.316	4	0.493	5	0.481
P_{230}	7	0.558	4	0.658	6	0.684

P_{270}	12	2.675	10	1.113	21	3.817
P_{300}	35	8.606	25	3.538	47	9.968
P'_{170}	2	0.053	3	0.281	2	0.132
P'_{200}	2	0.282	4	0.360	2	0.359
P'_{230}	3	0.498	4	0.459	5	0.664
P'_{270}	10	2.607	7	0.998	13	3.085
P'_{300}	34	8.368	18	3.316	41	9.324

It can be found that the molecular ion peak of small molecules gradually increased with the increase of reaction temperature and the peak area also increased accordingly, which indicates that the PAO oil samples decayed seriously in the oxidation reaction at high-temperature and the long-chain molecules of PAO oil broke to some alkanes, alkenes and other short-chain molecules. The degradation of PAO oil samples weakened significantly after adding antioxidants, while antioxidant itself consumed constantly and the peak area decreased gradually.

Table 2 shows that the oxidation products of PAO base oil at 170°C were 2 kinds of n-alkanes, 4 kinds of iso-paraffins and 5 kinds of olefins and the relative contents were 0.067%, 0.370% and 0.224% respectively. While enhanced the temperature to 300°C, the molecular chain structure was severely damaged to generate oxidation products of 35 kinds of n- alkanes, 25 kinds of iso-paraffins, 47 kinds of olefins and the relative contents were 8.606%, 3.538 % and 9.968%. Generally, the enhancement of temperature resulted in the deepen cracking of molecular and produced many small molecular compounds to increase the fluidity of lubricating oil itself. The kinematic viscosity decreased because intermolecular interaction weakened and intermolecular friction reduced.

When we added antioxidants into PAO base oil, the oil sample also went degradation along with the temperature increasing. But the categories and relative contents of the products after thermal oxidation reaction were less than that of PAO base oil without antioxidants, especially category and relative quality of olefins decreased greatly. There were 5 kinds of olefins in the products oxidized of PAO base oil with relative contents of 0.481% at 230°C; while adding antioxidants, there were 2 kinds of olefins with relative contents of 0.359% and reduction of 25%. Thus, antioxidants may effectively retard the thermal oxidation reaction of PAO lubricating oil and inhibit the formation of olefins

4 Conclusions

The present study revealed that antioxidants could enhance the thermal oxidation stability of PAO aviation lubricating oil as the decay process of the oil sample was significantly slowed down. Analyzed with IR and GC/MS instruments, we found that during the thermal oxidation step, PAO lubricating oil underwent a free radical chain reaction process, following pyrolysis and thermal oxidation reaction mechanism.

Acknowledgements

This work was subsidized by the Jiangsu Provincial Natural Science Foundation of China (Grant BK20161187), and Xuzhou Science and Technology Development Funds (Grant KC16SG269).

References

1. B. K. Sharma, A. J. Stipanovic, Thermochimica Acta **402,** 1-18 (2003).
2. J. S. Basta, Petroleum Science &Technology **28,** 1611-1619 (2010).

3. K. S. Pedersen, H. P. Ronningsen, Energy Fuels **17**, 321-330 (2003).
4. C. D. Gamlin, N. K. Dutta, N. R. Choudhury, et al., Thermochim Acta, **357**, 392-393 (2002).
5. H. Diamond, H. C. Kennedy, R. G. Larsen, Chemistry &Physics of Lipids, **176**, 99-104 (2013).
6. R. G. Larsen, R. E. Thorpe, F. A. Armfield. Ind. Eng. Chem. **34**, 183-193(2002).
7. G. Kreisberger, C. W. Klampfl, W. W. Buchberger, Energy Fuels, **10**, 1021-1025 (2016).
8. M. J. Abedin, H. H. Masjuki, M. A. Kalam, et al. Energy Conversion & Management, **85**, 173-189 (2014).
9. R. Shah, D. A. Pratt, J. Org. Chem., **81**, 6649-6656(2016).
10. R. G. Larsen, F. A. Armfield. Ind. Eng. Chem. **35**, 581-588 (2002).
11. M. J. Yanjarappa, S. Sivaram, Prog. Polym. Sci. **27**, 1347-1398 (2002).
12. J.C. O. Santos, I. M. G. Santos, A. G. Souza, et al., Fuel **83**, 2393-2399 (2004).
13. Y. Si, E.T. Samulski, Nano Lett. **8**, 1679-1682(2008).
14. A. Bagri, C. Mattevi, M. Acik, Y.J. Chabal, M. Chhowalla, V.B. Shenoy, Nat. Chem. **2**, 581-587 (2010).
15. D.W. Le, V.L. De Los Santos, J.W. Seo, J. Phys. Chem. B **114**, 5723-5728 (2010).

MATEC Web of Conferences **114**, 02003 (2017) DOI: 10.1051/matecconf/201711402003
2MAE 2017

Mechanical Behavior of 316L Stainless Steel after Strain Hardening

Kaishang Li[1,2], Jian Peng[1,2,a] and Jian Peng[1,2]

[1] *School of Mechanical Engineering Changzhou University, Changzhou, 213164, China.*
[2] *Jiangsu Key Laboratory of Green Process Equipment, Changzhou University, Changzhou, 213164, China.*

Abstract. The effects of strain hardening on the mechanical behavior of 316L stainless steel were studied in the paper. The original and different strain hardening materials were compared to investigate the mechanical behavior. The results demonstrate that the yield strengths increase with the magnitude of strain hardening significantly, but the ultimate strengths of the original and different strain hardening materials are closed. In addition, the plastic parameters of 316L stainless steel including fracture elongation and fracture surface shrinkage decrease with the magnitude of strain hardening. Finally, the Ramberg-Osgood equation is used to predict the stress-strain curves after strain hardening, and the results indicate that the predicted values agree with the experimental values.

1 Introduction

Austenitic stainless steel, used for important structure of pressure equipment, retains high strength and excellent ductility from cryogenic to elevated temperatures [1, 2]. 316L austenitic stainless steel, a common structural material, meets requirements for high-pressure vessels, pipes, valves, etc. [3]. But the mechanical behaviour of 316L stainless steel will change after strain hardening. The influences of strain hardening on mechanic behaviour including strength and plastic parameters of austenitic stainless steels have been studied in recent years. Tensile tests of 316H stainless steels with strain hardening up to 4%, 8% and 12% at room temperature in Ref. [4] found that the values of Young's modulus and the ultimate strengths are similar for different strain hardening, but the yield strengths increase with the magnitudes of strain hardening at different temperatures. Wang et al investigated fatigue properties and damage mechanisms of pre-strained plasticity steel, and found that the yield strengths increase with magnitudes of pre-strain [5]. Moreover, the relationship between stress and strain needs to predict according to constitutive equations. Ramberg and Osgood [6] proposed a technique of describing and predicting the relationship between stress and strain in metallic materials. Skelton and Maier [7] predicted the stress-strain relationship under variable loading conditions by the Ramberg-Osgood equation. Hight and Brandeau [8] using the Ramberg-Osgood relation, developed a mathematical equation of the stress-strain rate behavior. But the Ramberg-Osgood equation discussed above is without considering strain hardening, and cannot predict the tensile curve with strain hardening.

[a] Corresponding author: joepengjian@163.com

2 Specimen design and experimental procedure

Tensile tests were performed using commercially available 316L stainless steel. Figure 1 shows that the plate specimen used in the tests. Firstly, plate specimens were pre-deformed to 5% and 10% at room temperature. Then, tensile tests were operated at room temperature with a strain rate of 5e-5/s. Finally, the differences of mechanical behavior among original and 5% and 10% strain hardening materials was analyzed according to strength and plastic parameters. The fracture surface of original and 5% and 10% strain hardening specimens were examined using the scanning electron microscope (SEM). Figure 2 shows the tensile testing machine used in the tests.

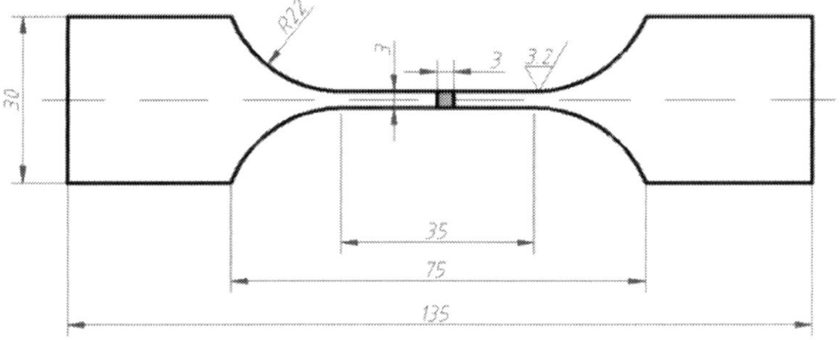

Figure 1. Dimension of tensile sample (unit: mm).

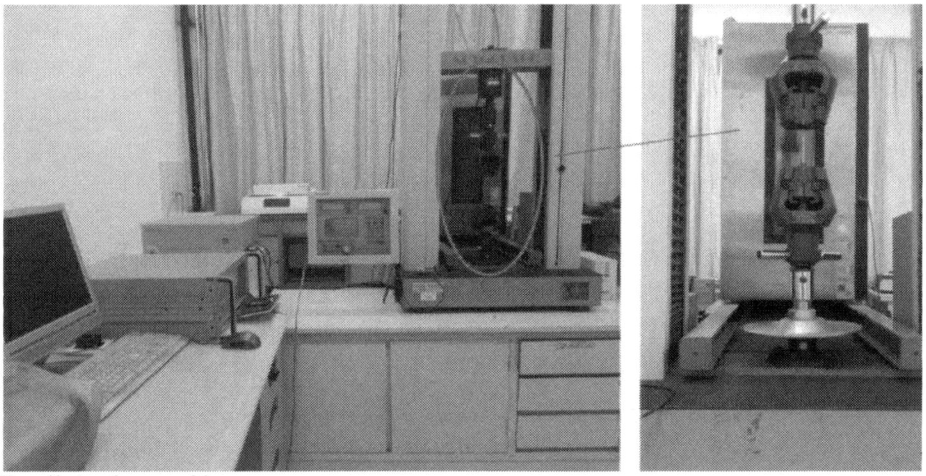

Figure 2. Tensile testing machine.

3 Results and discussion

3.1 Effect of strain hardening on strength parameter

Figure 3 shows the tensile stress-strain curves of original and 5% and 10% strain hardening specimens at room temperature, and the yield strengths and ultimate strengths are shown in Table 1. With comparison among the original material and strain hardening materials, the yield strength increase from 339 MPa to 370 Mpa with with the magnitude of strain hardening. But ultimate strengths are basically unchanged with the magnitude of strain hardening at room temperature.

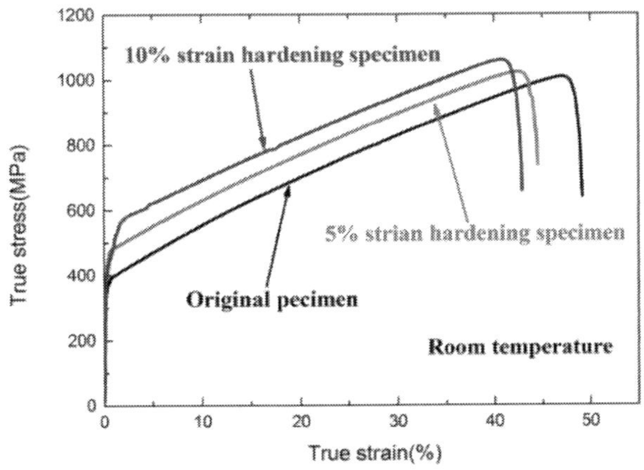

Figure 3. Tensile curves of original, 5% and 10% strain hardening specimens at room temperature.

Table 1. Yield strength and ultimate strength of original, 5% and 10% strain hardening specimens.

Specimens	Original	5% strain hardening	10% strain hardening
Yield Strength (MPa)	339	353	370
Ultimate Strength (MPa)	1008	1018	1024

3.2 Effect of strain hardening on plastic parameter

Figure 4 shows plastic parameters of 316L stainless steel including fracture elongation and fracture surface shrinkage with the magnitude of strain hardening at room temperature. Both fracture elongation and fracture surface shrinkage decrease with the magnitude of strain hardening. The fracture elongation and fracture surface shrinkage are approximately proportional to the magnitude of strain hardening. It indicates that strain hardening can increase the yield strength, but decrease the plastic parameter of 316L stainless steel.

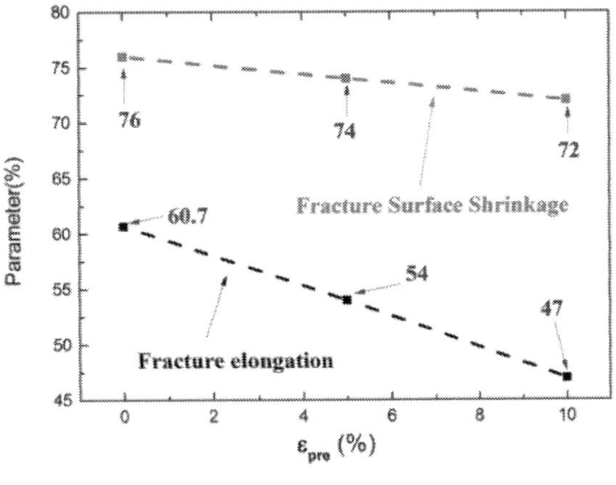

Figure 4. Fracture surface shrinkage and fracture elongation with different strain hardenings.

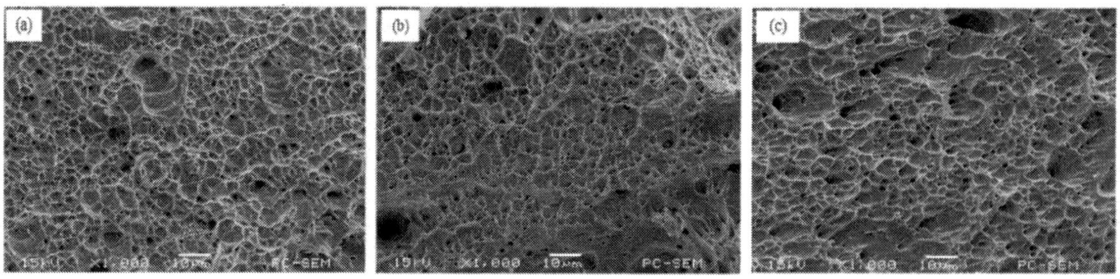

Figure 5. SEM images of tensile fracture surfaces of (a) original; (b) 5% strain hardening; (c) 10% strain hardening specimens at room temperature.

Figure 5 shows the SEM images of original, 5% and 10% strain hardening specimens at room temperature. The tensile fracture surfaces of specimens are filled with dimples, which are a typical ductile failure mode at room temperature. But the tensile fracture surfaces have no obvious change with the magnitudes of strain hardening. It indicates that strain hardening will not change the ductile failure mode of 316L stainless steel.

3.3 Analysis of strain hardening by the Ramberg-Osgood equation

Ramberg-Osgood is a theoretical equation to predict and describe the relationship between stress and strain of the metallic material near its yield point. In this paper, Ramberg-Osgood equation is used to analyze the change of stress index of 316L stainless steel after strain hardening.

The general form of the Ramberg-Osgood equation in tensile experiments is as follows:

$$\varepsilon = \sigma / E + \alpha \cdot \sigma_0 / E \cdot (\sigma / \sigma_0)^M \quad (1)$$

Where, σ_0 is the reference stress (MPa), E is the initial elastic modulus (MPa), α the material constants and its value is 3/7 normally, M is the stress index which indicates strengthening behavior of materials after yielding. By taking transposition and logarithm, Eq. (1) can get the expressions of the stress index:

$$\ln(\varepsilon - \sigma / E) = M \cdot \ln \sigma + \ln(\alpha / E) - (M - 1) \cdot \ln \sigma_0 \quad (2)$$

The relative plot $\ln\sigma$-$\ln(\varepsilon$-$\sigma/E)$ is used to fitting a straight line based on tensile curves and the initial elastic modulus (E). The stress index (M) values and the reference strain (σ_0) can be calculated according to the slope and intercept of the straight line. The fitting process is shown in Figure 6 and the values of stress index (M) are shown in Table 2, which shows that the values of stress index increase with the magnitudes of strain hardening. The phenomenon indicates that the hardening behavior is more and more obvious and the deformation capacity of 316L stainless steel decreases with the magnitudes of strain hardening.

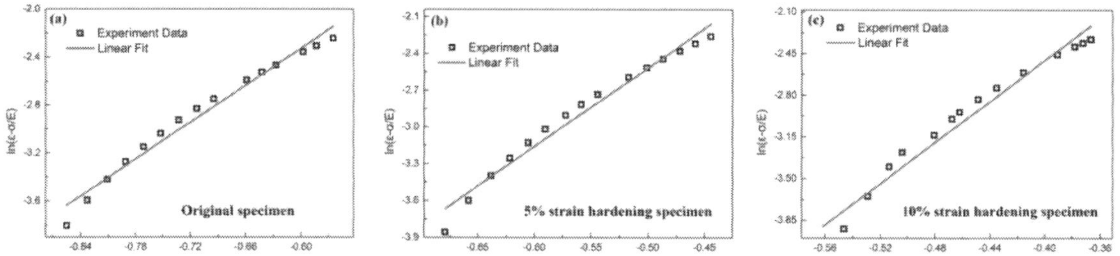

Figure 6. The linear fitting of stress index of (a) original; (b) 5% strain hardening; (c) strain hardening specimens.

Table 2. The values of stress index of original, 5% and 10% strain hardening specimens.

Specimens	Original	5% strain hardening	10% strain hardening
Stress Index	5.14	6.42	8.63

Figure 7 shows the plots of experimental values and the lines of predicted values by Ramberg-Osgood equation, and most of the experimental values are close to the predicted values. It illustrates that Ragberg-Osgood equation is feasible in tensile test of original, 5% and 10% strain hardening specimens. But it is obvious that the difference between the experimental data and the predicted values began to enlarge with the magnitudes of strain hardening.

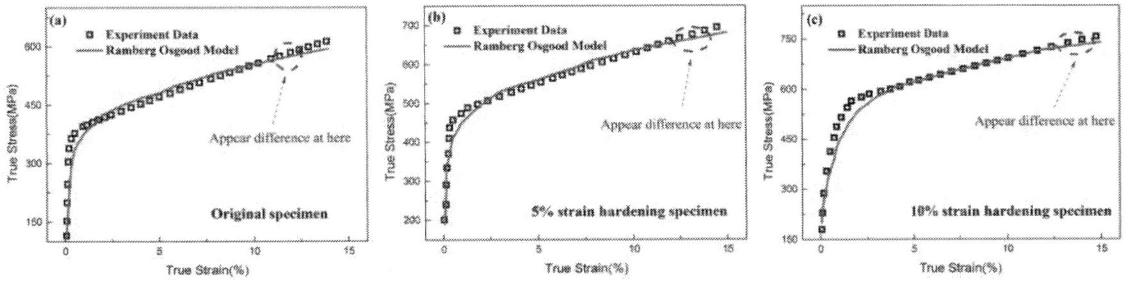

Figure 7. Comparison between the experimental data and predicted values by Ramberg-Osgood equation.

4 Summary

For the strength parameters of original, 5% and 10% strain hardening specimens at room temperature, the yield strength increases significantly but the ultimate strength is almost unchanged with the magnitudes of strain hardening. For plastic parameters of 316L stainless steel, the fracture elongation and fracture surface shrinkage have the same changed trend, which decreases with the magnitudes of strain hardening.

The Ramgerg-Osgood equation was applied to predict the stress strain curve of original, 5% and 10% strain hardening specimens, and the values of stress index increases with the magnitudes of strain hardening.

Acknowledgments

This work was financially supported by National Natural Science Foundation of China (Grant No. 51505041) and Natural Science Foundation of the Jiangsu Higher Education Institutions of China (Grant No. 16KJB460002).

References

1. Al-Mangour B, Mongrain R, Irissou E, et al. Improving the strength and corrosion resistance of 316L stainless steel for biomedical applications using cold spray [J]. Surface & Coatings Technology, 216(3):297-307 (2013).
2. Brnic J, Canadija M, Turkalj G, et al. Behavior of AISI 316L steel subjected to uniaxial state of stress at elevated temperatures [J]. 25(2):175-180 (2009).
3. Pham M S, Holdsworth S R. Dynamic strain ageing of AISI 316L during cyclic loading at 300 °C: Mechanism, evolution, and its effects[J]. Materials Science & Engineering A, 2013, 556:122–133.

4. Mehmanparast A, Davies C M, Dean D W, et al. Effects of plastic pre-straining level on the creep deformation, crack initiation and growth behaviour of 316H stainless steel[J]. International Journal of Pressure Vessels & Piping, 141:1-10 (2016).
5. Wang B, Zhang P, Duan Q Q, et al. High-cycle fatigue properties and damage mechanisms of pre-strained Fe-30Mn-0.9C twinning-induced plasticity steel[J]. Materials Science & Engineering A, 2017, 679:258-271.
6. Ramberg W, Osgood W R. Description of stress-strain curve by three parameters [J]. NACA Technical Note, 902 (1943).
7. Skelton R P, Maier H J, Christ H J. The Bauschinger effect, Masing model and the Ramberg–Osgood relation for cyclic deformation in metals [J]. Materials Science & Engineering A, 238(2):377-390 (1997).
8. Hight T K, Brandeau J F. Mathematical modeling of the stress strain-strain rate behavior of bone using the Ramberg-Osgood equation [J]. Journal of Biomechanics, 16(6):445-450 (1983).

MATEC Web of Conferences **114**, 02004 (2017) DOI: 10.1051/matecconf/201711402004
2MAE 2017

The Application of Modified Artificial Neural Network on Tensile Behaviour of 316L Austenitic Stainless Steel

Jian Peng[1, 2], Jian Peng[1, 2, a] and Kaishang Li[1, 2]

[1] *School of Mechanical Engineering Changzhou University, Changzhou, 213164, China.*
[2] *Jiangsu Key Laboratory of Green Process Equipment, Changzhou University, Changzhou, 213164, China.*

Abstract. In this paper, genetic algorithm (GA) was used to enhance the accuracy of back propagation artificial neural network (ANN) for the tensile behaviour of 316L. With GA optimization, the random weight and threshold between different layers in original ANN model can be updated constantly to describe the characteristics of 316L austenitic stainless steel accurately, just considering the nodes of hidden layer. According to the sample distribution, the data were divided into training and test. When the model was set to predict test data, the results showed coincident with the experimental ones, and errors of the data between prediction and experiment displayed the small values at room and intermediate temperatures.

1 Introduction

Constitutive models are usually treated as the mathematical description, which can reflect the macroscopic characteristics of materials. Conventional models are able to obtain the results, but can hardly calculate them efficiently and accurately, so the calculation of parameters will cost a lot of time.

In recent years, with the development of artificial intelligence, machine learning is becoming mainstream gradually. As one of the most popular methods of machine learning, artificial neural network was used widely in many fields [1], due to the simplified calculation, large scope of application and high accuracy. When the mechanical properties of many engineering materials were involved, the artificial neural network showed high practicability, which was superior to the models based on mathematical theories and physical phenomenon. In this regard, it was applied in flow stress prediction of metallic materials, such as Nb-Ti alloy steel [2], stainless steel [3], Cu-Al [4] and Al-Cu-Mg-Pb alloys [5]. Especially the back propagation artificial neural network (BP ANN), it seems like the brain of human beings, which can reduce errors by adjusting the weight and threshold autonomously. But the weight and threshold might fall into local optimal values easily without algorithm optimization.

Genetic Algorithm (GA) is considered as an optimized algorithm, which was studied by many researchers, including mechanical property [6], structural design [7] and reliability analysis [8]. With GA, the BP ANN constitutive model can be established to describe the tensile flow behaviour of 316L austenitic stainless steel accurately, and only a small amount of work should be accomplished, when the artificial neural network is considered as a constitutive model of 316L austenitic stainless steel.

[a] Corresponding author : joepengjian@163.com

2 Material and Experiment

From Fig.1 (a), 316L austenitic stainless steel was chosen as the material, and tensile experiment was executed on the tensile machine, which showed in Fig.1 (b).

According to the tensile experiments at different temperatures and strain rates, the relationship between flow stress and strain can be found, but the key point is how to quantitate the qualitative relation.

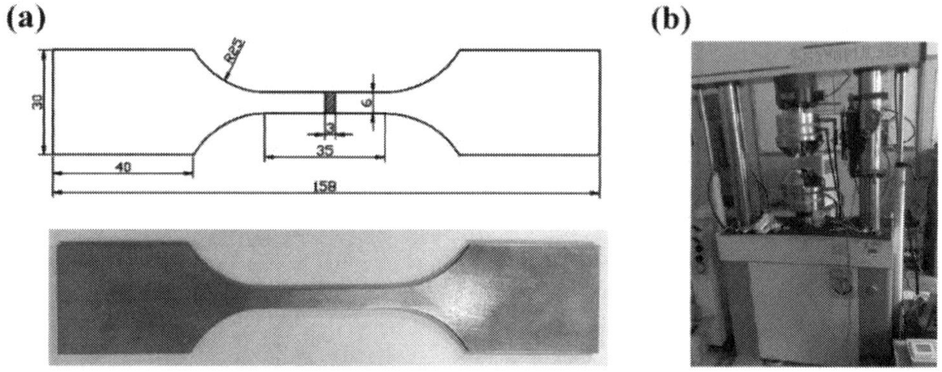

Figure.1 (a) Structure of 316L austenitic stainless steel and (b) tensile machine

3 Constitutive model

3.1 Artificial neural network

Back propagation (BP) neural network is one of the artificial neural networks, which imitates the biological brain learning, and the system of dealing with information consists of a lot of neurons. According to Fig.2, it takes the single hidden layer as an example. If the number of nodes in the input layer is m, and the number of nodes in the output layer is n. then the m dimensional to n dimensional mapping can be realized by the BP neural network of a single hidden layer in any cases [9]. In addition, the selection of node number in the hidden layer can affect the performance of neural network, the little nodes lead to the low ability of identifying sample, while too many nodes lead to more training time. The number of nodes in the hidden layer can not be determined by a regular way, but the range of nodes can be estimated with an empirical formula shown in Eq.1.

$$l = \sqrt{m+n} + a \qquad (1)$$

Where a represents the adjustment parameter range from 1 to 10, and l , m , n represent the range of nodes in hidden layer, input layer and output layer respectively.

The procedure of BP neural network can be divided into two parts. One is the forward transmission of information, and the other one is the backward transmission of errors. In the forward transmission of information, the current output can be obtained from the output of the former layer, the weight between current and former layer, the threshold of former layer, as well as the transfer function. However, in the latter part, which means the backward transmission of error, the error will reduce by changing the weight between input layer and hidden layer, as well as the weight between hidden layer and output layer, corresponding to the change of threshold in hidden and output layer.

 DOI: 10.1051/matecconf/201711402004

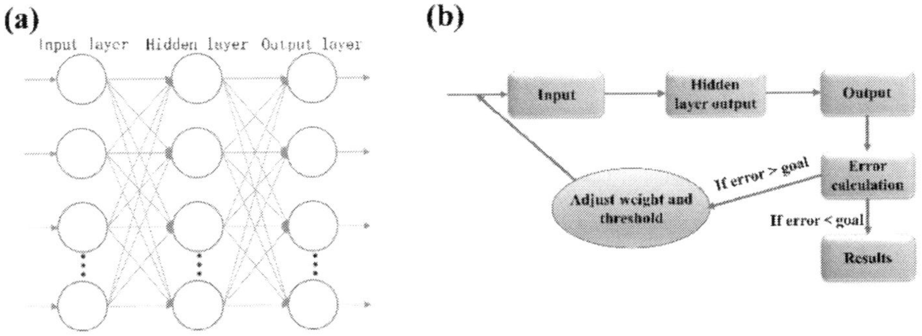

Figure 2. (a) Structure and (b) the procedure of back propagation artificial neural network

3.2 Genetic Algorithm (GA)

GA is one of the intelligence algorithm based on evolution, and the problem of local optimum can be solved efficiently with it. According to some rules about biology, including heredity, mutation and so on, the procedure of GA shows as Fig.3. According to Fig.3, it is found that, the first step is to code the original weight and threshold like biological gene code. Then the error between prediction and expectation data is considered as fitness function. The last step consists of selection, cross and mutation, in order to simulate the biological evolution.

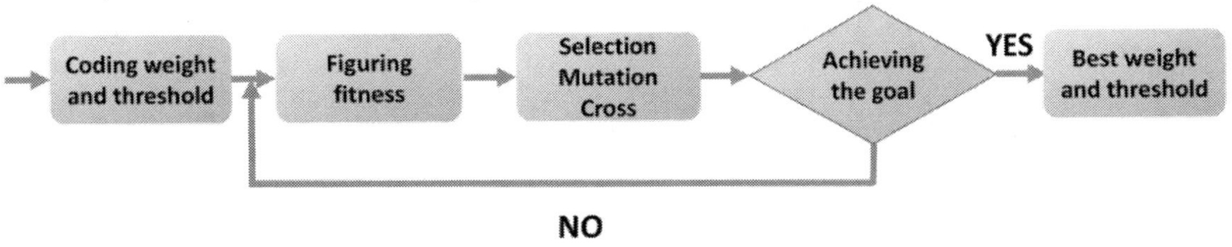

Figure.3 Procedure of genetic algorithm

4 Application of constitutive model

According to the theories above, the application of BP artificial neural network can predict the flow behaviour of 316L austenitic stainless steel, including the prediction of flow stress at room and intermediate temperatures. It is determined that input layer has 3 neurons (strain, temperature and strain rate), while the output layer has 1 neuron (flow stress). Depending on the formula shown in Eq.1, the number of hidden layer nodes can be estimated with the range from 3 to 12.

4.1 Normalization of sample

Due to the difference of strain rate, temperature and strain, the convergence rate and accuracy of the network can be influenced. Thus, the dealing of normalization is shown in Eq.2, and the normalized data can focus on 0.1-0.9.

$$x_n = 0.1 + 0.8 * \frac{x - x_{\min}}{x_{\max} - x_{\min}} \tag{2}$$

Where, x is the original input and output data, x_n is the data of normalized data, x_{min} , $x_{\max}$ are the maximum and minimum of one set sample, such as strain rate.

4.2 Training and test

The data at different temperatures and strain rates need to be divided into training data and test data. Table 1 gives the distribution of training data and test data, three strain rates (1e-2/s, 5e-3/s, 1e-3/s) were selected as test data at three temperatures (373K, 473K, 573K) respectively, and the rest are training data.

Table. 1 Distribution of sample data

	293K	373K	473K	573K
5e-4/s	train	train	train	train
1e-3/s	train	train	train	test
5e-3/s	train	train	test	train
1e-2/s	train	test	train	train

According to the training of some sample data, the error is minimal when the number of nodes in hidden layer is eight. Training parameters are shown in Table 2. Function 1 and function 2 represent activation function of hidden layer and output layer, and function 3 represents the training function of network. Figure 4(a) gives the comparison between experimental and prediction of training data. And Figure 4(b) gives the comparison between experimental and prediction data of test.

Table. 2 Training parameters of artificial neural network

Function 1	**Function 2**	**Function3**	**Epochs**	**Target error**	**Learning rate**
Sigmoid (tansig)	Linear (purelin)	Lm (trainlm)	1000	1e-5	0.01

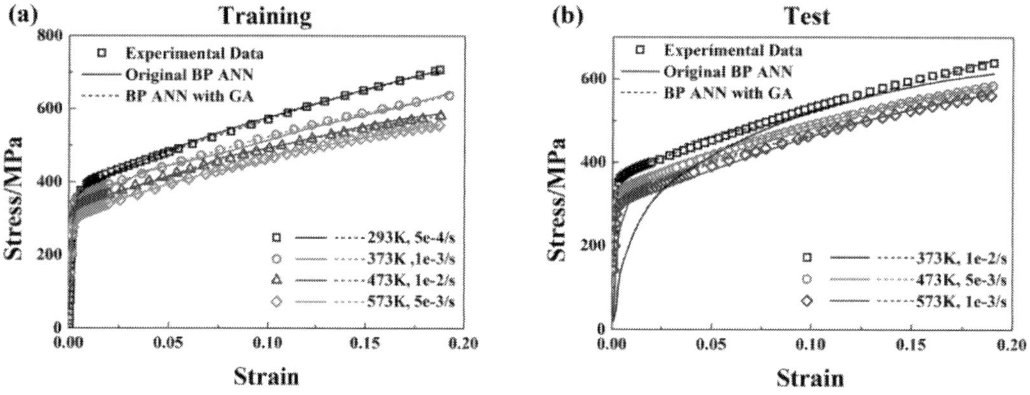

Figure. 4 Comparison between experimental and prediction data by (a) training and (b) test

Correlations and errors at the temperatures and strain rates are shown in Table 3. Among the results, it can be seen that correlation coefficients are all at high level, and the average errors of training data and test data are all at low level, which indicates the rationality of network training, and the prediction accuracy is well.

Table.3 Correlation coefficients and average errors at different temperatures and strain rates

Temperature (K)	**Strain rate(/s)**	**Correlation coefficients**	**Average errors**	**Correlation coefficients**	**Average errors**
		Original BP ANN		**BP ANN with GA**	
293 (Training)	**5e-4**	0.998	3.29	0.999	2.21
373 (Training)	**1e-3**	0.995	2.94	0.998	1.13
473 (Test)	**5e-3**	0.993	4.26	0.999	1.53
573 (Training)	**1e-2**	0.998	1.15	0.999	0.29

5 Conclusion

(1) A constitutive model based on BP artificial neural network and genetic algorithm was established, which can describe the flow behavior of 316L austenitic stainless steel at room and intermediate temperatures properly.
(2) According to the comparison between experimental data and prediction, not only the training data can get high accuracy, but the test data can be found in good performance by BP artificial neural network.
(3) When comparing the results of original BP artificial neural network with those of BP artificial neural network with genetic algorithm, it can be found that the prediction accuracy of the improved model is higher than that of original model.

Acknowledgements

This project is supported by National Natural Science Foundation of China (Grant No. 51505041) and Natural Science Foundation of the Jiangsu Higher Education Institutions of China (Grant No. 16KJB460002).

References

1. P. Li, L.K. Zeng and A.Z. Rui, Comp. Appl. Software, **150**, 184 (2008).
2. S. W. Wu, X. G. Zhou, G. M. Cao, Z. Y. Liu, G. D. Wang, Mat. Des, **15**, 2 (2017).
3. A. R. Sayyed, A. Fakhreddin, F. Saghar, Mater. Sci. Eng. A, **15**, 10 (2016).
4. M. Amirjan, H. Khorsand, M. H. Siadati, R. E. Farsani, J. Mater. Res. Technol, **4**, 2 (2013).
5. H. R. Rezaei Ashtiani, P. Shahsavari, J. Alloy. Compd, **5**, 12 (2016).
6. G. Petronea, V. Meruaneb, Compos Part A: Appl. Sci. Manufac, **94**, 3 (2017).
7. L. Wang, A. Koliosa, T. Nishinoa, P. L Delafina, T. Birdb, Compos. Struct, **153**, 10 (2016).
8. M. Shahnewaza, R. Machialb, M. S. Alamc, A. Rteilc, Eng. Struct, **107**, 1 (2016).
9. W.D. Zeng, Y. Shu and Y.G. Zhou, Rare. Metal. Mat. Eng, **10**, 33(2004).

Application of High Speed Serial Data Transmission System in Remote Sensing Camera

Zhang Ye [1,a], He Qiangmin[1] and Pan Weijun[1]

[1]Beijing Institute of Mechanics & Electricity, Beijing

Abstract. In order to meet the system needs of miniaturization, real-time, high-speed data and high reliability transmission in remote sensing camera, a high-speed serial protocol based on high-speed serial/parallel converter SerDes is proposed. For the SerDes high-speed serial data transmission chip for the physical layer, FPGA for the link layer of the communication system, design high-speed serial transmission communication protocol. Through the development of a simplified protocol for the upper user to provide a simple data interface. The system completed the data rate of 12.5Gbps point to point high-speed transmission. Verification is done by sending a pseudo-random code between boards. System work 3 hours, the measured bit error rate is less than 10^{-12} .

1 Introduction

Currently in many imaging systems and high-speed data processing systems, as the large amount of data, so the system's data transmission has high requirements. Space remote sensing instrument will get a lot of image data when working. In order to facilitate the latter part of the data processing and analysis applications, it needs for high-speed data acquisition and storage. As the amount of data is very large, the current high-speed serial differential mode is usually used for transmission[1]. This can greatly reduce the mutual crosstalk between signals and external noise interference. The adoption of new technologies to solve multi-channel and high-speed CCD / CMOS image data transmission have become an inevitable trend. High-speed differential serial bus interface technology to improve data transmission bandwidth becomes effective solution[2].

This paper introduces a new type of high-speed serial-to-parallel converter in remote sensing camera applications and SerDes high-speed serial data transmission chip Redundant XAUI Transceiver. According to the characteristics of the project designed the data transmission system of new SerDes high-speed serial data transmission chip.

2 Introduction to SerDes

The SerDes high-speed serial data transmission chip is a new high-speed transceiver device based on Serializer /DESerializer technology from Texas Instruments Inc which enables error-free transmission on 80-inch long standard backplane lines. Dual point-to-point serial transfer rate of up to 12.5Gbps. SerDes high-speed serial data transmission chip is using VML differential signal which has good anti-

[a] Corresponding author : yueyuan7763@sina.com.cn

jamming capability. It uses the self-synchronous communication with clock and data recovery technology instead of synchronous transmission of data and clock. This can resolved signal and clock offset problems. In addition, the serial communication technology to take full advantage of the transmission channel capacity, reduce the number of connector pins required and the number of chip peripheral pins. The equipment and cable wiring is more simple, and the system is more resistant to interference. SerDes high-speed serial data transmission chip is flexible, redundant XAUI serial transceiver, as shown in Figure 1, SerDes has A, B two identical XAUI port, with self-backup function.

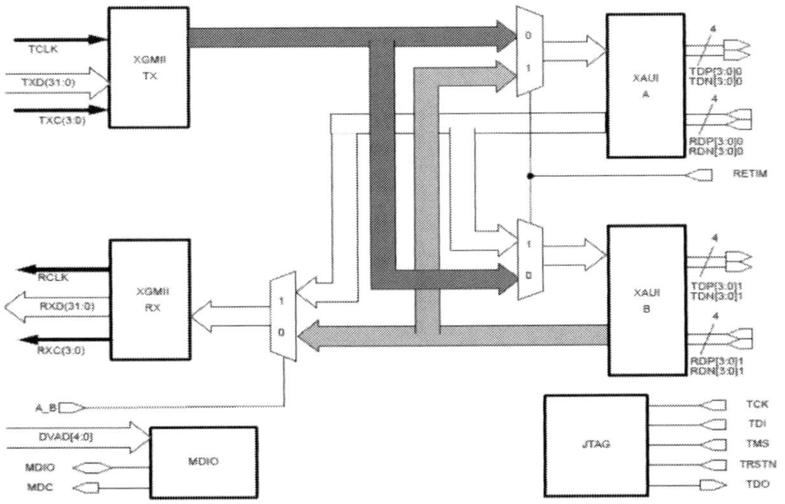

Figure 1. SerDes high-speed serial data transmission chip

The new SerDes uses full-time duplex point-to-point high-speed serial signal communication protocol. The parallel data transmitter and receiver are 4 channels. Each channel is 8 bit wide with a total of 32bit as a transmission unit.

3 System composition and processing flow

3.1 Design

The system is designed to achieve high-speed serial CCD image data transmission which mainly for the satellite payload and back-end data processing equipment between the communication interface. The high-resolution image data acquired from the satellite payload is transmitted to the back-end data processing module. The output of the SerDes is mainly for data transmission to complete the function of high-speed serializer[3].

In this paper, high-speed serial data transmission structure to replace parallel which can greatly reduce the number of interconnection lines and line interference. The number of the corresponding connector terminals and the transmitting and receiving terminal chips are greatly reduced, and the whole system is miniaturized in accordance with the current development trend. The parallel data loaded into the transmitter is transmitted to the receiver via a serial channel, which can be a coaxial cable, an impedance-controlled backplane, or a fiber-optic link [4-5].

SerDes high-speed serial data transmission interface consists of three parts, including the transmitter, transmission cable and receiver. The transmitting end includes transmitting data processing circuit, SerDes send circuit, crystal, clock management, electrical connectors and other parts. The receiving end includes receiving end data processing circuit, SerDes receiving circuit, crystal, clock management, electrical connectors and other parts.

SerDes interface using serial transmission mode, the clock frequency is 156.25MHz ± 200ppm. In the receiving end need to ensure that REFCKJ and TCLK is the same source clock, so pay special attention to the clock signal design. The clock management is used to provide a reference clock to the SerDes. The clock management is a high performance, low phase noise and low clock skew synchronizer that synchronizes the frequency of a voltage controlled crystal on a circuit board to an external reference clock. System block diagram is shown in Figure 2.

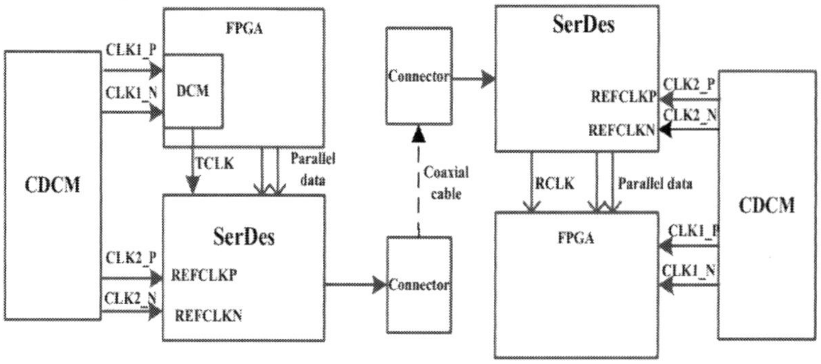

Figure 2. system block diagram

3.2 Transfer Protocol

SerDes high-speed serial data interface using point-to-point simplex communication. The data is passed on the channel as a data frame. In order to ensure that the data frame in the protocol channel in the correct transmission, you must transfer some control characters in the channel, and design a certain transmission protocol. The manner of this agreement is: After power-on or reset, both sides are out of synchronization (It requires the transmitter to reset, the receiving end to ensure that the receiver is already in the state). The sending end sends a sync character (idle character) for one millisecond. The system completes the synchronization process. After 1 millisecond, the transmitter starts the transmission of the data frame. The transmitting side is required to transmit not less than a certain number of synchronization characters before the transmission of each frame data is completed and before the transmission of the next data frame is started. And then send the next frame of data, the loop until the camera stops working. If the receiver side in the frame data transmission process out of sync, rely on data of the synchronization characters to re-establish the synchronization between the transceiver relations.

This protocol specifies that the frame data consists of frame header (S), frame end (T) and data. The start character of frame data is S (XGMII side is 0xFB), can only appear on Lane0. The termination character is T (0xFE on the XGMII side) and can appear on any Lane. The format is shown in Figure 3:

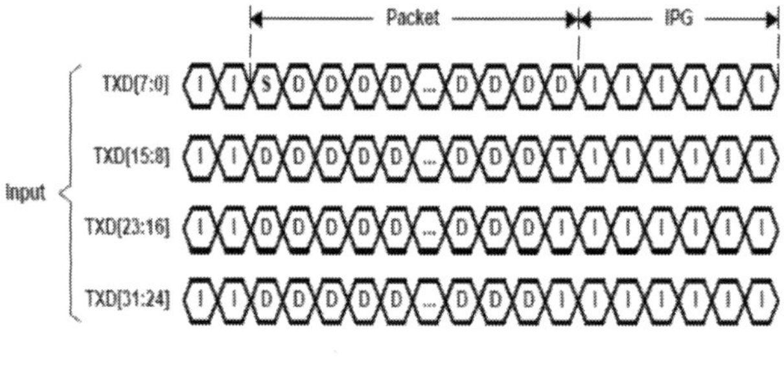

Figure 3. frame data format

The valid control characters are shown in table 1:The frame header is defined as: 07070707FFFFFFFB (64-bits converted to DDR 07070707, FFFFFFFB);The frame tail is defined as: 070707FD07070707 (64-bit converted to DDR 070707FD, 07070707);Idle characters are defined as: 0707070707070707;

 DOI: 10.1051/matecconf/201711402005

Table 1. Valid control characters

Primitive Description	K code	Control word TXC(3:0)	Data TXD[X:X-7]
SOP(S)	K27.7	FB	1
EOP(T)	K29.7	FD	1
Idle	K28.0	07	1
Valid characters	00~FF	00~FF	0

3.3 System Workflow

The SerDes high-speed serial data transmission system is first to complete the reset operation, and then enter the Loop-Back self-test mode. If the link has been synchronized to enter the next step, otherwise it has been in self-test state. After the link is synchronized, send the data sync word and receive the data sync word at the same time. The working process is in Figure 4.

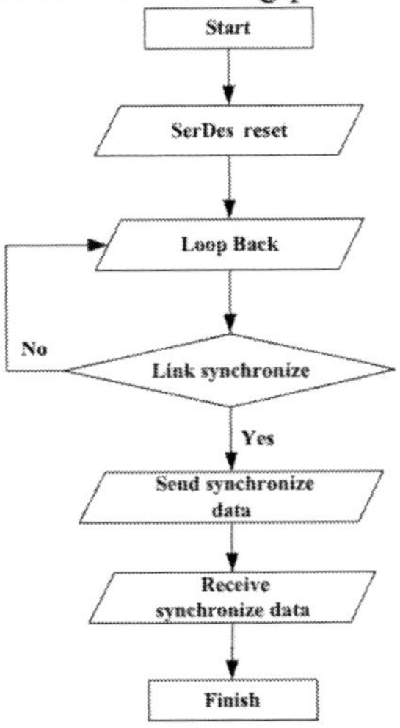

Figure 4. system workflow chart

The FPGA as the control center of high-speed serial transceiver is adjusting the control word of SerDes to configure its work state. Control word as shown in the Figure 5.

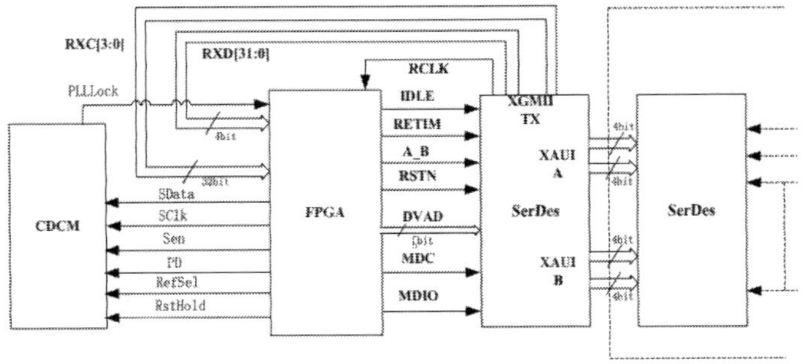

Figure 5. high-speed serial data transmission system

Among them: RSTN (Chip Reset) is a SerDes reset signal that requires at least 10us reset time. RETIM (Re-Timer Mode Enable) signal is pulled low to configured SerDes as redundant mode. IDLE signal is pulled low to mapping XGMII TX data of XAUI A and XAUI B port at the same time. By configuring A_B to choose output port. By configuring DVAD(0) to choose DTE (Device Address) or PHY(Physical Address). Whether SerDes working status is correct is judged by setting the read and write different address register values with MDIO. The read and write reference clock is provided by the MDC (Management Data Clock). Specific timing requirements are shown in Figure 6. Where tp is a minimum of 400ns, tsu is 10ns minimum, and th is 10ns minimum. So the frequency of MDC is 512ns.

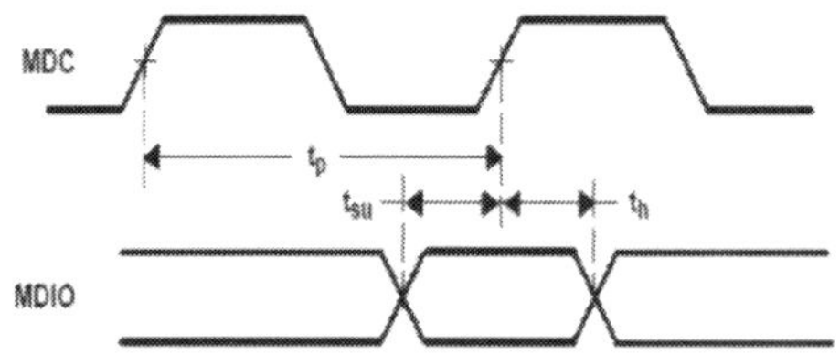

Figure 6. MDIO Read/Write Timing Diagram

First, SerDes is set to Loop Back mode as self-test mode. By writing address 0x0000 to data 0x6040, set this register to 0x6040 to check if SerDes has completed the function of spontaneous self-collection. If Self-test mode is correct, read 0x0008 address register data, check the transceiver link whether is normal. If the return value is 0x8000, then the link is normal, the link has been synchronized, if the return value of 0x8C00 said that the transceiver is not normal, if the return value of 0x8400 said the receiving link is not normal, if the return value of 0x8800 Indicates that the sending link is not working properly.

Through the state machine to control the register of read and write operations are shown in Figure 7. From the simulation waveform can be seen, STATE (1) and MDIO (1) for write register address, STATE (2) and MDIO (2) for write register data, STATE (3) and MDIO (3) for read register address. The actual application by reading and writing different address register values to check the working status of SerDes.

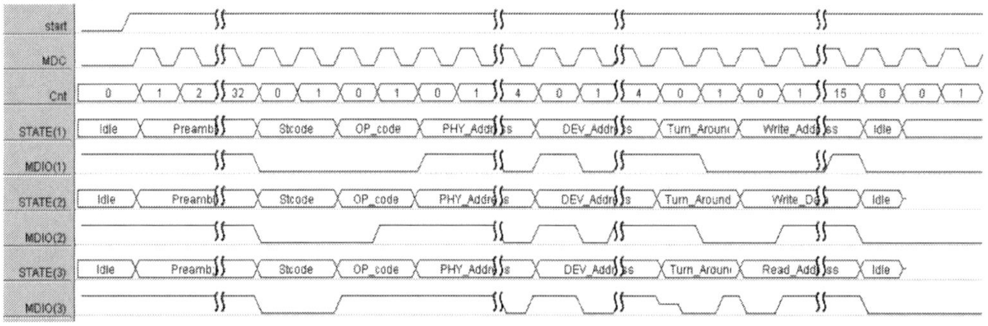

Figure 7. Data Input Timing Diagram

After checking the link is correct, data processing circuit will send parallel data with frame start and end. The SDR parallel data is converted to DDR data via ODDR, and the clock is sent to SerDes. The transmission data timing is shown in Figure 8.

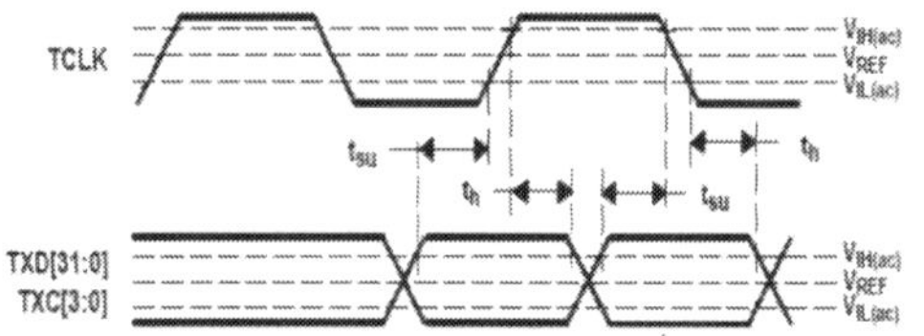

Figure8. Send clock timing diagram

Where tsu is at least 480ps and th is at least 480ps. It sends data at the sending clock signal along the transition edge. That is send the clock transition edge along the data center, the effective data in the clock after the shortest time to keep 480ps, the data valid to the clock transition edge needs at least 480ps to settle. So the frequency of tclk is 156.25M and the change frequency of data is 312.5M.

DOI: 10.1051/matecconf/201711402005

The receiver reads the data by receiving the SerDes recovery clock RCLK. The receiving circuit performs data extraction on the received data frame according to the alignment character. It converts 32-bit DDR data into 64-bit SDR data, remove the frame start and end, extract the valid data unit, and give the data flag. The received data timing is as shown in Figure 9.

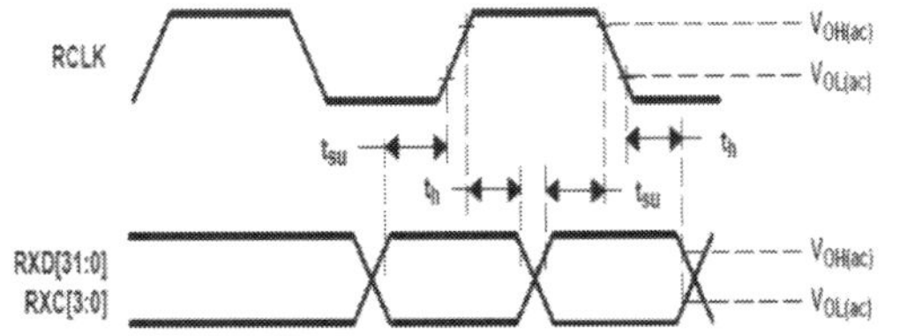

Figure 9. Restore the clock timing diagram

Where tsu is at least 960ps and th is at least 960ps. The clock edge of the receive clock signal reads the data, that is, the receive clock transition edge is aligned with the data center. Valid data is kept at a minimum of 960ps after clock hopping, and data settling time to clock hopping is as short as 960ps. SerDes's send clock and reference clock must be homologous clocks, so SerDes restores the clock based on the clock manager to give the reference clock according to SerDes which is 156.25MHz ± 200ppm.

4 Testing and verification

The reliability of the system is verified by testing the bit error rate by connecting the system to the BERT. In the test the sending end sends pseudo-random code, the receiving end monitor whether the received data is sent to the data. In the receiving end of the data processor FPGA for error counter, the error flag bits are counted, in order to statistics error rate.

System test 3 hours and sending clock is 156.25MHz. Receiver error flag bit is always high, indicating that the pseudo-random number sent and received the same, there is no error code.

5 Conclusion

In order to meet the mass data transmission requirements of high resolution camera, a SerDes serial transmission method is proposed. The system not only greatly reduces the complexity of high-speed image data transmission process, but also to ensure the reliability of signal transmission. Verification of the system compared to the traditional SerDes data transmission system transmission efficiency is higher, the transmission capacity increased by 1 times the transmission rate increased to 12.5Gbps. Single-chip new SerDes with A, B two redundant interface, comes with backup function, effectively saving space. Compared with the traditional serial conversion system, not only improve the transmission rate, but also reduces the cost of space. The test shows that the system is stable and reliable.

References

1. M. ZHANG, H. DU, H. GUAN, T. PAN, Design of Satellite High-speed Serial Data Interface Based on TLK2711, SPACECRAFT ENGINEERING, Vol.24, No.6, 13-19, 2015.
2. J.C. JIA, T. CHEN, Z. LI, High-speed serial transmission for data based on TLK2711, Application of Integrated Circuits, Vol.39, No.9, 47-49, 2013.
3. S.J. YU, J.Y. Wang, Application of Serializer/Deserializer in Remote Sensing Camera System, SPACECRAFT RECOVERY & REMOTE SENSING, Vol.33, No.6, 93-98, 2012
4. J.J. NI, T. LI, J.Y. WANG, Study of high-speed duplex communication serial data transmission based on TLK2711, Electronic Design Engineering, Vol.21, No.10, 76-80, 2013.

5. D. ZHANG, S.Y. XU, High−Speed Serial Transmission System for Image Data Based on TLK2711, Microcomputer information, Vol.25, No.1-2, 284-286, 2009.

DOI: 10.1051/matecconf/201711402006

Comparison of Machining Temperature in High Speed Grinding of Metallic Materials and Brittle Materials

Chongjun Wu[1,a], Beizhi Li[1], Jianguo Yang[1] and Steven Y. Liang[2]

[1] *Donghua University, Shanghai, China*
[2] *Georgia Institute of Technology, Atlanta, USA*

Abstract: Grinding has always been a high energy expenditure process, which generally produce massive grinding heat and possibly produce thermal damage in workpiece ground surface. Therefore, a good understanding of grinding temperature characteristics is important to lower the thermal effect on the workpiece quality. However, the grinding temperature characteristic differs when grinding of different materials, especially for those hard-to-machine materials (eg. Titanium alloys and ceramics). Therefore, this paper is devoted to investigate different grinding temperature characteristics for metallic materials and brittle materials. In order to detect the grinding temperature, a grindable thermocouple technique with NI-DAQ device is adopted in diamond wheel grinding of ceramics and CBN grinding of Titanium alloys. The results show that the temperature characteristics for ceramics is totally different with Titanium alloys. When the grinding wheel speed increases, the grinding temperature increases all the way when grinding of Titanium alloys. However, there is a temperature turning point for grinding of ceramics. When the wheel speed surpasses the turning point, the grinding temperature goes down and closes at a relatively low temperature.

1 Introduction

Grinding has been widely used as a precision machining method in various engineering application, such as aerospace, automobile and machinery industry [1-2]. However, grinding is also a complex machining process with dynamic active number, random grit topology and cutting angle. The interactive abrasive grits have typical negative rake angle [3], which will cause a high degree of material deformation and thus produce voluminous grinding heat and unpleasant residual stress [4]. Moreover, the grinding induced temperature is obviously different when machining different materials. It has been proven that a thermal shock condition of temperature 600-1500k and pressure 5–25 GPa for Silicon Carbide (SiC) will bring about a phase transition [5] and strength reduction [6]. While for the metallic materials, the ground surface is easy to get burnt to worsen the desired quality [7]. Therefore, a good understanding of grinding temperature for different materials would be helpful to minimize the thermal effect on grinding quality.

[a]Corresponding author:wcjunm@126.com

The high speed grinding process has been regarded as a high efficient method in precision machining of hard-to-machine materials. It is generally characterized by an increase of wheel velocity above 60m/s, which will substantially reduce the maximum undeformed chip thickness compared with conventional grinding of below 40m/s [8]. High speed machining was first proposed by German engineer Dr. Carl Salomon in the year 1931[9]. Based on some experiments, it was assumed that the grinding surface temperature will first increase with the increase of machining speed in high speed machining and then falls down after a turning point However, there is no consistent result on this assumption [10-13]. This is because that different machining methods and materials may cause different temperature characteristics. In recent years, lots of work has been done to investigate the grinding temperature. Thermocouples have been widely as a reliable temperature detecting method to obtain the machining temperature [14-16]. Based on grindable thermocouples, researchers have done experiments on high speed grinding of metallic materials [17-18] and brittle materials [13]. It was found that the temperature characteristics are totally different. Therefore, to understand grinding temperature of different materials, more efforts are still needed to be done.

This paper is intended to investigate different temperature characteristics when grinding different materials. The typical brittle materials ceramics and metallic material Titanium alloys will be used in this paper. High speed grinding experiments are conducted on both of these materials to detect the machining induced temperatures. Then a comparison of the process parameters on grinding temperature will be fully discussed.

2 Experiments

The grinding experiments are conducted on a high speed cylindrical grinder MGKS1332/H CNC from Shanghai Machine Tool Works Ltd. This grinder can reach up to a rotational speed up of 8000 rpm or linear velocity 160m/s. The workpiece specimens used in this paper are reaction-sintered SiC and Titanium Alloys. A vitrified diamond grinding wheel, model D91 V+2046J1SC C150 E, was used for SiC grinding. Another vitrified CBN grinding wheel, model CBN B91V175, was used for TC4 (Ti-6Al-4V) grinding. Two of these grinding wheels have a width of 22mm and diameter of 400mm. The experiments setup was given in the Fig. 1(TC4 as example). It can be found that a NIUSB-9213 was connected with a computer to detect the grinding temperature. The grindable embedded thermocouple was installed into the workpiece to detect the real time grinding temperature.

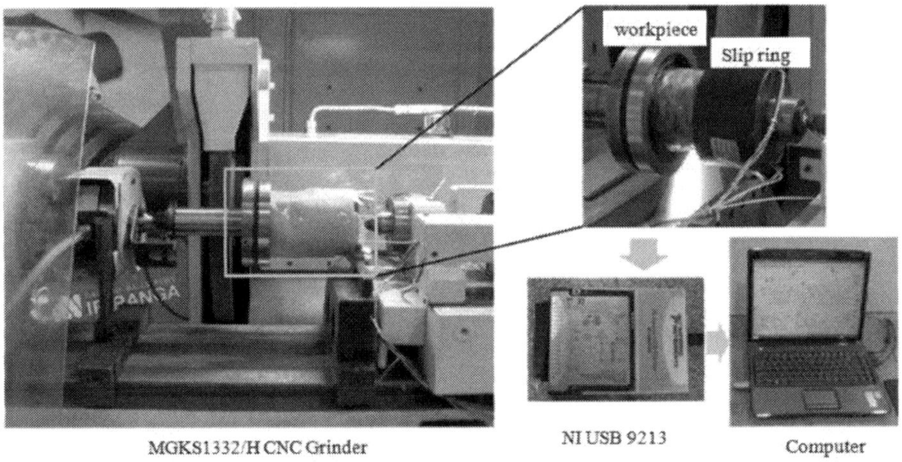

Figure1. Temperature experiments setup

3 Results and Discussion

3.1 Speed effect on grinding temperature

In high speed grinding process, the increased grinding velocity will greatly reduce the undeformed chip thickness and thus a reduction of grinding force. However, increase of grinding wheel velocity will cause a substantial increase of grinding power, which will produce more heat. The consumed grinding power can be expressed as:

$$P = F_t . V_s \qquad (1)$$

where P is the grinding power, F_t is the tangential force and V_s is the wheel liner speed. As the increase of wheel speed is higher than the reduction of tangential grinding force, thus the grinding power needed to remove a certain amount of material will increase. Thus, more heat will be produced and the grinding surface temperature is expected to increase with the increase of machining speed. However, it may not be the same in high speed grinding process. In a higher speed grinding process, the faster the machining is, the less time there will be for the heat to be absorbed by the workpiece. Moreover, the grinding chips can be helpful in taking away the produced heat. In high speed grinding, the heat can be taken away before a new chip was generated[19]. Therefore, the mechanics of grinding temperature is complicated in high speed grinding process.

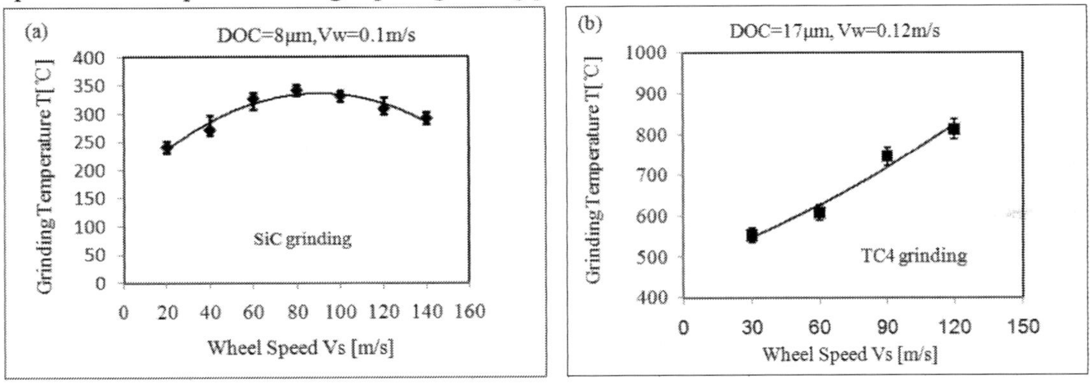

Figure 2. Speed effect on grinding temperature

Grinding experiments on SiC and TC4 was developed on a high speed grinding machine to study the actual features in grinding processes. Fig. 2 shows that grinding temperature trend under an elevation of grinding wheel speed while keeping the worktable speed and depth of cut constant. In this paper, the error bar gives the error of 10 groups of temperature data. Moreover, the solid curve was fitted by binomial model. It can be found that the temperature does not increase all the way as predicted. The temperature value for SiC is much lower than TC4. From the Fig. 2 (a), the grinding temperature of SiC first rises when the wheel speed is below 80m/s, then it goes down. However, the grinding temperature rises up all the way with the increase of grinding wheel speed (Fig. 2(b)). Therefore, it can be found that more heat was accumulated in TC4 grinding than SiC grinding. One is because of the heat dissipation for TC4 that is much slower than the SiC. The heat conductivity for SiC is about 83.6 W/(m.K) compared with 6.8 W/(m.K) for TC4, which means that the heat dissipation of SiC is 12 times faster than TC4. The other factor is more plastic deformation in TC4 grinding. As typical metallic materials, the TC4 can allow a higher elastic and plastic deformation, which is the main stage to produce machining heat. However, as the typical hard and brittle materials, SiC have a elastic modulus of more than 3 times of TC4, much stronger deformation resistance than TC4. That's why the grinding of SiC is much easier to lead a lower temperature and a shift of temperature trend.

In order to analyze the cooling efficiency in temperature effect on SiC, the grinding data was given in Fig. 3. In the Fig. 3, the comparison result of dry and wet grinding under 5% water-soluble metal cutting fluid was depicted. It can be found that the wet grinding helps to reduce the grinding temperature and it is just about 70% of the the dry grinding temperature. Moreover, the wet grinding shows the same temperature trend with the dry grinding in high speed grinding process.

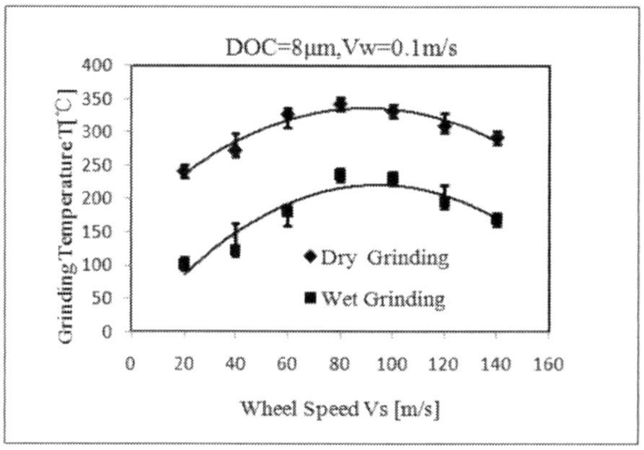

Figure 3. Cooling effect on grinding temperature

3.2 High speed efficient grinding under a lower temperature

Machining induced temperature is always one of the most significant factors that affect the machined surface quality. The material removal rate is a parameter to assess the machining economical efficiency. It can be defined as:

$$Q'_w = V_w.a_e \tag{2}$$

where Q'_w is the material removal rate, a_e is the depth of cut (DOC) and V_w is the worktable speed. The combined increase of depth of cut and worktable speed is always the main method to help increase the material removal volume. Fig. 4 gives the typical temperature trend when the wheel speed and depth of cut keep constant. From the Fig. 4 (a), it can be found that the temperature of SiC does not increase with the increase of worktable speed. However, the temperature for TC4 drops down when the worktable speed increases. The increase of worktable speed will lift the workpiece rotational speed, which will not only help to reduce the cutting time for a single grit, also improve the heat transfer rate into the air. Therefore, the increase of worktable speed is different from the increase of depth of cut, which has been proved to lead a liner increase of grinding temperature when increased [13]. Thus, in order to abtain a higher material removal volume, the increase of workpiece rotational speed will be more helpful in suppressing the grinding temperature.

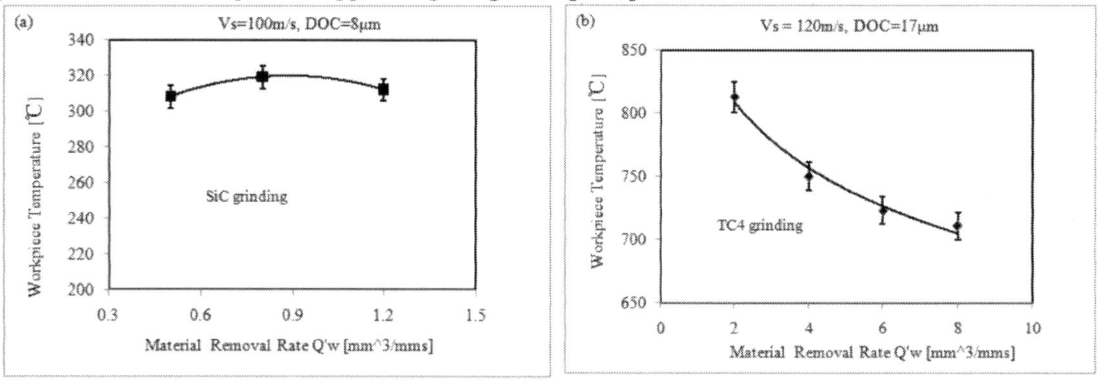

Figure4. Workpiece rotational speed's effect on temperature

In order to lower the grinding temperature, a fast shallow plunge grinding tests are deployed. The results for SiC and TC4 are given in the below Fig. 5. In the experiments, the material removal rate and wheel speed were set to be a constant. It can be seen from the data that the fast shallow plunge grinding will help substantially reduce the grinding temperature. Thus, in order to keep a stable material removal rate, a joint of the reduction of depth of cut and increase of workpiece speed will help to reduce the heat generation.

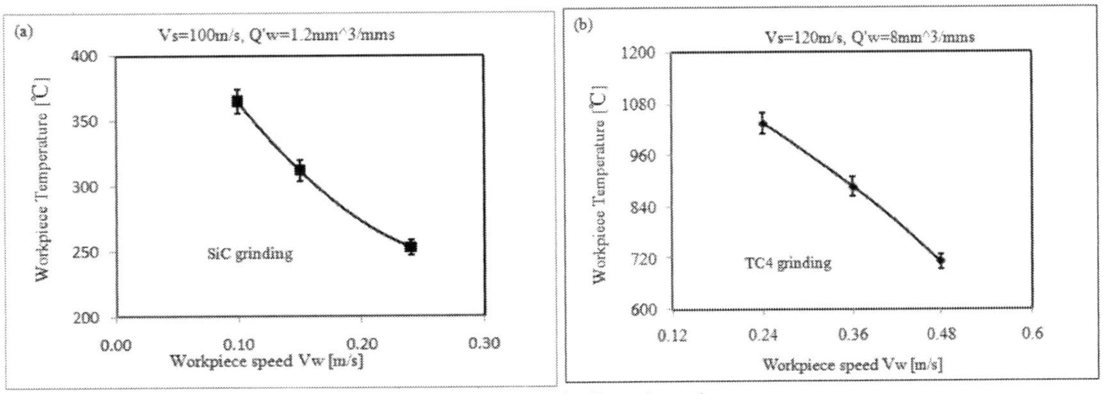

Figure5. Fast shallow plunge grinding for a lower temperature

4 Conclusions

This paper has discussed the different temperature characteristics in high speed grinding of typical brittle materials and ceramics in detail. The results show that the grinding temperature for different materials differs a lot. For the SiC, when the wheel speed is below 80m/s, the temperature rises and when it is higher than 80m/s the temperature goes down. However, the TC4 shows a liner increase all the way with the increase of wheel speed. This is more because that SiC has a higher heat dissipation rate and stronger deformation resistance. Moreover, coolants help to reduce about 30% of grinding temperature for SiC grinding. In order to achieve a higher material removal rate, the increase of the worktable speed brings out a lower temperature for the increase of heat transfer rate into the air. Finally, it is suggested that a fast shallow grinding helps to reduce both of the SiC and TC4 to suppressing a higher temperature.

Acknowledgements

This work is sponsored by the National Natural Science Foundation (51675096), National Major Projects (2013ZX04001-141), the Innovation Funds of Donghua University (CUSF-DH-D-2015100) and Morris M. Bryan Jr. Professorship for Advanced Manufacturing Systems. The authors hope to record their special thanks to the generous supports.

References

1. S. Agarwal, P. Venkateswara Rao, Int. J. Mach. Tool. Manu. 50 1077–1087(2010)
2. C.J. Wu, B.Z Li, J.Z. Pang, S. Y. Liang, J. Adv. Mech. Des. Syst. Manu. Vol.10, No.2(2016)
3. C.J. Wu, B.Z. Li, S.Y. Liang, Proce. Instit. Mech. Eng. B-J. Eng. Manu. Vol.230(8)1372-1380(2016)
4. C.J. Wu, B.Z. Li, Y. Liu, J.Z. Pang, S. Y. Liang, Int. J. Adv. Manuf. Tech. (to be published)
5. Y. Q. Zhu, T. Sekine, T. Kobayashi, E. Takazawa, J. Mater. Sci. 33, 5883– 5890(1998)
6. G.A. Schneider, [D]. Stuttgart: Universität Stuttgart (1989)
7. C.H. Zhan, W.Y. Yang, Int. J. Mech. Sci. Volume 119, Pages 125-143(2016)
8. H. Huang, K. Eng. Mater. Vol. 404, pp 11-22(2009)
9. C Salomon,Verfahren zur Bearbeitung von Metallen oder bei einer Bearbeitung durch schneidende Werkzeuge sich ähnlich verhaltenden Werkstoffen: DE, (1931)
10. H. Schultz, Proc. Win. Ann. Meet. ASME, New Orleans, , pp. 241–244(1984)
11. D. O'Sullivan, M. Cotterell, J. Mater. Process. Tech. 118 301–308(2001)
12. M. Hitchiner, Ind. Diam. Assoc. Amer.(2003)
13. C.J. Wu, B.Z. Li, S.Y. Liang, J.G Yang, Adv. Mater. Res.ols. 1120-1121, pp. 1251-1256(2015)

14. A.D. Batako, W.B. Rowe, M.N. Morgan, Int. J. Mach. Tool. Manu.45, 1231–1245(2005)
15. T. Kuriyagawa, K. Syoji, H. Ohshita, J. Mater. Process. Tech. 136,39–47 (2003)
16. A. Lefebvre, P. Vieville, P. Lipinski, C. Lescalier, Int. J. Mach. Tool. Manu. 46(14), pp. 1716-1726. (2006)
17. D.H. Zhu, B.Z. Li, J.Z. Pang, J.G. Yang, D. Zhang. ASME Int. Mech. Eng. Cong. Exp. Proceed., 9: 2035-2041(2010)
18. B.Z. Li, D.Z. Zhu, J.Z. Pang, J.G. Yang, Int. J. Adv. Manu. Tech. 54(9), pp. 931-940(2011)
19. J.M. Longbottom, J.D. Lanham, Int. J. Mach. Tool. Manu. Vol. 46, 14, pp1740-1747. (2006)

MATEC Web of Conferences **114**, 02007 (2017) DOI: 10.1051/matecconf/201711402007
2MAE 2017

Molecular Dynamics Study of Fluid Solid Interfacial Slip and Its Effect on Aerodynamic Drag

Asef Mohammed[a] and Jeetu S. Babu

Department of Mechanical Engineering, Amrita School of Engineering, Amritapuri, Amrita Vishwa Vidyapeetham, Amrita University, India

Abstract. Interfacial hydrodynamic slippage is controlled by two factors say physical structure and chemical composition. Various studies have been conducted experimentally which try to connect the physical structure of the surface and its chemical property on the interfacial wettability. One such example is the Tunable wettability in surface-modified ZnO-based hierarchical nanostructures [2]. In which vertically aligned Nanoneedles and Nanonails were employed as a platform to determine the effect of surface structure. According to which a variation in static contact angles were observed as the cap size the nanonails constantly increased. Starting with a contact angle of 104° the contact angle first increases and then decreases, which means that the slip length first increases and then decreases. The increase in slip length reduces the drag, which has immense application in the aerodynamic field. This paper investigates the relation between the chemical wettability and aerodynamic drag by performing MD simulations of couette flow with varying fluid-surface interaction.

1 Introduction

The fluid drag force acting on a moving solid body in the direction of the free stream flow of the fluid is termed as aerodynamic drag in aerodynamics. The forces due to pressure distribution over the surface result in drag. Reducing the drag plays a major role in increasing the stability of a moving body, It becomes significant when aerospace is the subject under consideration. You and Moin showed that the slip velocity present in hydrophobic surfaces could result in a significant drag reduction [1]. It could be concluded from their work that higher the slip length, higher will be the reduction in drag and rms lift. Since it is clear that slip and drag are directly related, in this work we investigate methods to manipulate slip in order to reduce drag with the help of surface modifications.

Various studies have been conducted experimentally as well as numerically which try to connect the physical structure of the surface and its chemical property on the interfacial wettability. One such example is the tunable wettability in surface modified ZnO-based hierarchical nanostructures by Li et. al[2]. In which vertically aligned nanoneedles and nanonails are used to study the deviation of surface structure, according to which a variation in static contact angles were observed as the size of the nanonail cap constantly increased. Starting with a contact angle of 104^{0} the contact angle first increases and then decreases, which means that the slip length first increases and then decreases. In a numerical work done by Bakli and Misra [3] on the comparability of chemical composition and

[a] Corresponding author: asefmohammed3@gmail.com

roughness of Nano channels in varying slippage of fluid at the interface through molecular dynamics (MD) simulations, by developing a typical water model the slippage of fluid through smooth and rough surfaces were analysed. The authors proposed a mathematical relation between static contact angle, surface roughness and slip length by determining the contact angles and slip length at different conditions. In a similar work done by Bakli and Chakraborty [4] they have studied the transition of slippery to sticky hydrophobic nano channels. They showed that a simple manipulation of the ionic and molecular positioning at the wall- fluid interface can be used to tune the actual wettability of the nano fluidic substrate. In this work they have opened up a new platform for studying the slip-stick behaviour by making the classification of hydrophobic and hydrophilic surfaces slightly imprecise, which they made possible by using colloidal and ionic inclusions in the fluid which considerably affected the dynamic contact angle.

The objective of this paper is to reduce drag by relating it with slip by applying surface modifications. In order to support the experimental work done by You and Moin [1] we have tried to mimic the idea of nano needles and nails by creating cylindrical and square shaped surface asperities by varying the contact angle. Further studies are also done by varying the wall-fluid interaction in order to change the contact angle and hence to manipulate slip length.

2 Methodology

This work is done using MD simulation [5]. MD can be explained as a computational tool where positions, velocities and orientation of molecules and atoms evolve over time. It is one of the most widely used atomistic simulation method. It can be considered as a computational microscope, where the behaviour of atoms and molecules are observed using a computational technique instead of using a conventional microscope. With the help of MD we can get atomistic insight into molecular structure and kinetics. It is a deterministic method, given a preliminary set of position and velocities the following time evolution is totally calculated so that the entire trajectory of the atoms could be obtained. MD follows laws of classical mechanics most notably Newton's law: $F = M \times a$, where F is the net force acting on the atom, M is the mass of the atom and a is the acceleration. The total force acting on the atom can be calculated from the interaction between the atom and its neighbouring atoms. This can be modelled using an interatomic potential function. The force acting on the atom is defined as the first derivative of the interatomic potential function with respect to position. Once the total force acting on the atom is obtained, the acceleration can be calculated and from the acceleration of the atom the positions and velocities can be obtained by time integration. So, in MD we are actually solving the Newton's law for obtaining the trajectory of the atoms. From the atom trajectory the properties of the system can be calculated with the help of statistical mechanics relation.

Hence, Running a simulation in MD can be summarised as follows

1. Set an initial system configuration [$\vec{r}$(t=0),$\vec{v}$(t=0)]
2. Get an interaction potential for the system v($\vec{r}$)
3. Find a way to integrate $F = M \times a$ and get new forces F
4. Solve the equation of motion numerically over a short step Δt
5. Get the new positions and velocities over the step Δt
6. Continue step 1-5 till t > t_{max}
7. Calculate the results and stop

The basic flowchart for MD simulation is shown in Fig. 1

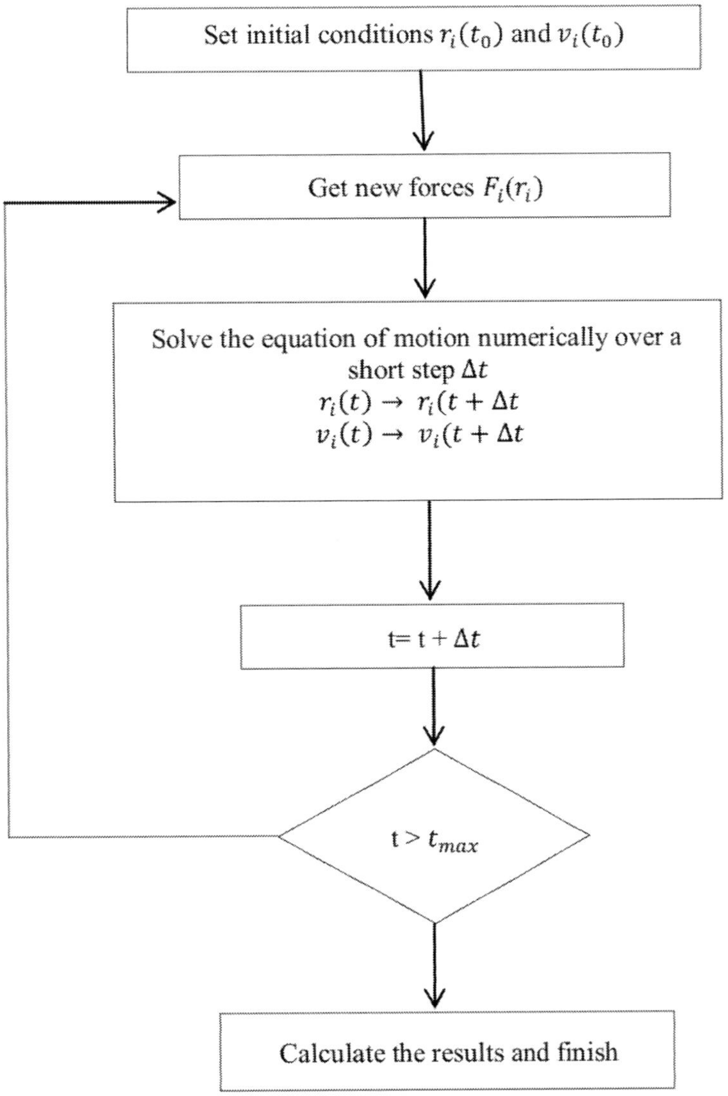

Figure 1. MD simulation flow chart

The MD simulation is implemented using LAMMPS open source library package [6,7] and the visualisation of the same is done using OVITO [8] which is also an open source software. The simulation system consists of a channel of height 1nm bounded by rigid solid walls in close pack face centred cubic (FCC) structure. Argon and platinum atoms constitute the fluid and wall respectively. The complete dimension of the system is 8x3.2x8 nm as shown in Fig. 2. The system is designed for Couette flow so that the upper plate moves with a small velocity V and the lower plate is kept stationary.

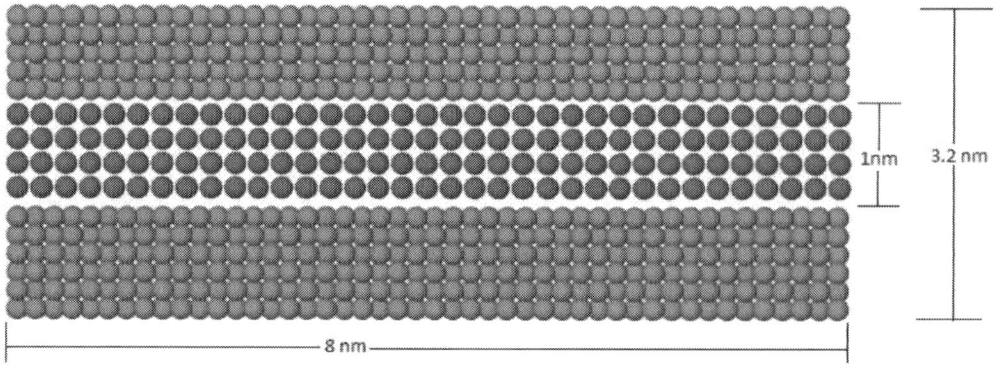

Figure 2. Couette flow simulation system in two dimensional view

Lenard Jones (LJ) interaction potential is used to model the wall-wall, fluid-fluid and wall-fluid interactions.

$$\phi LJ\,(r) = 4\varepsilon\left[\left(\frac{\sigma}{r}\right)^{12} - \left(\frac{\sigma}{r}\right)^{6}\right] \tag{1}$$

The interaction parameter between the atoms of the wall and fluid are as shown in Table-1 and ε_{wf} is varied to change the wall fluid interaction and hence the contact angle. The second part of the work is done by creating square and cylindrical surface asperity in the lower wall so us to vary the contact angle. Fig. 3 shows the simulation of couette flow with surface asperity. The complete dimension of the system is 8x3.2x8 nm.

Table 1. Interaction parameters

	ε(kcal/mole)	**σ(A°)**
Wall-wall	0.120183	2.475
Fluid-fluid	0.238	3.04
Wall-fluid	-	2.7575

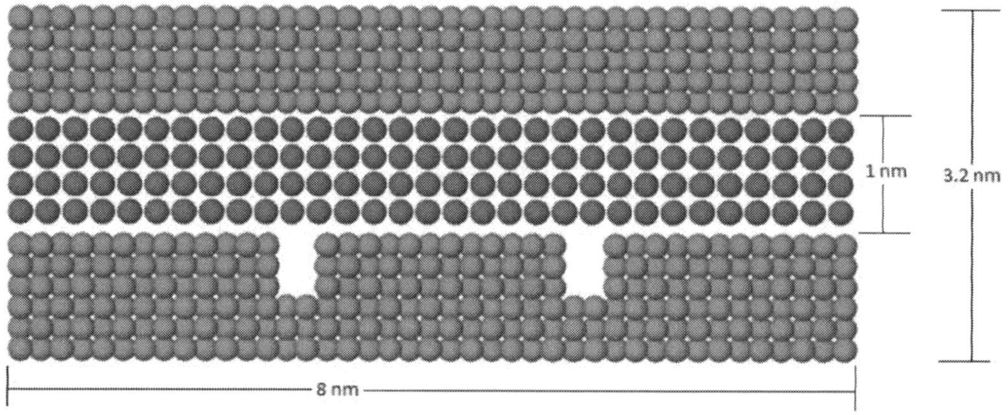

Figure 3. Couette flow simulation system with surface asperity in two dimensional view

Extrapolating the velocity profile obtained through MD simulation of Couette flow to meet the velocity of the wall will give the slip length. The numerical calculation of slip length was done by using the formula $V_s = L_s \times \dot{\gamma}$, where V_s is the slip velocity; L_s is the slip length and $\dot{\gamma}$ is the slope. The total energy exerted by the wall on the fluid slip layer is also calculated in every case. The relation between total energy and slip length is also studied.

3 Results and Discussions

All the values obtained are represented with respect to the values obtained for $\varepsilon_{wf} = 0.169$ kcal/mole. First part of the work was done by varying the wall fluid interaction in order to vary the contact angle and the variation of slip length observed is as shown in the Fig. 4 and the variation of total energy observed is as shown in Fig. 5

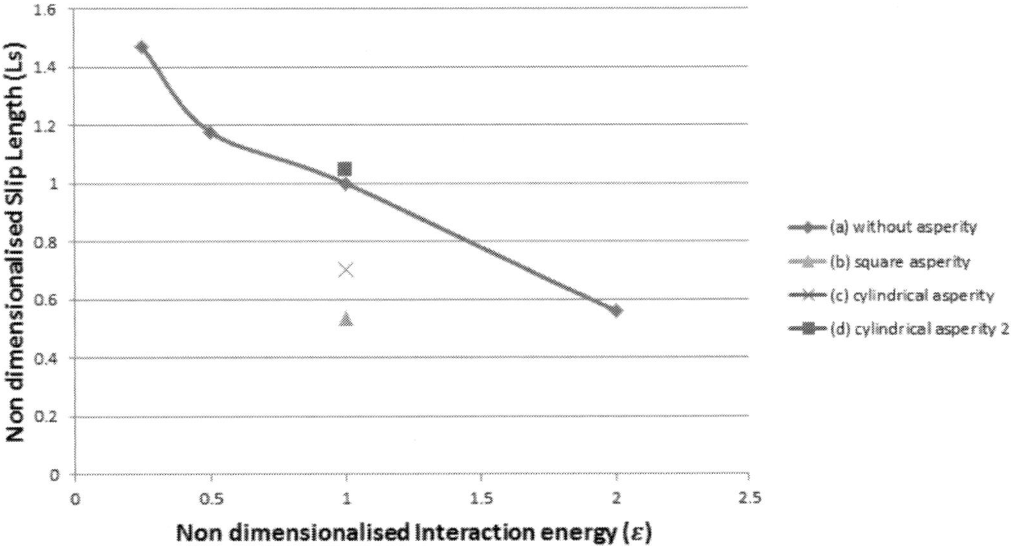

Figure 4. Variation of slip length with respect to changes in the wall-fluid interaction energy (♦), Changes in shape of the surface asperities [square (▲) and cylinder (×,■)] and also changes in the surface area of the asperity [×,■].

As shown in Fig. 5 when the wall-fluid interactions were reduced, the total energy exerted on the fluid atoms by the wall was found to reduce as well, as a result of which the fluid transformed from philic to phobic hence, the slip length increased. The increase in slip length with reduction in wall-fluid interaction (ε) is shown in Fig. 4. The result obtained when surface asperity was created in order to change the contact angle is also shown in the Fig. 4.

Figure 4 shows the variation of slip length with respect to flat plate, cylindrical asperity and square asperity for a particular value of ε. The slip length is found to reduce from what was obtained from flat plate, when asperities were created. This can be explained due to the change in contact angle such that when the contact angle increases the slip length reduces. A further reduction in slip length was found when the asperity shape was changed from cylindrical to square by assuring the same surface area. This effect can also be explained as a result of change in contact angle which can be supported by the work done by Li et. Al [2].

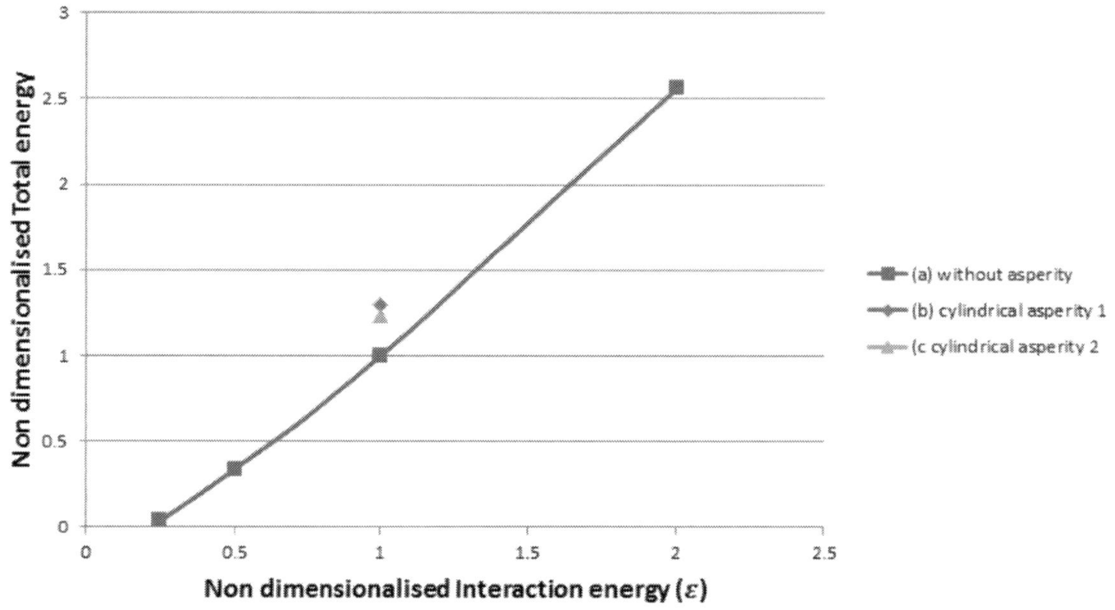

Figure 5. Variation of total energy with respect to change in the wall fluid interaction energy (■) and also change in the surface area of the asperity (♦, ▲).

In an extended analysis done by reducing the surface area of the cylindrical asperity, a further reduction in total energy and an increase in slip length was found. Due to the reduction in contact angle the total energy exerted on the fluid by the wall will increase as shown in Fig. 5 and as a result due to change in phobic to philic nature the slip length will reduce as shown in the Fig. 4.

4 Conclusion

Through MD simulation of Couette flow, the slippage of fluid through hydrophobic surfaces were studied. By measuring slip length for various conditions of varying contact angles the manipulation of slip was made possible. Varying the wall-fluid interaction energy as well as creating surface asperity were tried in order to change the contact angle. The manipulation of slip length can considerably effect the drag which could be used for aerodynamic applications.

References

1. D. You, P. Moin, Physics of Fluids **19**, 081701 (2007)
2. G. Li, T. Chen, B. Yan, Y. Ma, Z. Zhang, T. You, Z. Shen, H. Chen, T. Wu, Appl. Phys. Lett. **92**, 173104 (2008)
3. C. Bakli, A. Misra, Physics.flue-dyn, arXiv :1605.02479 (2016)
4. C. Bakli, S. Chakraborty, Nano Lett. **15**, 7497-502 (2015)
5. C. Rapaport, The Art of Molecular Dynamics Simulation, Cambridge university press New York, 2004
6. S. Plimpton, Fast Parallel Algorithms for Short-Range Molecular Dynamics, J Comp Phys, **117**, 19 (1995).
7. http://lammps.sandia.gov
8. A. Stukowski, Visualization and analysis of atomistic simulation data with OVITO - the Open Visualization Tool Modelling Simul. Mater. Sci. Eng. **18**, 015012 (2010)

MATEC Web of Conferences **114**, 02008 (2017) DOI: 10.1051/matecconf/201711402008
2MAE 2017

Effect of Rare-earth Element Addition on Structure of Heat-resistant Ti–6.5Al–4Zr–2.5Sn–2.4V–1Nb–0.5Mo–0.2Si Titanium Alloy

Svetlana Skvortsova[1], Ivan Grushin[1], Oksana Umarova[1] and Konstantin Speranskiy[1,a]

[1]*Moscow Aviation Institute (National Research University), 125993 Moscow, Russia*

Abstract. To guarantee the requirements for modern technology it is necessary to create new alloys with enhanced characteristics or upgrade existing ones. Especially it concerns high-tech industries, particularly aviation and rocket engineering. When designing and creating modern aircraft engines, a special place is devoted to heat-resistant titanium alloys. High strength characteristics and relatively low density make it possible to create alloys with outstanding specific parameters. However creation of new alloying systems is quite expensive and implies parallel development of technological process as well as various processing modes. There are several ways to solve this problem; one of them is modification of already existing alloy. Due to the unique combination of physical-chemical properties of rare earth metals adding them to existing alloying systems makes it possible to create alloys with increased characteristics. In particular, microalloying of titanium alloys with gadolinium leads to transformation of their microstructure and increase of some mechanical characteristics. Thus, microalloying of the high-temperature Ti-6.5Al-4Zr-2.5Sn-2.4V-1Nb-0.5Mo-0.2Si titanium alloy in as-cast state results in a change in the intragranular structure, but does not significantly affect the macrostructure of the ingots. At the same time, in deformed state increase of gadolinium content leads to refinement of the initial β-grain, growth and formation rates of the primary and secondary phases change, which in its turn affects the mechanical characteristics of the alloying system as a whole. The Rockwell hardness measurements carried out using the BUEHLER Macromet 5100T device show an increase of parameters in the alloy with gadolinium.

1 Introduction

To modify the existing alloys of different grades and create new alloys the lanthanides are used at present as additive agents at microalloying. Lanthanides or, as they are also called, rare-earth metals (REMs) [1-3] became widespread for microalloying due to the unique combination of physical-chemical properties they possess in full. They are obtained during reprocessing of the ore which contains minerals in form of mixtures consisting of REM oxides, phosphates, carbonates or silicates and also thorium, yttrium, scandium, titanium, niobium and tantalum similar to them by the properties [2-4]. REMs are many times more in the nature than zinc, copper, lead, tin and noble

[a] Corresponding author: speranski.konstantin@mail.ru

metals. Closeness of REMs in their physical-chemical properties is explained by the identical structure of the outer electron shells of their atoms. All of them possess the three outer-shell electrons; particular REMs differ in number of electrons on $4f$ level. These electrons are shielded from the common external influence by the surrounding closed shells and therefore have practically no effect on chemical activity of REMs. Slight differences in their properties are generally connected with the differences in the size of an atom nucleus [3-5].

All REMs possess an average refractoriness (cerium, Ce – 800°C; gadolinium, Gd – 1350°C; lutetium, Lu – 1700°C) [5-7]. In solid state they have typical for metals hexagonal closed-packed or cubic crystal structures [8]. The majority of them is polymorphous and differs by closeness of atomic and ionic radii (lanthanide compression) [8-10]. In their pure form REMs are plastic and can be obtained wire-shaped or sheet-shaped. REMs possess strongly negative electrochemical potential and therefore are good reducers. All of them are oxidized in air at the room temperature with different rapidity [8-10]. A set of such unique physical-chemical properties, close electronic and crystal structures makes it possible to predict improvement of parameters for titanium alloys. One of the main goals in development of heat-resistant titanium alloys is to increase operating temperatures for the components made of them as well as to improve lifetime, plastic fatigue strength and creep resistance. To solve these problems the required structure is needed that can be formed by alloying with rare-earth elements. Therefore in this work we considered the effect of gadolinium on structure and properties of heat-resistant alloy in as-cast and deformed conditions.

2 Materials and procedures

The ingots of a pilot Ti–6.5Al–4Zr–2.5Sn–2.4V–1Nb–0.5Mo–0.2Si alloy with different gadolinium content – without Gd (Composition 1), with 0.04% of Gd (Composition 2) and 0.18% of Gd (Composition 3) – were smelted to study the effect of Gd on structure formation and properties.

The pilot ingots were manufactured by triple remelting method. After a second remelting the ingots were subjected to mechanical treatment to remove corona, surface defects and pipe. The crystallizer with diameter of 160 mm was used for a third remelting which allowed improving chemical composition homogeneity and decreasing hydrogen and chlorides content. After a third remelting corona and pipe were removed, and then the ingots were turned for the further deformation.

Metallographic investigations were carried out using the Axio Observer A1m optical microscope with the Image Expert Pro3 digital software intended for acquisition and analysis of the images. X-ray diffraction analysis was carried out using the DRON-4 and Dron-7 diffractometers in filtered CuK_{α} radiation.

Hardness was measured according to the Rockwell method in HRC scale using the BUEHLER Macromet 5100T device by diamond cone with a vertex angle of 120° and the load of 1500 N.

3 Results and discussion

3.1 Structure and phase composition of a pilot alloy in as-cast condition.

Structures of casted alloys with the three compositions are presented in Fig. 1; the samples were cut from the upper part of the ingots. It was stated that the additional Gd alloying resulted in a change of the intragranular casted structure – gradual reduction of α-phase plate's length was observed and accompanied by gradual formation of a cage-netting structure similar to the Widmanstätten one with the growth of Gd content (Fig. 1).

 DOI: 10.1051/matecconf/201711402008

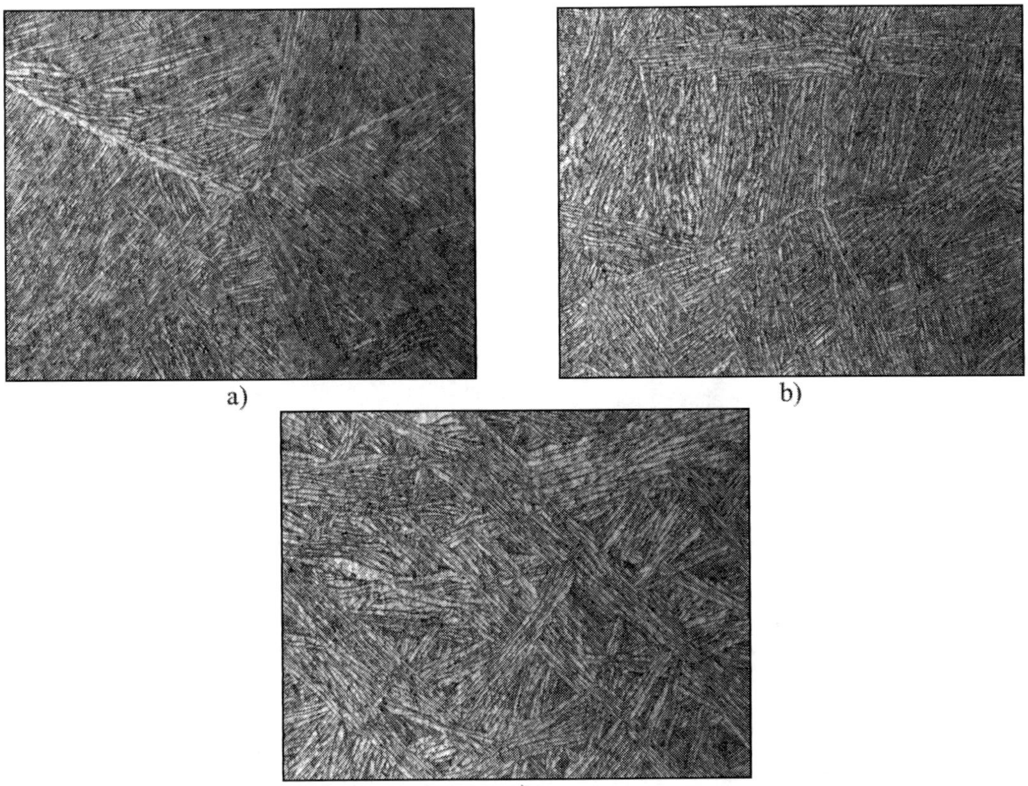

a) b)

c)

x100

Figure 1. Microstructure of the samples made of Ti–6.5Al–4Zr–2.5Sn–2.4V–1Nb–0.5Mo–0.2Si alloy without Gd (a), with 0.04% of Gd (b) and 0.18% of Gd (c).

X-ray diffraction analysis of the samples cut from the ingots with the different chemical composition showed that the phase composition corresponded to near-α-titanium alloys: the substrate was presented by α-matrix with small (less than 5%) volume fraction of β-phase. No additional reflections corresponding to the other phases were identified.

Despite of changes in the intragranular structure we did not observed any visible changes in macrostructure of the ingots (i.e. no changes of grain size) under increase of Gd content (Fig. 2).

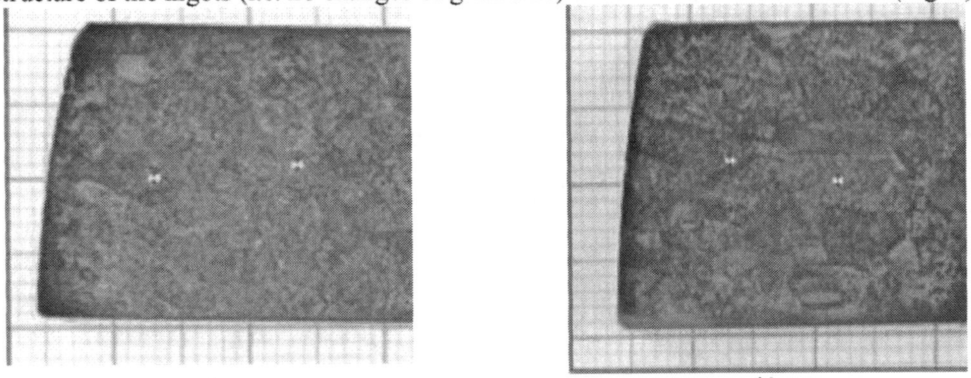

a) b)

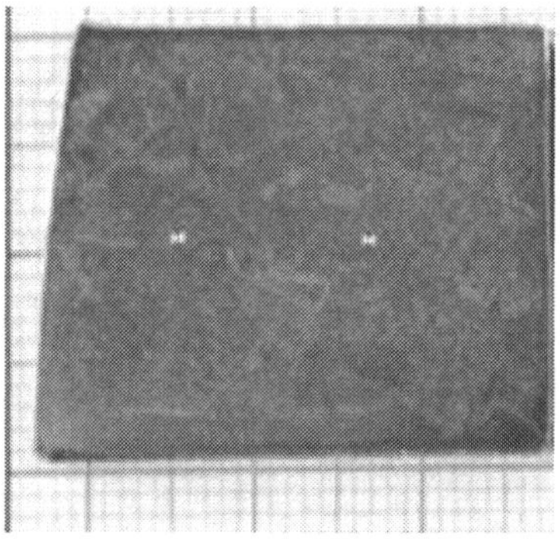

c)

Figure 2. Macrostructure of the ingots from Ti–6.5Al–4Zr–2.5Sn–2.4V–1Nb–0.5Mo–0.2Si alloy without Gd (a), and with 0.04% of Gd (b) and 0.18% of Gd (c).

To evaluate the influence of Gd content on the initial β-grain's size the samples cut from the upper part of the ingots were quenched from the temperature of 1060°C which is to be corresponded to a single β-phase region for all of the compositions according to preliminary evaluation based on reference data. The grain's size was evaluated by the Axio Observer A1m optical microscope with the use of the Image Expert Pro3 software; measurements were carried out in the two orthogonally related directions in more than 15 fields of view.

An average size of the initial β-grain equaled approximately 1700 μm in Ti–6.5Al–4Zr–2.5Sn–2.4V–1Nb–0.5Mo–0.2Si alloy without Gd; in the alloy with 0.04% of Gd the initial β-grains were greatly prolonged and had the average width of about 1550 μm and the length of more than 2000 μm; and at 0.18% of Gd the average size of the initial β-grain equaled about 1700 μm as in the alloy without Gd. Thus the conducted investigations showed that the additional alloying of Ti–6.5Al–4Zr–2.5Sn–2.4V–1Nb–0.5Mo–0.2Si alloy with Gd had no significant influence on the initial β-gran's size in as-cast conditions.

3.2 Investigation of deformed alloys

On the next stage of the work hot-rolled plates were made from the experimental ingots.

Rolling of the casted workpieces to the width of 13 mm was carried out at the temperature of 980°C after holding during 30 minutes at this temperature. Total degree of deformation equaled 70%. Rolled workpieces were air-cooled to the room temperature and subjected to mechanical treatment.

Samples cut from the rolled workpieces were used to determine polymorphic transformation temperature (PTT) by the trial quenching method. To this effect the samples were heated up to the temperature range of 960°–1060°C and water-cooled after isothermal holding during 30 minutes at these temperatures. Phase transition temperature was determined by metallographic method according to the occurrence of primary α-phase particles. The studies showed that PTT equaled 1000°C in Ti–6.5Al–4Zr–2.5Sn–2.4V–1Nb–0.5Mo–0.2Si alloy without Gd. Increase of Gd content resulted in gradual growth of PTT up to 1020° and 1040°C for the alloys with 0.04% and 0.18% of Gd, respectively (Figure 3).

970°C

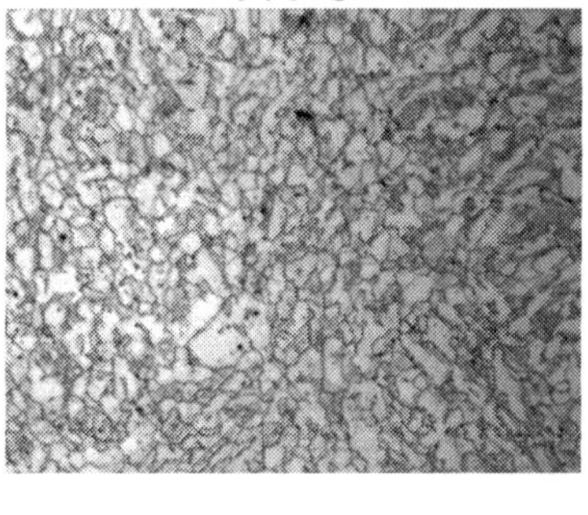

990°C

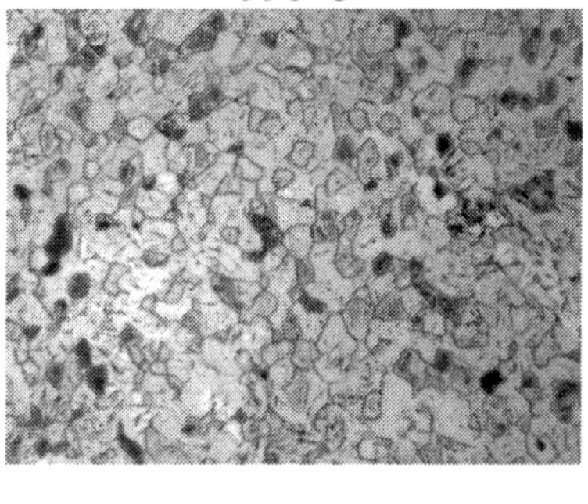

1010°C

×200

Figure 3. Microstructure of the samples made of the alloy with Composition 3 after different quenching temperatures

Determination of PTT allowed developing proper modes of rolling to obtain plates with optimal structure from casted workpieces with the different compositions.

Structure of the rolled plate with Composition 1 presented typical structure of heat-resistant near-α-titanium alloys after rolling near PTT (Fig. 4 a). Additional alloying with Gd resulted in decrease of the primary α-phase volume fraction and increase of the secondary α-phase formed in β-phase during cooling from rolling temperature down to the room temperature (Fig. 4 b and c).

Composition 1 Composition 2

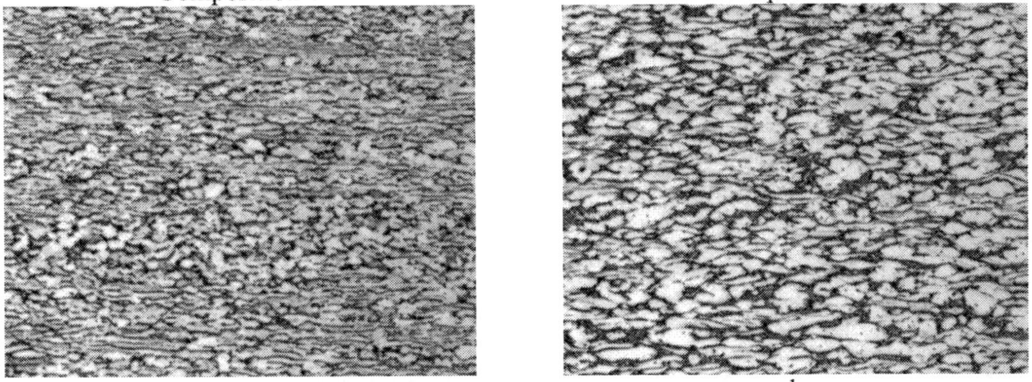

a b

Composition 3

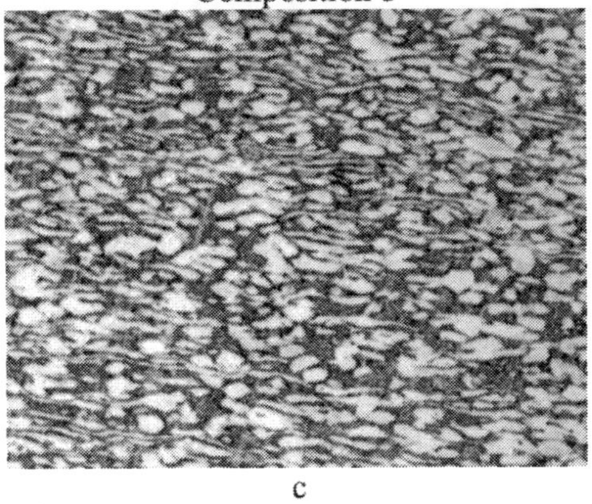

c

×200

Figure 4. Microstructure of plates made of the pilot alloys with different chemical compositions after rolling

Rolling of the workpiece with Composition 2 began at the temperature by 40°C lower than PTT, and for the workpiece with Composition 3 it was by 60°C lower than PTT. Therefore at rolling temperature structure of the alloys with Compositions 2 and 3 consisted of greater quantity of α-phase than the alloy with Composition 1. Evidently the difference in structure is related to the presence of Gd. Cooling to the room temperature after rolling of the workpiece with Compisition 1 results mainly in the growth of existing primary α-phase particles, but in case of Compositions 2 and 3 decomposion of β-phase occurs with precipitation of dispersed secondary α-phase. It is formation of more dispersed structure that contributes to increase of workpieces hardness after hot plastic deformation: from 37 HRC units for Composition 1 up to 39.5 HRC units for Composition 3 (Table 1).

Table 1. Hardness of rolled plates made of the pilot alloys depending on Gd content

Gd content	Hardness, HRC units
Composition 1 (without Gd)	37.0
Composition 2 (0.04% Gd)	38.0
Composition 3 (0.18% Gd)	39.5

As it was mentioned above the additional alloying of heat-resistant Ti–6.5Al–4Zr–2.5Sn–2.4V–1Nb–0.5Mo–0.2Si alloy with Gd did not lead to any significant change of the initial β-grain size in as-cast condition. Therefore on the next stage of the work we evaluated the initial β-grain size after hot plastic deformation. Samples cut thereto from the rolled plates of different compositions were quenched from the temperature of 10°C higher than the corresponding PTT, i.e. from 1010°, 1030° and 1050°C for Compositions 1, 2 and 3, respectively. Measurements were made in the two orthogonally related directions in more than 20 fields of view.

Investigation carried out showed that the addition of Gd had an impact on the β-grain size in deformed semi-finished products: in the alloy containing 0.18% of Gd the grain size was 3.5 times less than in the alloy without Gd.

4 Conclusions

1. Alloying of the heat-resistant Ti–6.5Al–4Zr–2.5Sn–2.4V–1Nb–0.5Mo–0.2Si titanium alloy with gadolinium has no influence on initial β-grain size in as-cast condition but contributes to transformation of intragranular structure and changes its type.
2. Addition of gadolinium results in gradual increase of polymorphic transformation temperature in the pilot alloy at the growth of Gd content: from 1000°C in the alloy without Gd up to 1040°C in the alloy containing 0.18% of Gd.
3. Addition of gadolinium to the Ti–6.5Al–4Zr–2.5Sn–2.4V–1Nb–0.5Mo–0.2Si alloy contributes to grain refinement in deformed state.

Acknowledgements

Equipment support from the Centre of Collective Use "Aerospace Materials and Technologies" of MAI is gratefully acknowledged.

References

1. Pod red. Grigor'eva I.S., Meilikhova E.Z.. *Physical quantities: Handbook*, Ed. Grigoryev I.S., Meilikhov E.Z.. // M.: Energoatomizdat, S.. – 1232 (1991)
2. Il'in A.A., Kolachev B.A., Pol'kin I.S.. *Titanium alloys. The composition, structure and properties. Handbook.* // M.:VILS-MATI,. 520 s. (2009)
3. Siemers C., Brunke F., Laukart J.,. Hussain M.S, Rösler J., Saksl K., Zahra B.. Rare earths, **1**, 281 – 292, (2012)
4. Holm M., Ebel T., Dahms M.. Materials and design.. Volume **51**. P. 943-948 (2013).
5. Brunke, Waalkes, Siemers. Engineering and Technology International Journal of Chemical, Molecular, Nuclear, Materials and Metallurgical Engineering.. Vol. 8. No.**11**. (2014)
6. K Xia, W Li, C Liu. Scripta Materialia.. Vol. 41. Issue **1**. P. 67-73. (1999)

7. Nie J.F., Majumdar A., Muddle B.C.. Materials Science and Engineering : A. 1 May. Vol. 179-180. Part **1**. P. 619-624, (1994)
8. Morteza Hadi, Mahmood Meratian, Ali Shafyei. Journal of Alloys and Compounds. Volume **618**. P. 27–32. (2015).
9. Skvortsova S.V., Grushin I.A., Mamontova N.A., Speranskii K.A., Slezov S.S. Nauchno-tekhnicheskii zhurnal «Titan». **№2 (52**). S.12-18. (2016).
10. Skvortsova S.V., Grushin I.A., Speranskii K.A., Demakov A.A., Mamontova N.A. Nauchno-tekhnicheskii zhurnal «Titan». **№1**. S. 4-9. (2017)

Research on nanosatellite thermal cycling test applicability

Xiyuan Li[1,2,a], Xiaoning Yang[2], Yonghong Shang[2], Yanqiang Bi[2], Jing Wang[2] and Guiping Lin[1]

[1]*Beihang University, Beijing, China*

[2]*Beijing Institute of Spacecraft Environment Engineering, Beijing, China*

Abstract: In order to verify the spacecraft performance in extreme temperature and vacuum, and to screen spacecraft early defect, generally spacecraft TV (Thermal Vacuum) test should be carried out before launch. Designed in small size and with low cost, nanosatellite is made from a large number of COTS (Commercial off the shelf) components; therefore, the test should be low-cost, simple and quick. With the intention of screen out early defects of the product in lower cost, nanosatellite developers usually use TC (Thermal Cycling) test to partially replace the TV test because TV test is more expensive. However, due to the air convection, TC test is different from TV test in heat transfer characteristics, which may be over-test or short-test in TC test. This paper aims to explore the applicability of different nanosatellites in TC/TV test. Using rule number analysis method, Heat Transfer model in vacuum and ambient environment has been built to analyse the characteristics of heat transfer under different temperature and characteristic length, and to deliver the recommended limits on using TC test instead of the TV test. The CFD and test methods are applied to verify the rule number analysis above.

1 Introduction

Limited by cost, time and other constraints, nanosatellites developers generally avoid long time TV (Thermal Vacuum) test in TV chamber. Some developers shorten the test by reducing the number of cycles or soak time; while others try to simplify the TV test by vacuum test, TC (Thermal Cycling) test and high and low temperature start-up test[1]. Firstly, a vacuum test, including system electrical performance test, is carried out in a small vacuum chamber (no 1: 2 limit) to verify the ability of the system working in vacuum and at high temperatures. Secondly, a TC test (including TC tolerance and TC performance) is applied to screen early defects of the system and to verify the system performance at

[a] Corresponding author: lxy_422@msn.com

low temperatures. Finally, a hot and cold start-up test is employed to verify the start ability of the system at high and low temperature [2]. To sum up, a large, complex, high-cost test could be cut into multiple small, simple, low-cost test in nanosatellite verification; as showed in Table 1, different tests meet different purposes of TV test.

Table 1. Test replacement

TV Test Purpose	Nanosatellite test
Screening early defects	TC test
Verify the ability to work at high temperatures coupled with the vacuum environment	vacuum test
Verify the ability to work at low temperatures coupled with the vacuum environment	TC test
Verify the ability to withstand extremely high and low temperature in non-operating state	TC tolerance test
Verify system startup capability at high and low temperature	hot/cold start-up test

As showed in the table above, some purposes can be achieved by a simple test: the TC test and TV test can achieve the same purpose in screening early defect; vacuum test can also perfectly achieve the purpose of verifying the ability of system working in the environment coupled with extremely high temperature and vacuum (via heating device in satellite); Hot/Cold start-up test can verify the system start in high and low temperature; and TC tolerance test can verify the ability of system survive extremely high and low temperature in non-operating state.

However, regarding "verifying system the ability to work at low temperatures coupled with the vacuum environment", there are certain differences between the TC and TV test. Initially, the TC test has no vacuum environment, which is a test condition and can only be tailored. Secondly, because of natural convection, heat transfer characteristics in the satellite should have some differences between TC and TV test. This paper describes the internal temperature distribution differences between TV test and TC test of nanosatellite via rule number analysis method, and validate the model by CFD simulation and test.

2 Rule number analysis

In international standards draft "Space Systems — Design Qualification And Acceptance Tests Of Small-Scale Satellite And Units Seeking Low-Cost And Fast-Delivery", the rule number analysis method was employed for nanosatellite TC applicability research, and a $50W/m^2$ heat flux threshold was set for TC test. It is mainly for a single external shell temperature (300K) and completely radiation heat transfer state[3], which has certain differences with the actual test environment. This paper constructs concentric sphere model to simulate heat transfer in satellite: the inside sphere is heat source, which has heat power Q(w), radius R1(m) and temperature T1 (k); while the outer sphere is satellite shell, which has radius R2(m) and temperature T2 (k). The model is showed in Figure 1:

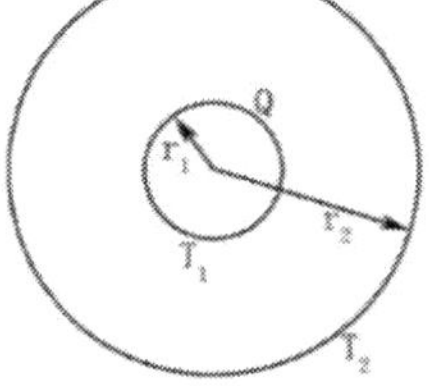

Figure 1. Heat transfer model

According to the basic formula of heat transfer, the heat flux through the inner shell in TV test can be expressed by the following equation,

$$\frac{Q}{A_1} = \frac{\sigma(T_1^4 - T_2^4)}{\frac{1}{\varepsilon_1} + \left(\frac{1}{\varepsilon_2} - 1\right)\frac{A_1}{A_2}} \tag{1}$$

Where A_1 is internal surface area(m^2), A_2 is the external surface area(m^2), ε_1 and ε_2 are the inner and outer surface emissivity. σ is Stefan-Boltzmann constant. The heat flux through the inner shell in TC test can be expressed by the following equation,

$$\frac{Q}{A_1} = \frac{\sigma(T_1^4 - T_2^4)}{\frac{1}{\varepsilon_1} + \left(\frac{1}{\varepsilon_2} - 1\right)\frac{A_1}{A_2}} + Nu\frac{k_{air}}{r_2 - r_1}(T_1 - T_2) \tag{2}$$

Where k_{air} is air thermal conductivity (W/m·K), *Nu* is the Nusselt number, can be expressed by the following equation[4],

$$Nu = 2 + \frac{0.589(GrPr)^{1/4}}{(1 + (0.469 / Pr)^{9/16})^{4/9}} = 2 + \frac{0.589(gPr)^{1/4}}{(1 + (0.469 / Pr)^{9/16})^{4/9}} \times \left(\frac{al^3\Delta t}{\nu}\right)^{1/4} \tag{3}$$

Due to the small changes of the Prandtl number *Pr* with temperature, when consider it as 0.7, the *Nu* can be simplified as follow,

$$Nu = 2 + \frac{0.589(gPr)^{1/4}}{(1 + (0.469 / Pr)^{9/16})^{4/9}} \times \left(\frac{al^3\Delta t}{\nu}\right)^{1/4} \approx 2 + 0.7344 \times \left(\frac{l^3\Delta t}{T\nu}\right)^{1/4} \tag{4}$$

Where l is characteristic length (m), Δt is temperature difference(℃), T is qualitative temperature(K), ν is kinematic viscosity (m^2/s). When qualitative temperature and kinematic viscosity changed little, Nusselt number can be considered only relieve with temperature difference Δt and characteristic length *l*, When assuming thermal conductivity changes little when temperature changed. Convection heat transfer can be considered only relieve with temperature difference Δt and characteristic length *l*, heat flux in TV and TC can be expressed as follow,

$$\frac{Q}{A_1} = \frac{\sigma(T_1^4 - T_2^4)}{\frac{1}{\varepsilon_1} + \left(\frac{1}{\varepsilon_2} - 1\right)\frac{A_1}{A_2}} \tag{5}$$

$$\frac{Q}{A_1} = \frac{\sigma(T_1^4 - T_2^4)}{\frac{1}{\varepsilon_1} + \left(\frac{1}{\varepsilon_2} - 1\right)\frac{A_1}{A_2}} + \left(2 + \left(\frac{0.7344}{T\nu}\right)^{1/4} \times l^{3/4}\Delta t^{1/4}\right)\frac{k_{air}}{r_2 - r_1}(T_1 - T_2) \tag{6}$$

Heat flux was selected for x-axis, temperature difference between inner and outer shell was selected for y-axis. Outer shell temperature was set to 258K according to the literature [2] low temperature test condition, and the inner shell radius was set to 0.05m. The temperature difference curve between TC and TV test with outer shell radius of 0.06, 0.1, 0.25m in 258K have been drew like Fig. 2-a, and curve in 300K have been drew in Fig. 2-b.

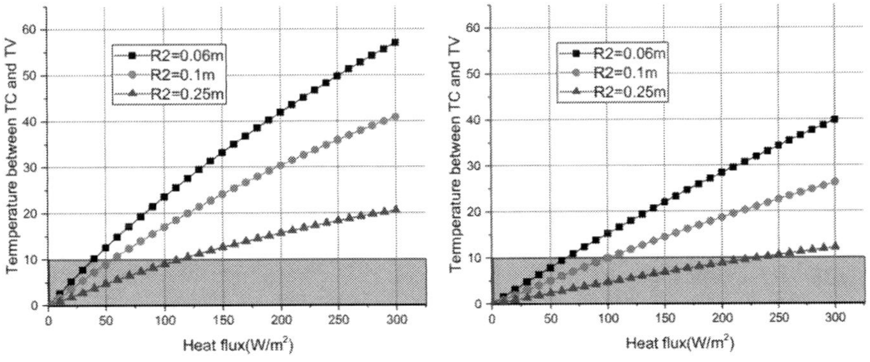

a Shell temperature=258K b shell temperature=300K

Figure 2. The difference between TC and TV test in various characteristic length

As showed in Figure 2, temperature difference between inner and outer shell increase with heat flux of nanosatellite heat source, and heat exchange increase when outer shell reduced in diameter. The temperature curve also has some differences between 258K and 300K. According to existing test standards [5,6], "If the worst-case temperature difference between the two environments is greater than 10°C, TV testing shall be performed ".To ensure the maximize temperature difference below 10°C,the heat flux should below 50w/m^2 when shell temperature is 300K, and below 40W/m^2 when shell temperature is 258K. TC test mentioned in Table 1 should attain the purpose of verifying the ability of system performance at low temperatures. The estimated heat flux limits should be in accordance with the temperature difference under cold conditions. Therefore, the heat flux of the heat source in nanosatellite should below 40W/m^2.Nanosatellite power density range is generally 0.3 ~ 2W/kg [7]. Consequently, not all the nanosatellites heat flux can meet the requirements of less than 40W/m^2.

As to a general nanosatellite, its heat source power is about 1-5W, and its heat transfer is mainly in radiation. the TC alternative test threshold should be 40W/m^2, or look up in Figure 2 using heat flux and characteristic length ratio.

When the internal heat source directly installed on a deck which was like a heat sink, it is assumed that one side of the device (20% area) has thermal conductivity with deck, as showed in Figure 3. The heat conduction will be more significant, which will not change between TC and TV test. Hence, the TC alternative test threshold should be much larger.

DOI: 10.1051/matecconf/201711402009

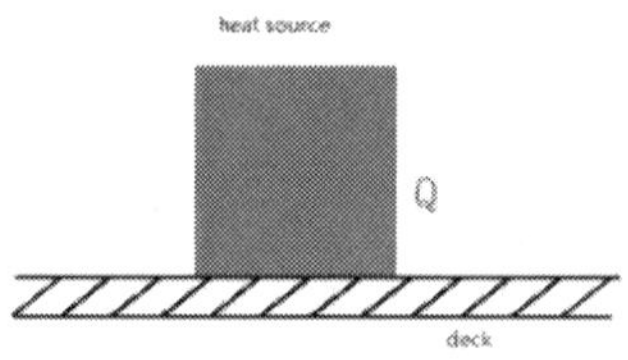

Figure 3. Heat source directly install on a deck

The contract was set to be dry contract. The deck was a honeycomb plate within 1cm thickness, and its equivalent thermal conductivity can be expressed by the following equation,

$$K = \frac{1}{\frac{1}{g} + \frac{l}{\lambda}} = \frac{1}{\frac{1}{50W / m^2 \cdot K} + \frac{0.01m}{1.5W / m \cdot K}} = 37.5W / m^2 \cdot K \tag{7}$$

where g is contact heat transfer coefficient(W/m^2·°C) and is set to 50W/m^2·°C for dry contract. Meanwhile, λ is honeycomb equivalent thermal conductivity (W/m·°C) and l is honeycomb plate thickness.

Heat flux in TV and TC test can be expressed by the following equation,:

$$\frac{Q}{A_1} = \frac{\sigma \ T_1^4 - T_2^4}{\frac{1}{\varepsilon_1} + \left(\frac{1}{\varepsilon_2} - 1\right)\frac{A_1}{A_2}} \times 0.8 + \ T_1 - T_2 \ \times k \times 0.2 \tag{8}$$

$$\frac{Q}{A_1} = [\frac{\sigma \ T_1^4 - T_2^4}{\frac{1}{\varepsilon_1} + \left(\frac{1}{\varepsilon_2} - 1\right)\frac{A_1}{A_2}} + (2 + (\frac{0.7344}{T\nu})^{1/4} \times l^{3/4} \Delta t^{1/4}) \frac{k_{air}}{r_2 - r_1}(T_1 - T_2)] \times 0.8 + \ T_1 - T_2 \ \times k \times 0.2 \tag{9}$$

where k is equivalent thermal coefficient (set to 37.5 W/m^2·°C). The curve of temperature difference between TV and TC test in different conditions is drew as Figure 4。

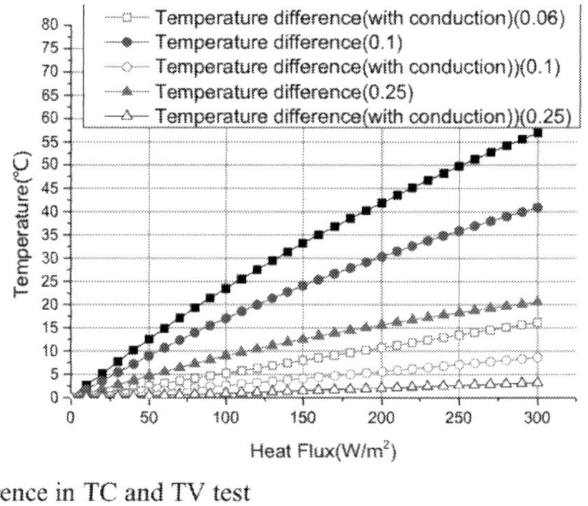

Figure 4. Temperature difference in TC and TV test

From the graph above, when the heat source set directly on the deck of nanosatellite, heat flux threshold can be enlarged to 100W/m^2 to ensure the difference between TV and TC test under 10°C. Concerning a higher characteristic length ratio or larger test article, heat flux threshold could be greater, which can be calculated by equ.9.

DOI: 10.1051/matecconf/201711402009

3 Validation

3.1. CFD simulation

A Fluent model was used to verify the rule number analysis, including four units (two amplifiers and two TWT, traveling-wave tube) and six heat pipes in cabin body. The model was a rectangular cuboid which dimensions are 600(L)×400(W)×300(H)mm, whose appearance and internal mesh showed below:

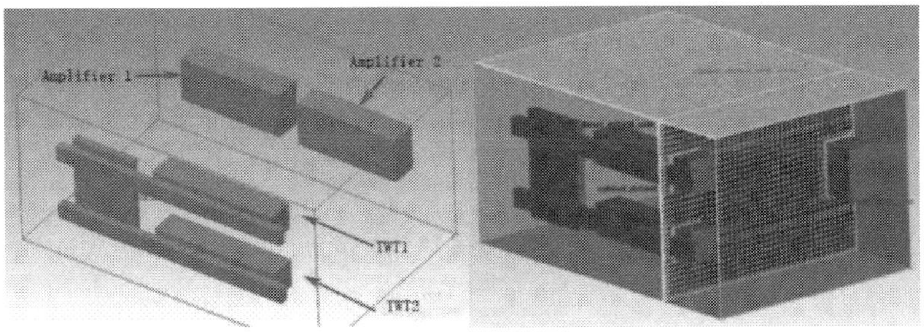

Figure 5. Simulation model and mesh

When the temperature of the outer shell is -20°C, the internal temperature distribution in TV and TC are given in Figure 6.

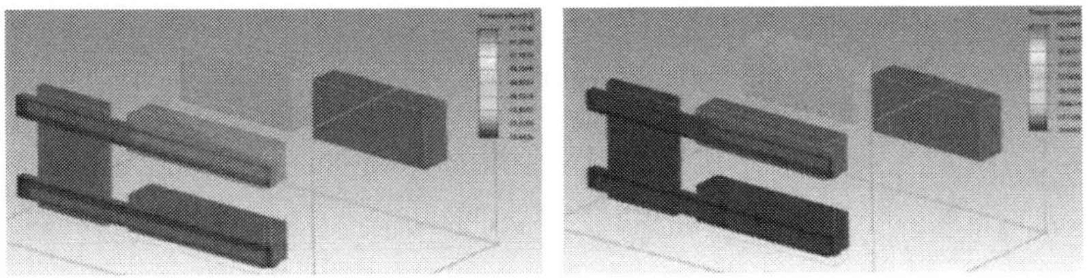

TC test TV test

Figure 6. Temperature distribution

The outer shell temperature was set 253K. When the amplifiers power was 0.5W, 1W, 2W, 5W and 10W, the model had been simulated under two different conditions - heat source contracting with wall or not contracting with wall. The temperature difference between TV and TC test can be seen in Table 2.

Table 2. Temperature comparison

Power(W)	Heat Flux(W/m^2)	Without Conduction			With Conduction		
		TV(°C)	TC (°C)	Difference (°C)	TV(°C)	TC (°C)	Difference (°C)
0.5	7.94	-16.97	-19.31	2.34	-19.19	-19.46	0.27
1	15.89	-15.23	-18.85	3.62	-18.50	-19.01	0.52
2	31.77	-11.75	-17.95	6.20	-17.11	-18.12	1.01
5	79.43	-1.90	-15.25	13.35	-12.96	-15.44	2.48
10	158.85	13.31	-10.78	24.08	-6.14	-10.99	4.86

The simulation results and the rule number analysis in table above follow same trend. When heat flux of heat source is greater than 40W/m^2 and without contracting with walls, there may be a significant temperature difference between TV and TC test(> 10°C). It may be in consequence short-test for article. The threshold should be much greater than 40W/m^2 for a unit directly contracting to deck and could be estimated by equ.9 using heat flux and characteristic length ratio.

3.2. Test

The comparative test was carried out in BISEE; specifically, the TC test was done in thermal cycling chamber and TV test was completed in thermal vacuum chamber. test article was a rectangular cuboid which dimensions are 600(L)×400(W)×300(H)mm. Inner structure of the article was same as the model in 3.1. The test article and test system are showed in Figure 7. There were four programmable power supplies and film heaters used to simulate the heat source. Also, sixteen thermal couples were set on test article to measure the temperature.

Figure 7. Test article and test system

When the shell temperature was 273K and 353K separately, the temperature difference between TC and TV test was calculated by equ.9. The curves are showed in Figure 8.

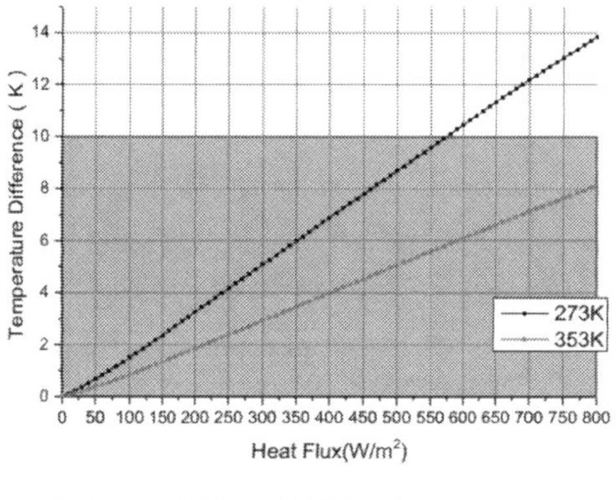

Figure 8. Temperature difference between TC and TV test

To verify the analysis, heat flux of the heat source was set to 750W/m^2. According to the analysis, temperature difference between TC and TV test will be less than 10°C in 353K case, and it will be more than 10°C in 273K case. The test result is showed in Table 3.

Table 3. Temperature difference in TC and TV test

Case	Temperature difference between shell and heat source(°C)		
	TV test	TC test	Delta
273K	43.89	20.37	23.52

353K	20.89	19.34	1.55

It can be seen that test result follow same trend with the rule number analysis. The temperature difference between TV and TC in low temperature was larger than in high temperature case. If a heat source thus set on deck directly, the heat flux threshold for alternative thermal cycling test can be calculated by the rule number analysis method.

4 Conclusion

In this paper, an internal heat transfer model coupled radiation, convection, conduction has been built by simplified nanosatellite model. Via rule number analysis method, it describes the applicability of using TC instead of TV for nanosatellite whose characteristic length and heat source size are different. Analysis results show it is feasible to use alternative TC test for nanosatellites with inner heat source flux less than $40W/m^2$. For a nanosatellite with internal heat source directly installed on deck, the threshold could be extended to $100W/m^2$, which can be calculated by rule number analysis method. The results of CFD and test method used to verification are consistent with the rule number analysis. The analytical method in this paper could be a positive choice in test method of nanosatellite in the future.

References

1. M. Cho, H. Masui, T. Hatamura, K.Date, S.Horii, S. Obata, Status of Micro/Nano satellite Environmental Tests Standardization Project: Test Campaign and Standard Draft. (Nanosatellite Symposium, La Seine, 2016)
2. M. Cho, H. Masui, T. Hatamura, K.Date, S.Horii, S. Obata, Overview of Nano-satellite Environmental Tests Standardization Project: Test Campaign and Standard Draft.(26th Annual AIAA/USU Conference on Small Satellites, Logan, UT, 2012)
3. ISO/TC 20 / SC 14 N 1004, Space Systems — Design Qualification And Acceptance Tests Of Small-Scale Satellite And Units Seeking Low-Cost And Fast-Delivery. (2014)
4. O.G. Martynenko, P.P. Khramtsov, Free-Convective Heat Transfer: With Many Photographs of Flows and Heat Exchange.(2005)
5. SMC-S-016A, Test Requirements For Launch, Upper Stage And Space Vehicle. (2014)
6. MIL-STD-1540C, Test Requirements For Launch, Upper Stage And Space Vehicle. (1994)
7. J. Bouwmeester, J. Guo, Survey of worldwide pico- and nanosatellite missions, distributions and subsystem technology, Acta Astronaut, **67**, 854-862(2010)

MATEC Web of Conferences **114**, 02010 (2017) DOI: 10.1051/matecconf/201711402010
2MAE 2017

Modular simulation software development for liquid propellant rocket engines based on MATLAB Simulink

Mahyar Naderi[1], Guozhu Liang[2a] and Hasan Karimi[3]

[1]*Ph.d Candidate, School of Aeronautics and Astronautics, 100191 Beihang University, China*
[2]*Professor, School of Aeronautics and Astronautics, 100191 Beihang University, China*
[3]*Associate Professor, Faculty of Aerospace Engineering, K.T. University of Technology, Iran*

Abstract. Focusing on Liquid Propellant Rocket Engine (LPRE) major components, a steady state modular simulation software has been established in MATLAB Simulink. For integrated system analysis, a new algorithm dependant on engine inlet mass flow rate and pressure is considered. Using the suggested algorithm, it is possible to evaluate engine component operation, similar to the known initial parameters during hot fire test of the engine on stand. As a case study, the reusable Space Shuttle Main Engine (SSME) has been selected and the simulation has been performed to predict engine's throttled operation at 109 percent of the nominal thrust value. For this purpose the engine actual flow diagram has been converted to 34 numerical modules and the engine has been modelled by solving a total of 101 steady state mathematic equations. The mean error of the simulation results is found to be less than 5% compared with the published SSME data. Using the presented idea and developed modules, it is possible to build up the numerical model and simulate other LPREs.

1 Introduction

In order to reduce the costs during design and enhancement phases, using software capable of precise modeling and simulating is favorable. For this purpose in this research a simulation tool is developed for steady state analysis of a RLPRE which can be further extended for simulating other types of LPREs.

In development of simulation software for LPRE analysis, many contributions have been done for example in 1995, C. Goertz suggested a modular method for static performance analysis of different LPRE cycles using Fortran (as it appears). [1]. In 2000, K. Liu, Y. Zhang developed a methodology for modularization modeling and transient simulation of LPREs and created an object oriented program using Microsoft Visual C++ 6.0 [2]. In 2009, C. R. Koppel et al.; used EcosimPro tool kit along with ESPSS libraries to implement and validate satellite propulsion systems modellings [3]. In 2011, F. D. Matteo and M. D. Rosa developed steady state library for ESPSS software for static simulation of liquid propellant engines [4]. In 2012, Y. Zheng et al. used the AMEsim software to simulate the dynamic behavior of unified spacecraft propulsion system [5]. In 2013, M. Naderi, A.R.Jalali et al. developed a GUI software for dynamic simulation of open cycle LPRE using FORTRAN 90 [6]. In 2015, L. Wei et al. used Modelica to establish componenet model libraries for LPRE [7]. As it is visible, by the evolution of computers, the researchers are tending towards

[a] Guozhu Liang : lgz@buaa.edu.cn

developing more general softwares similar to toolkits. In the current research in order to analyze different types of LPRE cycle performance, a modular simulation software has been developed for steady state analysis, using MATLAB Simulink. Here MATLAB is chosen due to its high accessibility, applicability, simple programming language, convinient usage, rich library and toolboxes for different mathematical solution methods and finally ease of operation for future researchers to enhance the current software. Despite some of the previous literatures, the presented algorithm uses engine inlet mass flowrate and pressure as the system input, similar to the known initial conditions during hot fire test. For case study, SSME, as one of the most complicated, successful and comprehensive LPRE built up to now, is simulated and the results are compared with published data.

2 Realization approach for modular simulation of LPRE

The process of modular simulation is classified into five major steps: 1) Developing modular models for LPRE main components (here in MATLAB Simulink); 2) Validating the modules; 3) Establishing the desired LPRE numerical flow model from the actual flow diagram; 4) Developing an algorithm for determining the sequential solution order and method, based on module's input/output data; 5) Establishing the integrated model.

Reviewing the major elements for different types LPRE cycles, it is found that LPREs are generally constructed of seven major modules. These include connecting pipes, combustion chamber, nozzle, pre-burner (gas generator), pumps, turbines and controling valves. In this paper the cooling jacket is also considered as a module therefor a total of eight madules have been considered (Table 1). Some of these modules can also be connected to form a subsystem such as turbo-pump assembly or thrust chamber. With modular form of simulation, it possible to use the same modules for other LPREs containing similar componenets. Here the modules are constructed of linear/nonlinear lamped parameter steady state equations. Despite frequently used algorithms for steady state analysis [8-10], the chamber or pre-burner pressure is not known, but itself is a parameter required to be simulated. For this pupose an algorithm is developed which is based on the known parameters at main feed line enterance (i.e. oxidizer and fuel mass flow and pressure at engine inlet), similar to the case of testing the engine on test stand. The solution method is based on Newton Raphson iterative loops. For interpolating between the required inputs, the shape-preserving piecewise cubic interpolation method was used as Matlab function and showed good accuracy.

Table 1. Main modules in LPRE flow diagram

Model/ Subsystem										
	CC	CJ	Nozzle	Pipe	PB	Pump	Turbine	Valve	TC	TP

3 Case study

In this paper, SSME is selected as a case study. This engine is not only a SCCLPRE but also a RLPRE and is among one the most successful LPREs ever built. Furthermore, due to its special design, it has most of the complexities that all other engines might have, namely: using cryogenic propellant, separate turbo-pump assemblies on each propellant line, low and high pressure turbo-pumps, hydraulic turbine and etc. A detail description of SSME can be found in [11]. In order to model the engine, with regard to its actual flow diagram, a numerical model using the eight modules already discussed is prepared (Figure 1). The model is constructed of 34 modules as described in Table 1. Each module is based on steady state linear/nonlinear lumped parameter model mathematical equation.

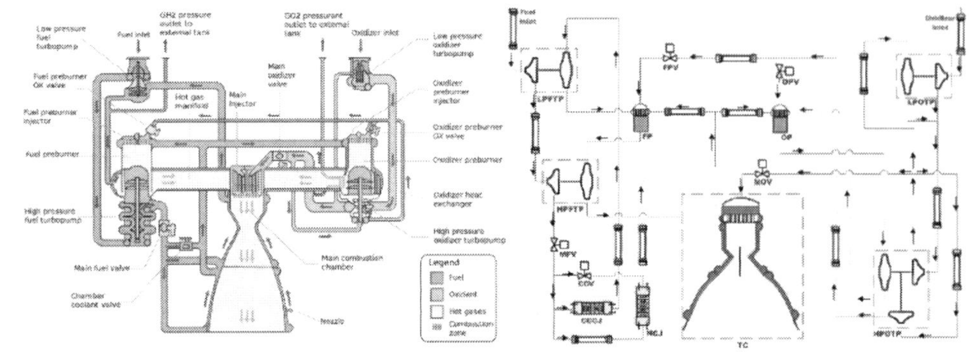

Figure 1. Comparison between SSME actual (left-[12]) and current simulated model flow diagram

The required input and produced output data for modules is described in Table 2 to Table 5 but due to page number limitations, only the modules on fuel line are mentioned.

The input variables are devided into three main catagories: Type "a" parameters: constant input values (e.g. LPFTP disk diameter) and in Figure 2 are depicted by pink color; Type "b" parameters: curves and graphs from which the desired value should be interpolated (e.g. combustion product gas temperature which is plotted versus propellant mixture ratio), depicted by orange color in Figure 2; Type "c" parameters: estimated in other blocks and will later be given to the current block as input (e.g. LPFTP pump outlet pressure is an input for HPFTP pump) depicted by cyan color in Figure 2.

Table 2. LPFTP turbo-pump input/output data

LPTFP Turbine Parameter	**Type**	**Cat.**	**LPFTP Pump Parameter**	**Type**	**Cat.**
LPFTP Shaft Rotational Speed	**Input**	c	**LPFTP Shaft Rotational Speed**	**Input**	c
CC Main Fuel Injector Pressure	**Input**	c	**CC Main Fuel Injector Pressure**	**Input**	c
CC Cooling Passage Pressure	**Input**	c	**Fuel Density**	**Input**	a
Fuel Vapor Gas Constant	**Input**	a	**LPFTP Characteristic Map**	**Input**	b
CC Cooling Passage heat ratio	**Input**	c	**LPFTP Pump Shaft Torque**	**Output**	-
LPFTP Turbine Polytropic Efficiency	**Input**	a	**LPFTP Pump Input mass flow rate**	**Output**	-
CC Cooling Passage Temperature	**Input**	a			
LPFTP Characteristic Map	**Input**	b			
LPFTP Turbine Shaft Torque	**Output**	-			
LPFTP Turbine mass flow rate	**Output**	-			

Table 3. MFV and CCV input/output data

MFV Parameter	**Type**	**Cat.**	**CCV Parameter**	**Type**	**Cat.**
HPFTP Pump Discharge Pressure	**Input**	c	**MFV Discharge Pressure**	**Input**	c
HPFTP Pump Discharge mass flow	**Input**	c	**Pre-Burners Main Fuel Supply Line**	**Input**	c
MFV Nominal Area Ratio Graph	**Input**	b	**CCV Nominal Area Ratio Graph**	**Input**	b
MFV Throttle Situation	**Input**	a	**CCV Throttle Situation**	**Input**	a
MFV Discharge Pressure	**Output**	-	**CCV Discharge mass fllow rate**	**Output**	-

 DOI: 10.1051/matecconf/201711402010

Table 4. Fuel pre-burner and combustion chamber input/output data

Fuel Pre-Burner Parameter	Type	Cat.	Combustion Chamber Parameter	Type	Cat.
Fuel Pre-Burner Fuel mass flow rate	Input	c	MOV mass flow rate	Input	c
Fuel Pre-Burner Oxid mass flow rate	Input	c	CC Fuel Injector mass flow rate	Input	c
Nozzle Cooling Passage Temperature inlet	Input	c	HPOTP Boost Pump mass flow	Input	c
Nozzle Cooling Passage mass flow	Input	c	HPFTP Pump mass flow	Input	c
Nozzle Cooling Passage Temperature	Input	a	CC Temperature-Mixture Ratio Graph	Input	b
CCV mass flow rate	Input	c	Product gama-Mixture Ratio Graph	Input	b
Fuel Pre-Burner Throat Diameter	Input	a	Gas Constant-Mixture Ratio Graph	Input	b
Pre-Burner Temperature-Mixure Ratio	Input	b	Combustion Chamber Throat Area	Input	a
Fuel Pre-Burner Pressure	Output	-	Combustion Chamber Pressure	Output	-
Fuel Pre-Burner Temperature	Output	-	Combustion Chamber Temperature	Output	-

Table 5. Nozzle cooling jacket and FP fuel line Inlet input/output data

Nozzle Cooling Jacket Parameter	Type	Cat.	FP Fuel line parameter	Type	Cat.
MFV Discharge Pressure	Input	c	Fuel Pre-Burner Pressure	Input	c
Nozzle Cooling Passage Pressure	Input	c	Pre-Burners Main Fuel Supply Line	Input	c
Nozzle Cooling Passage Friction Graph	Input	b	Nozzle Cooling mass flow	Input	c
Nozzle Cooling Passage Discharge mass flow	Output	-	CCV mass flow	Input	c
			Fuel Pre-Burner Fuel mass flow	Output	-
			Fuel Density Feeding the Pre-Burners	Output	-

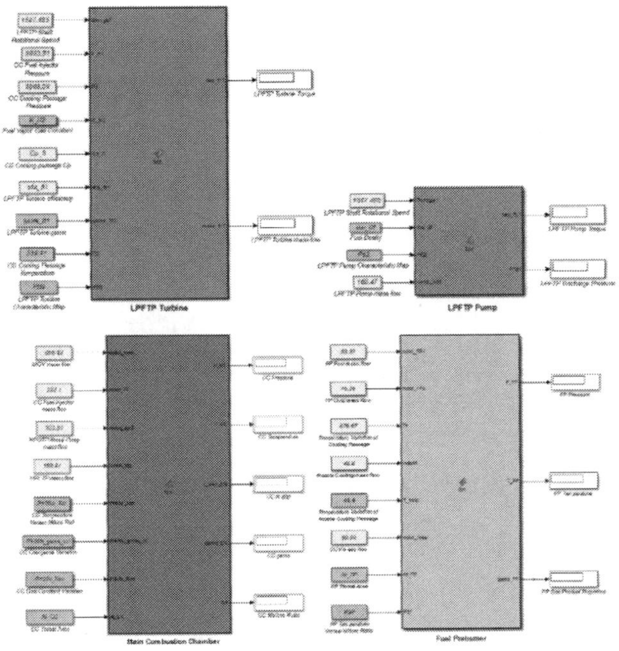

Figure 2. LPFTP turbine, pump, fuel pre-burner and combustion chamber Simulink module

After each module is verified, in order to form the integrated system model, the modules should be linked together with a correct logic according to the desired engine flow diagram (Figure 3).

Using given input data and nine initial guesses, the equations are solved by Newton Raphson iteration method. By correct connection of modules, an integrated model based on desired flow diagram can be formed. Considering SSME as our case study, the outer layer of the integrated Simulink model is shown in Figure 4.

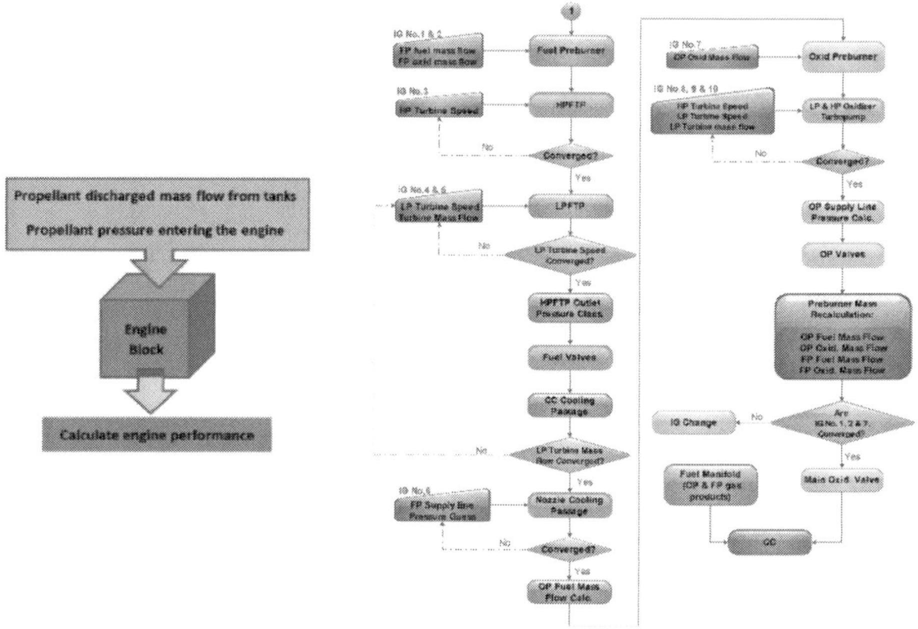

Figure 3. System simulation block diagram (left); SSME simulation algorithm (right)

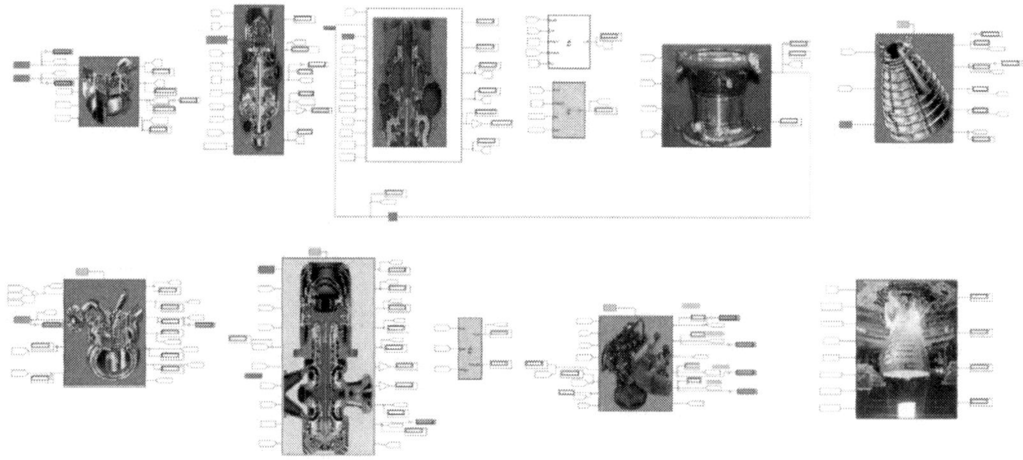

Figure 4. SSME integrated Simulink system simulation

4 Results

For validating the suggested algorithm and simulation software, SSME RLPRE data throttled at 109 percent of nominal operation condition is used from Rocketdyne report [13]. For this purpose using the suggested algorithm in Figure 3, the integrated model of the engine (containing 34 modules) is formed in MATLAB Simulink. The total system of equations consisting of 101 linear/nonlinear equations are solved with Newton Raphson method. The required system level input values can be found in Table 6.

Table 6. SSME System input parameters at 109% thrust level

Input Parameter	Value
Fuel tank discharge mass flow	72.79 kg/s
Oxidizer tank discharge mass flow	436.72 kg/s
Absolute pressure at the end of the fuel tank	2.04 atm
Absolute pressure at the end of the oxidizer	6.8 atm

The simulation results can be found at Table 7. As it can be seen from the results, the governing equations used for turbo-pump system has leaded to fairly good accuracy except for the LPFTP wahich was in lack of initial data. The outlet pressure which is of course quiet dependable on the pump performance has been affected by the torque and speed, and the results seem to be following the expected trend. The turbine mass flow rate analysis shows great convergence and the results are satisfactory. The pre-burner pressure are precise which confirm the usage of the introduced equation sets for this element. For the combustion chamber, using CEA software for computing the gas product thermodynamic properties, the results do not show good precision but using the suggested equation for computing the gas properties from [13], the simulation results for this element are far better.

Table 7. Comparing simulation results

Parameter		Simulation	Rocketdyne [13]	Error %
Shaft rotational speed, Ω (rpm)	HPFTP	36954.02	37354.67	-1.07
	HPOTP	31497.53	31133.36	1.17
	LPFTP	15235.47	15806.83	-3.61
	LPOTP	5505.54	5447.11	1.07
Torque, τ (N.m)	HPFTP	14389.05	14594.05	-1.40
	HPOTP	6489.89	6350.64	2.19
	HPOBTP	449.72	438.58	2.54
	LPFTP	1238.39	1327.43	-6.71
	LPOTP	2305.37	2271.32	1.50
Pump outlet pressure (atm)	HPFTP	463.92	476.61	-2.66
	HPOTP	361.1	350.62	2.99
	HPOBTP	590.79	573.95	2.93
	LPFTP	16.05	17.63	-8.96
	LPOTP	30.82	30.10	2.39
Turbine mass flow rate, $\dot{m}_t$ (kg/s)	HPFTP	72.79	73.32	-0.72
	HPOTP	30.01	28.61	4.89
	LPFTP	15.1	15.73	-4.01
	LPOTP	79.35	78.73	0.79
PB pressure (atm)	FP	404.02	401.30	0.68
	OP	401.96	399.30	0.67
CC pressure (atm)	Exp. Gas prop.	226.39	227.00	-0.27
	CEA	280.96	227.00	23.77

5 Conclusion

A modular simulation software for steady state simulation of RLPRE was developed based on MATLAB Simulink. Eight general simulation modules for simulating engine main components was established. In the current simulation, despite commonly used algorithms, the simulation is not dependant on engine down stream parameters. As a case study, SSME as the most successful RLPRE ever built, was selected and the engine flow diagram was simplified and modularized to 34 component modules. A total of 101 linear/nonlinear equations were solved using Newton-Raphson iteration method. Comparing with the actual results for the engine's throttled operation at 109 percent of the nominal thrust value, it is concieved that the simulation has an acceptable average precision of more than 95 percent. As future work, the current precision can be enhanced by introducing more detailed initial data especially for the LPFTP assembley and the software can be further verified with other engine cycles.

References

[1] C. Goertz, "A modular method for the analysis of liquid rocket engine cycles," 1995.
[2] K. Liu and Y. Zhang, "A study on versatile simulation of liquid propellant rocket engine systems transients," 2000.
[3] J. M. C. R. Koppel, R. Perez , J. Steelant, P. Omaly, "A Platform Satellite Modelling With EcosimPro: Simulation Results," presented at the 45th AIAA/ASME/SAE/ASEE Joint Propulsion Conference & Exhibit, Denver, Colorado, 2 - 5 August, 2009.
[4] F. Di Matteo and M. De Rosa, "Steady State Library for Liquid Rocket Engine Cycle Design," 2011.
[5] X. H. P. Z. Yan, Y. Q. Cheng, J. J. Wu,, "System Dynamic Characteristic Simulation of Spacecraft Propulsion System Based on AMESim," *Advanced Materials Research,* vol. 605-607, pp. 679-683, 2013
[6] A. R. J. M. Naderi, H. karimi, "Modeling And Simulation Of Open Cycle Liquid Propellant Engines," *Journal of Science and Engineering,* vol. Vol. 1, 2013.
[7] C. L. L. Wei, X. Gang, D. Jing, Z. Haiming, Y. Hao, "Modeling and Simulation of Liquid Propellant Rocket Engine Transient Performance Using Modelica," presented at the Proceedings of the 11th International Modelica Conference, Versailles, France, September 21-23, 2015.
[8] M. Chitsaz, E. A. Tahmasebi, and H. Karimi M, "A Method for Simulation of a Cryogenic Liquid Propellant Engine," *Applied Mechanics and Materials,* vol. 110-116, pp. 2370-2376, 2011.
[9] D. H. H. Dieter K Huzel, *Modern Engineering for Design of Liquid Propellant Rocket Engines*. Washington DC, USA: American Institute of Aeronautics and Astronautics, 1992.
[10] H. L. Humble, *Space Propulsion Analysis and Design* USA: McGraw-Hill Inc., 2007
[11] "Space Shuttle Main Engine Orientation," in "BC98-04," Rocketdyne Propulsion and Power, BoeingJune 1998.
[12] (March, 2016). www.wikipedia.com.
[13] N. Dynamics, "SSME Model Dynamic Characteristics Related to POGO," Rocketdyne Division Rockwel International, 30th September, 1973.

Abbreviation

CC	Combustion Chamber	**LPRE**	Liquid Propellant Rocket Engine
CCV	Cooling Control Valve	**MFV**	Main Fuel Valve
CJ	Cooling Jacket	**MOV**	Main Oxidizer Valve
FP	Fuel Pre-Burner	**OP**	Oxidizer Pre-Burner
FPV	Fuel Pre-Burner Valve	**OPV**	Oxidizer Pre-Burner Valve
HPFTP	High Pressure Fuel Turbo-pump	**PB**	Pre-Burner
HPOTP	High Pressure Oxidizer Turbo-Pump	**RLPRE**	Reusable Liquid Propellant Rocket Engine
IG	Initial Guess	**SCCLPRE**	Stage Combustion Cycle LPRE
LPFTP	Low Pressure Fuel Turbo-pump	**TC**	Thrust Chamber
LPOTP	Low Pressure Oxidizer Turbo-Pump		

Air dehumidification by membrane with cold water for manned spacecraft environmental control

Shang Yonghong[1,a], Wang Jing[1], Bi Yanqiang[1],Jian Yabin[1],Pei yifei[1], Yuan Weixing[2] and Yang Bo[2]

[1]*Beijing Institute of Spacecraft Environmental Engineering, Beijing, China*
[2]*School of Aeronautic Science and Engineering, Beihang University,Beijing, China*

Abstract. The traditional condensation dehumidification method requires additional gas-liquid separation and water recovery process in the manned spacecraft humidity control system, which would increase weight and complexity of systems. A new membrane dehumidification with cold water is proposed, which uses water vapor partial pressure difference to promote water vapor transmembrane mass transfer for dehumidification. The permeability of the membrane was measured and the experimental results agree well with the theoretical calculations. Based on the simulation of dehumidification process of cold water-membrane, the influence of module structure and working condition on dehumidification performance was analyzed, which provided reference for the design of membrane module construct. It can be seen from the simulation and experiments that the cold water-membrane dehumidification can effectively reduce the thermal load of the manned spacecraft.

1 Induction

Humidity control is takes an important role in Environmental Control and Life Support System for manned space missions. An astronaut will release almost 1.5 kg/d moisture by metabolism[1]. High humidity will not only influence the thermal comfort and health, but also is harmful to electronic equipment, which may cause many security problems[2].An active environmental control method of condensing dehumidification has been researched and utilized in many manned spacecraft such as Space Shuttle, ISS and Shenzhou[3]. Warm and humid air is cooled below dew point by Condensing Heat Exchanger(CHX) with humidity and temperature decreasing. The condensate water is absorbed by hydrophilic coating on the surface of CHX fins and is separated from air-water mixture by a mechanical rotary separator under microgravity environment. But CHX requires much energy consumption to cool and reheat air before send back to cabin, and to run air/water separator. Moreover, the reliability and lifespan of the movement assembly may limit the system performance[4].

2 Membrane dehumidification

Membrane dehumidification technology is based on the principle that water vapor can transfer through membrane pores by partial pressure difference of water vapor. A novel cold water and

[a] Corresponding author: yonghongshang@126.com

membrane assembly for air dehumidification is researched. The membrane dehumidifier is a typical shell-and-tube construct as shown in Figure1. A bundle of hollow fibre membranes is packed in the shell with terminations epoxy sealed to create the tube and shell side. The warm and humid air enters entrance and flows at the shell side, and the water vapor permeates through membrane pores into the tube side, because cold water flows in the tube side bring condensing water away to realize air dehumidification. The water at the tube side cannot leak out from pores because of the hydrophobic nature of membrane. There are no moving parts or electronics within the system.

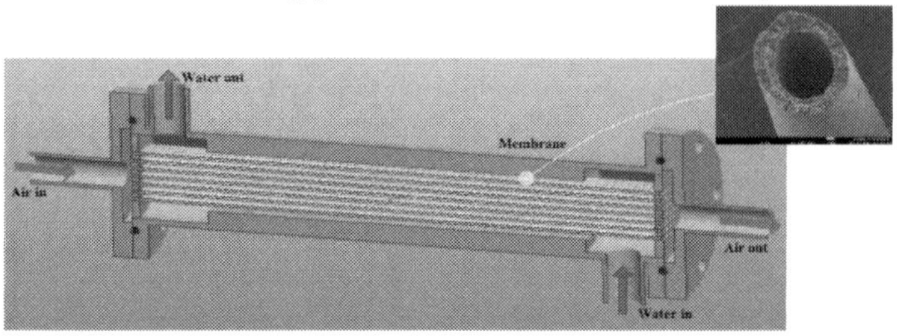

Figure 1. The hollow fibre membrane assembly

Spacesuit Water Membrane Evaporator have been developed by NASA to provide heat rejection capability in Mars[5]. Bundles of Nafion membrane are assembled inside vacuum shell as humidity control subassembly for the Boeing CST-100 Starliner spacecraft[6].A semi-permeable membrane material is used to remove water vapor from humid air. Various materials such as Polyvinylidene Fluoride(PVDF), Polytetrafluoroethylene(PTFE) and Cellulose Acetate(CA) were used and compared.

3 Methodology

Dehumidification process is a combination of heat and mass transfer. Heat transfers serially through the thermal boundary layer at the air-side, and then conducted by membranes and transfers through the thermal boundary layer at the water-side. The convective transport boundary layer theory is applicable to the mass transfer as well. The mass transfer resistance of water-side is considered to be negligible, only the mass transfer resistance of air-side should be taken into account.

3.1 Membrane diffusion

A kind of hydrophilic-hydrophobic membrane was adopted in this research. The inner layer of membrane is PVDF porous structure of as hydrophobic support. The outer layer of membrane is thin hydrophilic PVA with dense structure. The parameters of material are listed in Table 1.

Table 1. Physical parameters of material

Layer	Parameter	Symbol	Unit	Value
Inner	Thickness	δ_1	μm	110
	Average pore size	d_p	μm	0.15
	Porosity	ε	-	0.65
	Tortured	τ	-	3
Outer	Thickness	δ_2	μm	40
	Diffusion coefficient	D_{ws}	m^2/s	3.2×10^{-10}

	Density	ρ_s	kg/m^3	1280

The average pore size of the support layer is 0.15μm, which is close to the average molecular free path (0.09μm). According to the diffusion model, the diffusion mechanism of water vapor in the support layer includes Knusson diffusion, viscous flow and molecular diffusion[7,8]. Where $C_a = 3.203 \times 10^{-4}$, $v_v = 20.1$, $v_a = 12.7$; R is the gas constant; M_v is the molar mass of water vapor; M_a is the molar mass of air; μ is the kinetic viscosity of water vapor; p and T are the average pressure and temperature in the pores.

$$D_K = \frac{d_p \varepsilon}{3\tau}\sqrt{\frac{8RT}{\pi M_v}} \tag{1}$$

$$D_P = \frac{d_p^2 \varepsilon p}{32\tau\mu} \tag{2}$$

$$D_{va} = \frac{C_a T^{1.75}}{p\left(v_v^{1/3} + v_a^{1/3}\right)^2}\sqrt{\frac{1}{M_v} + \frac{1}{M_a}} \tag{3}$$

According to the diffusion resistance model, the diffusion coefficient of the support layer is

$$D_{vm1} = \frac{1}{1/D_K + 1/D_{va}} + D_P \tag{4}$$

The diffusion coefficient of water vapor in the outer layer is calculated by the equivalent of the diffusion coefficient of water, where ρ_a is air density; ψ is partition coefficient equals to 15.

$$D_{vm2} = \frac{\rho_s}{\rho_a}\psi D_{ws} \tag{5}$$

So the total diffusion coefficient of water vapor is

$$D_{vm} = \frac{\delta_1 + \delta_2}{\delta_1/D_{vm1} + \delta_2/D_{vm2}} \tag{6}$$

According to above, the diffusion coefficient of water vapor through membrane is 6.28×10^{-6} m^2/s.

3.2 Convective heat and mass transfer

According to the similarity theory, the mass transfer can be compared with the convective heat transfer. The relation of the mass transfer number *Sh* and the convective heat transfer number *Nu* will satisfy with the Chilton-Colburn correlation[9], $Sh = NuLe^{1/3}$,Where *Le* is the Lewis number, $Le = \frac{Pr}{Sc}$; *Pr* is the Prandtl number, $Pr = \frac{\mu c_p}{\lambda}$; *Sc* is the Schmidt number, $Sc = \frac{\mu}{\rho D_{va}}$; ρ, c_p, λ and μ are respectively the density, specific heat capacity, thermal conductivity, and dynamic viscosity of humid air.

3.3 Control equation

The sensible heat transfer will occur between humid air and cold water, and the temperature of humid air will be decreased. While the moisture content of humid air will decrease because of transmembrane mass transfer of water vapor. Cold water absorbs the sensible heat from warm humid air, while the latent heat will be produced in the process of moisture absorption, so the temperature of cold water will increase.

$$dQ + rdJ = \dot{m}_w c_{pw} dt_w \tag{7}$$

The rate of heat and mass transfer will be calculated as follows, where h_t and k_t are the total heat and mass transfer coefficient.

$$dQ = h_t\left(t_a - t_w\right)dA \tag{8}$$

$$dJ = \rho_a k_t (\omega_a - \omega_w) dA \tag{9}$$

3.4 Membrane permeability experiments

In order to determine the diffusion coefficient of water vapor through membrane by experiment, a shell-and-tube membrane assembly was made, which was filled with six hollow fibre membrane tubes and with the total membrane area of $19.1cm^2$. The micro-flow test bed was built, including circulating water and gas loop to provide 10mL/min water flow and 10L/min air flow. The experimental diffusion coefficient of water vapor in membrane was $6.44\times10^{-6}m^2/s$. The experiment results were consistent with the theoretical model result of $6.28\times10^{-6}m^2/s$, verifying the reliability of the theoretical model.

4 Dehumidification performance analyses

4.1 Flow form

The performance analysis of the counter-flow or cross-flow of the shell-and-tube heat exchangers is carried out. The convective mass transfer resistance of the counter flower is larger than that of the cross-flow at the same Reynolds number as shown in Figure 2. However, the heat transfer resistance of counter-flower is also larger than cross-flow as shown in Figure 3. When the Reynolds number is 300, the heat transfer resistance of counter-flow ($0.0457m^2K/W$) is 6 times of cross-flow (0.0076 m^2K/W) and the mass transfer resistance of counter-flow (80s/m) is 1.8 times of cross flow (44.8s/m). In order to prevent the dewing on fibre outer surface as result of the faster heat transfer rate than the mass transfer, the form of counter-flow and the cylindrical structure of shell are adopted. This configuration is also more favourable for the flow uniformity.

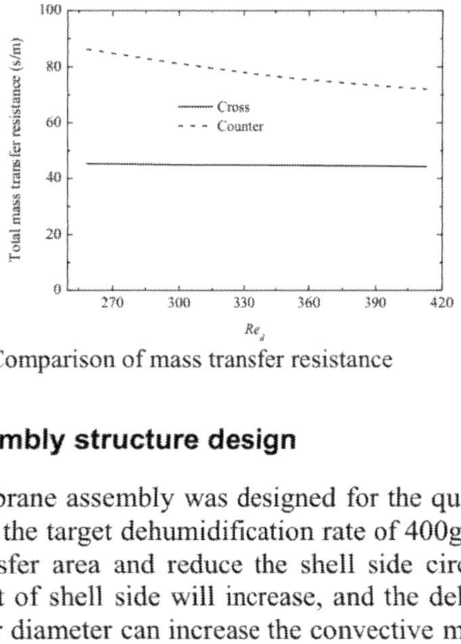

Figure 2. Comparison of mass transfer resistance

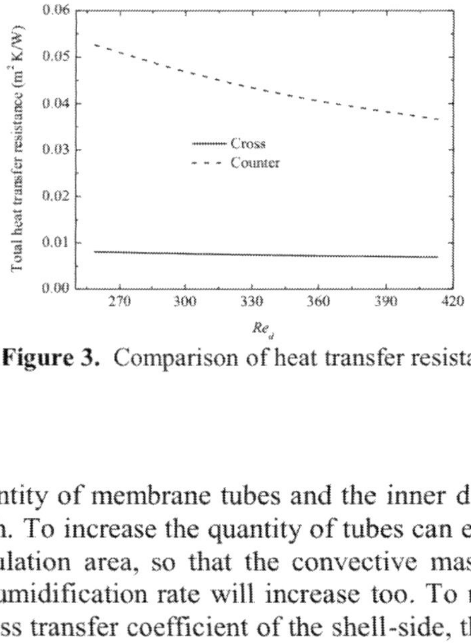

Figure 3. Comparison of heat transfer resistance

4.2 Assembly structure design

The membrane assembly was designed for the quantity of membrane tubes and the inner diameter of shell with the target dehumidification rate of 400g/h. To increase the quantity of tubes can enlarge the mass transfer area and reduce the shell side circulation area, so that the convective mass transfer coefficient of shell side will increase, and the dehumidification rate will increase too. To reduce the shell inner diameter can increase the convective mass transfer coefficient of the shell-side, thereby the dehumidification rate will increase. Improve the mass transfer performance will enhance the heat transfer performance at the same time, which will lead to the relative humidity of the outlet air increase and the risk of water condensation on the tube surface.

According to the typical operating conditions of the manned spacecraft humidify control requirements, the inlet air temperature is 26°C and the relative humidity is 60%, the inlet cold water temperature is 5°C, the flow rate of air and cold water are both 0.1kg/s and the tube length is 300 mm. The design range of the membrane tube quantity and shell inner diameter is calculated as shown in

Figure 4. In order to achieve dehumidification rate is more than 400g/h, the reasonable quantity of membrane tube and shell inner diameter should be designed below the solid line and above the dotted line. On the other hand, the smaller shell diameter, the greater pressure drops on the air-side as shown in Figure 5.

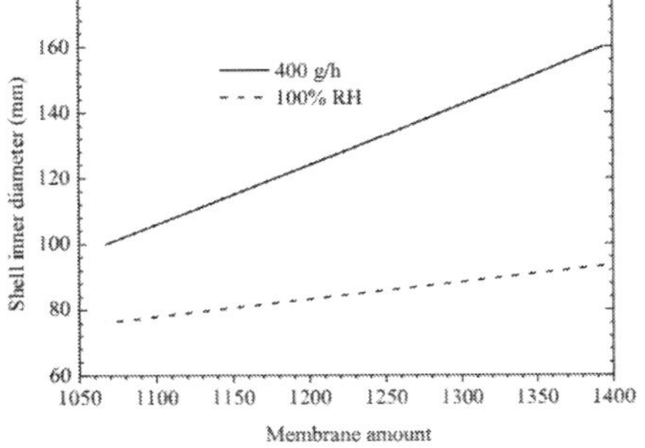

Figure 4. Design range(high-humidity)

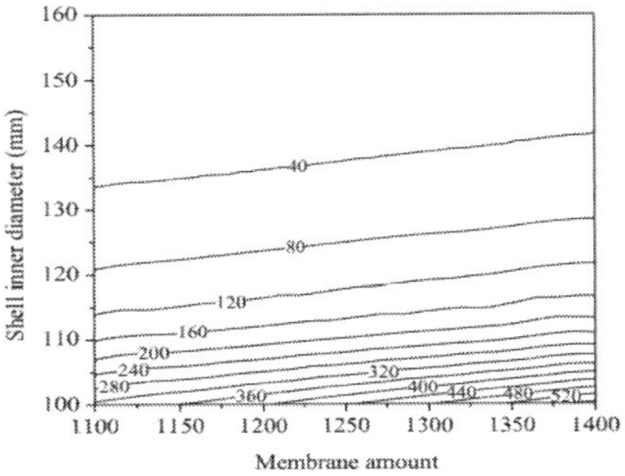

Figure 5. Air-side pressure drop (high-humidity)

Under the operating condition that the inlet air temperature is 22.5 °C and the relative humidity is 50%, the tube length is 400 mm, and the other condition as above. The design range of tube quantity and shell inner diameter is calculated as shown in Figure 6. The quantity of tubes required in the low humidity condition is significantly increased compared with Figure 4, and the inner diameter of the shell is also increased. The design range under low humidity conditions is significantly narrower, that is to ensure dehumidification rate, and the risk of condensation is increasing. The smaller shell diameter, the greater pressure drops on the air-side as shown in Figure 7.

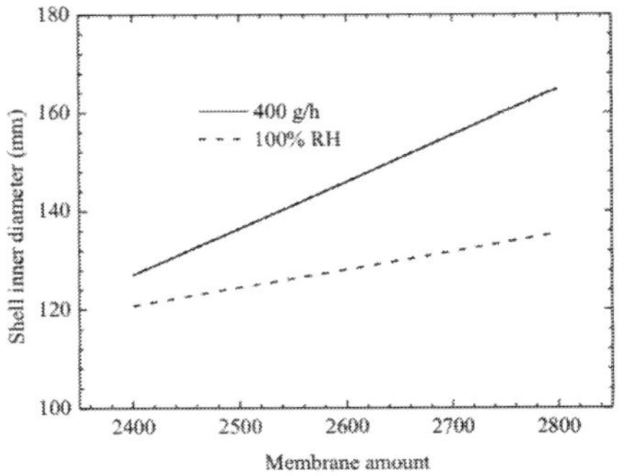

Figure 6. Design range (low-humidity)

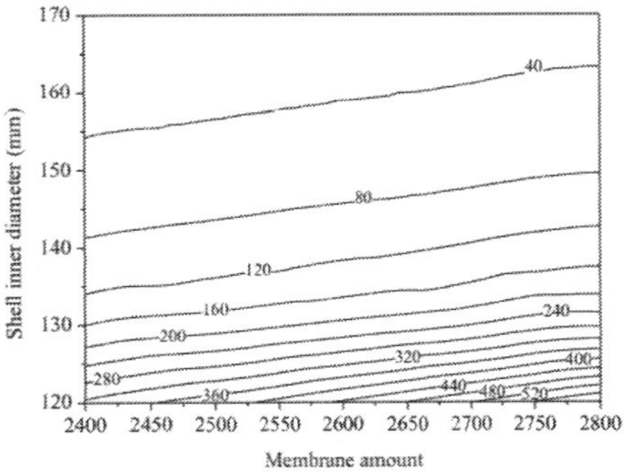

Figure 7. Air-side pressure drop (low-humidity)

4.3 Operating condition

When the manned cabin conditions of temperature and humidity are keep constant, the dehumidification rate is adjusted by changing the air flow and temperature of cold water. The greater the air flow, the greater the mass transfer coefficient on the air side, the easier moisture content of air is maintained at higher value, and the greater mass transfer potential difference, both of which will contribute to the dehumidification performance. And on the other hand, the lower temperature of cold water can effectively reduce partial pressure of water vapor, thereby improving the dehumidification performance.

Figure 8 shows the design results of different dehumidification performance of operating conditions, where the air temperature and humidity are maintained at 26 °C and 60%, cold water flow is 0.1kg/s. In order to maintain the dehumidification capacity, the air flow needs to be enlarged when the temperature of cold water increases. Since the relationship between partial pressure of water vapor and cold water temperature is non-linear, and the magnitude of partial pressure of water vapor will increase more and more with the water temperature increases, which will result in a sharp decrease in mass transfer potential difference, so the increase extent of air flow will be more and more.

From the above simulation analysis, the membrane assembly can meet the dehumidification requirements even the temperature of cold water rose to 10°C. Compared with condensing dehumidification's cold water operating temperature is 5°C, cold water-membrane dehumidification system has lower requirements on the cooling capacity, which has better thermodynamic properties.

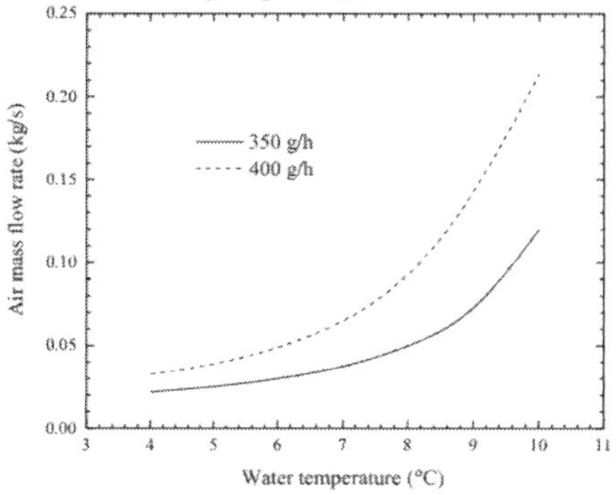

Figure 8. Operating condition design under different dehumidification demands

4 Results and discussion

A novel cold water-membrane dehumidification method is proposed for manned spacecraft. The heat and mass transfer model was established for the cold water-membrane dehumidification process, and the membrane permeability was measured by experiment. The process of cold water-membrane dehumidification assembly was simulated and the counter-flow form was determined. The structure and operating conditions were designed for certain dehumidification requirements. Compared with condensation dehumidification, cold water-membrane dehumidification has advantages:(1) achieves water recovery without additional gas-liquid separation processes;(2) assembly consisting of polymer materials are lighter in weight;(3) has better thermodynamic properties.

References

1. Huang JR and Fan HL. Numerical simulation of the thermal and humid environment in manned spacecraft habitation cabin. Chinese Space Sci Technol 2004; 24: 7–13
2. Michael Izensonl, Weibo Chen. Evaporative Cooling and Dehumidification Garment for Portable Life Support Systems. 43rd International Conference on Environmental Systems, July14-18, 2013.
3. Westheimer D., Tuan G. Active Thermal Control System Considerations for the Next Generation of Human Rated Space Vehicles. AIAA 10-13 January 2005
4. David W., Jason D., Gregory G. International Space Station Environmental Control and Life Support System Status for the Prior Year:2010-2011. 42nd International Conference on Environmental Systems. 2012
5. Tony JR, John WS, Grant B. Performance of a Water Recirculation Loop Maintenance Device and Process for the Advanced Spacesuit Water Membrane Evaporator. AIAA, 2013
6. Christie Iacomini, Josh Hecht, Jim Harrell, John Lumpkin. Qualification of the Boeing Starliner humidity control subassembly, 2016-322
7. Tomaszewska M., Gryta M., Morawski A. Mass Transfer of HCl and H_2O across the Hydrophobic Membrane During Membrane Distillation. Journal of Membrane Science, 2000, 166(2): 149-157
8. Zhang L.-Z., Huang S.-M., Pei L.-X. Conjugate Heat and Mass Transfer in a Cross-Flow Hollow Fibre Membrane Contactor for Liquid Desiccant Air Dehumidification. International Journal of Heat and Mass Transfer, 2012, 55(25): 8061-8072
9. Zhang L.-Z., Li Z.-X., Zhong T.-S., et al. Flow Maldistribution and Performance Deteriorations

in a Cross Flow Hollow Fibre Membrane Module for Air Humidification. Journal of Membrane Science, 2013, 427: 1-9

MATEC Web of Conferences **114**, 02012 (2017) DOI: 10.1051/matecconf/201711402012
2MAE 2017

Integrated analysis of aerodynamic/stealth characteristics of blended-wing-body lifting vehicle

Ye Tao[1], Jianquan Ge[1], Qiangqiang Xu[1] , Hao Huang[1] and Tao Yang[1,a]

[1] *College of Aerospace Science and Engineering, National Univ. of Defense Tech.*, 410073, *Changsha, P.R. China*

Abstract. In this paper, a blended-wing-body lifting vehicle is designed, with the cross section described by Class/Shape transformation method (CST), the outline controlled by an exponential function, and tail regulated by B-spline curve. Rapid estimation procedures of aerodynamic based on plane method and of radar cross section (RCS) based on physical optics are developed and validated. A frame of integrated design and analysis of aerodynamic and stealth characteristics is constructed and applied to parameter sensitivity analysis. The results show that the parameters that affect the aerodynamic differ from those of RCS, which means the vehicle could be good at both aerodynamic and stealth, and the study followed should be focused on it.

1 Introduction

The design of hypersonic glide vehicle (HGV) has been a hot field in vehicle research for a long time. It involves disciplines like aerodynamic, thermal, structural, trajectory, guidance, stealth and so on, which have strong coupling effects between each other. The coupling effects should be fully considered in the overall design process. However, the design of HGV is general guided by the aerodynamic characteristics currently, which induced the appearance of many high lift-to-drag ratio shapes[1] in the laboratory and paper, it is doubtful that whether these shapes possess practical significance. Against this problem, some scholars have proposed that introducing new constraints or evaluation criteria like the loading rate[2], structure[3], heat[4], stability[5] and so on. But none of these are related to the shape directly. This paper argues that, in the shape design of HGV, the capability of stealth should be treated as equal as aerodynamic. Not only because the shape feature is the most important technology to improve the stealth performance, but also the low observable technology is the most efficient technique to against the rapid development of supersonic aircraft defense system and improve the penetration ability of assault weapons.

Lots of work has been carried out about the integrated design of aerodynamic and stealth. Chu[6] develops a mixed RCS calculation method to calculate the radar scattering performance of a kind of cruise missile. Liu[7] introduces three control factors to accelerate the multi-layer fast multi-level algorithm and ensure the appropriate accuracy meanwhile. In order to reduce the computational cost, the optimization process is divided into two levels by Wu[8], the first is the coupling optimization of aerodynamic and stealthy, and the effect of structure is considered later. These are intended to improve

[a] Corresponding author: Ty93pical@163.com

the precise and accelerate the computation of RCS. There are also works focus on the structure of integrated modeling[9-12], and the main purpose of them is to acquire an efficient operation system of multi-discipline.

In summary, the aerodynamic and stealth integrated design technology of low speed of fighter planes, unmanned aerial vehicles and cruise missile is pretty mature, but the corresponding research on HGV is relatively few. In this paper a simple parameterization method for hypersonic **B**lended-wing-body **L**ifting **V**ehicle (BLV) is proposed, and then an integrated platform established with rapid estimation modules of aerodynamic and RCS, an experiment is designed to obtain the sensitivity of shape parameters on the aerodynamic or stealth properties, which lay the foundation for shape optimization study subsequently.

2. Parameterization for blended-wing-body lifting vehicle

Geometric modeling is the prerequisite for performance analysis. The shape parameter system is the basis of geometric modeling. A kind of lifting vehicle with blended wing is proposed, which is based on the parameterization of Class/Shape transformation method (CST), exponential function method and B-spline curve method. The aircraft is mainly divided into two parts: the fuselage and body-merged tail. The cross section of fuselage is described with CST method, the outline is controlled by an exponential function, and the tail is regulated by B- spline curve. The vehicle designed is shown in figure 1.

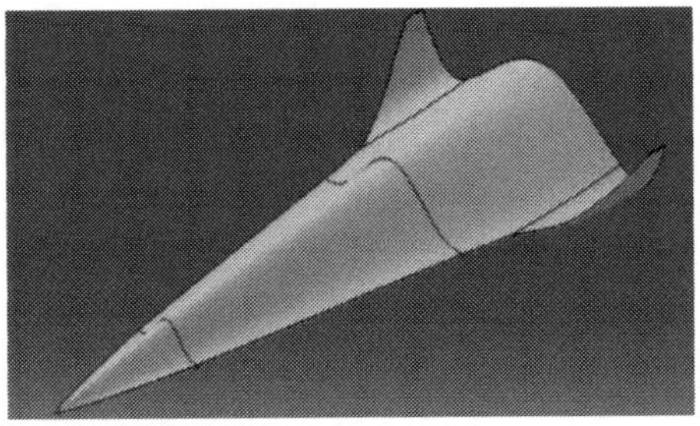

Figure 1. The blended-wing-body lifting vehicle

2.1 Parameterization

The Class/Shape Transformation Method is proposed by Kulfan and Bussoletti[13-15], which use the combination of Class Function and Shape Function to express the geometric Shape change. The formula for CST curve is

$$\eta(\psi)=C(\psi)S(\psi)+T(\psi) \tag{1}$$

In the formula, ψ is a normalized length, $C(\psi)$ is the Class Function, which represents the geometric property of the curve, $S(\psi)$ is the shape function, which represents the shape of the curve precisely. $T(\psi)$ is the thickness function. In the design of the cross section of the fuselage, the thickness function is assumed as $T(\psi)=0$, and the cross section is divided into the upper and lower surfaces to be controlled by two curves, as shown in figure 2.

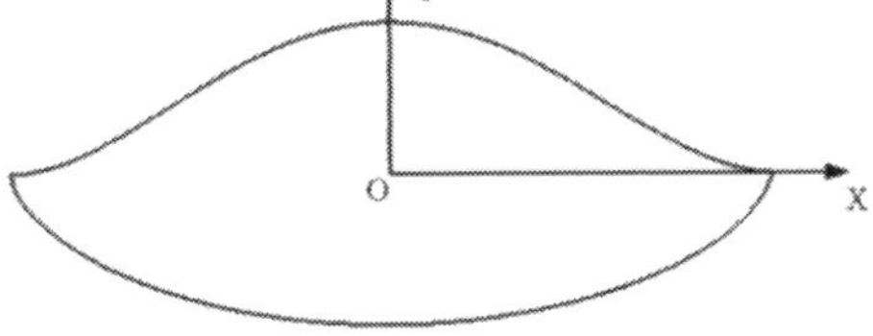

Figure 2. CST cross section curve

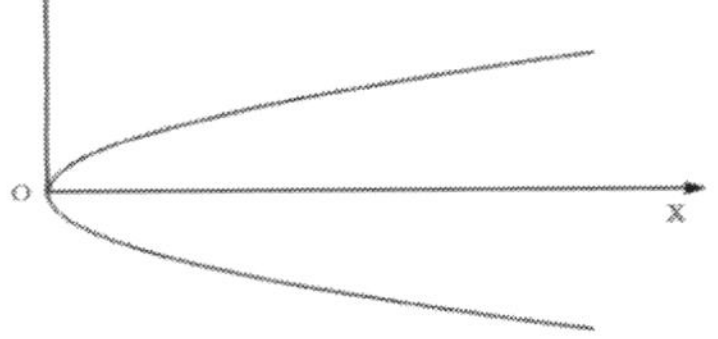

Figure 3. Exponential function control line

The width of the section is set to W, so the range of x is $[-W/2, W/2]$, and the height of the curve is set to H, so the range of y is $[0, H]$. With $C(\psi)=\psi^{Nc1}(1-\psi)^{Nc2}$ and $S(\psi)=2^{2Nc}$, the control curve of the upper and lower surface can be expressed as

$$y_u=H_u \cdot (x/W+0.5)^{Ncu} \cdot (0.5-x/W)^{Ncu} \cdot 2^{2Ncu} \quad (2)$$

$$y_l=-H_l \cdot (x/W+0.5)^{Ncl} \cdot (0.5-x/W)^{Ncl} \cdot 2^{2Ncl} \quad (3)$$

The subscript u and l are denoted as upper and lower curve parameters. Thus, each section will be controlled by five parameters, which are H_u, H_l, Ncu, Ncl and W.

The outline of the fuselage is controlled by an exponential function, which is $y=0.5W \cdot (x/L)^n$, where W is the maximum width of the model, L is the total length, n is the index to control variation of the curve. An example of the outline generated by exponential function shown in Figure 3.

A B-spline curve is controlled by a series of key point, the more of control points, the finer the curve is. Set $B_{i,p}$ as P order B-spline basis function, P_i as the coordinates of the main point, $A(t)$ as a parametric curve. The mathematical expression of $A(t)$ is:

$$\mathrm{A(t)}=\sum_{i=1}^{n+1} B_{i,p}(t)P_i \quad (4)$$

Figure 4 explains how to use the B-spline curve to control the tail. Point A is adhered the longitudinal section of the body and fixed, due to the limitation of the overall width, the coordinate value of the direction X of point D is fixed too. Thus, the main control points are point B and C. In order to ensure the rationality of the two control points, the heights are fixed too, and only the values of the coordinate X are changeable.

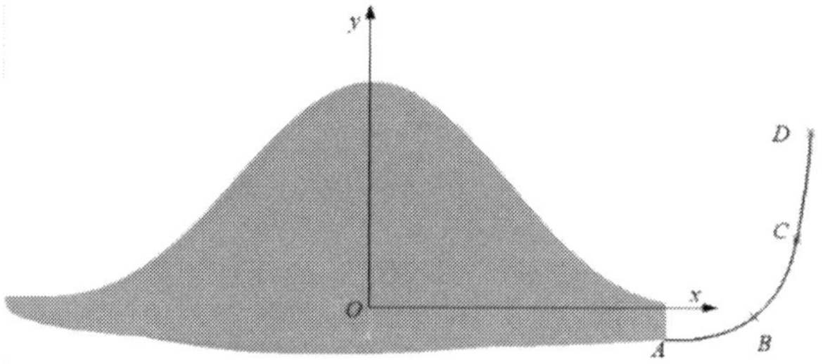

Figure 4. Curve controlled by B-spline method

The tail could be formed by using multi-section solid technique guided by the B-spline mentioned above. The sections used are the longitudinal section in point A and normal planes in points B, C and D, which are generated by narrowing 0.66 times in sequence based on the section in point A.

2.4 Main parameters of the BLV

The fuselage is controlled by three sections, and there are six parameters (H_{ui}, H_{li}, Ncu_i, Ncl_i, l_i and W_i) for each section, along with n the index of exponential function and overall length L, there are 20 parameters to generate the body in summary. Besides, the tail is simplified to be controlled by the coordinates X of point B (a_1) and C (a_2), plus the total width W, a total of 23 parameters make up the system of BLV's parameterization. All parameters used to control the shape of BLV are presented in table 1, along with their symbol, initial value and range.

Table 1. Main parameters of the BLV

No.	Symbol	Initial Value	Range	No.	Symbol	Initial Value	Range
1.	L	6500 *mm*	—	13.	H_{u2}	650 *mm*	(550, 700)
2.	W	2960 *mm*	—	14.	H_{l2}	120 *mm*	(100, 250)

3.	l_1	1500 *mm*	—	**15.**	N_{cu2}	4	(1.32, 5)
4.	l_2	3000 *mm*	—	**16.**	N_{cl2}	0.3	(0.3, 0.62)
5.	l_3	2000 *mm*	—	**17.**	W_3	2500 *mm*	—
6.	n	0.8	(0.5, 0.85)	**18.**	H_{u3}	750 *mm*	(700, 900)
7.	W_1	800 *mm*	(600, 900)	**19.**	H_{l3}	150 *mm*	(150, 250)
8.	H_{ul}	320 *mm*	(250, 450)	**20.**	N_{cu3}	4	(1.32, 5)
9.	H_{ll}	120 *mm*	(100, 250)	**21.**	N_{cl3}	0.3	(0.3, 0.62)
10.	N_{cul}	4	(1.32, 5)	**22.**	a_1	1400	(1350, 1410)
11.	N_{cll}	0.3	(0.3, 0.62)	**23.**	a_2	1450	(1410, 1470)
12.	W_2	1800 *mm*	(1650, 1800)				

3 Method of aerodynamic computation

In the conceptual design of a new vehicle, a large number of aerodynamic characteristics prediction is needed. Due to the huge expenditure of time, manpower and money of numerical computation, it is necessary to develop a fast approximation method in the initial stage.

It is difficult to calculate the aerodynamic characteristics in hypersonic flows precisely for a complex object. But for a simple facet, things are different, the expenditure of time could be much less and the precise could be relatively high with Plane Method. So it would be a good idea to create a method by dividing complex curved surface into small triangle surface and calculate the pressure coefficient on each face.

The plane method of aerodynamic is based on the modified Newton method, which is aimed at the situation presented in figure 5, the pressure coefficient of plane A is $C_p=C_{pmax}\sin^2\theta$, where C_{pmax} is the maximum pressure coefficient, and in the hypersonic flows, C_{pmax} is calculated on the stationary point of the normal shock, which is:

$$C_{p\max}=\frac{2}{\gamma M_1^2}\left\{\left[\frac{(\gamma+1)M_1^2}{4\gamma M_1^2-2(\gamma-1)}\right]^{\frac{\gamma}{\gamma-1}}\left(\frac{1-\gamma+2\gamma M_1^2}{\gamma+1}\right)-1\right\} \tag{5}$$

As shown in figure 6, **n** is the normal vector and G is the geometric center of a facet, $\mathbf{V}_\infty$ is the velocity of the free stream, which would be **V** after the impact with the facet, and the included angle between $\mathbf{V}_\infty$ and **V** is Θ:

$$\Theta=\frac{\pi}{2}-arc\cos\left(\frac{\mathbf{V}_\infty\cdot\mathbf{n}}{|\mathbf{V}_\infty|\cdot|\mathbf{n}|}\right) \tag{6}$$

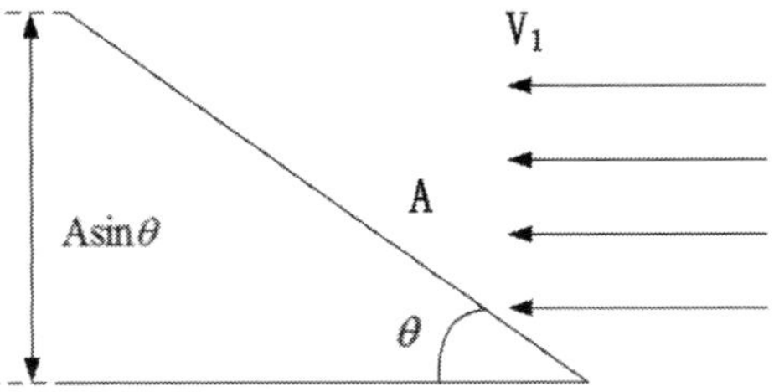

Figure 5. Sketch of the modified Newton method

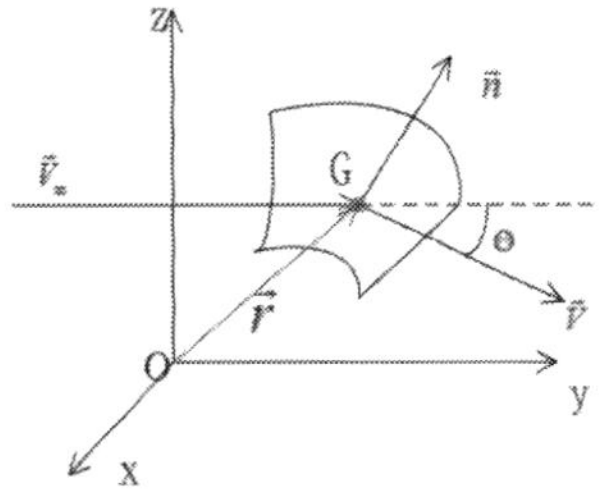

Figure 6. Sketch of the plane method

With C_p, **n**, G and the area A of every facet, the overall aero force is expressed as followed:

$$\mathbf{F}=-\int_s p\mathbf{n}ds=-\sum_{i=1}^{n_e}p_in_{ix}ds_i\mathbf{i}-\sum_{i=1}^{n_e}p_in_{iy}ds_i\mathbf{j}-\sum_{i=1}^{n_e}p_in_{iz}ds_i\mathbf{k} \tag{7}$$

where, $p=C_p{\cdot}A$, and n_e is the number of facets.

In order to verify the rapid estimation procedure of the aerodynamic characteristics developed in this paper, eight angles of attack (1°, 2.5°, 4°, 5°, 5.5°, 6.5°, 8°, 10°) were calculated for the BLV model with the non-stick CFD program Cart3D in the free stream whose Mach number is 7, the height is 25km, the reference area is 1m^2, and the reference length is 6.5m.

The lift-to-drag ratios of estimation procedure and Cart3D are compared at the angle of attack from 1 to 10 degrees, and plotted in figure 7 for conspicuous. The results shows that the estimation procedure is dependable for that the trends for the two methods are similar and the numerical error between are regular and easily corrected, and the relative error in the maximum lift-to-drag ratio is only 5.2%, which is pretty acceptable for an estimation method.

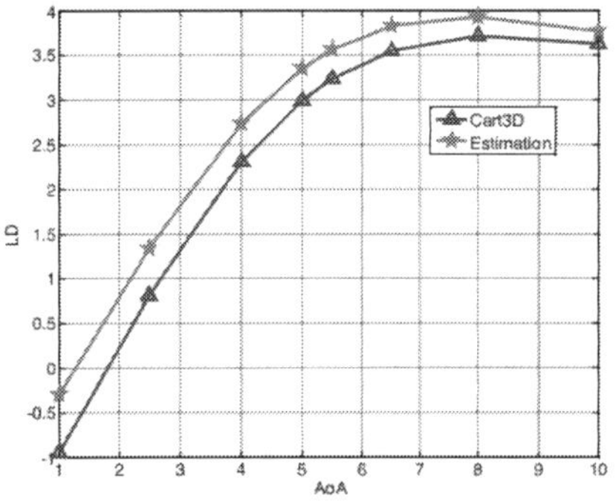

Figure 7. Sketch of aerodynamic compare

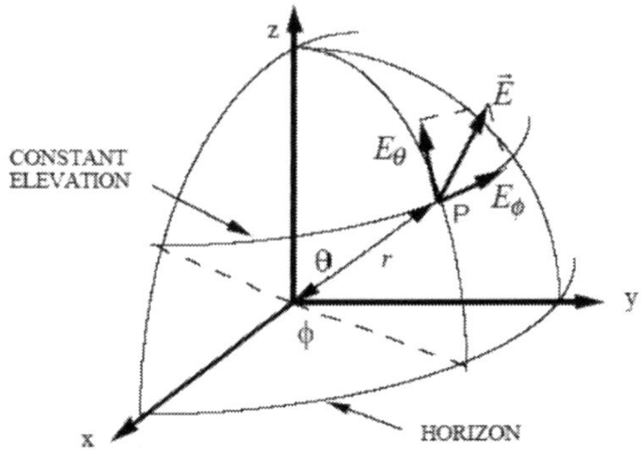

Figure 8. Coordinate system of physical optics

4 Method of electromagnetic scattering calculation

4.1 RCS definitions

The scattering energy of radar target can be expressed by the product of an effective area and the incident power density, which is called radar cross section σ, and the basic theoretical definition formula is:

$$\sigma = \lim_{R\to\infty} 4\pi R^2 \frac{|\mathbf{E}_s|^2}{|\mathbf{E}_i|^2} = \lim_{R\to\infty} 4\pi R^2 \frac{|\mathbf{H}_s|^2}{|\mathbf{H}_i|^2} \tag{8}$$

Where $\mathbf{E}_s$ and $\mathbf{H}_s$ is the energy intensity of the scattering electric field and magnetic field at the point whose distance to the goal is R, $\mathbf{E}_i$ and $\mathbf{H}_i$ is the energy intensity of the incident electric field and magnetic field. Considering that the RCS varies greatly, it is usually expressed in logarithmic form, σ_{dBsm}=10lg(σ), and its unit is dBsm, meaning the decibel of the square meter.

RCS is a function of the target shape, material, electromagnetic wave frequency, polarization type and target attitude. Normally, RCS is designated as a symbol $\sigma_{pq}(\theta, \varphi)$ after the target is determined. Where p and q are the polarization type of the wave, θ and φ is the relative angle to the line of sight[16].

4.2 Physical optics method

The physical optics method is for electrically large targets in the high frequency, the source of the scattered field is replaced by the induced current on the surface in the theory, and then the surface induced current is integrated to obtain the scattering field. Assuming that the radius of curvature of the surface of the scattering is much larger than that of the wavelength, it is considered that the influence of the scattering from the other surface of the object can be neglected during the solution, so that the RCS of each surface element can be solved independently.

As shown in Figure 8, suppose there is a radiation source in the origin, the scattering field in P is:

$$\mathbf{E}_s(r,\theta,\varphi) = E_\theta(r,\theta,\varphi)\hat{\theta} + E_\varphi(r,\theta,\varphi)\hat{\varphi} = \frac{-jKZ_0}{4\pi r} e^{-jkg} \iiint_V \mathbf{J} e^{jkg} dv' \tag{9}$$

where, r is the distance between point P and the origin, θ and φ is the Euler transition angle, $\mathbf{J}$ is the induced current, Z_0 is the inherent impedance of surrounding space, $k=2\pi/\lambda$ (λ is the wavelength) is the wave number, g is defined as $g=\mathbf{r}\cdot\mathbf{r}'$. When the scattering center is a triangular element, the volume integral can be converted into surface integral, which is

$$\mathbf{E}_s(r,\theta,\varphi) = \frac{-jKZ_0}{4\pi r} e^{-jkg} \iint_A \mathbf{J}_s e^{jkg} ds_p \tag{10}$$

Where $\mathbf{J}_s$ is the surface current, A is the area of the facet. Using the integral formula in all surface element, and the total scattering field of the target could be acquired by summing them up. In order to ensure the accuracy of calculation, the length of grid should generally set to 1/10 ~ 1/3 of the wavelength.

4.3 Results and verification

The monostatic RCS characteristics of a cylinder and the BLV are calculated and compared with the commercial software **CST** and the experimental data.

Cylinder: length 71.1cm, diameter 14cm, frequency 5.4GHz. Compares of the results are shown below, and the experimental data could find in the reference book wrote by Yingzheng Ruan[17].

BLV: total width 2.96m, total length 6.5m, frequency 9.0GHz.

As shown in figure 9, the result of cylinder of the self-developed program fits the experimental data quite good with the same fluctuating characteristics in detail and the same trend on the whole, but the value is generally small, it could be due to the lack of edge diffraction. For the vehicle designed in this paper, result presented in figure 10 reflects the symmetrical features of the vehicle and is very close to the result of commercial software. In conclusion, the RCS estimation procedure could predict the electromagnetic scattering exactly, and possess high practical value.

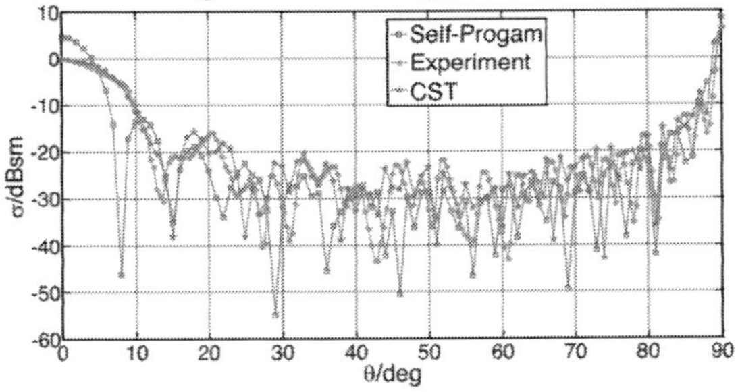

Figure 9. Monostatic RCS compare of the cylinder with different data

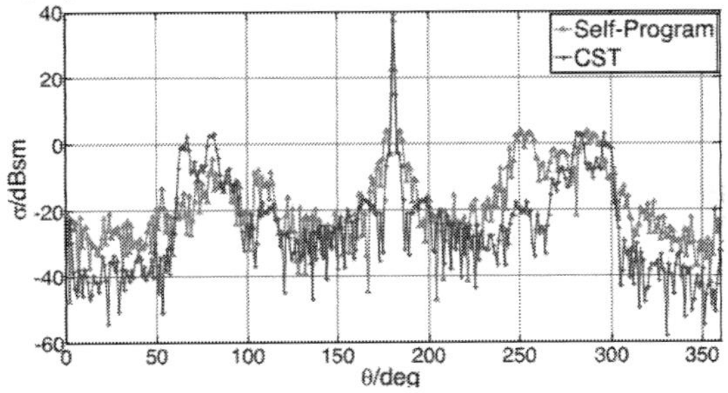

Figure 10. Monostatic RCS compare of the BLV with Self-Program and **CST** data

5 Integrated design of aerodynamic and stealth

The technology of 3D geometric modeling, rapid grid generation, read and transformation of grid data, rapid RCS calculation, and rapid aerodynamic calculation is involved in the integrated design. In this paper, integrated platform of the hypersonic BLV could link the output from the aircraft shape control parameters with the aerodynamic characteristics and RCS characteristic data automatically.

As shown in Figure 11, when the aircraft shape control parameters are changed, the CATIA module updates the model and the display automatically, and then mesh module according to grid needs of aerodynamic and RCS calculation generates two sets of grid files, aerodynamic branch read grid files for aerodynamic and calculate for the aerodynamic coefficient, and the RCS branch reads the RCS grid to obtain the average RCS in the forward direction of the vehicle.

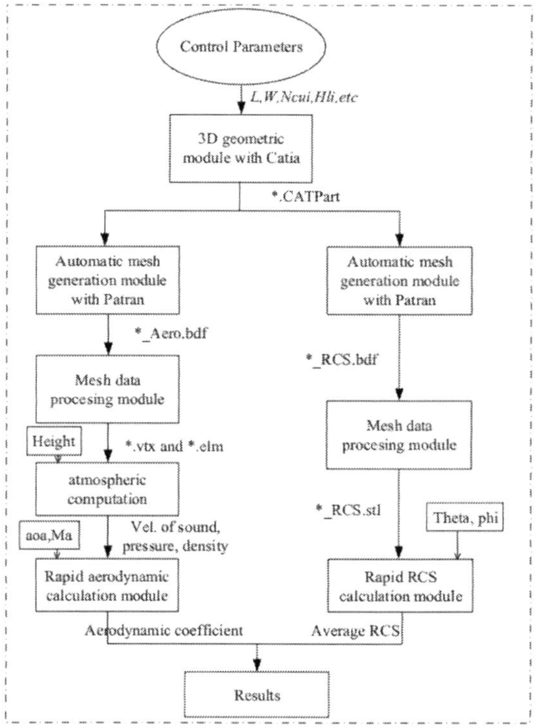

Figure 11. The work flow of the aerodynamic/stealth integrated platform

6 Sensitivity analysis of control parameters

Sensitivity analysis is a branch of mathematical statistics, which is one of the most important statistical methods in optimization nowadays. With appropriate sensitivity analysis before optimization, twice as much can be accomplished with half the expenditure.

The Orthogonal Arrays (OA) is an efficient, rapid and economical method for fractional factorial analysis. It is a set of combination of multiple factors at multiple levels which are balanced to ensure that all levels of all factors are considered equally. Dr. Genichi Taguchi has designed a group of matrix filled with OA of 2 or 3 levels, with which the experiments could be applied more scientific.

As there are 23 parameters of the blended-wing-body lifting vehicle, but only 17 of them is alterable, an experiment of 17 factors at 3 levels (the low level is the lower limit, the high level is the upper bound, and the middle level is the average of bounds.) is carried out for sensitivity analysis on RCS and aerodynamic performance. The results are presented below:

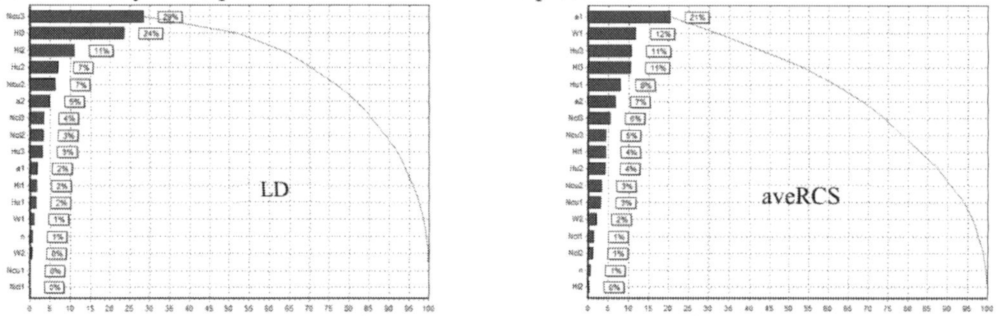

Figure 12. The results of sensitivity analysis on LD and average RCS.

The influence of control parameters on LD or average RCS is expressed as percentage shown in figure 12, assume that value greater than 7% indicates a significant effect, thus the variables have a remarkable impact on LD are N_{cu3}, H_{l3}, and H_{l2}, and on RCS are a_1, w_1, H_{u3}, and H_{l3}. It's easy to figure out the reason why N_{cu3} and H_{l3} is notable to LD, both N_{cu3} and H_{l3} are the control parameters of the biggest cross section, which means the greatest impact on the shape in the group of N_{cui} and H_{li}.

The effects of shape control parameters on RCS is vital, as it is difficult to master the influence of shape on RCS. Both LD and RCS are impacted definitely by H_{l3}, which means H_{l3} should be carefully considered in further design. According to the result of sensitivity analysis of control parameters, there is no great conflict between the aerodynamic and stealth, which shows that the overall performance of the vehicle could be better after optimization.

7. Conclusion

In this paper, a blended-wing-body lifting vehicle is designed by mixing three kinds of curve control method, meanwhile a platform for integrated design and analysis is built. With the help of the platform, parameters which are prominent in the aerodynamic and radar reflection characteristics are mastered by sensitivity analysis, thus it becomes clear that how to design a vehicle with particular aerodynamic and radar needs. The further work would be concentrated on raising and solving the MOP of aerodynamic and stealth, and improving the precision of each module to make the results more reliable.

Acknowledgments

This work is supported by Aeronautical Science Foundation of China (Grant No. 2016ZA880003).

References

1. T. Smith, K. Bowcutt, J. Selmon, et al. HIFiRE 4: A Low-Cost Aerodynamics, Stability, and Control Hypersonic Flight Experiment [C]. *17th AIAA International Space Planes and Hypersonic Systems and Technologies Conference.* (San Francisco, California. 2011).
2. L. Zhang. Rapid aerodynamic shape optimization design of aircraft based on surrogate model [D]. Changsha, Hunan, P.R. China: National University of Defence Technology (2015).
3. J. Hu, R.H. Wang, W.J. Wang, et al. Multidisciplinary Optimization of Transport Wing Aerodynamic/Structural Integrated Design [J]. J. Nanjing Univ. Aero. & Astro. **44**, 458-463 (2012).
4. X. Chen, L. Liu, Z.J. Yue. A Reduced Order Modeling for Aerothermodynamic of Hypercsonic Vehicles Based on Surrogate Method [J]. Tran. Beijing Ins. Tech. **36**, 340-348 (2016).
5. Q. Gao, J.B. Zhao, Q. Li. Study on Lateral-Directional Stability of HTV-2 Like Configuration [J]. J. Astro. **36**, 657-662 (2014).
6. X.Y. Chu. Integrated Design Method of Cruise Missile Including Configuration, Aerodynamics and Radar Stealth [D]. Nanjing: Nanjing University of Aeronautics and Astronautics. (2004).
7. Z.H. Liu, P.L. Huang, X. Gao, et al. Multi-frequency RCS Reduction Characteristics of Shape Stealth with MLFMA with Improved MMN [J]. Chin. J. Aero. **23**, 327-333 (2010).
8. W. Di, L. Teng, L.Y. Lin et al. Aero-Structure-Stealth Coupled Optimization for High Aspect Ratio Wing Using Adaptive Metamodeling Method [C]. *15th AIAA/ISSMO Multidisciplinary Analysis and Optimization Conference*. (Atlanta, GA. 2014).
9. J.X. Su, Y. Lu, Z.Y. Zheng et al. Fast analysis and optimal design of metasurface for wideband monostatic and multi-static radar stealth [J]. J. Appl. Phys. **120**, (2016).
10. X.W. Jiang, Q.J. Zhao, G.G. Zhao et al. Integrated optimization analyses of aerodynamic/stealth characteristics of helicopter rotor based on surrogate model [J]. Chin. J. Aero. **28**, 737-748 (2015).

11. J. Hu, T. Liu, P. Zhang et al. Aerodynamics/Structure/Invisibility Integrative Design and Multidisciplinary Design Optimization of Loitering Weapon [J]. Aero Weap. **3**, 9-13 (2016).
12. Z.H. Jiao, F. Deng, H. Liu et al. Aerodynamic and Stealthy Optimization Design Method of Hypersonic Vehicle [J]. J. Astro. **39**, 1031-1040 (2016).
13. K. Brenda. A universal parametric geometry representation method – "CST" [C]. *45th AIAA Aerospace Sciences Meeting and Exhibit*. (Reno, Nevada. 2007).
14. K. Brenda. Universal parametric geometry representation method [J]. J. Airc. **45**, 142-158 (2008).
15. K. Brenda, B. John. "Fundamental" parametric geometry representations for aircraft component shapes [C]. *11th AIAA/ISSMO MAOC*. (Portsmouth, Virginia. 2006).
16. C. Filippos. Development of code for a physical optics radar cross section prediction and analysis application [D]. California: Naval Postgraduate School. (2004).
17. Y.Z Ruan. Radar Cross Section and Stealth Technology [M]. *Beijing:National Defence Industry Press*. (1998).

MATEC Web of Conferences **114**, 02013 (2017) DOI: 10.1051/matecconf/201711402013
2MAE 2017

Effects of TiO_2 Addition on the Electrochemical Performance of the Overlay Welded High-manganese Steel Cladding

Jianglong Yi[1,a], Kai Wang[2], Ben Niu[1], Xinxin Wang[1],Jianyi Wang[1] and Chunlin Dong[1]

[1] *Guangdong Welding Institute (China-Ukraine E. O. Paton Institute of Welding) Guangdong Provincial Key Laboratory of Advanced Welding Technology,*
[2] *Foshan University*

Abstract. Series of overlay flux-cored wires with different content of TiO_2 were fabricated, and different claddings of high-manganese steel are welded by flux cored arc welding (FCAW). The influences of TiO_2 addition on the electrochemical behaviour of the FCAW claddings over a high-manganese steel were studied. The results demonstrate that all the LPR values of the overlay welded claddings are higher than that of the base metal. However, the LPR values did not increase with the addition of TiO_2. The EIS simulation results indicate that the addition of titanium dioxide improves the corrosion resistance of the high-manganese steel cladding, and the trends of the Rp values is highly consistent with the change of LRP value. At the same time, the grain refinement was achieved by the TiO_2 addition. In conclusion, the proper content of the titanium dioxide in high-manganese steel cladding contributes to its high LRP and Rp values, in this paper, the suitable content of the TiO_2 addition is 20g (0.2wt. %).

1 Introduction

It is a well-known fact that overlay welding is an excellent technique to improve the surface properties of a substrate. There are many remanufacturing technologies used to accomplish the overlay weld, such as flux cored arc welding, plasma arc welding, electroslag welding and Laser welding [1,2]. For example, a cladding weld of high-manganese steel overlaid on the surface of a carbon steel can strengthen the hardness, wear resistance and crack resistance, which is widely used in many fields such as machinery, metallurgy and mining equipment. However, because of the harsh working condition, those high-manganese steel claddings were easy to damage caused by corrosion, fatigue, and stress corrosion cracking.

In order to improve the corrosion resistance, many elements have been added into the base metal of high-manganese austenitic steels. Such as Al, Cr, Cu, and Mo, which improve the tendency to form passive films on the steel surface [3-5]. However, there are few researches focused on the effects of TiO_2 addition, especially for the electrochemical performance of the overlay welded high-manganese steel cladding. Many researchers have reported that the addition of titanium dioxide can affect the characteristics of the high-manganese steel cladding. This paper aimed to investigate the influence of TiO_2 addition on electrochemical behavior of the high-manganese steel.

In this study, we fabricate series of overlay flux-cored wire with different content of TiO_2, and made different claddings of high-manganese steel using FCAW. The linear polarization for different

[a] Corresponding author: yijl@gwi.gd.cn

deposited metals on the Q235A steel were tested, aim to investigate the effect of the TiO_2 addition on the corrosion properties of the overlay weled high-manganese steel claddings.

2 Experimental

Titanium dioxide powder of -200μm mesh (99.7% purity level) was added to the base flux-cored wire in the range from 10g (0.1wt. %) to 40g (0.4wt. %) %, the major chemical composition of the weld metal of base overlay flux-cored wire was listed in Table 1. The Q235A steel according to Chinese standard GB/T11253-2007 was used as the experimental base metal with the size of 200 mm × 100 mm × 10 mm , and the chemical composition was listed in Table 2. The different claddings were obtained by FCAW with the following welding parameters as shown in table 3. Using the EWM (Phoenix 521 forceArc puls DW) as the welding machine and use the mixture of 80% Ar+20%CO_2 as protecting gas. The plate was cladded up to a thickness of 15 mm. The prepared claddings were cut to different pieces for microstructure observation and electrochemical measurement by linear cutting. The specimens for metallographic examination were mechanically polished using standard metallographic procedure and etched with a 4 vol% nital solution.

After cladding the welded samples were cut to prepare the electrochemical test with the size of 10 mm × 10 mm × 0.5 mm. The LP and EIS testes were carried out by an Autolab PGSTAT302N electrochemical workstation. And a three-electrode system was used to test the EIS corrosion properties. Using the cladded sample as the working electrode, the Pt plate as the auxiliary electrode, the reference electrode was Ag/AgCl (3 mol/L KCl), and use the 3.5 wt.% NaCl water solution as the test electrolyte. The working electrode was soaked in the electrolyte for 30min before electrochemical test, and set the open circuit potential (OCP) as the EIS mode over a frequency range of 10^4~10^{-2} Hz. While the EIS test was finished, all samples were immediately charactered by SEM analysis (JEOL, JXA-8100).

3 Results and discussion

3.1 Microstructure characterization

The micrograph of cladding metals made by the addition of TiO_2 and without addition of TiO_2 are presented in Fig.1. The microstructures of the entire clad layer revealed austenite structure. It is observed that the claddings fabricated by the addition of 20g TiO_2 (0.2 wt. %) and 30g TiO_2 (0.3 wt. %) resulted in fine equiaxed grains compared to the others. When increasing the TiO_2 addition to 40g (0.4 wt. %), the fine equiaxed grains disappear, and the grain size also become larger, which might be contributed by the generate of fine particles such as TiC, TiN [6].

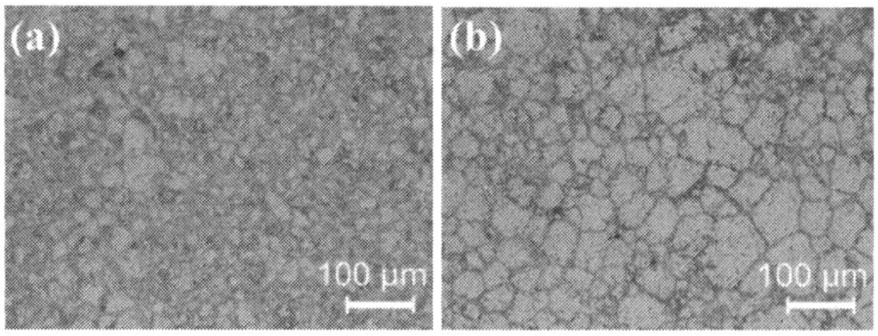

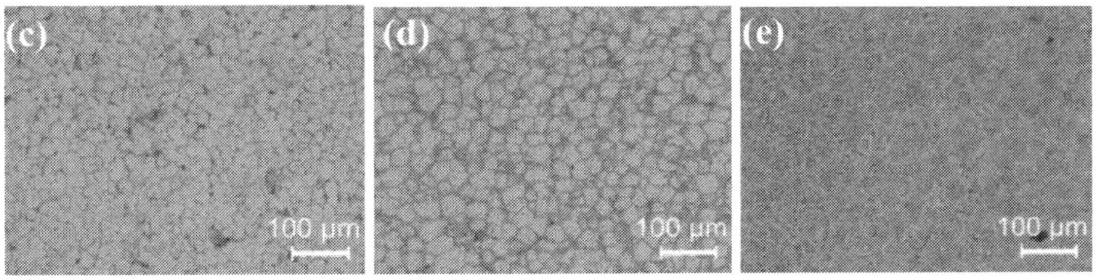

Figure 1. Optical microstructure of the five clad layer a) 0g TiO_2 (0 wt. %) addition; b) 10g TiO_2 (0.1 wt. %) addition; c) 20g TiO_2 (0.2 wt. %) addition; d)30g TiO_2 (0.3 wt. %) addition; e) 40g TiO_2 (0.4 wt. %) addition;

3.2 Electrochemical analysis

The linear polarization results of the different TiO_2 addition claddings are shown in Fig. 2. The LP lines of the clad layer the addition of 30g TiO_2 (0.3 wt. %) is the most stable with the most positive self-corrosion potential, compared with those of the overlay welded metals. The linear polarization resistances (LPR) values are shown in Tab.4, and the results demonstrate that all the LPR values of the overlay welded claddings are higher than that of the base metal. However, the LPR values are not increased with the addition of TiO_2.The LPR value of the clad layer of 1# at 10g TiO_2 addition is 11356.35 Ohm.cm^2, while the 2# presents a similar LRP value of 10467.56 Ohm.cm^2 to the LPR value of 1#. Both of them are extremely higher than the others. As the TiO_2 increased to 40g (0.4 wt. %) and 50g (0.5 wt. %), the LRP value decreased sharply to 6273.13 and 5024.23 Ohm.cm^2, respectively.

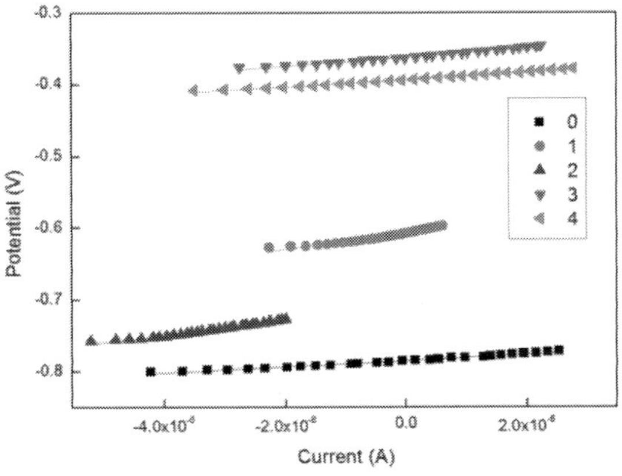

Figure 2. Linear polarization curves of the different clad layer

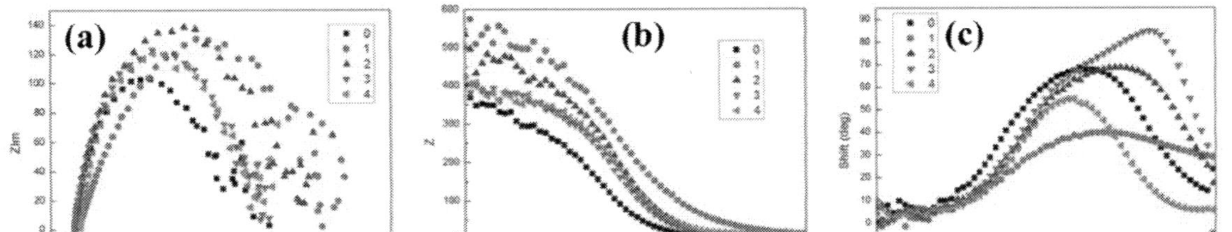

Figure 3. (a) Nyquist, (b) Bode and (c) Bode-phase plots of the five clad layer at open-circuit potential. a) 0g TiO_2 (0 wt. %) addition; b) 10g TiO_2 (0.1 wt. %) addition; c) 20g TiO_2 (0.2 wt. %) addition; d)30g TiO_2 (0.3 wt. %) addition; e) 40g TiO_2 (0.4 wt. %) addition;

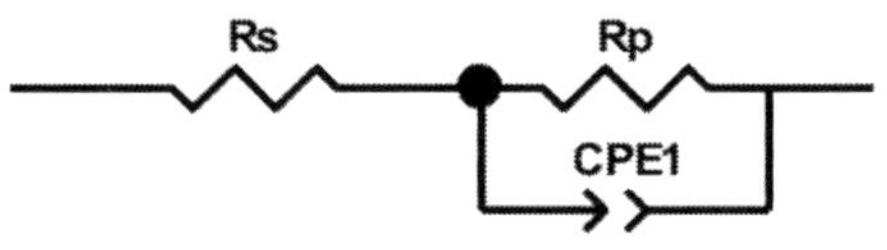

Figure 4. Best equivalent circuit used to model the experimental EIS data

Table 1. Chemical composition of the weld metal of base overlay flux-cored wire %

C	Si	Mn	Cr	Ni	V	Mo	Al
0.032	0.063	12.39	11.28	1.23	0.38	0.79	0.085

Table 2. Chemical composition of Q235 steel %

C	Si	Mn	P	S
0.14~0.22	≤0.30	0.30~0.65	≤0.045	≤0.050

Table 3. Welding parameters of the welding process

Parameter	Value
Current (A)	240-260
Voltage (V)	24-26
Electrode polarity	DCEP
Filler wire diameter (mm)	1.6
Interpass temperature (°C)	150
Welding speed (mm·s-1)	6.8

Table 4. The linear polarization resistances of the weled metal at different TiO2 Adddition

No.	TiO_2 Adddition	LPR (Ohm.cm2)
0#	0g	4418.89
1#	10g	11356.35
2#	20g	10467.56
3#	30g	6273.13
4#	40g	5024.23

Table 5. Variations in the impedance parameters of the EIS tested samples

	0#	1#	2#	3#	4#
	Value	Value	Value	Value	Value
Rs ($\Omega \cdot cm^2$)	1.27	1.12	7.17	0.33	11.78
Rp($\Omega \cdot cm^2$)	323.9	437.4	554.6	343.2	359.9
CPE1T(mf)	5.57×10-4	2.09×10-4	3.66×10-4	1.66×10-4	2.49×10-4
CPE1-P	0.82	0.81	0.54	0.91	0.81

The EIS measurements were conducted for cladded samples in a 3.5% NaCl solution in order to evaluate the corrosion protection performance of the overlay welded high-manganese steel. Figures 3 show the results for Nyquist, Bode, and Bode-phase plots. It was possible fit well with all the impedance spectra using the equivalent circuit model shown in Fig. 4.

The Nyquist plots of all the clad layer show similar semicircles, represented as the capacitive arc. The diameter of the semicircle of the overlay welded claddings with TiO_2 addition was bigger than that of the base metal without TiO_2 addition. The 1# and 2# sample show similar and also the biggest semicircle diameter which reflected the best passivation behavior. Moreover, the imperfect semicircle diameter for the clad layer of 1# and 2# is higher than that of the 3# and 4#, which agrees well with the LPR values.

As seen in Fig. 3b, the Bode plots show higher electrical resistance value at the low-frequency range, which represent the stronger corrosion resistance and consistent with the result of the above

Nyquist plots. One time-constant was observed on the bode plots of all samples (only one maximum phase lag at the middle frequency range). As can be seen from the Bode-phase curves, except the 3# all the clad layer exhibit certain constant plots at the middle frequency range [7]. Also, the phase angles values remained far below to 80° except the 3#, which indicated the formation and growth of the corrosion pits at middle frequency [8, 9]. And the maximum phase angles values of 3# was over 80°, which revealed the formation of a passive film. In other words, it means a pure or ideal capacitive response [10].

Figure 4 shows equivalent circuit best describing the electrochemical processes [11]. In this equivalent model, Rs is the solution resistance; Rp stands for the resistance of the sample in the solution, and CPE1 counts for the constant phase element corresponding to the capacitance of the corrosion pits. As the most important characteristic of all the EIS value, the polarization resistance Rp represent the resistant ability to against the corrosion [12, 13]. Table 5 shows all the values of equivalent circuit elements of the cladding samples in 3.5% NaCl solution. The highest value of Rp was reached to 554.6 Ω•cm2, which indicated the 2# with 20g TiO_2 (0.2 wt. %) possesses best anti-corrosion ability of the entire clad layer. The simulation results demonstrate that the addition of titanium dioxide improves the corrosion resistance of overlay welded high-manganese steel cladding.

3.3 Corrosion morphologies analysis

Fig. 5 shows SEM images of the corrosion morphologies of the high-manganese steel cladding with different TiO_2 addition after the electrochemical tests in a 3.5% NaCl solution. It is apparent that, the rust films formed at all the test samples. The rust films became increasingly dense and uniform as the increase of the TiO_2 addition. Fig. 4a shows the morphology of the rust presented in a loose porous structure, and obvious cracking phenomena can be easily seen on the clad layer surface. It is obvious that a dense passive film without crack can be observed on the surface of the cladded sample with 20g TiO_2 (0.2 wt. %) addition, which prevents the corrosive penetration trend of the experimental solutions. Meanwhile, this phenomenon well explained and consistent with the close LRP value and EIS measurements result. It is found that there are more corrosive crack appears on the surfaces, as increase the addition amount of the TiO_2. It will accelerate the corrosive solution penetration trend of the experimental solutions [14].

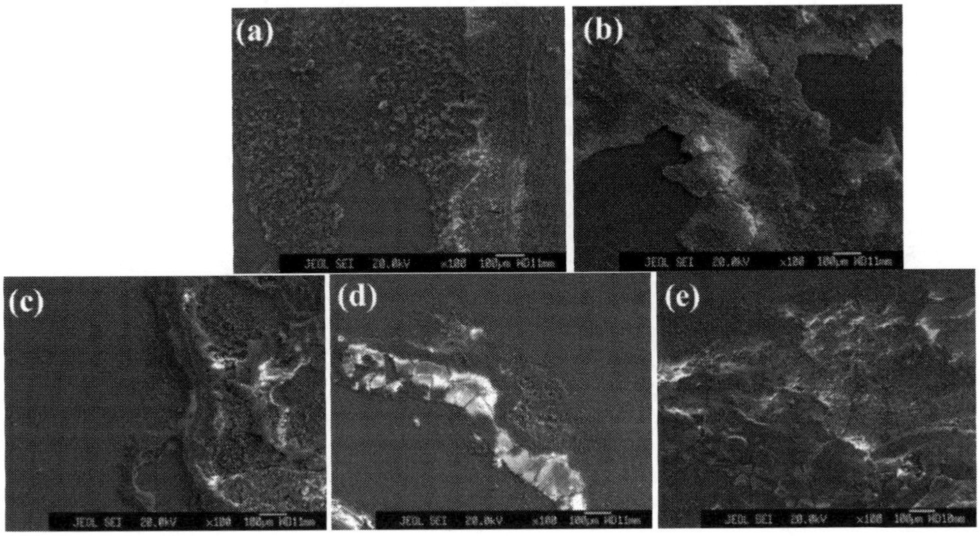

Figure 5. The corrosion morphology of the five claddings after the electrochemical tests ×500a) 0g TiO_2 (0 wt. %) addition; b) 10g TiO_2 (0.1 wt. %) addition; c) 20g TiO_2 (0.2 wt. %) addition; d)30g TiO_2 (0.3 wt. %) addition; e) 40g TiO_2 (0.4 wt. %) addition;

DOI: 10.1051/matecconf/201711402013

4 Conclusion

The results demonstrate that all the LPR values of the overlay welded claddings are higher than that of the base metal. However, the LPR values do not increase with the addition of TiO_2. To simulate the measured impedance EIS data, the equivalent circuit was used, the simulation results indicate that the addition of TiO_2 improves the corrosion resistance of the high-manganese steel cladding and the trends of the Rp values is highly consistent with the change of LRP value. At the same time, the grain refinement is achieved by the TiO_2 addition. In conclusion, the proper content of the TiO_2 in high-manganese steel cladding contributes to its high resistance values and good corrosive surface morphology, as in this paper, the suitable content of the TiO_2 addition is 20g (0.2wt. %). As resulted in the increase of the polarization resistance to present a similar electrochemical behavior as the high-manganese steel cladding.

Acknowledgements

This research was supported by the technical project of Guangdong (2014B050503004, 2013B050800028, 2015B050502005, 2016B090918120, 2012A061400011), the Guangzhou Science and Technology research project (201510010101, 201508030024, 2015B050502003), The Scientific Program of GDAS (2016GDASPT-0205) and the National Natural Science Foundation of China (No. 51601043).

References

1. N.V. Rao, G.M. Reddy, S. Nagarjuna, Mater. Des. **32**,4 (2011).
2. Metals hand book. *Welding, brazing and soldering*, 9th ed. vol. 6. Metals Park (OH, USA): ASM; (1983)
3. M.B. Kannan, R.K.S. Raman, and S. Khoddam, Corros. Sci., **50**,10 (2008)
4. Y.S. Zhang, X.M. Zhua, and S.H. Zhong, Corros. Sci., **46**, 4 (2004)
5. T. Dieudonne, L. Marchetti, M. Wery, F. Miserque, M. Tabarant, J. Chene, C. Allely, P. Cugy, and C.P. Scott, Corros. Sci. **83**,7 (2014)
6. G. Mallaiah, P. R. Reddy, A. Kumar, Procedia Mater. Sci. **6**, (2014)
7. J. Soltis, K.A. Lichti,. Corros. Sci. **68**, (2013)
8. L.R. Bhagavathi, G.P. Chaudhari, S.K. Nath, Mater. Des., **32**,1 (2011)
9. A. Fattah-alhosseini, S. Vafaeian, J. Alloys Compd., **639** ,(2015)
10. H. Y. Ha, M. H. Janga, T. H. Lee, J. Moon, Corros. Sci. **89**, (2014)
11. F. Pastorek, K. Borko, S. Fintova, D. Kajsnek, B. Hadzima, Coating, **6**,4 (2016)
12. M. Mhaede, F. Pastorek, B. Hadzima, Mater. Sci. Eng., C, **39**, (2014)
13. E. A. Avila, L. A. Rocha, Mater. Sci. Forum, **492**, (2005)
14. G. Q. Su, X. H. Gao, L. X. Du, et al. Int J Electrochem Sc, **11**, 11, (2016)

Effect of Oxidation Degree on Valence Change and Distribution of Octahedral Fe Element in Biotite

Li Ziqian[1], Huang Zhiliang[1,a] and Wei Yu[1]

[1]*School of Material and Engineering, Wuhan Institute of Technology, Wuhan 430074, China*

Abstract. In this paper, the valence change and distribution of iron elements in octahedral vacancies of biotite were studied in the oxidation process. The biotite and saturated barium nitrate solution were mixed in dilute hydrochloric acid under hydrothermal reaction conditions, the solid after reaction was used as the test sample. Firstly, the remainder potassium and iron content were measured by atomic absorption spectrometry(AAS). Secondly, the state of iron along with oxidation degree increased was analyzed, in addition, the phase composition and the change of layer spacing in samples was detected by X-ray diffraction(XRD). Thirdly, The variation mode of Si-O bond were characterized by Fourier transform infrared spectroscopy(FT-IR).This research was adopted hydrogen ions in diluted hydrochloric acid and nitrate ions in barium nitrate to provide oxidation environment for reaction, and the oxidation degree was controlled by adjusted the amount of hydrogen ion introduced. We found out that the amount of hydrogen ion is positively correlated with oxidation degree in biotite, and the deeper oxidation degree in biotite, the lower electronegativity of singer layer. Potassium and iron element would be release out of micaceous structure during the biotite oxidation. The higher the oxidation degree is, the greater the releasing happen. The charge density combining Fe oxidation and releasing firstly increased then decreased with the oxidation degree turned greater. During the oxidation, the Si-O vibrated would change from parallel layer vibration model to vertical vibration model.

1 Introduction

Layered silicate is a good template material to prepare organic composite. [1-2] Mica, as an important member of layered silicates, has been attached more and more attention. It is almost inevitable that Fe element occupies part of octahedral sites in natural mica. And the content and valence of Fe element contribute a responsibility to the charge density of aluminosilicate sheet which is very important to the utilization of mica.

Many papers about oxidation of mica have been published. Some of them are concentrated on the potassium releasing in the view of potassium supplement in soil. [3] Their research about the structure of mica after oxidation is mostly focused on the study of the hydroxyl in mica. [4-5]The research about valence change of Fe element and the distribution of Fe element with different valence at different oxidation degree has not been exposed too much. [6]

[a] Corresponding author : hzl6455@126.com

In this paper, the effect of oxidation degree in biotite on potassium located between aluminosilicate sheets and iron element located in octahedral sites was studied, aiming to introduce a sight of the valence change of Fe element and the distribution of Fe element with different valence at different oxidation degree.

2 Experimental

2.1 Reagents and instruments

The materials used in the research were listed in Table 1:

Table 1. Raw materials

Name	Remark
Biotite	High purity, purchase from Lingshou town of Hebei Province
$Ba(NO_3)_2$	AR
HCl	AR, diluted to 0.5 mol/L
H_2O	Homemade in lab

The experiment involved these instruments: CPA1245 analytical balance; DF-101S temperature control magnetic stirring apparatus; HH-S smart thermostat water bath; SHZ-DIII air pump filtration; DZ-2BC drying tunnel.

2.2 Methods

Quantitative biotite powder and barium nitrate were weighted accurately into an Erlenmeyer flask. Then different amounts of hydrochloric acid (V_{HCl}) and distilled water (V_{H20}) were added according to a series of setting solid to liquid ratios. The total volume of the final solution was 100 mL. the untreated raw biotite sample was named as S-0. The Erlenmeyer flask was thereafter putted into thermostat water bath at 60℃ for 5 hours. After that, a filtration was performed immediately. The residue was dried at 90 ℃for 12 hours, then put in a new clean flask. The above steps were repeated 4 times. The final residue was saved for characterization. The final sample were marked as S-1 that the V_{HCl}:V_{H20} was 0.0 : 100.0, S-2 that the V_{HCl}:V_{H20} was 2.0 : 98.0, S-3 that V_{HCl}:V_{H20} was 4.0 : 96.0, and S-4 that the V_{HCl}:V_{H20} was 8.0 : 92.0.

2.3 Characterization

Germany Ru-200 X-ray powder diffraction (Cu Target K alpha, 40kV tube voltage, current of 100mA tube, scanning step 0.04 degree per second), Fourier transform infrared spectrometer (Niocolet impact of 420), MS-500 Mossbauer spectrum (testing temperature of 300K and testing time of 60 hours)was used to characterize the crystal structure, chemical bonds vibration, and valance of iron element of the final sample, separately.

The content of potassium and iron values in filter residue were determined by atomic absorption spectrometry(AAS).

3 Results and discussion

Firstly, we have to explain that the oxidation environment in reaction solutions was supplied by the hydrions in diluted hydrochloric acid combined with the NO_3^- in barium nitrate. It's obvious that the NO_3^- in solution is abundant, so we control the degree of oxidation by controlling the amount of hydrochloric acid added.

3.1 AAS analysis

The content of potassium and iron values in samples were showed in Table 2.

Table 2. K, Fe content of different samples

Samples	S-0	S-1	S-2	S-3	S-4
V_{HCl}:V_{H2O}	—	0.0:100.0	2.0:98.0	4.0:96.0	8.0:92.0
K (wt%)	6.81	5.36	3.12	1.65	0.59
Fe (wt%)	10.34	10.21	7.68	5.52	2.18

The K content of oxidized biotite (S-1, S-2, S-3, and S-4) were gradually declining with the increase of V(HCl), as well as the Fe content, as shown in Table 2. This means potassium and iron element were released out of micaceous structure during the oxidation, and the higher the oxidation degree is, the greater the releasing happens.

3.2 Mossbauer spectral analysis

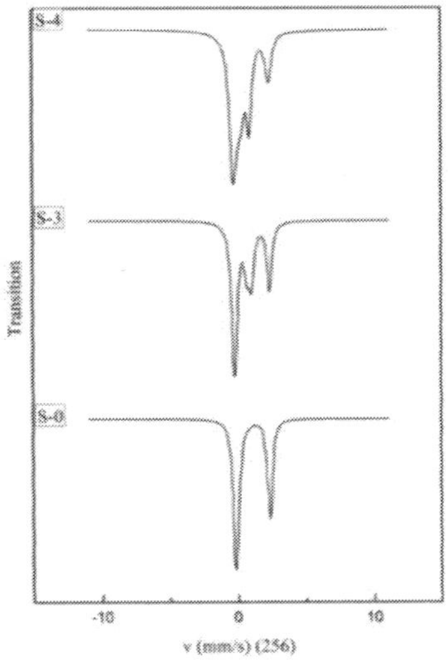

Figure 1. Mossbauer spectrums of S-0, S-3, and S-4

Figure 1 shows the Mossbauer spectrum of S-0, S-3, and S-4. The corresponding data are shown in Table 3. M1 and M2 refer the two kinds of octahedral sites in micaceous structure. As shown in Figure 4-a, if the OH ions are situated at adjacent corners of an octahedron the site is called M2, whereas if they are at opposite corners the sites called M1. [7]

Table 3. Mossbauer spectrum parameters of S-0, S-3, and S-4

Samples	Fe^{2+}							Fe^{3+}						
	M2			M1			A (%)	M2			M1			A (%)
	C.S (mm/s)	Q.S (mm/s)	A (%)	C.S (mm/s)	Q.S (mm/s)	A (%)		C.S (mm/s)	Q.S (mm/s)	A (%)	C.S (mm/s)	Q.S (mm/s)	A (%)	
S-0	1.12	2.56	80	0.94	2.41	20	73				0.01	0.12	100	27
S-3	1.08	2.59	100				41	0.49	1.15	57	0.26	0.9	43	59
S-4	1.37	1.93	100				28	0.33	1.23	55	-0.03	0	45	72

As shown in Table 3, the content of Fe^{3+} in S-4 is greater than that in S-3, and both of them are greater than that in S-0. The content of Fe^{2+} shows a reverse trend. The means that the oxidation degree of biotite increased as the V(HCl) increased. Furthermore, in S-3 and S-4, Fe^{2+} is more likely to occupy in M2 vacancy and Fe^{3+} in M1 vacancy, which demonstrates Fe^{2+} in M1 vacancy was prior to been oxidized. The ratio of Fe^{3+} in M2 and M1, which in S-3 is 1.3 and in S-4 is 1.2, is close, demonstrating the over Fe^{3+} content in M2 would be partly released to keep the charge balance.

3.3 XRD analysis

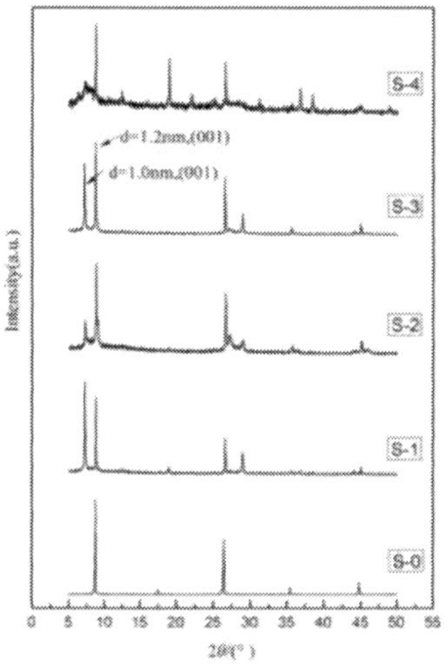

Figure 2. XRD pattern of raw ore and treated biotite

Figure 2 is the XRD patterns of samples. The specific diffraction peaks of mica were clear, meaning the micaceous structure retained during the oxidation. A new peak whose d value is about 1.2 nm emerged in oxidized biotite samples (S-1, S-2, S-3, and S-4). This is due to the hydrated barium had occupied part of the potassium site after Ba^{2+}/K^{+} exchange. [8]

The strength of the new peak went up first then down as oxidation degree turned greater. This is due to the change of charge density of aluminosilicate sheet. At the beginning of the oxidation, the charge increase of Fe^{2+} was over the charge loss of Fe releasing. The charge density of the aluminosilicate sheet would go down, weakening the electronical force and resulting in the sheets expanded. As the oxidation stronger, the charge change was reversed, leading to the charge density went up and the sheets shrank.

3.4 Infrared analysis

DOI: 10.1051/matecconf/201711402014

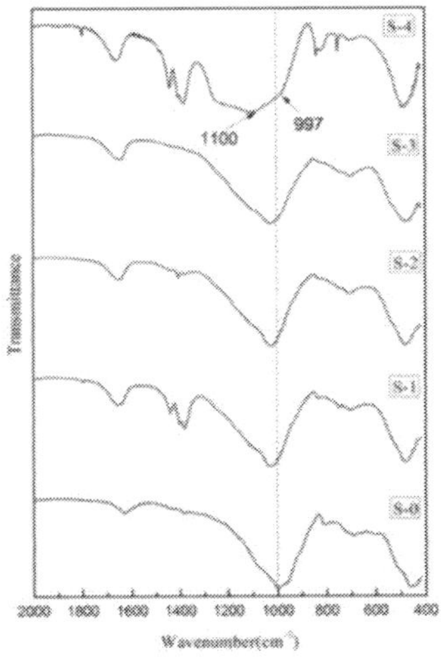

Figure 3. FT-IR spectrogram of raw ore and treated biotite

As showed in Figure 3, the Si-O vibration band which is at 1000 cm^{-1} around shifted towards high frequency(1100 cm^{-1}). According the infrared spectroscopy study in clay minerals by Farmer V C, the infrared absorption peak with a wave number of 1000 cm^{-1} belongs to the layer vibration(mode E_1^1), and the infrared absorption peak of 1100 cm^{-1} belongs to the vertical vibration(mode A_1^1) [9]. The frequency of Si-O vibration band in vertical vibration(A_1^1) is higher than that in parallel layer vibration(E_1^1), because Si-O tetrahedron force field deflects from trigonal symmetry. [10] The symmetry deflects of the silicon tetrahedron structure results in an infrared absorption peak shift that can be demonstrated using the Figure 4-a.

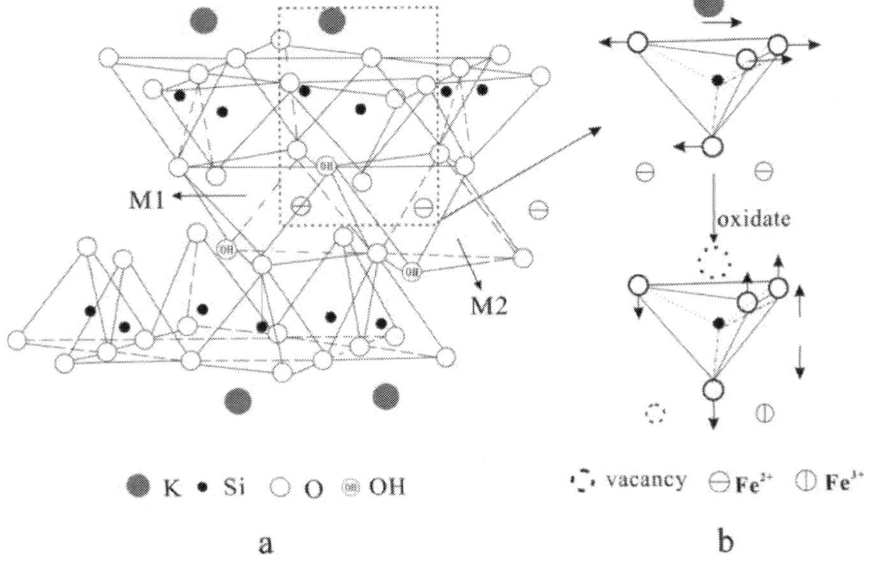

Figure 4. Diagram of the biotite structure (a), change of Si-O vibration mode after oxidation (b)

We can see the environment of Si-O tetrahedral clearly from Figure 4-a, under K^+, upper iron ions. The Figure 4-b show the change of Si-O tetrahedral environment. In original ore (S-0), Si-O vibrates as parallel layer vibration. After oxidized, the octahedron site and potassium site would partly be vacancies, the vacancy defects cause the symmetry of silicon tetrahedron structure deflects, resulting the Si-O vibrated as vertical vibration. Therefore, the frequency of Si-O vibration band would shift to higher position after oxidized.

4 Conclusion

Potassium and iron element would be release out of micaceous structure during the oxidation. The higher the oxidation degree is, the greater the releasing happens. The new emerged peaks of 1.2 nm turned stronger then weaker, and back and forth. This is due to the charge density combining Fe oxidation and releasing firstly increased then decreased with the oxidation degree turned greater. Many vacancies left after oxidation and K/Ba exchange, resulting in the change of Si-O symmetry. During the oxidation, the Si-O vibrated would change from parallel layer vibration model to vertical vibration model.

Foundation item

This research is supported by National Natural Science Foundation of China (51374155), Hubei Province Key Technology R&D Program of China (2014BCB034).

References

1. Agag T, Taepaisitphongse V, Takeichi T. Reinforcement of polybenzoxazine matrix with organically modified mica[J]. Polymer Composites, 2007, **28**(5):680–687.
2. YuMin Chen. Organic modification and intercalation of synthetic fluorine mica with poly(oxyalkylene)-amines[J]. 2004.
3. Kathsutoshi Tomita, Mitsuhiko Dozono. Formation of an interstratified mineral by extraction of potassium from mica with sodium tetraphenylboron. Clays and Clay Minerals, 1972, (2)225-231.
4. Beran A. The occurrence of OH absorptions in phenakite—an IR spectroscopic study[J]. Mineralogy and Petrology, 1990, **41**(2):73-79.
5. Rouxhet P G. Hydroxyl stretching bands in micas: quantitative interpretation.[J]. Clay Minerals, 1970, **8**(4):375-388.
6. Rebbert C R, Partin E, Hewitt D A. Synthetic biotite oxidation under hydrothermal conditions[J]. American Mineralogist, 2015, **80**(3-4):345-354.
7. Chandra U, Lokanathan S. A Mossbauer study of the effect of heat treatment on biotite micas[J]. Journal of Physics D Applied Physics, 1982, **15**(11):2331-2340.
8. Peng Meng, Ziqian Li, et al. Extraction of potassium from Biotite by Ba^{2+}/K^{+} ion exchange and the structural transformation. Physics and Chemistry of Minerals, 2016 (43), 387-393.
9. Farmer V C. Infrared Spectroscopy in Mineral Chemistry[M]// Physicochemical Methods of Mineral Analysis. 1975:357-388.
10. Lantenois S, Bény J M, Muller F, et al. Integration of iron in natural and synthetic Al-pyrophyllites: an infrared spectroscopic study. Clay Minerals, 2007, volume **42**(1):129-141(13).

MATEC Web of Conferences **114**, 02015 (2017) DOI: 10.1051/matecconf/201711402015
2MAE 2017

Preparation of SiC Powders by Carbonthemal Reduction Method at Low Temperature

Zhao Wang[1], Xiao-Yuan Dai[1], Shao-Peng Xu[1]and Man Xu[1, 2, a]

[1] *School of materials science and Engineering. Wuhan Institute of Technology. Wuhan, Hubei 430073*
[2] *Energy saving materials and Membrane Technology Institute.Wuhan Institute of Technology. Wuhan, Hubei 430073*

Abstract: The stable slurry was prepared by ball-milling with a certain proportion of silica fume as silicon source,glucose as carbon source and metal niobium (Nb) as additives mixed with alcohol water. After the slurry was dried and pulverized, put the powder into a tubular furnace heated to 650°C for 2h under nitrogen atmosphere to prepare precursor, the heating rate was 5°C/min. Treated the precursor under vacuum by carbothermal reduction method to prepare silicon carbide (SiC) powder. The effect of temperature and additive content on the morphology of synthesised silicon carbide and the impurity removal order on product purity were explored. After firing at 1300°C,1400°C,1500°C for 2 h, the 3C-SiC powders are detected, and as the temperature increased, the crystallinity of the product become better. When the content of the additive is 1% of the quality of the silica fume, the particle size of the silicon carbide synthesized at the temperature of 1500°C is even and the dispersion is better. As for the impurity removal order, removed SiO2 first, then removed C can effectively remove the impurities in the product.

1 Introduction

Silicon carbide (SiC) is one of the most important non-oxide ceramics due to its unique characteristics such as high melting point, excellent oxidation resistance, high chemical inertness, high thermal conductivity, good microwave absorbing ability, wide energy band gap, and high mechanical strength. It is considered as a promising material in many fields including aerospace structures, biomaterials, high temperature semi-conducting devices, membrane supports, and catalysis in harsh environments [1–6].

The carbothermal reduction method has the advantages of high purity, simple preparation process and low cost of raw materials, so it has broad prospects for development of SiC powder. But the traditional carbonthermal reduction is often simply using quartz mixed with carbon black, which makes the reaction time too long and the synthesis temperature too high. In addition, the effect of the impurity removal order on product purity are seldom reported in the literature. So this paper explores the influence of reaction temperature and additive content on SiC powder preparation system and the effect of impurity removal order of reaction product purity on the base of carbonthermal reduction

a Corresponding author.Tel.:18707123506
E-mail address:1733784513@qq.com

reaction of the original preparation of SiC powders.

In this paper, SiC powder was prepared by carbothermal reduction method with silicon powder as silicon source, glucose as carbon source and Nb as additive.The effect of temperature and additive content on the synthesis of SiC morphology and the impurity removal order on product purity were explored.

2 Experimental

2.1 Experimental preparation

Took a certain amount of silica fume powder, glucose and Nb with the carbon silicon ratio of 1:4 (amount of substance ratio), amount of Nb was the quality of silica fume powder respectively 1 %, 2%, 3%. The glucose was completely dissolved in the mixed solution of distilled water and absolute ethyl alcohol first, then the silica fume powder was added into the mixed liquid, and finally, an appropriate amount of Nb was added, then ball milled the slurry at high speed for 4h. Then put the slurry in the oven under 100°C for 24 h. After the slurry was dried and pulverized, put the powder into a tubular furnace heated to 650°C for 2h under nitrogen atmosphere to prepare precursor, the heating rate was 5°C/min. The precursor was heated to reaction tempeture under vacuum for 2h. The synthesized powders were washed with hydrofluoric acid (HF) to remove unreacted silica and metal impurities, then heated the powders to 650°C in atmosphere for 2h to remove the residual carbon in the product. Finally, the SiC powder could be obtained by washing, filtering and drying.

2.2 Sample characterization

The phase composition and crystallinity of the materials were characterized by powder XD-5A x-ray diffraction（XRD）which was produced by Japan Island Company[7].

The investigation for chemical bonds was performed via a Fourier transform infrared（FT-IR）spectrometer which was produced by British Renishaw Company[7].

Microstructures of the products were observed on a JSM-5510LV Scanning Electron Microscope （SEM）which was produced by Japan[7].

3 Results and discussion

3.1 Effect of Nb on the formation silicon carbide

The samples with the Nb addition amount of 0%, 1%, 2%, 3% are fired at 1300℃ and 1500℃ respectively, and the phase analysis was carried out. The results were shown in Fig.1 and Fig. 2.

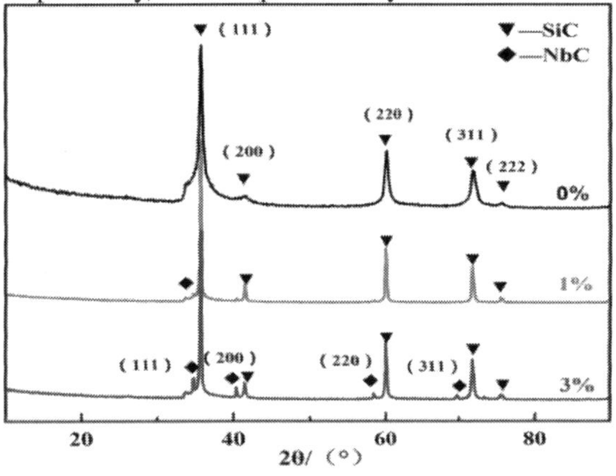

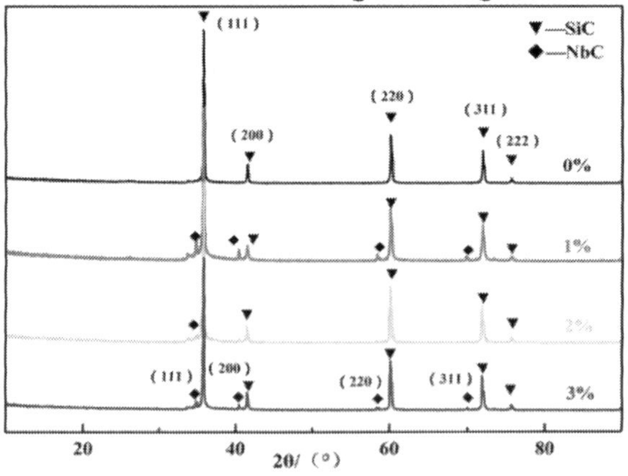

Figure 1. Effect of different content of Nb on the products reaction at 1300°C

Figure 2. Effect of different content of Nb on the products reaction at 1500°C

As shown in Fig.1, without adding Nb firing at 1300°C for 2h, the product is SiC, and the diffraction peak is wide, indicating that the crystallization degree of SiC is not high. When the addition amount of Nb is 1%, the characteristic diffraction peak of NbC appears in addition to the characteristic diffraction peak of SiC. With the increase of Nb addition amount, the diffraction peak intensity of SiC is enhanced, the results showed that the crystallinity of SiC increased, indicating that the addition of Nb could promote the formation and growth of SiC. When the content of the metal niobium powder was 3%, the characteristic peak of NbC increased obviously and the peak of SiC in the product was not further enhanced, which indicated that the amount of NbC was increased and the amount of Nb is reasonable when the addition amount is 1% . It can be seen from Fig. 2 that when the amount of Nb is added, the diffraction peak of NbC appears and the diffraction peak of SiC increases, and with the addition of Nb increasing, the diffraction peak of SiC is not further enhanced, which is the same as Fig.1 shows.

3.2 Effect of impurity removal order on product purification

The products of 1400°C were treated in three different ways （①: the product was first removed SiO_2 then removed C; ②: the product was first removed C then removed SiO_2; ③: the product do not do any processing） . Infrared and Raman tests were used to analyze the effect of impurity removal order on the purity of the product, and the results were shown in Fig. 3 and Fig. 4.

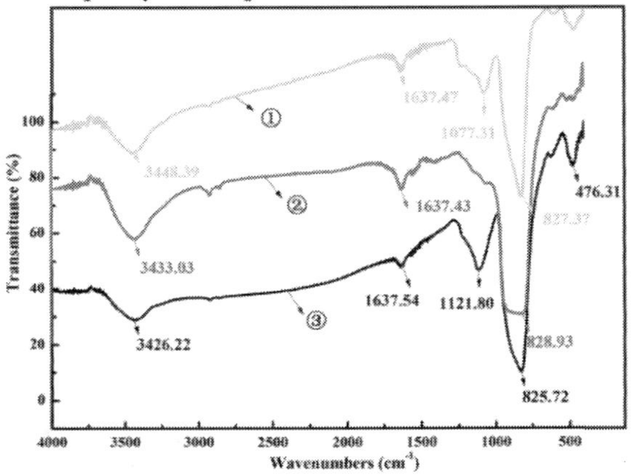

Figure 3. The Infrared test of the 1400°C products after different treatment

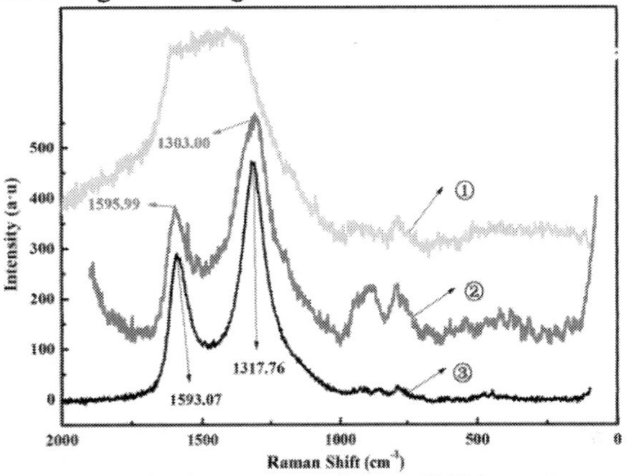

Figure 4. The Raman testof the 1400°C products after different treatment

In accordance with the FTIR spectrum indicated in Fig. 3③, the bands at 3426.22 and 1637.54cm^{-1} represent hydroxyl groups. There is a band at 1121.80 cm^{-1} which corresponded to siloxane bonds (Si-O-Si) which are formed by unreacted silica fume. A band presented at 825.72 cm^{-1} corresponded to (Si-C) bonds that originated from the products containing silicon carbide. As shown in Fig.3②, after hydrofluoric acid treatment, the silicon dioxide disappeared. As shown in Fig.3①, the product was first removed SiO_2 then removed C , there is a band at 1077.31 cm^{-1} which corresponded tosiloxane bonds (Si-O-Si) which are formed by product oxidized. In accordance with the Raman spectrum indicated in Fig.4②③, the bands at 1593.07,1317.76,1595.99 and 1303.00cm^{-1} which corresponded (C-C) bonds. As shown in Fig.4①, there is no C-C bonds, it can be concluded that removed SiO_2 first, then removed C can effectively remove the impurities in the product.

3.3 Effect of temperature on Preparation of SiC powder

XRD patterns of the as-received products after the carbothermal reduction at different temperatures are shown in Fig.5. [7]

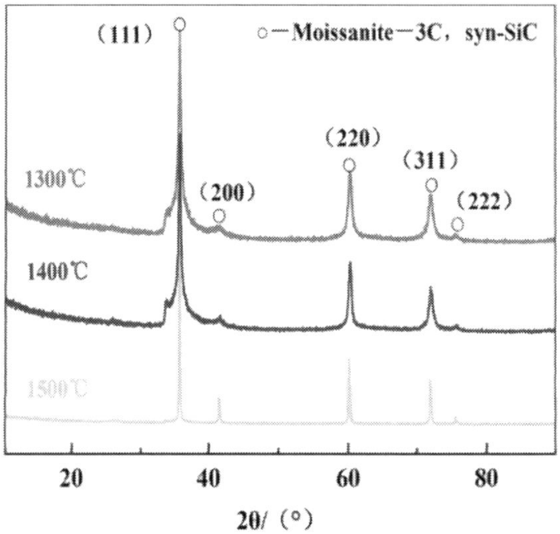

Figure 5. XRD analysis of products at different temperatures

Comparing XRD patterns shows that the crystalline phases of the products synthesized at 1300-1500°C are quite similar, and the crystallinity of the products increases slightly with increasing reaction temperature. From the XRD patterns, it can be easily observed that there are four sharp diffraction peaks located at 35.6°(111), 41.4°(200), 60.0°(220), 71.8°(311) and 75.5°(222), all of which are attributed to β-SiC(3C-SiC). No other obvious diffraction peak is found. Therefore, it can be concluded that the products synthesized at 1300-1500°C are relatively pure β-SiC(3C-SiC)[7].

3.4 Morphology of synthetic products

During the experiment, the results of the SEM analysis of the vacuum sintered products of the powder which does not contain the additive of the Nb at 1300°C, 1400°C and 1500°C are used to illustrate the influence of the temperature on the SiC, which are shown in Fig. 6.

Figure 6. SEM images of SiC powder at different temperatures: (a) 1300°C; (b) 1400°C and (c) 1500°C

As shown in Fig.6 (a), when the reaction temperature is 1300°C, the SiC is granular, and the single particle diameter is about 250nm, when the temperature is 1400°C, the SiC has irregular particle distribution, particle agglomeration together and poor dispersion, when the temperature is 1500°C, SiC morphology generated large and small particles together, and compared with 1300°C and 1400°C, the particle size increased, this is due to that the increase of temperature can promote the crystallization of SiC, which is consistent with XRD analysis results.

The effect of Nb content on the morphology of the synthesized SiC powder was investigated by SEM analysis of different samples at the same temperatures.The amount of Nb used in the four groups was 0%, 1%, 2%, 3% (the mass ratio of silica fume). The SEM images are shown in Fig.7 and Fig.8.

0% 1% 2% 3%

Figure 7. SEM images of SiC powder at 1300 °C with different content of Nb vacuum sintering

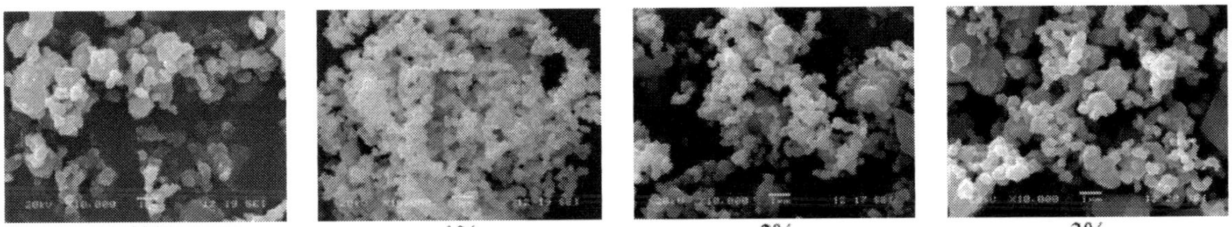

0% 1% 2% 3%

Figure 8. SEM images of SiC powder at 1300 ℃ with different content of Nb vacuum sintering

As shown in Fig.7, at 1300°C, when the Nb is not added, the SiC powder is granular, and the particle diameter is about 250nm, when the content of Nb is 1%, the particle size of SiC powder is slightly increased, the dispersion is improved, and the development of SiC crystal is perfect, when the addition amount of Nb is 2%, the SiC particles increase obviously, the particle diameter is about 400nm, when the content of Nb is up to 3%, the SiC with large particles can be seen more and more, and the SiC will gradually grow into rods. It can be inferred that with the increase of the content of Nb, the size of the SiC powder will increase.

As shown in Fig.8, at 1500°C, when the powder content of Nb is 1%, the generated SiC powder has uniform size and good crystal growth, and compared to 1300°C, the dispersion of product is better. This is due to the high temperature is beneficial to the nucleation and growth of SiC, thus relatively homogeneous particles are formed. With the increase of Nb content, the product gradually increased and the homogeneity become worse.

4 Conclusions

(1) Silica fume powder and glucose as raw material, under vacuum conditions at 1300 °C for 2h can produce 3C-SiC.
(2) The morphology and crystallinity of SiC can be significantly improved by adding Nb powder, and the particle size of SiC increases with the increase of niobium content at the same temperature.
(3) In this experiment, the best process for the synthesis of SiC is that under vacuum conditions at 1500°C for 2 h, the content of Nb is 1% of the mass of silica fume powder.
(4) The removal of impurities in the product can be effectively removed by using the method of removing SiO_2 first then removing C.

Acknowledgments

This work was supported by the Postgraduate Innovation Education Fund (CX2016013), Hubei Province Science and Technology Support Project (Grant No.2014BAA102) and Project of Technology Innovationin Hubei Province (Grant No.2016ACA160).

References

1. A. Najafi, F.Golestani-Fard, H.R.Rezaie, N.Ehsani, J.Sol-Gel Sci.Technol. **59**, 205-214(2011).

2. V.G. Lutsenko, ActaMater. **56**, 2450-2455(2008).
3. H.P. Martin, R. Ecke, E. Müller, J.Eur.Ceram.Soc. **18**, 1737-1742(1998).
4. B. Li, C.R. Zhang, H.F. Hu, Y.B. Cao, G.J. Qi, R.J. Liu, J.Mater.Eng.Perform. **16**, 775-778(2007).
5. P.V. Prabhakaran, K.J. Sreejith, B. Swaminathan, S. Packirisamy, K.N.Ninan, J.Mater.Sci. **44**, 528-533(2009).
6. A. Najafi, F.Golestani-Fard, H.R.Rezaie, N.Ehsani, PowderTechnol. **219**, 202-210(2012).
7. B.Li, Y.C.Song, C.R.Zhang, J.S.Y, J.Ceram. Int. **40**, 12613-12616(2014) .

Progress in Treatment of Oily Wastewater by Inorganic Porous Ceramic Membrane

Xiaoyuan Dai[1], Zhao Wang[1] and Man Xu[1, a]

[1]*School of Materials Science and Engineering, Wuhan Institute of Technology, Wuhan 430074, People's Republic of China*

Abstract : The composition and complexity of oily wastewater contains many solid particles, free oil, emulsified oil and so on.It brought about a series of environmental pollution problems when oily wastewater was directly discharged into rivers, lakes and other water bodies. Therefore, researchers are committed to study how to deal with oily wastewater to deal with oily wastewater to apply it to meet the requirements of water injection.Inorganic porous ceramic membrane has excellent properties among many filtering methods. For example, high temperature and high pressure resistance, resistance to acid and alkali, low energy consumption, no pollution to the environment and has a good prospect in the field of oily wastewater treatment, which has attracted the attention of many scholars not only at home but also on abroad. This article describes the present situation of the research on the treatment of oily wastewater by ceramic membrane in recent years, and expounded the significance of the treatment of oily wastewater to people's lives and makes an expectation for the development of inorganic porous ceramic membrane in the future.

0 Introduction

Porous ceramic membrane is a kind of ideal filter material. In recent years, many scholars at home and abroad have paid more attention to it in recent years. .Porous ceramic membrane has been being studied for more than 20 years in China. Both the production process and technical level have been greatly improved, which did a great favor to expand the scale of production. And the ceramic membrane undoubtedly takes an important part in the filters nowadays.

In the 50s of last century, China began to study ion exchange membrane and an experimental study on desalination of seawater by electrodialysis was carried out. In 1980s,inorganic membrane was developed rapidly and played an important role in membrane separation technology. After then, the application of porous ceramic membrane in gas-solid separation come into scholars' sight. The application of inorganic ceramic membrane have been used in China, in the removal of the bacteria, the fermentation liquid and in the collection of small molecular substances. Since 21st century, the ceramic membrane has been widely used in the oily wastewater, emulsion wastewater treatment, printing and dyeing, papermaking wastewater treatment, brine refining and so on. Sang H Hyun[1]used the self-made composite membrane to separate the emulsion, and the removal rate could

[a] Corresponding author: 1733784513@qq.com

reach 100%.Lin[2]used the polymer membrane to treat the wastewater containing vegetable oil, when the operating pressure was 2.1MPA,operating temperature is 313K,membrane surface velocity is 220L/h,the membrane flux is about 26.8 L/（$m^2 \cdot h$）.Thus, membrane filtration turns out to be a trend in the future.

For oily wastewater, the traditional separation technologies contained flotation expanded graphite adsorption, ultrafiltration, Electric-Fenton method[3].In order to avoid the shortcomings of traditional separation technology, people turn to research the ceramic membrane separation technology. Organic ceramic membrane technology is relatively mature while it has many disadvantages such as poor stability, poor acid and alkali corrosion resistance, non renewable, easy to occur two pollution, etc. Inorganic membrane[4]has been studied and used by scholars owing to its high temperature resistance, chemical corrosion resistance, good mechanical strength, strong anti microbial ability, high permeate flux and long service life. Therefore, the inorganic film is absolutely the key research direction in the ceramic membrane.

In this paper, we have made an introduction that the treatment of oily wastewater by inorganic ceramic membrane. The treatment of oily wastewater by inorganic ceramic membrane takes an important in people's life and environment, and provides a better research direction for the researchers who hammers at a more mature technology.

1 The traditional treatment of oily wastewater

Biochemical process is one of the treatment of oily wastewater in Daqing Oilfield:

Sedimentation + microorganism + membrane technology. But this method is used of organic film, which have some shortcomings like poor acid and alkali corrosion resistance, poor stability, easy to produce secondary pollution. Flocculation method also has a wide range of applications in recent years. The flocculant was added to the oily wastewater which then was hydrolyzed into colloid, and flocculation occurred after a series of physical and chemical actions. The chemical reagents were added to make the surface of the water charge neutralization reaction, and the oil droplets were gathered. The polymer material was added to carry out flocculation through the bridge. Finally the oil droplets are removed by sedimentation or air floatation. This method has a removal rate for the removal of oil droplets, but the addition of reagents will produce secondary pollution, which will have a lot of adverse effects on the treatment system. Biofilm method made aerobic microorganism attached to the surface of the solid filler, which can form the biological membrane to adsorb the oil and dissolved oxygen in the wastewater. Biofilm was also undergoing metabolic processes and could be recycled.

2 Treatment of oily wastewater by inorganic ceramic membrane

2.1 Hydrophilic alumina ceramics

Alumina ceramic membrane is the direction of most scholars, Chun Lei Ren proposed that the porous alumina membrane was used to treat the oily wastewater, and the α-Al_2O_3 with average particle size of 0.78μm was prepared by phase inversion casting sintering technology. In the treatment of oily wastewater, the water permeability and oil permeability test should be carried out, the results show that the flat membrane meets the requirement of treating oily wastewater. The specific practice of this experiment is to mix the octane and water in the proportion of 30:70 to the formation of oily wastewater and oil red and methyl blue staining to make the experimental results more obvious. As shown in Figure 1.The flat membrane is fixed on the glass tube at the bottom, and then put the oil-water mixture to the glass tube, we can see that the flow of water with methyl blue is linear through the flat film, and the oil droplets with oil red are all trapped above the flat membrane. The results show that the alumina ceramic membrane has the properties of hydrophilic and oil repellent, and has good effect on oil and water separation.

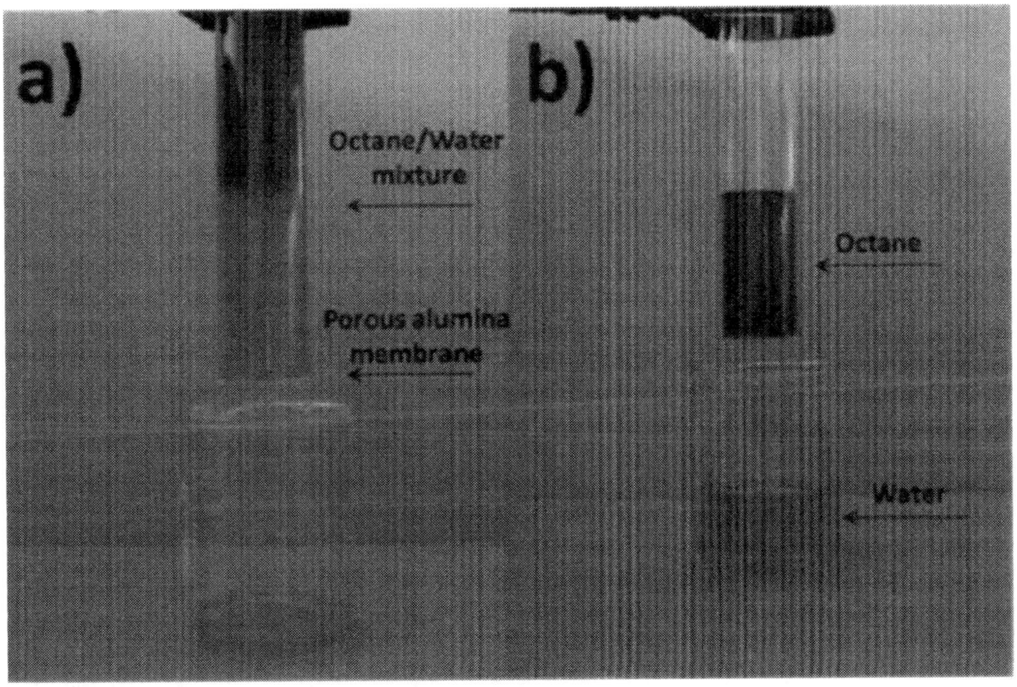

Figure 1. Oil-water separation of porous alumina membrane

2.2 Hydrophilic zirconia ceramic membrane

Chun Mei Wang studied the effects of various factors on the membrane flux in the treatment of oily waste water with a pore size of 0.2μm investigated, including the operating pressure, the film surface flow rate, the operating temperature, and so on. Fig. 2 flow chart of experiment.

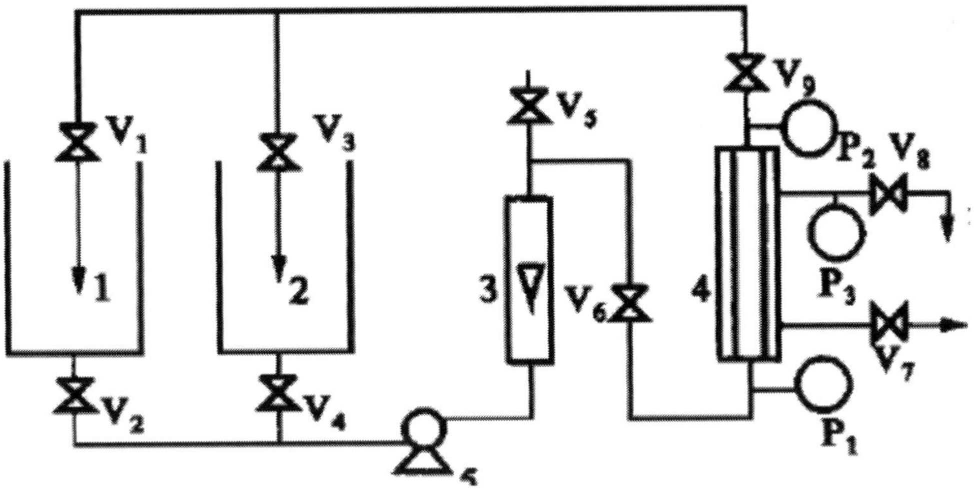

Figure 2. Flow chart of experiment. 1. storage tank; 2. cleaning tank; 3. rotor flow meter; 4. membrane components; 5. centrifugal pump; V1-V9 valve

Oily wastewater after being deal with the 0.2μm zirconia ceramic membrane, the operating differential pressure is 0.1MPA, the membrane surface velocity is 5.05 m/s, the operating temperature is 308k, the stability of the membrane flux can reach 250 L/(m^2/h),after treatment, the oil content of waste water is less than 10mg / L,which can meet the requirements of treatment, and the operation and process of the equipment are relatively simple. It is an effective process for treating oil-containing wastewater.

2.3 Hydrophilic silicon carbide ceramic membrane

There are few reports on silicon carbide ceramic membranes. Compared with alumina and zirconia ceramic membrane, it is more hydrophilic, acid and alkali corrosion resistance is stronger, and that is a kind of ideal filter material with the highest flux.

Shi Wei Ye used 0.1μm silicon carbide ceramic membrane to treat wastewater that oil content is 7.564mg/L-9.568mg/L and water temperature is about 20°C and the suspended matter content is 35mg/L-45 mg/L and pH is 6,with the method of dead end filtration.The experiment shows that when the temperature is 20°C, suspended solids and oil removal rate can reach more than 98% and 88%, when the temperature is 40°C, suspended solids and oil removal rate can reach 97% and 87% respectively; when the temperature is low, the filtering effect will be better.

2.4Treatment of oily wastewater by other membrane materials

In addition to studying the treatment of oily wastewater with inorganic ceramic membranes, the scholars also studied the organic/inorganic composite ceramic membrane, considering the properties of the two kinds of film together, not only can improve the surface hydrophilicity of [5-6] and toughness, but also can fight pollution.Wang Shu had a ceramic base membrane of polyvinylidene fluoride (PVDF) ultrafiltration membrane surface modification, prepared with polyamide/PE pure organic/inorganic composite membrane.The ceramic film deal with GS-1 type high speed vacuum pump oil with an average particle size of 2.365μm under the operating pressure of 0.4MPA.The experimental results show that the water flux is 120 L/（m^2·h）,the oil concentration after treatment is 5mg/L. The test results are not satisfactory, and need to be further improved. Xue Dong Zhang used to treat the sand filtration water in the eight sewage treatment plant of Da Qing oil production plant, using the modified PVDF ultrafiltration membrane with inorganic nano particles. The results showed that the content of suspended matter of three kinds of modified ultrafiltration membrane effluent was below 0.4mg/L,the removal rate was above 95%, oil content decreased from 12.52mg/L-84.42mg/L down to 1.0mg/L, the removal rate was above 90%, COD in the water form the 450.17mg/L-1280.55mg/L down to 80mg/L-90mg/L, the removal rate was 80% -95%,The treatment effect is very good, and the treated liquid can be discharged directly to meet the experimental requirements. Jin Ming Wang used the alumina microfiltration membrane modified by nano ZrO_2 treatment of oily wastewater, the modified membrane flux can reach a maximum of 295 L/（m^2·h）, the oil removal rate was 99%, and the oil content that treatment was less than 10mg/L, reached the national emission standard.

3 Conclusions and prospects

In this paper, we have summarized the various experimental methods made by researchers in recent years that the inorganic ceramic membrane is used to treat the oily wastewater, such as ceramic ceramics, zirconia ceramic film, silicon carbide ceramic film and organic / inorganic composite ceramic membrane, the filtering effect has unpredictable good prospects in many areas. Table 1 summarizes the results of the treatment of oily wastewater by these 3 kinds of inorganic ceramic membrane.

Table 1. Effect of the treatment by three kinds of inorganic ceramic membrane

ceramic membrane	effect
alumina ceramic membrane	Oil water separation

	efficiency was 96%
zirconia ceramic membrane	The oil content of the filtered water is less than 10mg/L, and the stable flux can reach 250 L/（m2·h）
SiC ceramic membrane	The removal rate of suspended solids and oil can be reached 97% and 87%

There are both advantages and disadvantages in the treatment of oily wastewater in each kind of inorganic ceramic membrane, and which has a very important impact on people's production and life. It is worth mentioning that in the treatment of oily wastewater should also ensure that no pollution to the environment, no waste of energy. But researchers also are faced with the problems that how to clean the oil droplets when the ceramic membrane pore blockage, to control membrane pollution degree, to guarantee the performance of the film does not have too big change after cleaning._How to control the pore size is whether suitable to be used for oil-water separation or not if the size of the oil droplets is too small? It could be seen that there is still a long way to explore in the treatment of oily wastewater by using inorganic ceramic membrane.

References

1. S.H. Hyun, G.T. Kim, Sep.Sci.Technol **32**, 2927 (1997)
2. L. Lin, K.C. Rhee, S.S. Koseoglu, J.Membrane.Sci **134**, 101 (1997)
3. M. Inagaki, H. Konno, M. Toyoda, Desalination **128**, 213 (2000)
4. T.A. Mann, E.Mendoza, T.Stemphenson. Wat.Res **33**,1108 (1999)
5. R.S. Faibish, Y.Cohen, Colloids.Surfaces.A **191**,27 (2011)
6. B. Tansel, J. Regula, R. Shalewitz. Desalination **102**, 301 (1995)

Design of All Digital Flight Program Training Desktop Application System

Yu Li[1,a], Gang An[2,b] , Xin Li[3,c]

[1] System Integration Center, FAI AVIC,Xi'an, CN
[2] School of management, Xi'an Jiaotong university, Xi'an, CN
[3] Landing Gear System Research Institute, FAI AVIC, Xi'an, CN

Abstract. All digital flight program training desktop application system operating conditions are simple. Can make the aircraft aircrew learning theory and operation training closely. Improve the training efficiency and effectiveness. This paper studies the application field and design requirements of flight program training system. Based on the WINDOWS operating system desktop application, the design idea and system architecture of the all digital flight program training system are put forward. Flight characteristics, key airborne systems and aircraft cockpit are simulated. Finally, By comparing flight training simulator and the specific script program training system, The characteristics and advantages of the training system are analyzed in this paper.

1. Introduction

Modern aircraft set a variety of complex systems and high-tech in one. The function and performance improvement are also used to aircrew put forward higher requirements. The traditional training mode has many problems, mainly reflected in: backward technology, single method, scattered resources, sharing difficulties, training is not flexible, intuitive, systematic and operability, low degree of automation, the lack of effective supervision and evaluation, and to deal with the new situation of the training needs. With the rapid development of computer, network, multimedia, digital simulation and modeling technology, by means of information to compensate for the lack of traditional training mode, carry out the resources development, management and training, improve the training quality and efficiency, has become the inevitable trend of development of modern flight training[1].

All digital flight program training desktop application system running on the WINDOWS OS, the minimum operating and training units for a single desktop computer or portable PC. And can flexibly set up the training room through the high speed Ethernet environment. Support single machine training mode, double pilot collaborative training mode, distributed synchronization teaching mode. For the aircrew learning theory, cognition, cockpit equipment operating procedures and flight training, maintenance training etc.

2. System architecture design

[a] kingliyu@sina.com; [b]An_gang@163.com; [c] 370713464@qq.com

2.1. Functional composition

The function of the all digital flight program training desktop application system is shown in figure 1.

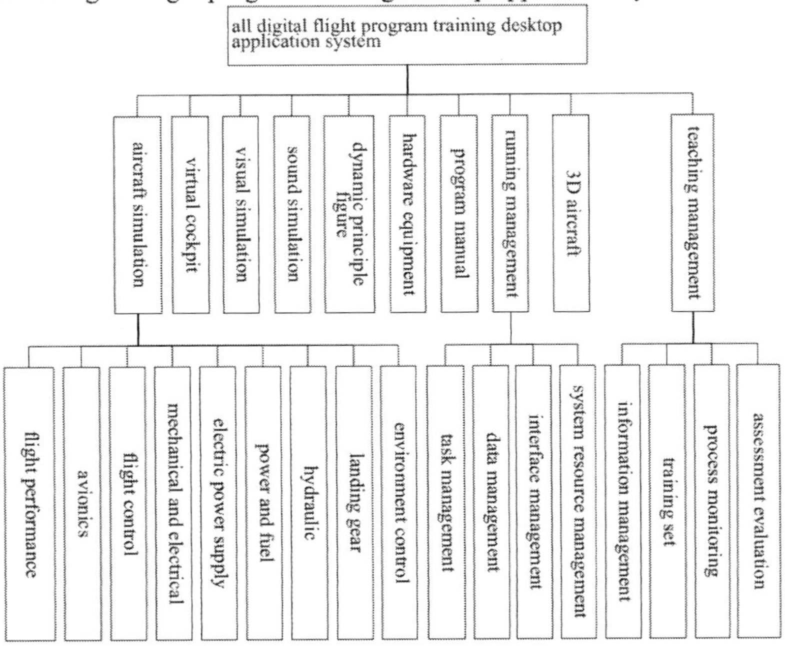

Figure 1. Functional composition

Aircraft simulation: including flight simulation and airborne system simulation. Mainly completes the flight performance simulation. Functional simulation of airborne system, working logic simulation and Simulation of cross link between systems[2].

Virtual cockpit: Construction of the 3D virtual cockpit environment consistent with a real aircraft. For the cockpit and cabin equipment cognition, aircrew flight operation training, emergency procedure training, power checking operation on ground, operation testing etc..

Visual simulation: the use of computer image generation technology, produce a realistic external scene, the simulation of the airport, route and flight base, including the corresponding airport runway, route terrain and terrain, etc..

Sound simulation: the simulation of the aircraft during the flight can be heard in a variety of sounds, including a variety of environmental noise, equipment noise, system tips and alarm voice, etc..

Dynamic principle figure: The simulation of the airborne system principle. The cross-linking relationship simulation of the different airborne systems. For the crew to learn and master the working process of the system.

Hardware equipment: the smallest unit of software running for a single desktop computer or portable PC, and can be through the server, high speed Ethernet to achieve virtual classroom or multi machine network training.

Running management: system scheduling, management and operation control, including task management, data management, interface management and system resource management.

Program manual: the training management system to load the aircraft's flight class and maintenance related to the electronic technical manuals, and the current program associated with the system training.

3D aircraft: The 3D aircraft environment that is consistent with a real aircraft, including the fuselage, wing, cargo hold, floor, and system layout. Through the training management system scheduling, and the simulation can be synchronized with the program, the implementation of different training operating procedures.

Teaching management: information management, training seting and process monitoring, assessment evaluation etc. To achieve the purpose of training class setting, system initialization settings, a variety of flight parameters, airport conditions, the status of the aircraft display and settings, aircraft failure and special circumstances set, flight training process monitoring, etc..

2.2. Software architecture

The all digital flight training program desktop application system based on operation management system as the core, the integrated simulation system, virtual cockpit system, visual simulation, sound simulation, dynamic principle and teaching management system, realize the whole system software and hardware resource management, data exchange, task scheduling and management. The system architecture is shown in figure 2. The internal function of the system is highly cohesive and loosely coupled, and the system can transfer and exchange the data flow and control flow through the common data pool, so as to achieve the flexible resolution and assembly of the software[3].

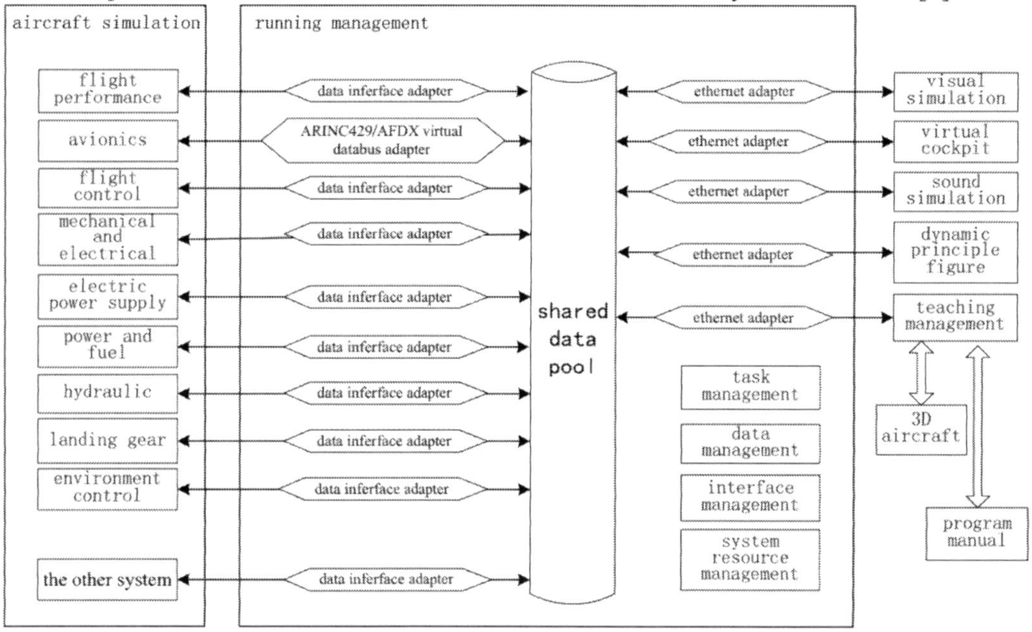

Figure 2. Software architecture

Operation management system mainly includes: public data pool, is responsible for the data exchange between the system; task management, task scheduling, task group is responsible for the rate allocation and management, process / thread allocation and management; data management, responsible for public data pool establishment, data update management, data read / write operation management interface; the management of the data exchange interface conversion, computer network communication interface for each subsystem and the public data pool management system; resource management, management of the operating system software resources and computer hardware resources such as CPU, memory, hardware interface etc.. Run management system to meet the WINDOWS desktop application system task management, resource management, interface management, data management, software architecture, network architecture, etc..

The cross linking of the avionics system simulation and operation management system. Data exchange between public data pool and the avionics system through ARINC429 and AFDX virtual bus data adapter software. visual simulation, virtual cockpit, dynamic principle, training management and operation management system linking, through the Ethernet interface adapter software and the public data pool data exchange; conversion flight simulation and flight control system simulation, mechanical and electrical system simulation, power system simulation, power/fuel system simulation, hydraulic system simulation, simulation, gear ring, door control simulation simulation system through the software interface adapter to exchange data and the public data pool. The 3D plane and procedure manuals and training management system directly connected by Ethernet, the main transmission training information and training process control instructions and procedures manual information to realize the training process of synchronous scheduling and operation of related subjects and the corresponding three-dimensional plane information.

2.3. Network teaching

The virtual training classroom can be flexibly set up through the high speed Ethernet environment, as shown in figure 3. The virtual training room is composed of one system server, one teacher computer, one training administrator computer, one system administrator computer and several student training computers.

The training computer's user interface is formed by the 4 display of the computer, and the display content can be independently arranged on each screen. The commonly used display mode in the training process:

- No. 1 display: attitude, participation, representation of a page.
- No. 2 display: Avionics navigation information.
- No. 3 display: by operating the dynamic way roaming the entire virtual cockpit.
- No. 4 display: you can choose to display the training manual, dynamic principle diagram, three-dimensional plane etc..

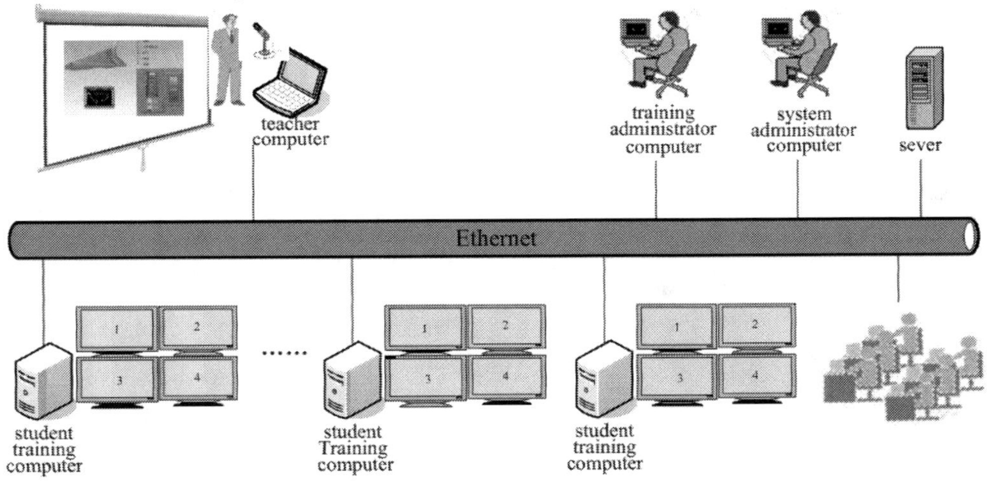

Figure 3. Virtual training classroom

3. System simulation design

3.1. Aircraft simulation modeling

The principle of aircraft simulation is shown in figure 4. Flight simulation includes: Aerodynamic module, mass features module, atmospheric environment module, wind module, turbulence module, shear module, reposition module, ground control calculation module and equation of motion[4].
Airborne system simulation includes[5]:

- Function simulation modeling of airborne system.
- Simulation of the interaction between cross linked systems.
- Simulation cockpit control panel for the control of the system and control box lights.
- Simulate common fault, alarm and display.
- Simulation of flight inspection procedures and display control.

3.2. Virtual cockpit simulation

Cockpit simulation is one of the most important systems simulation of the whole digital flight program, Requirements for the simulation of particle size to reach the component level. The effective area of the system simulation of the entire cockpit to ensure real aircraft cockpit simulation reappearance. Simulation of all the control and display equipment in the cockpit through modeling method[6]. A variety of display, indicator, instrument, switch, control box, switch, handle, button, indicator, knob, adjustment device, operation instructions and response in line with the actual aircraft. Shape, size, color, brightness, etc. with the real aircraft. Through the cabin glass to watch the scene outside the cabin, through the mouse and keyboard operation on the virtual cockpit to promote / zoom / translation / rotation and other roaming operations.

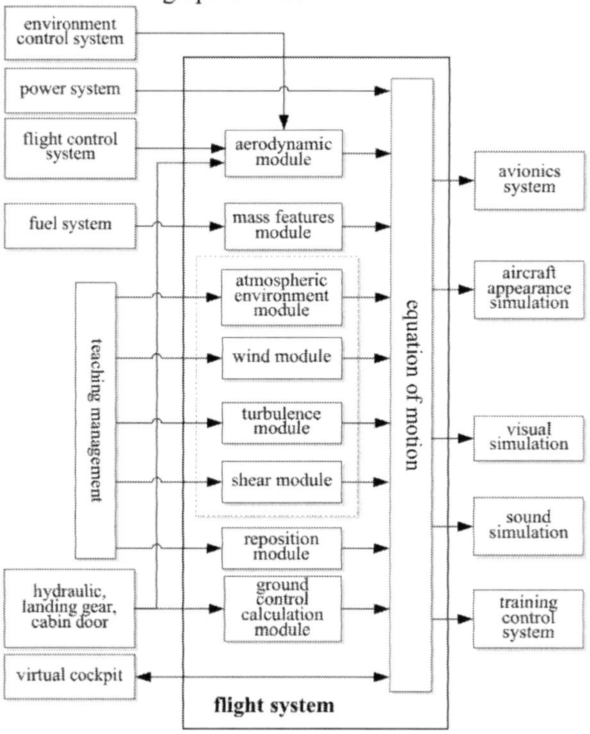

Figure 4. Aircraft simulation principle

According to the characteristics of "WINDOWS OS + desktop computer" with keyboard and mouse as the main input device, the cockpit operating mode (such as: rod, wheel, pedal, button, pulling the handle, with a protective cover switch, a trackball, keyboard and so on), to optimize the design of different equipment and operating characteristics, through the realization of common the keyboard and mouse to complete the cockpit all operations, and meet the design requirements of reasonable operation, intuitive and simple.

Trajectory ball operation switching model optimization: through the virtual cockpit software to re development of the avionics system, the trajectory of the ball function, so that the original independent avionics software is fully integrated into the virtual cockpit. As long as the students in the virtual cockpit view through the "CTRL + mouse" can make the Windows ordinary mouse operation fast and flexible switch for the control system of the track ball operation.

The throttle trim, handle operation optimization: throttle four engines can be operated at the same time, also can operate independently, through simple mouse control is difficult, therefore, the navigation window set in virtual cockpit in the auxiliary control work area, add a slide bar to achieve four engines control.

3.3. Teaching management design

The teaching management system for teaching and training of the whole process of information management, training set, process monitoring and assessment evaluation etc..

Information management functions, including:

- Teachers, students access to login and basic information management.
- The system can automatically establish and maintain the teaching and training of teachers and students, the examination of the file.
- The system supports the basic information of the teaching staff, training information and assessment results of the query.
- Record system status information, such as: technical state, system version change, etc..

Training set function, including:

- Course setting.
- System parameter setting.
- Aircraft initialization setting.
- Flight environment initialization setting.
- System fault setting.

Process monitoring functions, including:

- System operation control, such as start/restart, pause, stop.
- Automatically record the student assessment process, including the status of the aircraft system, operating procedures, the location of snapshots and simulation operation and other information.
- The teacher management student assessment authority, monitoring students' training and evaluation process.
- The teacher to maintain management system.

The execution of the whole system is completed by the user, the training computer and the server. The system operation process mainly includes: the server starts, the virtual training system start, the edition automatic renewal, the user registers, the teaching / training pattern establishment, the training and the movement control[7]. The timing logic of the operation is shown in figure 5.

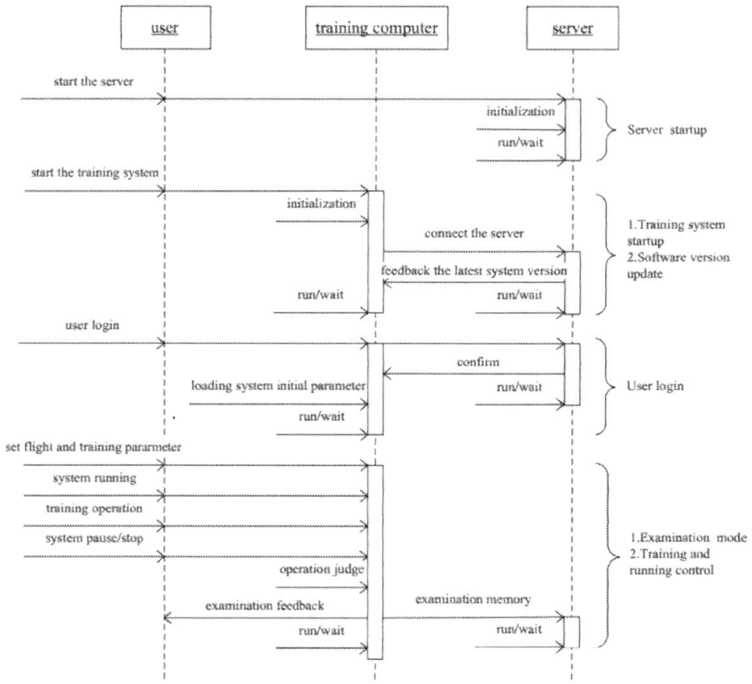

Figure 5. Operational process sequence logic

3.4. Teaching mode design

The design of teaching mode in high speed Ethernet environment is shown in Figure 6.

Single machine training mode consists of a single training computer. Students can be on the computer virtual instructors guide training, autonomous training and evaluation.

Double pilot collaborative training mode from two sets of training computer composed of a complete set of two driving units. Students can be on the computer virtual instructors guide training, autonomous training and evaluation.

Distributed synchronization teaching mode, one computer training as a teacher teaching, student learning as other computer systems, to provide different permissions for identification of the teacher when the flight operations performed in synchronous computer students.

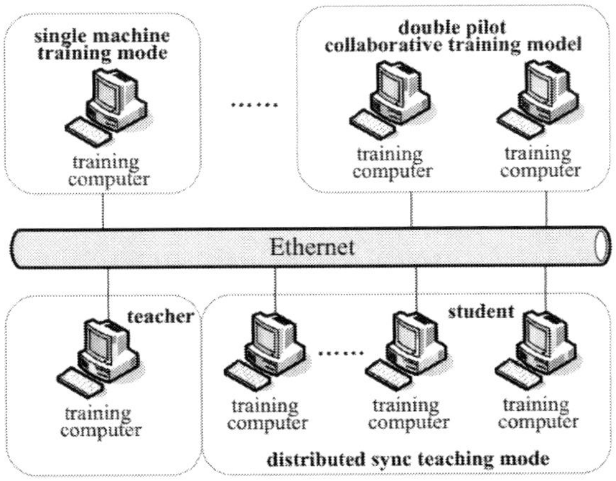

Figure 6. Teaching mode

4. Conclusions

The comparison of the all digital flight program training desktop application system and the flight training simulator: The flight training simulator is more suitable for the training of the pilot's flying skills, it is the advanced training based on the solid theoretical foundation and skilled operating skills. Due to the high cost of manufacturing, use and maintenance, limiting the number of equipment and training time. Flexibility difference is difficult to meet the requirements of autonomous learning and differentiated training.

The all digital flight training program desktop application system suitable for aircrew learning theory, cognitive equipment and flight operations training program. The system runs in WINDOWS OS, supports single machine or network teaching and training, the hardware resources and the use condition request is simple. The system can meet the training of self training, classroom teaching, automatic evaluation and other aspects of the system, and the training cost is low. All digital flight simulation technology and simulation software can be easily and efficiently transplanted or reused to the aircraft simulation platform, simulator, virtual reality training system, which has a good value for popularization and application.

References

[1] Huang Zhongrong, Liang Yunduan. Flight training system and development of trainer. training plane, 2011.
[2] Fang Zhenping, Chen Wanchun, Zhang Shuguang. Aircraft flight mechanics. Beijing: Beihang University press, 2012.
[3] Zhi Chaoyou, Li Zhenshui, Xue Feng. Research on virtual test system framework based on model component. Computer measurement and control, 2011.
[4] Jia Qiuling, Yuan Dongli. Simulation, analysis and design based on MATLAB7.x/Simulink/Stateflow system. Northwestern Polytechnical University press, 2008.
[5] Wang Yong, Yu Hongkun. Airborne computer system. Beijing: National Defense Industry Press. 2008.
[6] Yao Xudong. Design and implementation of avionics system simulation platform based on virtual instrument. Shanghai Jiao Tong University, 2012.
[7] Zhi Chaoyou, Tang Changhong. Development of modern aircraft system virtual test verification technology. Aviation science and technology, 2010.

Optimal Scheduling of GEO On-orbit Refuelling with Uncertain Object Satellites

Jing Yu[1], Yong-gang Yu[1], Jiang-tao Huang[1,a], Xiao-qian Chen[2] and Hong-yang Liu[1]

[1]*China Aerodynamics Research and Development Center, Mianyang 621000, China*
[2]*College of Aerospace Science and Engineering, National University of Defense Technology, Changsha 410073, Hunan Province, China*

Abstract. The mission planning of Geostationary Earth Orbit (GEO) on-orbit refuelling with uncertain object satellites is studied in this paper. Three key problems should be resolved here: Object Satellite (OSc) selection, mission sequence planning and transfer trajectory optimization. The optimization goal is to select the OSc and find the optimal servicing sequence with the best optimization objective value. Three optimization objectives, i.e. fuel cost, the total number of the OScs and the sum of the OSc priority value are taken into account. Considering this mission as a hybrid optimal control problem, a mathematical model is proposed. Multi-objective particle swarm optimization is employed to address the model. Several mission scenarios are compared and discussed. Numerical experiments indicate that 1) the method developed in this paper could address the problem efficiently; and 2) OSc distribution and varieties in optimization objective emphasis have considerable influences on the optimization results.

1 Introduction

Refuelling the satellites offers a great alternative for lengthening the lifetime of the constellation, as well as increasing the constellation's profitability and survivability. Furthermore, a multispacecraft rendezvous refuelling strategy can contribute significantly to improving the economic returns of on-orbit service missions [1, 2].

In the past few years, many studies focused on OOR mission planning, i.e. developing optimal sequence and trajectories for SSc refuelling multiple targets. Shen [3] studies the scheduling of servicing multiple satellites in a circular orbit. Ouyang et al. [4] transformed the refuelling problem into a travel salesman problem (TSP). Zhang et al. [2, 5] focused on the LEO long-duration multi-spacecraft rendezvous mission. Bo et al. [6] presented a new refuelling pattern based on formation flying, and proposed two strategies to address the problem. Tsiotras and Dutta [1, 7-10] focused on OOR missions using peer-to-peer (P2P) strategy. In the investigations mentioned above, the object satellites are always definite, and much work focuses on fuel cost only. Considering the capacity limitation of the SSc, the planning problem of GEO OOR mission with uncertain OScs is studied in this paper. The OScs are undetermined before planning, and multiple relevant optimization objectives

[a] Corresponding author : hjtcyf@163.com

 DOI: 10.1051/matecconf/201711403001

will be taken into account. We consider the scheduling of GEO OOR mission as a HOC problem, and use Hybrid Automaton to model and address it.

2 Mission scenario

The mission studied here can be stated as follows. One SSc running on GEO belt is required to service some of the fuel-insufficient GEO satellites (see Figure 1). Since the propellant capacity of the SSc is limited, it could not service all of the satellites in a mission. The goal of the mission planning is to select the OScs that would be refuelled and find the optimal mission sequence for the SSc. In order to simplify the problem, we take two-impulse phasing maneuver and Hohmann transfer to perform orbital rendezvous, and sufficient time is given for the mission.

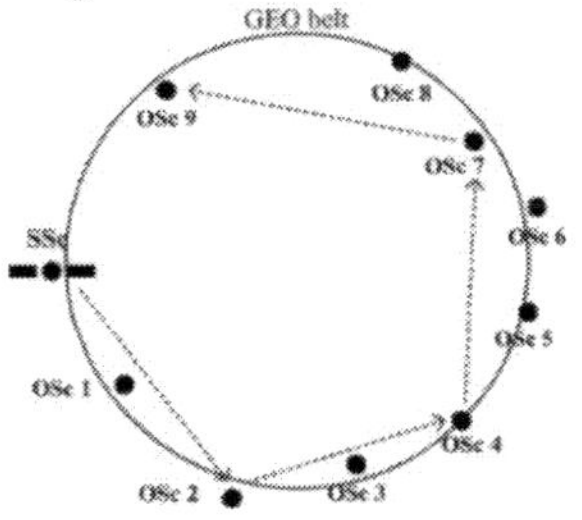

Figure 1. Mission scenario.

3 Mathematical description of the HOC problem

Recently, hybrid optimal control (HOC) theory has been applied to the solution of space mission planning [11-17]. The common way to model a HOC problem consists of the steps of categorical state space modelling; continuous-time dynamics modelling; continuous-valued state and control spaces modeling; discrete events modelling; and cost functions modelling. In this section, we will detail these steps.

3.1 Categorical/Discrete state space

Supposing there are N fuel-insufficient satellites, N_Q out of them are selected to be refuelled. Then the categorical state space for the problem is $Q = \{q_1, q_2, \ldots, q_{NQ}\}$. A directed graph (or called "digraph") can describe this space. Figure 2 shows a digraph for an example problem in which there are three possible categorical states that can be combined in some order to qualitatively describe the mission plan. The categorical state space is graphically depicted, as are the allowed transitions.

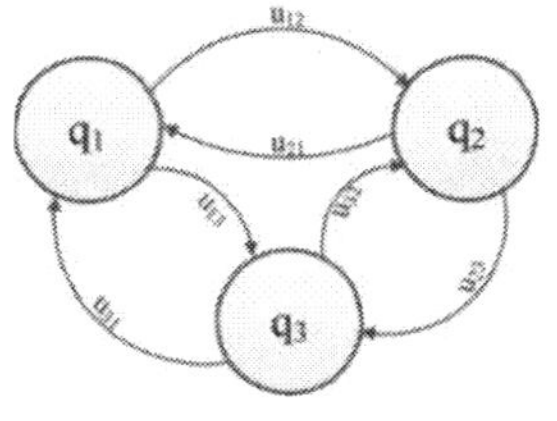

Figure 2. Digraph for the OOR mission.

In Figure 2, $N_Q = 3$, and each vertex q_i is the state corresponding to the SSc flying at the slot of the i-th OSc with zero thrust. Each edge represents a switch of the SSc's dynamics from one vector field to another. $\mathbf{u}_{ij}$ denotes the system input that triggers the switch from state q_i to q_j. A mission plan is described qualitatively by an event (or maneuver) sequence, e.g. $\mathbf{q} = (q_2, q_1, q_3)$, with events chosen from the categorical state space. This mission plan sequence describes that starting from the original location, the SSc refuels the OScs numbered 2, 1 and 3 one by one. The real trajectory of the SSc is $\mathbf{q}$

= (q_0, q_2, q_1, q_3), where q_0 denotes the initial fly state of the SSc. In this paper, we tacitly approve that q_0 is the original state of the SSc, and q_0 won't appear in the mission plan.

3.2 Continuous-time dynamics

Associated with each state $q \in Q$ is a continuous-time controlled dynamical system: $\dot{\mathbf{x}} = f(\mathbf{x}, \mathbf{u}, t, q)$. At each discrete state, we have the dynamics of SSc with $\mathbf{x} = [\boldsymbol{r}, \boldsymbol{v}]$. $\boldsymbol{r}, \boldsymbol{v}$ respectively denotes the radius vector and velocity vector of the SSc. Perturbations are not taken into account [18][19] in this paper.

$$\begin{cases} \dfrac{d\boldsymbol{v}}{dt} = -\dfrac{\mu}{r^3}\boldsymbol{r} \\ \dfrac{d\boldsymbol{r}}{dt} = \boldsymbol{v} \end{cases} \tag{1}$$

3.3 Continuous-valued state and control spaces

For the problem investigated in this paper, we should consider constraints as follows.
(1) The number of satellites in the mission sequence should not be more than the total number of the fuel-insufficient satellites, i.e. $N_Q \leq N$.
(2) The fuel mass that the OSc receives should not be more than the mass it requires.
And (3) when the mission is completed, the remaining fuel of the SSc should be more than a defined value $fuel_{\min}$.

3.4 Discrete events

Let $(\mathbf{x}, \mathbf{u})$ and $(\mathbf{x}', \mathbf{u}')$ denote the continuous-valued state and control variables associated with any two vertices $q, q' \in Q$. The generalized switching set, which may be empty, is called the event set $\mathbf{E}(q, q')$ and is defined (when nonempty) by means of an inequality constraint on a function $\mathbf{e}(\cdot, q, q')$.

$$\mathbf{E}(q,q') = \{(\mathbf{x}, \mathbf{u}, \tau, \mathbf{x}', \mathbf{u}', \tau') : \mathbf{e}^L \leq \mathbf{e}(\mathbf{x}, \mathbf{u}, \tau, \mathbf{x}', \mathbf{u}', \tau', q, q') \leq \mathbf{e}^U\} \tag{2}$$

$\mathbf{e}^L$ and $\mathbf{e}^U$ are the lower and upper bounds on the values of the function $\mathbf{e}(\cdot, q, q')$, respectively, τ is the time corresponding to $\mathbf{x}$ and $\mathbf{u}$. The function $\mathbf{e}(\cdot, q, q')$ is called the event function associated with the discrete states q and q'.

3.5 Cost functions

In this paper, we consider three optimization objective functions.

(1) J_f : fuel cost is as small as possible.

For impulsive transfers, the following expression holds between the mass of the satellite just before ($m(t_k^-)$) and just after ($m(t_k^+)$) an impulse at time $t = t_k$:

$$m(t_k^+) = m(t_k^-)e^{-\Delta v_k/(I_{sp}g_0)} \tag{3}$$

Where Δv_k is the gain (or loss) of the velocity due to the impulse at $t = t_k$.

When the SSc is running on the discrete state, there is no thrust, so the velocity increment is zero. Associated with any pair $(q,q') \in Q \times Q$, the Δv consists of two components, the plane change and the in-plane transfer for the rendezvous. In this paper, we apply phasing maneuver and Hohmann transfer to in-plane transfer.

(2) J_N: the number of satellites that will be serviced should be as large as possible.

As the SSc could not service all of the fuel-insufficient satellites, it is required that the fuel-insufficient satellites should be refueled as many as possible.

(3) J_p: the sum of the priority value in the mission sequence should be as large as possible.

Suppose that each fuel-insufficient satellite has a priority value. It is required that the satellite with high priority value should be serviced first.

3.6 Summary of the HOC problem

In summary, the On-orbit refueling mission researched in this paper could simply be described as a HOC problem, and can be formulated as:

$$\textit{find} \quad \mathbf{q},\ \mathbf{u} \tag{4}$$

$$\text{minimize} \quad J_f = \sum CFuel(q^i, q^{i+1}) \tag{5}$$

$$\text{maxmize} \quad J_N = Num(\mathbf{q}) \tag{6}$$

$$\text{maxmize} \quad J_p = \sum Prio(q^i) \tag{7}$$

s.t.

$$\dot{\mathbf{x}} = f(\mathbf{x}, \mathbf{u}, t, q) \tag{8}$$

$$q \in \mathbf{q},\ q \in Q \tag{9}$$

$$Num(\mathbf{q}) \le N \tag{10}$$

$$0 < AFuel(q) \le RFuel(q) \tag{11}$$

$$Fuel_s(\mathbf{q}) > fuel_{\min} \tag{12}$$

$$X(t_0) = X_{q_0} \tag{13}$$

In Eq. (4), $\mathbf{q}$ denotes the mission sequence of the SSc, and $\mathbf{u}$ gives the optimal input that could trigger the whole rendezvous trajectory with optimal cost. Equations (5)-(7) represent three optimization objectives. Equations (8) and (9) describe that each discrete state should satisfy continuous-time dynamics. Equation (10) gives the constraint that the number of satellites in the mission sequence should be not more than the total number of the fuel-insufficient satellites. Equation (11) means that the fuel mass that the OSc receives should be not more than the mass it requires, where $AFuel(q)$ denotes the fuel mass the OSc receives, and $RFuel(q)$ gives the fuel mass the OSc requires. Equation (12) describes that when the mission is completed, the remaining fuel of the SSc should be more than $fuel_{\min}$. Equation (13) represents the initial state constraint of the SSc.

4 Solution via particle swarm optimization

To address the HOC model above, three key problems should be considered and resolved:

(1) OScs selection: deciding which satellite will be refuelled.

(2) Mission sequence: deciding the servicing order of the SSc.

(3) Trajectory generation: optimizing the orbital rendezvous trajectories.

Multi-Objective Particle Swarm Optimization (MOPSO) algorithm is employed to deal with the model described above. The main algorithm is provided in [15]. Here we will detail the encoding method and the fitness computation process of MOPSO.

4.1 Encoding

In the MOPSO algorithm used in this paper, each particle is a row matrix with N columns, where N denotes the total number of the fuel-insufficient satellites. Each particle contains N real numbers $d_i \in (1,2)$ $i \in (1,2,...N)$. Sorting these numbers in ascending order, the task arrangement then could be obtained. For example, if a particle is described as $[1.5, 1.3, 1.7]$, it means that the SSc should service satellites 2, 1 and 3 in order.

4.2 Fitness computation

When the particle is given, the task arrangement can be decoded, and the fitness value can be computed as follows (see Figure 3).

Algorithm: Fitness Computation
Input: $task\ arrangement = \{q^1, q^2, \ldots, q^N\}$
Output: $\mathbf{q}$, J_f, J_N, J_p

$\mathbf{q} = \varnothing$; $flag = 1$; $i = 1$; $\mathbf{q}' = \varnothing$
while $flag = 1$ **&&** $Num(\mathbf{q}) \le N$
  $\mathbf{q}' = \mathbf{q} + \{q^i\}$;
  if $Fuel_s(\mathbf{q}') > fuel_{min}$
    $\mathbf{q} = \mathbf{q}'$; $flag = 1$; $i = i + 1$;
  else
    $flag = 0$;
  end if
end while
$J_N = Num(\mathbf{q})$; $J_p = \sum_{i=1}^{J_N} Prio(q^i)$; $J_f = \sum_{i=0}^{J_N} CFuel(q^i, q^{i+1})$

Figure 3. Fitness computation algorithm.

5 Simulations

5.1 Configurations

To demonstrate the model and the solution method, 15 objects of GEO belt were selected as fuel-insufficient satellites (see Table 1).The inclination and right ascension data are from literature [20]. Satellite dry mass, fuel requirements, and satellite priority value are given randomly.

Table 1. Parameters for GEO objects.

Satellite number	name	inclination (°)	RAAN (°)	True anomaly (°)	Fuel requirement (kg)	Priority
1	Morelos 2	0.9128	90.595	30.171	600	1
2	Intelsat VA F15	1.7156	87.1401	100.814	600	4
3	TDRS 4	2.3935	82.265	78.521	700	6
4	Gstar 1	2.6347	80.6943	240.339	800	4
5	IUS-13 SRM-2	4.1882	43.4184	330.478	700	6
6	IUS-15 SRM-2	6.3096	70.5983	200.158	660	1
7	Intelsat V F5	7.5253	49.9744	150.229	720	4
8	Intelsat V F2	8.2849	46.1791	300.554	750	4
9	Intelsat 501	8.7881	44.4057	175.874	650	1
10	GOES 5	9.9086	39.2166	286.178	850	1
11	DSCS II F-16	10.0680	41.2597	45.231	600	4
12	IUS-2 SRM-2	11.2379	37.5653	123.456	600	4
13	Leasat 2	11.8924	40.1735	91.221	700	6
14	FLTSATCOM F4	12.2709	31.7337	210.369	800	4
15	DSCS II F-13	12.6462	33.5862	350.109	850	1

It is assumed that SSc and all satellites are in circular orbits with the orbit altitude of $r_G = 35798$ km, and the SSc is originally located at the position where the inclination is 0 deg, the RAAN is 0 deg and the initial true anomaly is 0 deg. The initial dry mass of the SSc is 2000 kg, while the fuel mass it initially carries is 3000 kg. The specific impulse of the SSc is 300 s, and $fuel_{min}$ here is 1 kg. As sufficient time is provided for on-orbit refueling mission, each rendezvous could be achieved by phasing maneuver or Hohmann transfer. The time upper bound for each Phasing maneuver is 30 days. Four mission scenarios are given here to be compared and discussed. The SSc should return to its original location after mission in all these scenarios unless noted otherwise.

Scenario 1: all satellites and the SSc share the same orbit altitude r_G, the same inclination 0 deg and the same RAAN 0 deg.

Case 1-1: fuel demand of each satellite is 0 kg.

Case 1-2: fuel demand of each satellite is among [600, 900] kg, as depicted in Table 1.

Case 1-3: fuel demand of satellite 3 is defined to be 1800 kg.

Scenario 2: all satellites and the SSc share the same orbit altitude r_G, while the other parameters are as illustrated in Table 1.

Case 2-1: fuel demand of each satellite is 0 kg; and after mission, the SSc is not required to return to its original location.

Case 2-2: fuel demand of each satellite is among [600, 900] kg, as depicted in Table 1.

Case 2-3: fuel demand of satellite 3 is defined to be 1800 kg.

Scenario 3: all satellites share the same inclination 0 deg and the same RAAN 0 deg; the orbit altitude difference between two adjacent satellites is 300 km, i.e. satellite 1's altitude is r_G, while satellite 15 is $r_G - 4200$ km.

Case 3-1: fuel demand of each satellite is 0 kg; the orbit altitude of the SSc is initially r_G; and after mission, the SSc is not required to return to its original location.

Cade 3-2: fuel demand of each satellite is 0 kg; the orbit altitude of the SSc is initially $r_G - 4200\ km$; and after mission, the SSc is not required to return to its original location.

Case 3-3: fuel demand of each satellite is depicted in Table 1; and the orbit altitude of the SSc is initially r_G.

Case 3-4: fuel demand of each satellite is depicted in Table 1; and the orbit altitude of the SSc is initially $r_G - 4200\ km$.

Scenario 4: the orbit altitude of each satellite is the same as scenario 3; and the other parameters are the same as illustrated in Table 1.

Case 4-1: fuel demand of each satellite is 0 kg; the orbit altitude of the SSc is initially $r_G - 4200\ km$; and after mission, the SSc is not required to return to its original location.

Case 4-2: fuel demand of each satellite is depicted in Table 1; and the orbit altitude of the SSc is initially $r_G - 4200\ km$.

5.2 Results and discussions

When the fuel demand is 0 kg, J_N and J_p are senseless, as all satellites can be serviced. The phase distribution of all satellites is illustrated in Figure 4. Selecting cases which take fuel demand out of account, the optimization results are shown in Table 2.

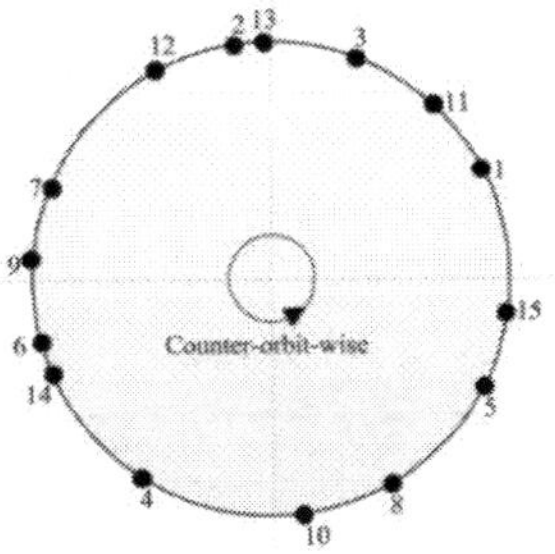

Figure 4. Phase distribution of all satellites.

Table 2. Optimization results for cases without fuel demand.

Case	J_f (kg)	Optimal mission sequence
1-1	146.799	(1, 11, 3, 13, 2, 12, 7, 9, 6, 14, 4, 10, 8, 5, 15)
2-1	1210.622	(1, 2, 3, 4, 5, 6, 7, 9, 8, 10, 11, 12, 13, 15, 14)

3-1	280.234	(1, 2, 3, 4, 5, 6, 7, 8, 9, 10, 11, 12, 13, 14, 15)
3-2	272.282	(15, 14, 13, 12, 11, 10, 9, 8, 7, 6, 5, 4, 3, 2, 1)
4-1	1242.609	(1, 2, 3, 4, 5, 6, 7, 8, 9, 10, 11, 12, 13, 14, 15)

As shown in Table 2: (1) the optimization result of case 1-1 is in accordance with Shen's research [3], (2) the result of case 2-1 is in accordance with Alfriend's [20] research; (3) the results of cases 3-1 and 3-2 demonstrate that when all satellites are coplanar but not share the same altitude, the SSc always service the nearest (altitude difference is the smallest) one first; and (4) the result of case 4-1 once again verifies Alfriend's conclusion. Table 2 also verifies that the solution approach proposed in this paper can resolve the problem efficiently. Now, taking fuel demand into consideration, the non-dominated frontiers for each scenario are shown in Tables 3-9.

Table 3. The non-dominated frontier for case 1-2.

Index	Mission sequence	J_f	J_N	J_p
1	(15, 5, 8, 10)	34.726	12	4
2	(11, 3, 13, 2, 5)	66.773	26	5
3	(11, 3, 13, 2, 12)	61.597	24	5
4	(15, 5, 11, 3, 13)	57.032	23	5
5	(15, 1, 11, 3, 13)	47.559	18	5
6	(1, 11, 3, 13, 2)	49.960	21	5

Table 4. The non-dominated frontier for case 1-3.

Index	Mission sequence	J_f	J_N	J_p
1	(1, 3, 13)	47.69	13	3
2	(1, 11, 13, 3)	48.43	17	4
3	(1, 11, 13, 2, 3)	52.17	21	5
4	(5, 11, 13, 2, 3)	72.18	26	5
5	(15, 5, 8, 10)	34.73	12	4
6	(15, 5, 11, 13, 3)	57.93	23	5
7	(11, 13, 2, 12, 3)	65.97	24	5
8	(15, 5, 1, 11, 3)	50.29	18	5
9	(15, 5, 8, 1, 11)	49.19	16	5

Table 5. The non-dominated frontier for case 2-2.

Index	Mission sequence	J_f	J_N	J_p
1	(5, 8, 7, 3)	606.37	20	4
2	(5, 8, 7, 2)	605.62	18	4
3	(1, 2, 3, 4)	255.68	15	4
4	(1, 2, 3, 5, 4)	371.01	21	5
5	(2, 3, 5, 13)	717.20	22	4

Table 6. The non-dominated frontier for case 2-3.

Index	Mission sequence	J_f	J_N	J_p
1	(2, 4, 5, 3)	449.41	20	4
2	(5, 8, 13, 3)	761.81	22	4
3	(1, 2, 5, 3)	401.75	17	4
4	(2, 3, 5)	314.07	16	3
5	(1, 3, 2)	195.61	11	3
6	(1, 2, 4, 3)	282.68	15	4
7	(2, 3, 4)	258.76	14	3

Table 7. The non-dominated frontier for case 3-3.

Index	Mission sequence	J_f	J_N	J_p
1	(2, 3, 5, 7, 13)	232.61	26	5
2	(1, 2, 3, 4, 5)	92.41	21	5
3	(2, 3, 4, 1)	85.20	15	4
4	(2, 3, 4, 5, 7)	121.09	24	5

Table 8. The non-dominated frontier for case 3-4.

Index	Mission sequence	J_f	J_N	J_p
1	(14, 13, 12, 11, 5)	198.96	24	5
2	(15, 14, 12, 11, 13)	94.37	19	5
3	(14, 13, 12, 11, 8)	149.45	22	5
4	(15, 14, 13, 12)	71.66	15	4
5	(13, 12, 11, 5, 3)	259.92	26	5

Table 9. The non-dominated frontier for case 4-2.

Index	Mission sequence	J_f	J_N	J_p
1	(4, 3, 2, 1)	546.21	15	4
2	(2, 3, 4, 5)	622.16	20	4
3	(2, 3, 5, 13)	979.64	22	4

Conclusions can be obtained from Table 3-Table 9 that:

(1) When priority is given to optimization objective J_p, satellite which has high priority value would be serviced first. Taking 1-2(2), 1-3(4), 2-2(5), 2-3(2), 3-3(1), 3-4(5) and 4-2(2) as examples, if it is possible, the SSc always service satellite 3, 5, and 13 first.

(2) When priority is given to optimization objective J_N, satellite which demands small fuel mass and consumes small in orbital transferring would be serviced first. Taking 1-2(4) and 1-3(6) as examples, in order to fuel one more satellite, the SSc services satellite 12, 5, 11, 13 first, and then satellite 3. Since the fuel demand of satellite 3 is large, if it is serviced first, there would not be sufficient fuel to satisfy other 4 satellites.

(3) When priority is given to optimization objective J_f, in most cases, satellite that demands large fuel mass would be serviced first. That is due to the fact that the smaller the J_N is, the less the orbital transfer would consume (see 1-1(1), 2-3(5) and 2-3(7)).

(4) When orbital plane differences are considered, the optimal servicing order is still mainly determined by orbital-plane differences (see tables 7, 8, and 9).

6 Conclusions

Considering the practical fact that the SSc capacity is limited and not all of the fuel-insufficient satellites could be refuelled in a mission, the problem of optimal scheduling of GEO OOR mission with uncertain OScs is studied. Specifically, the SSc is considered to be initially on the GEO belt. Numerous fuel-insufficient satellites request OOR at the same time. The SSc is unable to service all of these satellites. It just rendezvouses with some selected ones and refuels them. Phasing maneuver and Hohmann transfer are employed for each rendezvous. The optimization goal is to select the servicing targets and find the optimal servicing sequence with the best optimization objective value. Three optimization objectives, i.e. fuel cost, the total number of the servicing targets and the sum of the OSc priority value are taken into account. Considering this mission as a hybrid optimal control problem, a mathematical model is proposed. MOPSO is employed to address the model. Several OOR scenarios are compared in this paper. Numerical experiments indicate that 1) the method developed in this paper could address the problem efficiently; and 2) varieties in optimization objective emphasis and satellite distribution have considerable influences on the optimization results, and these influences appear regularly.

Acknowledgment

This work was supported by National Natural Science Foundation of China under Grant No. 71601179.

References

1. Tsiotras, P., and Nailly, A., Comparison Between Peer-to-Peer and Single Spacecraft Refueling Strategies for Spacecraft in Circular Orbits, Infotech at Aerospace Conference, AIAA Paper 2005-7115,Arlington, VA, Sept. 2005
2. J. Zhang, G.T. Parks, Y.Z. Luo, et al. Multispacecraft Refueling Optimization Considering the J2 Perturbation and Window Constraints. Journal of Guidance, Control, and Dynamics. 37(1) (2014): 111-122.
3. H. Shen, P. Tsiotras, Optimal Scheduling for Servicing Multiple Satellites in a Circular Constellation, in: AIAA/AAS Astrodynamics Specialists Conference and Exhibit, (Monterey, CA), Aug. 2002.
4. Q. Ouyang, W. Yao, X.Q. Chen, Mission Programming of On-orbit Re-fueling for Geosynchronous Satellites (in Chinese), J. Astronaut 31 (2010) 2629–2634.
5. J. Zhang, Y.Z. Luo, G.J. Tang, Hybrid planning for LEO long-duration multi-spacecraft rendezvous mission, Sci. China Tech. Sci. 55 (2012) 233-243.
6. X. Bo, Q. Feng, Research on constellation refueling based on formation flying, Acta Astronautica 68 (2011) 1987-1995.
7. Dutta A, Arora N, Russell RP. Peer-to-Peer Refueling Strategy Using Low-Thrust Propulsion [J]. Journal of Spacecraft and Rockets, 2012, 49(5): 944-954.
8. Shen, H., and Tsiotras, P.,"Peer-to-Peer Refueling for Circular Satellite Constellations,"Journal of Guidance, Control, and Dynamics, Vol. 28, No. 6, 2005, pp. 1220-1230. doi:10.2514/1.9570
9. Dutta, A., and Tsiotras, P.,"Egalitarian Peer-to-Peer Satellite Refueling Strategy,"Journal of Spacecraft and Rockets, Vol. 45, No. 3, 2008, pp. 608–618. doi:10.2514/1.31299
10. Dutta, A., and Tsiotras, P.,"Network Flow Formulation for Cooperative Peer-to-Peer Refueling Strategies,"Journal of Guidance, Control, and Dynamics, Vol. 33, No. 5, 2010, pp. 1539-1549
11. B.A. Conway, C.M. Chilan, B.J. Wall, Evolutionary Principles Applied to Mission Planning Problems, Celestial Mechanics and Dynamical Astronomy 97(2) (2007): 73-86.
12. I.M. Ross, C.N. D'Souza, Hybrid Optimal Control Framework for Mission Planning, Journal of Guidance, Control, and Dynamics 28(4) (2005): 686-697.
13. C.M. Chilan, Automated Design of Multiphase Space Missions Using Hybrid Optimal Control. Ph.D. diss., University of Illinois at Urbana-Champaign, 2009.
14. B.J. Wall, B.A. Conway, Genetic Algorithms Applied to the Solution of Hybrid Optimal Control Problems in Astrodynamics, J. Glob. Optim. 44 (2009): 493-508.
15. Jing Y, Chen X, Chen L. Biobjective planning of GEO debris removal mission with multiple servicing spacecrafts [J]. Acta Astronautica, 2014, 105(1): 311-320.
16. Jing Yu, Xiao-qian Chen, Li-hu Chen, Dong Hao. Optimal scheduling of GEO debris removing based on hybrid optimal control theory [J]. Acta Astronautica, 2014, 93: 400-409.
17. Jing Yu, Xiao-qian Chen, Li-hu Chen. Optimal planning of LEO active debris removal based on hybrid optimal control theory [J]. Advances in Space Research, 2015, 55(11): 2628-2640.
18. H. L. Shi, Y. B. Han, L. H. Ma, et al., Beyong Life-Cycle Utilization of Geostationary Communication Satellites in End-of-Life, Satellite Communications (2010) 323-365.
19. L. J. Friesen, A. A. Jackson, H. A. Zook, et al., Analysis of Orbital Perturbations Acting on Objects in Orbits Near Geosynchronous Earth Orbit, Journal of Geophysical Research 97 (1992) 3845-3863.
20. Alfriend KT, Lee D, Creamer NG. Optimal Servicing of Geosynchronous Satellites. Journal of Guidance, Control, and Dynamics, 2006, 29(1): 203-206.

MATEC Web of Conferences **114**, 03002 (2017) DOI: 10.1051/matecconf/201711403002
2MAE 2017

The Capturing of Space Debris with a Spaceborne Multi-fingered Gripper

Yin Zhang[1] and Qiang Zhan[2,a]

[1] *Robotic institute, Beihang University, 100191 Beijing, China*
[2] *Robotic institute, Beihang University, 100191 Beijing, China*

Abstract. With the massive launching of spacecraft, more and more space debris are making the low Earth orbit (LEO) much more crowded which seriously affects the normal flight of other spacecrafts. Space debris removal has become a very urgent issue concerned by numerous countries. In this paper, using SwissCube as a target, the capturing of space debris with a spaceborne four-fingered gripper was studied in order to obtain the key factors that affect the capturing effect. The contact state between the gripper fingers and SwissCube was described using a defined contact matrix. The law of momentum conservation was used to model the motion variations of the gripper and SwissCube before and after the capturing process. A zero-gravity simulation environment was built using ADAMS software. Two typical kinds of capturing processes were simulated considering different stiffness of fingers and different friction conditions between fingers and SwissCube. Comparisons between results obtained with the law of momentum conservation and those from ADAMS simulation show that the theoretical calculations and simulation results are consistent. In addition, through analyzing the capturing process, a valuable finding was obtained that the contact friction and finger flexibility are two very important factors that affect the capturing result.

1 Introduction

Space debris is the collection of defunct man-made objects in orbit around the Earth, which includes old satellites, spent rocket stages, fragments, erosion, and collisions, et al [1]. A lot of space debris has been generated during the past 60 years of human space activities. Most of the space debris spreads within the Low Earth Orbit (LEO) region, resulting in the density of space debris on orbit is approaching the safety threshold [2,3]. It is well recognized that the existence of space debris seriously threatens the safety of human space activities and the orbiting spacecrafts, so it is essential for the whole world to remove the LEO space debris.

The space environment is very complex with micro-gravity, high vacuum, strong radiation and weak illumination. Space debris belongs to non-cooperative target and has different shapes and sizes. It's significantly challenging to capture and remove space debris. Space agencies in many countries have proposed some projects and plans on space debris removal, which can be broadly categorized as follows: 1) "Robotic Arm & Gripper", such as the FREND project [4], the DEOS and the EPOS project [5] and the e.Deorbit project [6]; 2) "Flexible Net", such as the e.Deorbit project [6]; 3)

[a] Corresponding author : qzhan@buaa.edu.cn

"Harpoon", such as the e.Deorbit project [6]; 4) "Electro-dynamic Tether", such as the ProSEDS plan [7]. Some other methods have also been proposed and studied, such as the laser method [8], etc. Although many studies have been done on space debris removal, however, to find a safe, reliable and efficient method still has a long way to go.

In order to explore and verify the key technologies of the LEO space debris removal, EPFL (École Polytechnique Fédérale de Lausanne) and Swiss Space Center proposed the "CleanSpace One, CSO" plan [9,10]. Based on the information communicated with EPFL, we have studied the capturing of target SwissCube with a four-fingered spaceborne gripper in order to check the reliability and factors influencing the grasping scheme.

2 Spaceborne gripper capturing scheme

2.1 Brief introduction of SwissCube

The capturing target SwissCube was launched in 2009 by Swiss Space Center. Some of its main characteristic parameters are provided in Table 1. It was originally designed working for 4-12 months, however, it has been working for over 7 years to date. Some components have begun to age and therefor it is chosen as the experimental target to remove from the orbit [9].

Table 1. Characteristics of SwissCube.

Parameters	Values
Dimensions (Length×Width×Height)/ mm×mm×mm	113.50×100.00×100.00
Mass/kg	1.00
The maximum rotation speed /deg·s^{-1}	50.00
The maximum moment of inertia /kg·m^2	2.45×10^{-3}

2.2 Spaceborne gripper capturing system

In this paper, we studied a spaceborne gripper system that regards SwissCube as the target for capturing. Compared with the flexible net and other removal methods, this gripper capturing system can conduct the capturing and releasing process more easily, and it can repeat the removal tasks. According to the robot grasping theory, the number of fingers is typically greater than or equal to three. Because the cracks between fingers are larger of the three-fingered gripper, the possibility of target escape is larger. The structure of the five-fingered gripper is complex and difficult to control. Therefore, considering factors such as structural complexity, grasping stability and others, we select four-fingered gripper as the spaceborne capturing device scheme. The gripper is composed of a base and four arc-shaped fingers. The gripper can adjust its own posture according to the moving state of SwissCube by controlling the closing and opening of the fingers, the gripper can realize the capturing and releasing of the target. As we mainly focus on the capturing process, we just consider the system model comprising mother-satellite, spaceborne gripper, and SwissCube.

3 Modeling and analysis of the capturing process

The spaceborne capturing system will be nearly on the same orbit as the SwissCube after completing the tracking and closing steps. The angular velocity difference around the Earth could be ignored. When the distance between the capturing system and SwissCube is small, the linear speed difference between the two objects could be also neglected. Therefore, the final capturing process could be simplified as one where a spaceborne gripper captures SwissCube in a zero gravity environment.

3.1 Description of the contact status

DOI: 10.1051/matecconf/201711403002

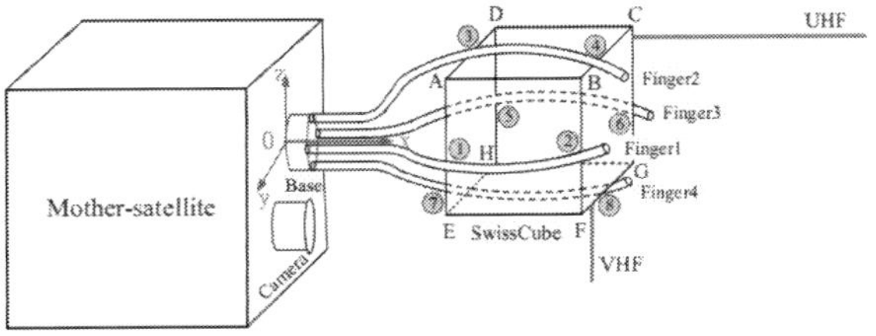

Figure 1. The grasping model of spaceborne gripper capturing SwissCube.

Fig. 1 shows the grasping model. SwissCube has two antennas, six feature planes, and twelve different edges. When the fingers contact with SwissCube, contact forces will be applied on SwissCube through the contact points and a group of point contact constraints will be formed. The maximum number of contact points between one finger and one edge of SwissCube is two. We defined a contact matrix $\boldsymbol{M}_{F_i,L_j}$ to describe the contact status among fingers and SwissCube,

$$\boldsymbol{M}_{F_i,L_j} = \left[\delta_{i,j}\right] \quad i=1,\ldots,4, \quad j=1,\ldots,12 \tag{1}$$

Where, F_i represents the ith finger of the gripper, F_i=Finger$i$; L_j represents the jth edge of SwissCube, $L_1=AB$, $L_2=BC$, $L_3=CD$, $L_4=DA$, $L_5=EF$, $L_6=FG$, $L_7=GH$, $L_8=HE$, $L_9=AE$, $L_{10}=BF$, $L_{11}=CG$, $L_{12}=DH$. $\delta_{i,j}$ denotes the contact status between the ith finger and the jth edge. If the ith finger contacts with the jth edge, $\delta_{i,j}=1$; else, $\delta_{i,j}=0$. When SwissCube is moving in the envelope region, different contact points will be formed and the contact matrix $\boldsymbol{M}_{F_i,L_j}$ will be different.

3.2 Theoretical analyses of the capturing process

The complex contact and collision process can be seen as an impulse effect. We defined a contact force matrix $\boldsymbol{F}_{F_i,L_j}$ to describe the contact forces between the fingers and SwissCube,

$$\boldsymbol{F}_{F_i,L_j} = \left[f_{i,j}\right] \quad i=1,\cdots,4, j=1,\cdots,12 \tag{2}$$

Where, $f_{i,j}$ denotes the contact force between the ith finger and the jth edge of SwissCube. If the ith finger contact with the jth edge, $\delta_{i,j}=1$ and $f_{i,j}\neq 0$; else, $\delta_{i,j}=0$ and $f_{i,j}=0$.

In the space environment, all the linear momentum and angular momentum of the whole system are both constants. We can utilise the law of momentum conservation to analyze the motion variation of the system before and after the capturing process. The initial linear momentum and the angular momentum of the whole system are shown in Eq. 3 and Eq. 4 respectively.

$$\boldsymbol{P}_I = \boldsymbol{P}_M + \boldsymbol{P}_S = \boxed{\sum_{i=1}^{n} m_i \boldsymbol{v}_i} + \boxed{m_S \boldsymbol{v}_S} \tag{3}$$

$$\boldsymbol{L}_I = \boldsymbol{L}_M + \boldsymbol{L}_S = \boxed{\sum_{i=1}^{n} (\boldsymbol{r}_i \times m_i \boldsymbol{v}_i)} + \boxed{\boldsymbol{r}_S \times m_S \boldsymbol{v}_S} \tag{4}$$

Where $\boldsymbol{P}_M$ and $\boldsymbol{L}_M$ denote the linear and angular momentum of the mother-satellite with gripper respectively, m_i , $\boldsymbol{v}_i$ and $\boldsymbol{r}_i$ denote the mass, the line speed, and the radius vector of each corresponding components. $\boldsymbol{P}_S$, $\boldsymbol{L}_S$, m_S , $\boldsymbol{v}_S$ and $\boldsymbol{r}_S$ denote the linear momentum, the angular momentum, the mass, the line speed and the radius vector of SwissCube respectively.

The initial line speed and angular velocity of the mother-satellite are both set to 0. The initial line speed and angular velocity of each finger are set to 0 and $\boldsymbol{\omega}_{f_k}$. The initial line speed and angular velocity of SwissCube are set to 0 and $\boldsymbol{\omega}_{S_z}$. Thus, $\boldsymbol{P}_I = 0$ and $\boldsymbol{L}_I = \sum_{k=1}^{4} J_{f_k} \boldsymbol{\omega}_{f_k} + J_S \boldsymbol{\omega}_{S_z}$, where,

DOI: 10.1051/matecconf/201711403002

J_{f_k} and J_S denote the inertia moment of each finger and that of SwissCube respectively. Since the four fingers are even-distributed along one circle, so $\sum_{k=1}^{4} J_{f_k}\boldsymbol{\omega}_{f_k} = 0$, and $\boldsymbol{L}_I = J_S\boldsymbol{\omega}_{S_z}$.

We use $\boldsymbol{P}_E$ and $\boldsymbol{L}_E$ to denote the final linear and angular momentum of the whole system respectively. According to the law of momentum conservation, the following equation is obtained,

$$\begin{bmatrix} \boldsymbol{P}_I \\ \boldsymbol{L}_I \end{bmatrix} = \begin{bmatrix} \boldsymbol{P}_E \\ \boldsymbol{L}_E \end{bmatrix} = \begin{bmatrix} 0 \\ J_S\boldsymbol{\omega}_{S_z} \end{bmatrix} \tag{5}$$

The final linear speed is zero, $\boldsymbol{v}_E = 0$, and the final angular velocity $\boldsymbol{\omega}_E$ is shown in Eq. 6.

$$\boldsymbol{\omega}_E = \begin{bmatrix} \omega_{Ex} & \omega_{Ey} & \omega_{Ez} \end{bmatrix}^T = \begin{bmatrix} 0 & 0 & J_S\boldsymbol{\omega}_{S_z} \end{bmatrix}^T / J_E \tag{6}$$

Where ω_{Ex}, ω_{Ey} and ω_{Ez} denote the angular velocity about the axes x, y and z respectively, J_E denotes the inertia moment of the whole system including SwissCube.

We use the following values to explain the changes before and after the capturing process, namely, $\sum_{i=1}^{n} m_i = 30\ \text{kg}$, $m_S = 1\ \text{kg}$, $J_S = 2.45\times10^{-3}\ \text{kg•m}^2$ (shown in Table 1), $J_E = 0.53483\ \text{kg•m}^2$, $\boldsymbol{\omega}_{S_z} = 50\ \text{deg/s}$. Thus, we can obtain the following results, $\boldsymbol{\omega}_{Ex} = \boldsymbol{\omega}_{Ey} = 0$ and $\boldsymbol{\omega}_{Ez} = 0.229\ \text{deg/s}$.

It indicates that if the gripper can capture SwissCube successfully, the gripper, the mother-satellite and SwissCube will rotate about their common axis at a same small angular velocity together.

4 Simulation and analysis of the capturing process

ADAMS software was used to build a zero-gravity simulation environment, some parameters are set as follows: the mass of the mother-satellite and SwissCube are 30kg and 1kg respectively, the initial angular velocity of SwissCube is 50 deg/s. If the contact belongs to frictionless point contact, it is difficult to capture SwissCube successfully. We mainly focus on the other two typical capturing cases with different stiffness of fingers and different friction conditions. In order to display the simulation results conveniently, we adopt the following abbreviations: *MS-Microsatellite, SC-SwissCube, P-Position, V-Velocity, A-Acceleration, AV-Angular Velocity, AA-Angular Acceleration.*

4.1 Simulation and analysis of "Rigid fingers & Contact friction"

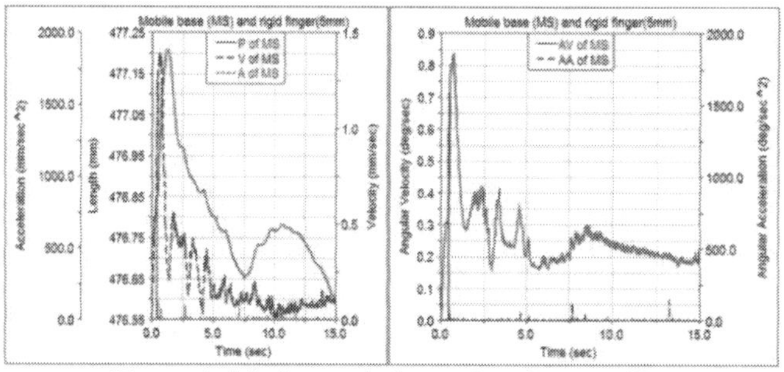

Figure 2. Motion variations of mother-satellite

Fig. 2 and Fig. 3 show the motion variations of the mother-satellite and SwissCube after capturing when the acting force is a force with friction. From these figures, we can find that when the fingers first contact with SwissCube, the mother-satellite has a notable acceleration, its line speed reaches a high value in a short time and then gradually decreases to zero. The mother-satellite deviates from its initial position slightly and then gradually returns to the initial position. The final angular velocity of SwissCube and the mother-satellite will be a same smaller one (≈0.2 deg/s) about their common equivalent axis. These results are consistent with those obtained from the theoretical analysis. Finally, the angular acceleration will eventually be zero. The gripper can capture SwissCube successfully.

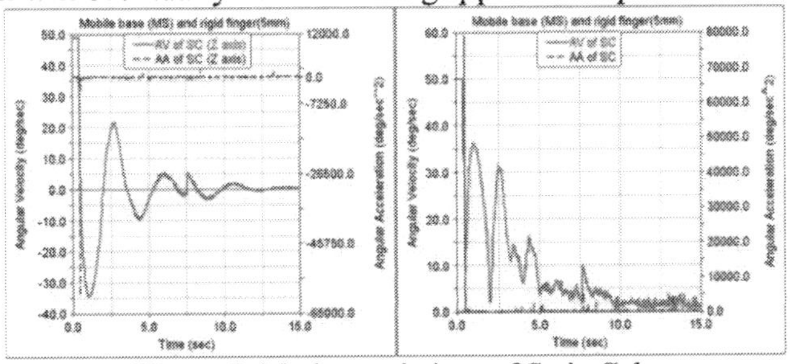

Figure 3. Motion variations of SwissCube

Fig. 4(a) shows the initial state of the whole system before capturing. When the contact belongs to point contact with friction, some kinetic energy of SwissCube could be "consumed" and converted into other forms of energy. SwissCube moves and rotates in the envelope region formed by the fingers and it gradually reaches a stable state (see Fig. 4(b)). As there exists motion coupling between the mother-satellite and SwissCube, the final angular velocity of SwissCube and the mother-satellite will be not zero, but a same smaller one about their common equivalent axis.

(a) Initial state (b) After capturing

Figure 4. Capturing result with contact friction and the fingers are rigid

4.2 Simulation and analysis of "Flexible fingers & Contact friction"

In this section, we'll explore the capturing characteristic of the system when the fingers are flexible bodies and there exists contact friction between fingers and SwissCube. It means that the rigid fingers are converted to elastic bodies using FEA software and replaced by flexible fingers in ADAMS software. Thus, a whole system consists of rigid parts and flexible fingers is built. Flexible fingers can deform under force. The contact between the fingers of the gripper and SwissCube also belongs to point contact with friction.

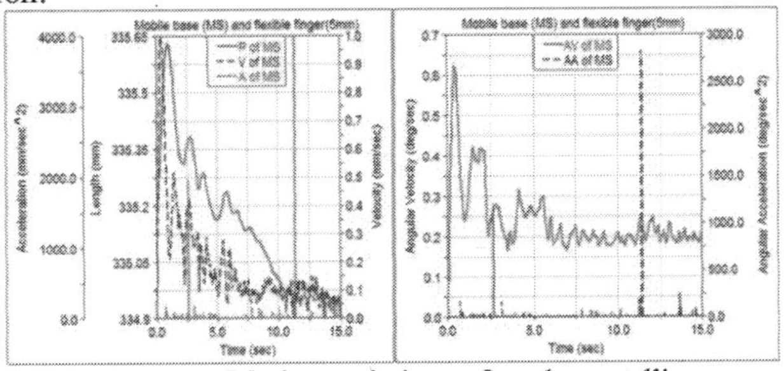

Figure 5. Motion variations of mother-satellite

When the simulation is completed, we get the motion variations of the mother-satellite and SwissCube (shown in Fig. 5 and Fig. 6). Because there exists motion coupling between the fingers and SwissCube, the mother-satellite deviates from its initial position slightly at a small line speed and it rotates about its own axis. Through comparing with the motion variations of the mother-satellite and SwissCube when the fingers are rigid bodies, we can find that the flexible fingers can make the capturing process more smooth and continuous than the rigid fingers, and the whole system can reach a stable state more quickly. This is because when the fingers are flexible bodies, some kinetic energy of SwissCube could be "consumed" and converted into other forms of energy, so the mother-satellite and SwissCube can reach the co-speed (≈ 0.2 deg/s) rotation state more smoothly and quickly.

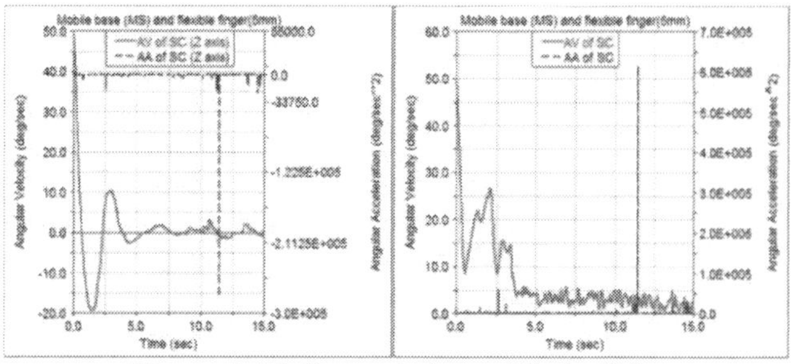

Figure 6. Motion variations of SwissCube

5 Conclusions

In this paper, we studied the capturing process and its influencing factors when a four-fingered spaceborne gripper captures a target space debris SwissCube. We can conclude the following:

1. Considering the structural complexity, control difficulty, grasping stability and other factors, the four-fingered gripper is an ideal capturing device scheme. If this spaceborne gripper can approach the approximate capturing region, it can capture the space debris target successfully.

2. Contact friction is an important and essential factor affecting the space debris capturing effect. It's helpful to "consumed" the kinetic energy of the object and converted it into other forms of energy. When designing a capturing device, the contact friction should be considered.

3. The flexibility of finger is another significant factor affecting the capturing effect. Flexible fingers have better performance than rigid fingers, it's helpful to make the capturing more smooth and quickly. When designing a capturing device, the flexibility of the gripper should be taken into account.

Acknowledgements

This work was supported by the Fundamental Research Funds for the Central Universities (YWF-14-JXXXY-025).

References

1. M.H. Shan, J. Guo and E. Gill, Review and comparison of active space debris capturing and removal methods, Prog Aerosp Sci. 80 (2016) 18-32.
2. J.-C. Liou, N.L. Johnson and N.M. Hill, Controlling the growth of future LEO debris populations with active debris removal, Acta Astronaut. 66 (2010) 648-653.
3. C. Bonnal, J.M. Ruault and M.C. Desjean, Active debris removal: recent progress and current trends, Acta Astronaut. 85 (2013) 51-60.
4. G. Creamer, The SUMO/FREND project: technology development for autonomous grapple of geosynchronous satellites, Adv Astronaut Sci. 128 (2007) 895-909.
5. T. Boge, T. Wimmer, O. Ma and M. Zebenay, EPOS-A robotics-based hardware-in-the-loop simulator for simulating satellite RvD operations. In: 10th International symposium on artificial intelligence, robotics and automation in space, Sapporo, Japan, 2010.
6. J.A.F. Deloo, Analysis of the rendezvous phase of e.Deorbit: guidance, communication and illumination. PhD Thesis, Delft University of Technology, NL, 2015.
7. S. Kawamoto, T. Makida, F. Sasaki, et al, Precise numerical simulations of electrodynamic tethers for an active debris removal system, Acta Astronaut. 59 (2006) 139-148.
8. C.R. Phipps, K.L. Baker, S.B. Libby, et al, Removing orbital debris with lasers, Adv Space Res. 49 (2012) 1283-1300.
9. M. Richard, L. Kronig, F. Belloni, et al, Uncooperative rendezvous and docking for MicroSats, In: 6th International conference on recent advances in space technologies, Istanbul, Türkiye, 12-14 June 2013.
10. O.A. Araromi, I. Gavrilovich, J. Shintake, et al, Rollable multisegment dielectric elastomer minimum energy structures for a deployable microsatellite gripper, IEEE/ASME Transactions on Mechatronics. 20 (2015) 438-446.

Noise Generation of a Bionic Airfoil Based on Owl Wings

Changjiang Ge[1,a]

[1] *College of Automotive Engineering, Jilin University, Changchun, 130022, P. R. China*

Abstract. A bionic airfoil is used to investigate a preliminary understanding of the noise mechanism generated by the flow around the owl wing. The flow indicates a transitional separation bubble on the surface and the turbulent boundary layer reattaches at the downstream end of the bubble. This causes turbulent boundary layer trailing-edge scattering noise corresponding to the broadband noise of owl wings at low Reynolds numbers. Furthermore, a large-scale vortex detaches from the bubble and then drifts downstream. Thus, a tone will be generated by the vortex shedding. However, the computations of the acoustic field for both cases, where the on-body data surface and the off-body data surface are used respectively, predominantly verify the broadband nature of the noise that well agrees with the noise scattering caused by the turbulent boundary layer passing by the sharp trailing edge. Furthermore, the significant decrease of the spectra around 2k Hz is consistent with that of silent owl wings. This fact suggests that the numerical simulations are capturing the basic physical mechanism responsible for the noise radiated by the natural owl wing. This might be able to provide a reference for engineering applications of the owl technology.

1 Introduction

"Owl technology" is a promising means of turbulent boundary layer trailing-edge noise reduction for applications such as fans, rotors and propellers [1~4], so bionic treatments, inspired by three wings and plumage: leading-edge serrations, trailing-edge fringes and a soft and elastic downy upper surface of the feathers [5], are used to attenuate noise caused by trailing-edge scattering. Furthermore, howe [6~7], using aeroacoustic theory, showed how serrated or porous trailing edge may be made quieter than a straight and solid trailing edge. Despite this accumulated knowledge of airfoil trailing-edge noise, actually alleviating it is difficult. Particularly, experimental results of full-scale models associated with trailing-edge serrations showed noise abatement in the low frequency band, but noise enhancement in the high frequency band [8]. According to howe's theory, significant noise reductions should be only expected in the high frequency region. This suggested that the trailing-edge noise reduction problem is complex, and involves a variety of opposing constrains.

Previous investigations were conducted by a flat plate or technical airfoils [9~13], such as NACA airfoils. Obviously, they are different from profiles extracted by nature owl wings [14~15]. It is well known that modification of the noise-producing characteristics of the trailing-edge turbulent boundary layer is possible through airfoil shape. This may be the reason for the discrepancy between the experiments and theory of noise reduction with the bionic treatments. In addition, Reynolds number of

[a] Corresponding author : gcj@jlu.edu.cn

less than 2×10^5 typical of flying owls implies the laminar flow on the surfaces of owl wings. However, measurements suggested that the pronounced broadband noise is generated by turbulent boundary layer scattering [16]. Thus, fundamental researches are needed to shed more additional light on the owl-based airfoil noise mechanism for further investigations of noise reduction technology.

The objective of this numerical study is to gain a preliminary understanding of the aerodynamic noise mechanism generated by the unsteady flow around a bionic owl wing airfoil. The bionic airfoil selected by the validity is representative of the characteristics of owl wings [17]. The flow field around the airfoil is simulated by Computational Fluid Dynamics (CFD), and the noise source is identified by the distribution of vorticity. The far-field Sound Pressure Level (SPL) is determined by inputting highly resolved unsteady near-field data obtained from converged CFD solutions into two-dimensional Ffowcs Williams and Hawkings equation (FW-H equation) in order to estimate noise levels at a radial distance of ten chords from the airfoil.

2 Materials and method

2.1 Grid topology

The computational domain of the bionic airfoil extends to 20c in the upstream direction, 24c in the downstream direction and 30c in the crosswise direction so as to dissipate reflections from the boundaries (Figure 1). The leading edge is located at the origin. The outflow boundary condition is a pressure far field, thus minimize effectively any feedback effect in the incompressible flow regime, the viscous wall boundary condition is specified for the bionic airfoil surface, and interface boundary condition is applied at the internal faces. Quadrilateral grids with C-type are employed for the simulation because they generate less numerical diffusion than triangular grids. Typical grids used for the two-dimensional simulations are 250×200 (Figure 2). The grid size Δ x is small enough to capture the interesting turbulence length scales, and to make the numerical diffusion smaller than the subgrid-scale turbulence viscosity. Further node savings are achieved by specifically clustering nodes near the wall. For instance, the node distribution at the wall yields non-dimensional wall distances $y^+<0.4$ everywhere and it can simulate the effect of laminar sub-layer. Moreover, the sufficiently dense grids around the areas of interest are also required to ensure adequate propagation of waves and restrict artificial diffusion.

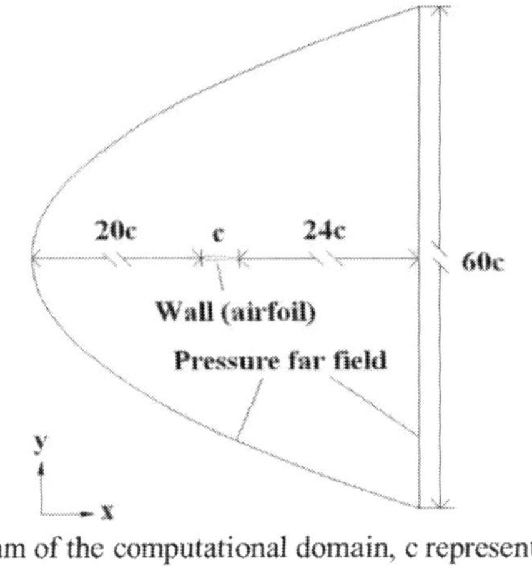

Figure 1. Schematic diagram of the computational domain, c represents the chord length of 0.1m.

Figure 2. Grid distribution around the bionic airfoil.

2.2 Numerical method

The Large Eddy Simulation (LES) is based on the incompressible Navier-Stokes (N-S) equations spatially filtered with the dynamic subgrid scale model. Therefore, broadband aeroacoustic noise can be predicted, because the LES resolves all eddies with scales larger than the grid scale. The central differences are used to discretize the spatial derivatives yielding overall second-order accuracy in Ansys Fluent V6.3. Initial conditions are provided by a Reynolds Averaged Navier-Stokes (RANS) simulation. The numerical method is Pressure Implicit with Splitting of Operators (PISO) algorithm on basis of the pressure-velocity formulation employing a predictor-corrector pressure scheme. The velocity u for the pressure far field is 8m/s, which is typical for the flying owls, and the corresponding Reynolds number based on the chord length c is about 5.5×10^{-4}. The angle of attack is $0°$, where the resulting noise source is related to the turbulent pressure fluctuations in the airfoil boundary layer. The time step Δt is 1×10^{-5} to obtain the interesting frequency band making CFL<0.6, calculated by $CFL=u\times\Delta t/\Delta x$. Once the flow simulation reaches a dynamically steady state, the accuracy for predicting SPL is usually dependent on the number of time steps used. Thus, the number of time steps is 1×10^{6} for a better result.

The far-field acoustic signals are achieved by solving two-dimensional FW-H equation to gain the correct features of the noise radiated from unsteady flow around the two-dimensional bionic airfoil surface. The source correlation length for the two-dimensional simulation is 1m. Both the data surfaces for noise are depicted in Figure 3. The propagated noise would be overpredicted due to two-dimensional method implying perfect correlation in the third dimension. However, the paper is focused on the near-field flow and the corresponding noise mechanism, so any effects associated with the overprediction of noise caused by two-dimensional approximations are obviated by the qualitative distribution trend in noise level.

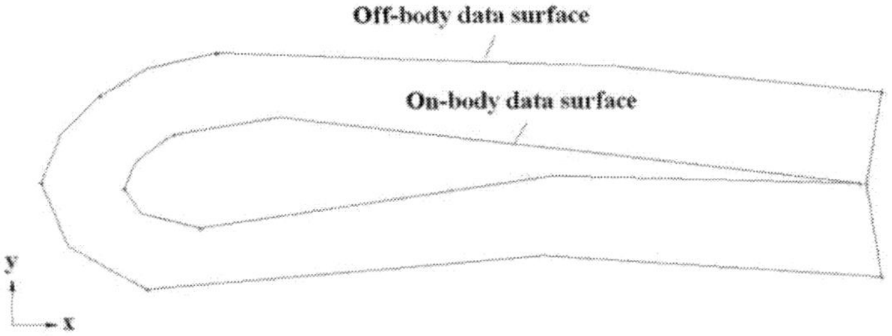

Figure 3. Schematic diagram of the data surfaces.

2.3 Experimental setup

Experiments are done in a low-speed wind tunnel with a cross section of 150×150 mm^2 and a length of 800mm. The wind tunnel is used as a smoke wind tunnel by installing laser sheet generator. The smoke lines are generated by engine oil dropping on an electrically heated 0.1mm nichrome wire moving downstream with the air to visualize the flow field. A wing model with the chord of 150mm and the span of 150mm is built to avoid the spanwise flow. The freestream velocity of 8m/s and the angle of attack of 0° are considered. The measurement equipment is wind tunnel LW-9117 manufactured by Long Win science and Technology Corporation.

3 Results and discussion

3.1 Instantaneous flow field

The instantaneous flow around the bionic airfoil indicates the leading-edge separation due to the adverse pressure gradient, as shown in Figure 4. After separating, the free shear layer rolls up forming a vortex by Kelvin-Helmholtz (K-H) instability. It re-energizes the low-velocity flow on the near wall, and thereby reattachment occurs with the turbulent boundary layer. This kind of separation is called the "long bubble", and its extent quickly increases as angles of attack increase until the presence of the deep stall at high angle of attack. At the downstream end of the transitional separation bubble, a large-scale vortex detaches from the bubble and moves further downstream, as depicted by Burgmann et al [18]. Simultaneously, the similar evolution is also shown on the lower surface. The reason for this is that the strong adverse pressure gradient is caused by the high curvature of the lower surface near the leading edge. In addition, the experimental flow visualized in low turbulence wind tunnel is in close agreement with the flow computed by Ansys Fluent V6.3 at the same conditions, as shown in Figure 5.

Figure 4. Distribution of velocity vector and static pressure (pa).

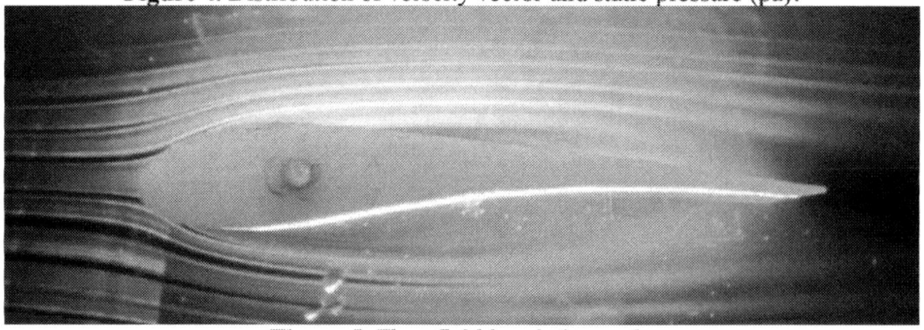

Figure 5. Flow field in wind tunnel.

The pressure coefficients Cp varying with the streamwise positions verifies the finding of the visualized flow field, as depicted in Figure 6. The distribution of the upper surface shows a stable process downstream of the first peak at the chordwise position about between x/c=0.2~0.5, which corresponds to the recirculation region downstream of the leading edge in Figure 4. Subsequently, the second peak confirms the rolling up of a vortex in the bubble, and then the pressure valley indicates that the leading-edge separation is formed by the reattachment of the transitional free shear layer. Furthermore, the third (maximum) pressure peak is consistent with the large-scale vortex separated from the bubble. The following pressure fluctuation close to the trailing edge suggests the turbulent boundary layer.

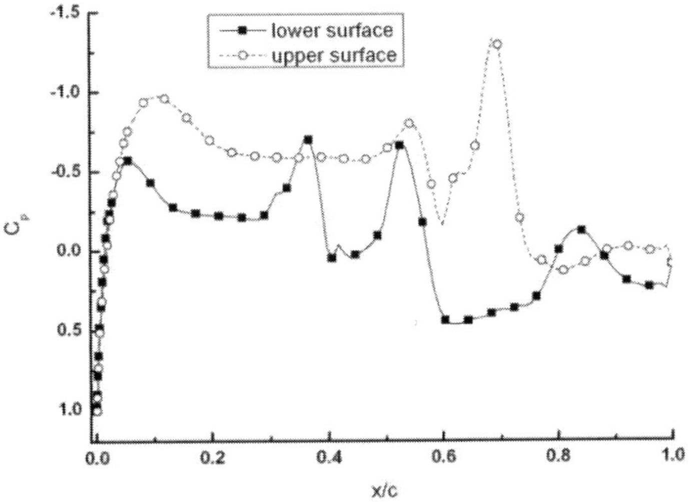

Figure 6. Distribution of pressure coefficient Cp, x/c is the normalized chordwise coordinate, where x is the chordwise coordinate.

The contours of vorticity also substantiate the finding of the flow field: the leading-edge separation, the roll-up process of the free shear layer, and the formation of the bubble, as shown in Figure 7. The large-scale vortex departing form the bubble and passing by the trailing edge will generate the vortex shedding responsible for the tonal loading noise. It is obvious that the K-H instability of the separated free shear layer is receptive to the acoustic perturbations caused by the trailing-edge vortex shedding and the process can produce an aeroacoustic feedback loop. Interesting, the turbulent boundary layer, which reattaches on the wall, will cause the broadband scattering noise due to the interaction of the turbulent eddies within the boundary layer with the sharp trailing edge of the bionic airfoil. This can be used to explain the noise mechanism of natural owl wings at the low Reynolds numbers.

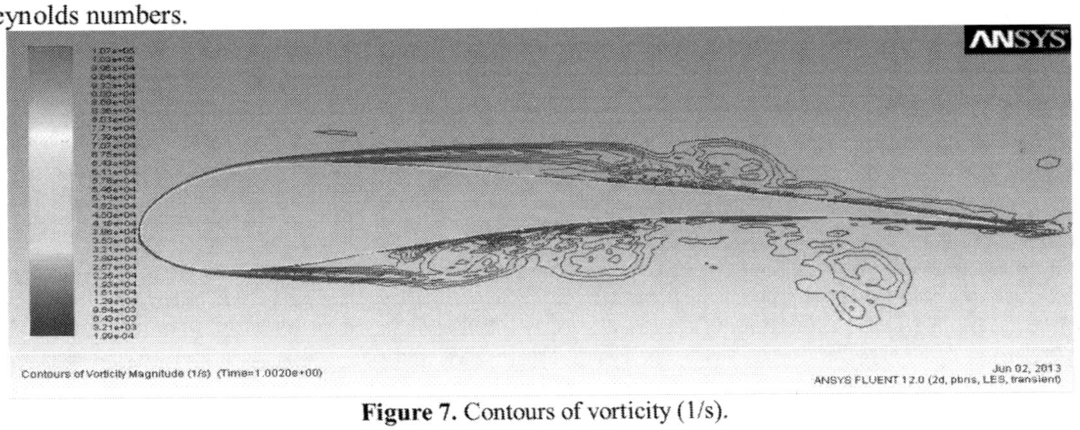

Figure 7. Contours of vorticity (1/s).

3.2 Acoustic field

Figure 8 shows a comparison of the acoustic spectra obtained at a distance of 10c directly under the leading edge for cases, in which the on-body data surface and the off-body data surface are used to compute the far-field noise. The humps agree well with the broadband noise nature of owls on the approach. The lack of the strong peak means that vortex-shedding noise is less than turbulent boundary layer trailing-edge noise that is the dominant source in the current simulations. The significant discrepancy of the SPLs between both the cases suggests that the on-body data surface can only capture the scattering source caused by wall pressure fluctuation in the turbulent boundary layer. In contrast, the off-body can also consider quadrupole effect of turbulence in the boundary layer by the permeable data surface. In particular, the collapse depicted in the SPLs is expected around 2k Hz, because landing owls can dramatically decrease noise above 2k Hz [2~4]. This indicates that the acoustic prediction is believable, although amplitudes for the on-body data surface are significantly lower than those for off-body data surface.

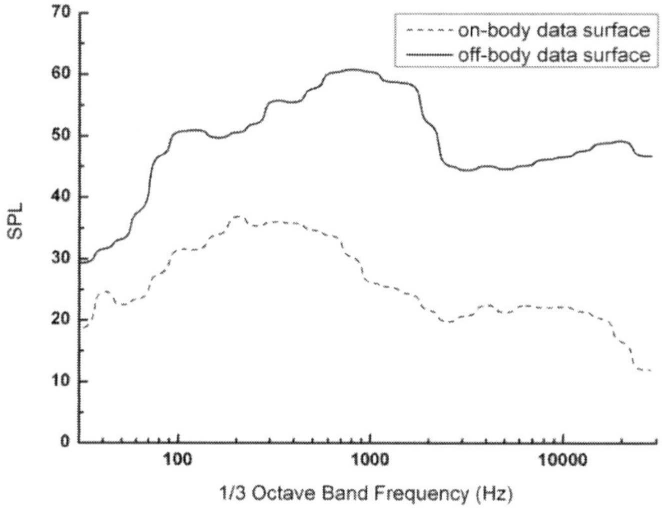

Figure 8. Sound pressure level at the observer position displaced 10c directly under the leading edge.

4 Conclusion

The unsteady simulation of the bionic airfoil is completed in order to investigate the noise related to owl wings. The flow field around the airfoil shows that the leading-edge boundary layer on the surface is separated, and later the free shear layer rolls up forming a vortex due to K-H instability. As a result, a bubble comes into being by the reattachment of the transitional free shear layer. The resulting turbulent boundary layer is responsible for the noise mechanism of owl wings. Furthermore, a large-scale vortex detaches from the bubble and drifts downstream. Thus, the aeroacoustic feedback loop is achieved by the interaction of trailing-edge vortex shedding with the free shear layer.

The flow from the unsteady LES calculation is used to provide the necessary information on the on-body and off-body data surfaces. The preliminary results indicate the broadband nature of the noise, which is consistent with the scattering caused by flying owls on the approach. Compared to the on-body data surface applied at the bionic airfoil, the off-body data surface, which contains the dominant vortexes on the basis of vorticity field, can account for the non-linear effect of sound scattering by interaction between turbulence and sharp trailing edge. In addition, both the spectra generated by the airfoil dramatically decrease around 2k Hz, as in the spectra of silent owl wings. This similarity suggests that the simulation for the unsteady flow around the bionic airfoil can predict the scattering noise mechanism generated by natural owl wings.

Acknowledgements

The authors would like to thank the Key Laboratory of Bionic Engineering for their assistance with the specimen and the equipments. This research is supported by the National Natural Science Fund of China (No. 51505182) and Jilin Scientific and Technological Development Program (No. 20160520065JH).

References

1. R.A. Kroeger, H.D. Gruschka, T.C. Helvey, Air Force Flight Dynamics Laboratory Technical Report Low speed aerodynamics for ultra-quiet flight., 71-75 (1971).

2. G.M. Lilley, A study of the silent flight of the owl. AIAA Paper, 98-2340 (1998).
3. G.M. Lilley, The Prediction of Airframe Noise and Comparison with Experiment. Journal of Sound and Vibration, **239** (4), 849–859 (2001).
4. D.P. Lockard, G.M. Lilley, The Airframe Noise Reduction Chanllenge. NASA/TM-2004-213013 (2004).
5. R.R. Graham, The silent flight of owls. J. R. Aeronaut., **38**, 837-843 (1934).
6. M.S. Howe, Aerodynamic noise of a serrated trailing edge. J. Fluids and Structures, **5**, 33–45 (1991).
7. M.S. Howe, On the added mass of a perforated shell, with application to the generation of aerodynamic sound by a perforated trailing edge. Proc. R. Soc. Lond. A, **365**, 209–233 (1979).
8. S. Oerlemans, M. Fisher, T. Maeder, K. Kogler, Reduction of wind turbine noise using optimized airfoils and trailing edge serrations. AIAA J., **47**(6):1470–1481 (2009).
9. M. Herr, New Results in Numerical and Experimental Fluid Mechanics V: Experimental Study on Noise Reduction through Trailing Edge Brushes, *Springer, Berlin Heidelberg*, 365-372 (2006).
10. M. Herr, W. Dobrzynski, Experimental investigations in low-noise trailing-edge design. AIAA J., **43**(6), 1167–1175 (2005).
11. M. Herr, On the design of silent trailing edges. New Res. in Num. and Exp. Fluid Mech. VI, NNFM 96, **96**, 430–437 (2007).
12. T.Geyer, E. Sarradj, C. Fritzsche, Porous airfoils: Noise reduction and boundary layer effects. AIAA Paper, 2009-3392 (2009).
13. T. Geyer, E. Sarradj, C. Fritzsche, Measurement of the noise generation at the trailing edge of porous airfoils. Exp. in Fluids, **48**, 291–308 (2010).
14. T.S. Liu, K. Kuykendoll, R. Rhew, S. Jones, Avian wing geometry and kinematics, AIAA J., **44**, 954-963 (2006).
15. S. Klan, T. Bachmann, M. Klaas, H. Wagner, W. Schroder, Experimental analysis of the flow field over a novel owl based airfoil. Exp. in Fluids, **46**, 975-989 (2009).
16. R.A. Kroeger, H.D. Gruschka, T.C. Helvey, Low speed aerodynamics for ultra-quiet flight. Air Force Flight Dynamics Laboratory Technical Report, 71-75 (1971).
17. C. Ge, Z. Zhang, P. Liang, C. Zhang, L. Ren, Prediction and control of trailing edge noise based on bionic airfoil. SCIENCE CHINA Technological Sciences E, **57**(7), 1462-1470 (2014).
18. Burgmann, S., Dannemann, J., Schroder W. Time-resolved and volumetric PIV measurements of a transitional separation bubble on an SD7003 airfoil. Exp. in Fluids, **44** (4), 609-622 (2008).

A New Method for Determining Horizontal Impact Load Based on Rotational Speed of Aircraft Wheel in Landing Gear Drop Test

Wang Huaitao[1], Wu Dafang[2,a] and Wang Feng[3]

[1,2,3] *School of Aeronautic Science and Engineering, Beihang University, No. 37 Xueyuan Road, Haidian District, Beijing, China.*

Abstract. The maximum horizontal impact load experienced by the aircraft landing gear wheel during landing is an important parameter in landing gear safety design and performance analysis. Using a non-contact photoelectric testing method, this paper experimentally obtained the dynamic data of the instantaneous rotational speed for an aircraft wheel when it contacts the platform. Moreover, the kinetic relationship between the transient rotational speed of the aircraft wheel and the force of horizontal impact was established. According to the change data measured from the transient rotational speed of the aircraft wheel, the maximum horizontal impact load at the instant of aircraft contact with the platform was calculated. To verify the accuracy of this method, in which changes in rotational speed determine the horizontal impact load, verification equipment without lateral constrains was established to test the horizontal impact load of the aircraft wheel. The experimental results showed that the horizontal impact load determined by the transient rotational speed of aircraft wheel is consistent with the results obtained by self-built measurement equipment without lateral constrains.

Introduction

The landing gear is an extremely important component of an aircraft structure. In the moment of the aircraft landing, the landing gear hits the ground with a high speed and the wheels immediately transform from a static state to a high-speed rotation state. During this interval, the change of impact load in the horizontal direction experienced by the aircraft wheel is rapid and complex. In addition, a direct relationship exists between changes in the horizontal impact load on the aircraft wheel and the design of key components, such as the aircraft wheel brakes, bumpers, and shimmy dampers for the rockers, all of which also affect performance studies, such as the strength and stiffness of the landing gear, brake damping efficiency, power absorption, and aircraft wheel spin-up and spring-back[1].Therefore, determining the horizontal impact load of the landing gear wheel upon contact with the ground through landing gear drop test, plays a very important role in the strength analysis and safety design of landing gear systems.

At present, the wheel horizontal impact load is generally measured by the force sensor. In aircraft landing gear drop tests, a traditional method in which force sensors are horizontally mounted on the end of a drop test platform is employed[2-4]. The aircraft wheel horizontal impact load upon contact

[a] Corresponding author : wdf1950@163.com

with platform is recorded by the horizon tally mounted force sensor at the side of drop test platform. However, the spherical connection mode of the vertical support column of the landing platform causes the restriction of the free movement of the landing platform in the horizontal direction, which has an influence on the accuracy of horizontal load measurement. The inertia force of the platform is considered by Shi Haiwen et al. and the spherical connection mode between the drop test platform and the vertical sensor is replaced with a sliding connection mode, which can improve the accuracy of the measured result to some extent[5]. The serious lateral shake will occur when the aircraft wheel touches the platform, so lateral friction between the slide rails restricts the free movement of the drop test platform in the horizontal plane. Thus, the accuracy of the horizontal impact force of aircraft wheel upon contact with the platform is affected. As a result, researchers have been searching for a better method for measuring the landing gear horizontal impact force.

This paper proposes a new non-contact measuring method with practical application that determines the landing gear horizontal impact load in drop tests by measuring and recording the change of aircraft wheel transient rotational speed. First, the dynamic relationship between the transient speed change and the horizontal impact force is established, and then the maximum horizontal impact load of the aircraft wheel is calculated based on the measured data of the aircraft wheel transient rotational speed. The proposed method can avoid the interactive effects caused by the design of landing gear drop platform, which is used to measure the different direction impact loads.

1 Determining the horizontal impact load

First, the kinetic relationship between the horizontal impact load and the transient rotational speed of the aircraft wheel must be established. Next, the change in the transient rotational speed of the aircraft wheel at the moment of impact with the ground is obtained through the drop test apparatus and the aircraft wheel transient rotational speed measurement system. Then, the size of the horizontal impact load at the instant of aircraft wheel contact with the ground is established according to the kinetic relationship between the transient rotational speed of the aircraft wheel and the force of horizontal impact.

1.1 Kinetic relationship between the horizontal impact load and the transient rotational speed of the aircraft wheel

The horizontal force experienced by the aircraft wheel during high-speed rotation in the drop tests, when dropped into contact with the platform, is shown in Figure 1.

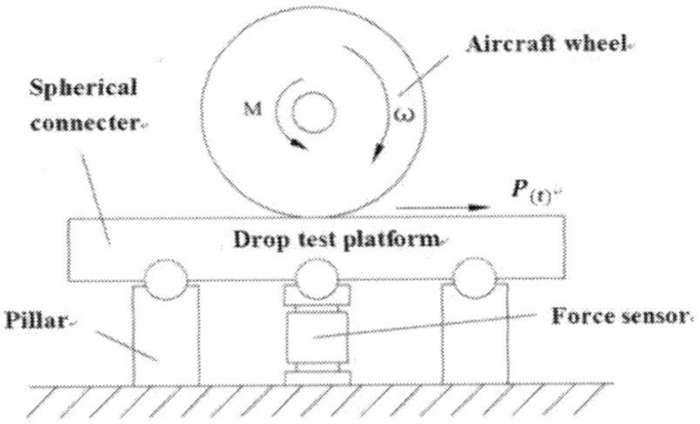

Figure 1 Mechanical model of an aircraft wheel in contact with the platform

The Figure 1 shows the horizontal impact load is related to the friction between the aircraft wheel and platform, as well as the friction between the aircraft wheel and axle. Given ω as the rotational speed of the aircraft wheel, $\frac{2\pi}{60}\omega$ is the angular velocity of the aircraft wheel. According to the theorem of the moment of momentum, the acquirable kinetic relationship between the horizontal impact load and the transient rotational speed of the aircraft wheel is

$$\frac{d(\frac{2\pi}{60}\omega I)}{dt} = P(t)R(t) + M \tag{1}$$

In the formula, $P(t)$ is the horizontal impact load of the aircraft wheel, t is time, R is the radius of the aircraft wheel, M is the friction torque of the aircraft axle, and I is the moment of inertia of the aircraft axle. Under the condition of this experiment, the above formula can be converted into the following form

$$\frac{2\pi}{60}\frac{d\omega}{dt}I = P(t)R + M \tag{2}$$

In the formula, $\frac{2\pi}{60}\frac{d\omega}{dt}$ is the angular acceleration while aircraft wheel rotating.

The calculation formula for the horizontal impact load of the aircraft wheel, $P(t)$, is

$$P(t) = (\frac{2\pi}{60}\frac{d\omega}{dt} - \frac{2\pi}{60}\frac{d\omega_0}{dt})I / R \tag{3}$$

Analysis of Formula (3) shows the amount of change over time of the measured transient rotational speed, ω, in the drop test when the aircraft wheel touches the platform and the amount of attenuation change over time in the aircraft wheel rotational speed, ω_0, in a state of free rotation. Substitution into Formula (3) can be done to immediately calculate the changes over time of the horizontal impact load of the aircraft wheel, $P(t)$.

1.2 Drop test apparatus and measurement of the transient rotational speed of an aircraft wheel

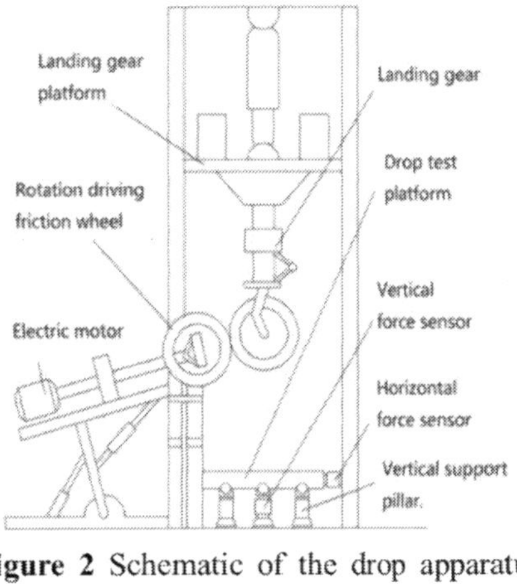

Figure 2 Schematic of the drop apparatus

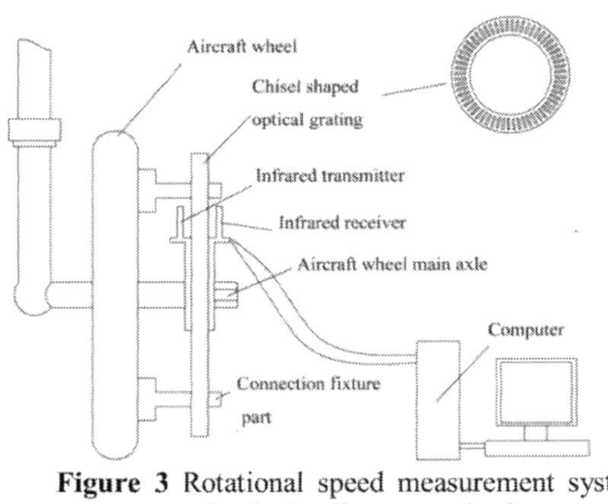

Figure 3 Rotational speed measurement system test for the landing gear wheel

The basic structure of the traditional column-style landing gear drop test apparatus is shown in Figure 2. It is mainly composed of three components, including a landing gear platform that can slide freely up and down, an apparatus for driving the rotation of the aircraft wheel, and a drop test force-measuring platform. In the drop test process, the sudden change in the rotational speed of the aircraft wheel upon contact with the ground is enormous. Within a few tenths of a second, the rotational speed of the aircraft wheel can be reduced from several thousand revolutions per minute to tens of revolutions per minute. The high-speed impact experienced by the landing gear at the instant of contact with the ground causes the landing gear to generate a wide range of elastic impact displacements in three-dimensional space; these displacements range from tens of millimeters to hundreds of millimeters and are accompanied by intense vibration, making the testing environment adverse. For this reason, a non-contact apparatus for measuring the rotational speed of the aircraft wheel, composed of a chisel shaped optical grating, an infrared transmitter, an infrared receiver, a connection fixture part, and a computer, was designed and installed on the landing gear wheel, as shown in Figure 3.

The process of measuring the rotational speed of the aircraft wheel is as follows: when the landing gear wheel drops at a rotational speed of several thousand revolutions per minute, a relative rotational displacement is generated between the infrared transmitter and receiver installed on the landing gear main axle and the grating disc installed on the aircraft wheel. Because the grating is an equidistant light transmission grating, the infrared receiver can receive the optical pulse signal associated with the changes in aircraft wheel rotational speed and then convert the optical pulse into an electrical pulse. Concurrently, the computer records the time interval between each adjacent pulse received by the infrared receiver. Because the chisel shaped optical grating contains equally spaced planes, the angle between the adjacent holes in the optical grating is a fixed constant; therefore, the change in the rotational speed of the aircraft wheel in the impact process can be computed from the recorded relationship between the rotation angle and the pulse interval time[6].

The horizontal impact load of the aircraft wheel can be determined by substituting the measured rotational speed of the aircraft wheel and related parameters into Formula (3).

2 Determining the horizontal impact load and experimental verification

2.1 Test apparatus

To determine the horizontal impact load of the aircraft wheel in the landing gear drop test and to verify the accuracy of the method used to determine the horizontal impact load from the measured rotational speed of the aircraft wheel, a drop test simulation apparatus was designed and built, as shown in Figure 4.

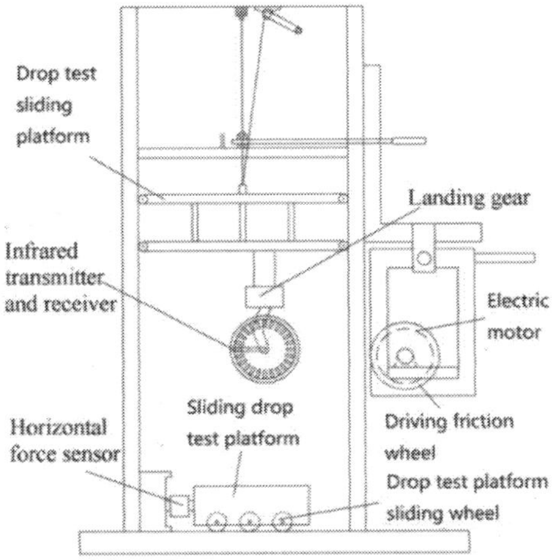

Figure 4 Drop test simulation and verification apparatus

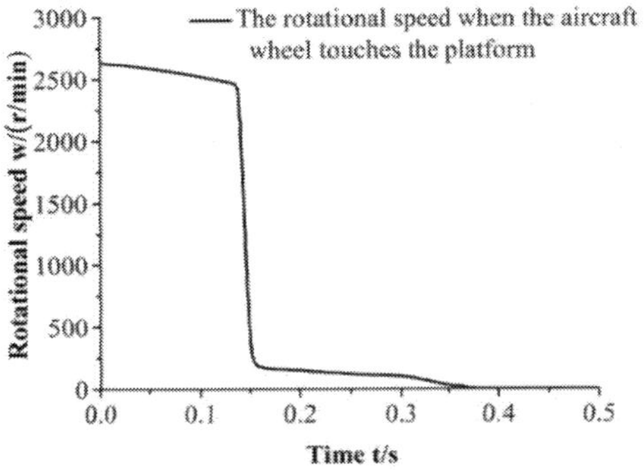

Figure 5 Curve of the rotational speed when the aircraft wheel touches the platform

A sliding drop test platform composed of a rolling bearing was set up at the base platform of the test simulation apparatus. This allowed the platform to move freely in the horizontal direction. A horizontal force sensor was installed at one end of the sliding drop test platform. One side of the sensor was fixed and connected to the support column of the test apparatus by a bolt, and there is no lateral constrained structure for the sliding drop test platform in the lower part of the test device. The horizontal friction generated between the aircraft wheel and the platform when the aircraft wheel is dropped and makes contact with the platform while under high-speed rotation is then transmitted to the horizontal force sensor through the sliding platform. Because this test apparatus only uses one horizontal force sensor, it avoids the mutual restraint and coupling effect in installation and measurement between the sensors that measure multiple dynamic strength parameters in large drop tests. The data measured by the horizontal force sensor represent the horizontal impact load of the rotating aircraft wheel upon contact with the platform. The apparatus is equipped with a rotational speed measurement system focused on the aircraft wheel; this system can simultaneously record the change in rotational speed when the aircraft wheel is allowed to drop and contact the platform.

2.2 Measurement results on the transient rotational speed of the aircraft wheel, ω

To collect the measurements, first, the landing gear wheel was driven by an electric motor to rotate at a high speed; then, the release apparatus was initiated to drop the landing gear into a vertical free-fall while the computer collected the electric pulse signals generated by the rotation of the chisel-shaped optical grating. The computer then calculated and completed the measurement while recording the transient rotational speed of the aircraft wheel.

Figure 5 is a curve of the measured change in the transient rotational speed of the aircraft wheel during the test. Because the optical grating was arranged tightly, the number of optical gratings could reach 180 or higher per revolution; therefore, the curve of the collected change in the rotational speed of the aircraft wheel when the aircraft wheel touched the platform was smooth. According to Figure 8, the rotational speed of the aircraft wheel suffers a certain amount of attenuation due to the effect of the friction torque of the axle from the drop until prior to touching the platform. Within approximately 0.15 seconds after release, the rotational speed of the aircraft wheel decreased from 2,630 r/min at the time of separation from the rotation driver to 2,460 r/min at the time of contact with the platform; in total, the rotational speed decreased by approximately 6.5%. At the instant of contact with the platform, the intense friction between the aircraft wheel and the platform caused a sudden decrease in the rotational speed of the aircraft wheel; hence, within an extremely short time, i.e., less than 0.03 seconds, the rotational speed rapidly decreased from 2,460 r/min to 165 r/min, representing an approximate decrease of 94%. After this point, the rotation of the aircraft wheel stopped at 0.4 second.

2.3 Measurement results of the rotational speed, ω_0 ,of the aircraft wheel in a state of free rotation attenuation

To obtain the rotational speed, ω_0 , during free attenuation of the aircraft wheel under the effect of the friction torque of the axle, M , in Formula (3), the landing gear wheel was driven to spin at the release rotational speed; then, the rotation-driving friction wheel was separated from the aircraft wheel, although the aircraft wheel was not allowed to drop, so that the aircraft wheel freely attenuated only under the effect of the friction torque of the axle. Meanwhile, the rotational speed measuring system was triggered to collect and record the rotational speed of the aircraft wheel in the free attenuation process.

The moment of inertia, I , of the aircraft wheel needed in Formula (3) can be determined using a torsion pendulum method[7].

2.4 Determining the horizontal impact load from the measurement results of the measured rotational speed of the aircraft wheel, ω and ω_0

The measurement results of the transient rotational speed of the aircraft wheel, ω , and the rotational speed of the aircraft wheel in a state of free attenuation, ω_0 , the moment of inertia of the aircraft wheel, and the value of the aircraft wheel radius were substituted into Formula (3). Through computation, the curve identifying the change in the horizontal impact load over time when the aircraft wheel touched the platform was obtained as shown in Figure 6.The figure indicates that within 0.012 second after the aircraft wheel made contact with the platform, the horizontal impact load rapidly increased from zero, reaching a maximum value at 0.146 second. Subsequently, the horizontal impact load continued to decrease and tended toward zero.

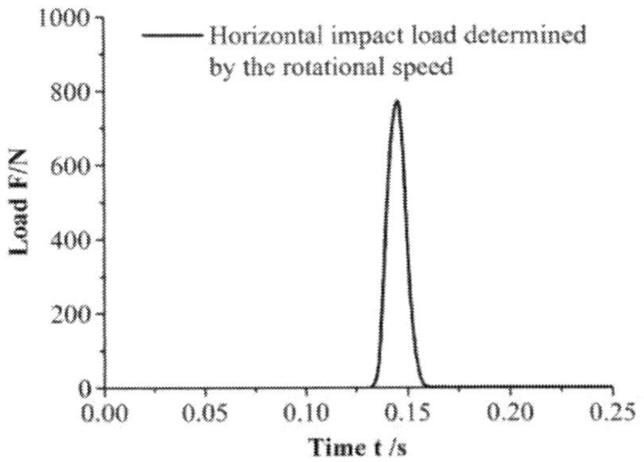

Figure 6 Curve of horizontal impact load determined by the measured rotational speed of the aircraft wheel

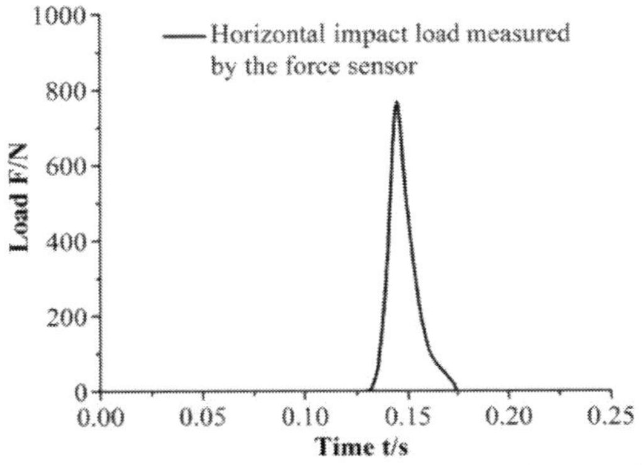

Figure 7 Curve of the horizontal impact load as measured by the force sensor

2.5 Results of horizontal impact load measured by force sensor

During the experiment, the landing gear wheel was first driven to rotate at a high speed by an electric motor, and then, the aircraft wheel was driven to rotate by friction to approximately 2,700 r/min. The release apparatus was started to induce a vertical free-fall of the aircraft wheel, while the computer completed the measurement and recording of the transient rotational speed of the aircraft wheel and the horizontal impact load.

The measurement results of the horizontal force sensor are shown in Figure 7. The curve in the figure reflects the changes that occurred in the horizontal impact load when the aircraft wheel contacted the platform. At the instant that the aircraft wheel made contact with the sliding platform, due to the rapid reduction in rotational speed from the large frictional force, the horizontal impact load increased significantly, reaching a maximum value of 744.36 N at approximately 0.012 second after the time of contact between the aircraft wheel and the platform. Subsequently, the horizontal impact load reduced to 110.23 N within 0.016 second, representing a decrease of approximately 85%. After another 0.017 second, the horizontal impact load neared zero.

2.6 Comparison of the horizontal impact load results determined by the rotational speed and the direct measurement results obtained with the force sensor

To compare the horizontal impact load determined by the measured rotational speed of the aircraft wheel with the horizontal impact load measured by the horizontal force sensor, Figure 8 shows the curves of the horizontal impact load obtained by both methods in the same experiment. According to the figure, both methods obtained the same trends in the changes of the horizontal impact load. The peak value of the horizontal impact load determined by the measured rotational speed of the aircraft wheel was 772.42 N, and the peak value of the horizontal impact load measured by the horizontal force sensor was 744.36 N; the relative error between the two was 3.77%, which indicates that this approach is able to meet the requirements of actual engineering application. The results of the above comparison indicate that the method in which the horizontal impact load is determined based on the measured transient rotational speed of an aircraft wheel is effective and feasible.

According to the friction deceleration principle, the different contact materials and different surface qualities between the aircraft wheel and the ground generate different sizes and courses in the horizontal impact load of the aircraft wheel. Moreover, the process of change in different horizontal impact loads corresponds to the process of change in the rotational speed of the aircraft wheel. By establishing the kinetic relationship between the force of the horizontal impact and the transient rotational speed of the aircraft wheel, this paper measured the dynamic change in the rotational speed of the aircraft wheel at the instant of aircraft contact with the ground and, from this, was able to determine the size of the horizontal impact load of the aircraft wheel.

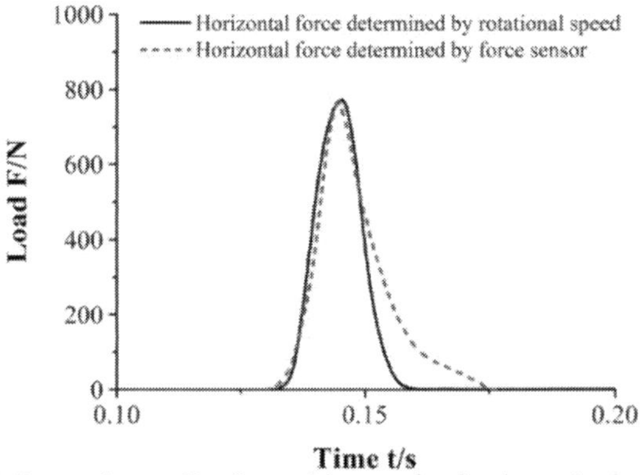

Figure 8 Comparison of horizontal impact load values obtained by two different measurement methods

Conclusion

By establishing the kinetic relationship between the transient rotational speed of the aircraft wheel and the force of the horizontal impact, This paper proposes a method that the horizontal impact load of the aircraft wheel can be determined based on the measured dynamic change in the rotational speed of the aircraft wheel at the instant at which the aircraft contacts the ground. The maximum values obtained by the method of using the measured rotational speed of the aircraft wheel to determine the horizontal impact load proposed in this paper match well with the measured horizontal impact load recorded by using horizontal force sensors. This demonstrates the accuracy and usability of this method in which the measured transient rotational speed of the aircraft wheel determines the horizontal impact load.

Acknowledgements

This work was supported by the National Natural Science Foundation of China under Grant No. 11427802.

References

1. Besch, Hans Martin. Large aircraft landing gears–a brief overview. *6th International Munich Chassis Symposium 2015*. Springer Fachmedien Wiesbaden, 2015:685-715.
2. Wei,W.U."Study On Load Measurement And Calibration Test For Airplane Landing Gear." Journal of Mechanical Strength (2003).
3. Zhang, Daqian. "The present situation and development of chinese drop test measurement and test system." Journal of Shenyang Institute of Aeronautcal Engineering (2002).
4. Xue Caijun, et al. "Optimization of the cushion properties of amphibious aircraft landing gear by simulation and drop test." Journal of Vibration & Control 19.19(2013):1753-1760.
5. SHI Hai-wen, et al."Parameters Study at Touchdown Instant of the Nose Undercarriage of Fighter Airplane." Acta Aeronautica et Astronautica Sinica 24.6(2003):517-520(In Chinese).
6. WU DaFang, LI YongTing, WANG YongHai, SONG Hao. Measurement method of the transient rotational speed of landing gear on drop test. Journal of Experimental Mechanics, 2008
7. W. Bin, M. GuiXian, P. MinYan. A Measurement Method of Moment of Inertia for the Artificial Satellite. Journal of Astronautic Metrology and Measurement, 2008, 28: 21-23 (In Chinese).

MATEC Web of Conferences **114**, 03005 (2017) DOI: 10.1051/matecconf/201711403005
2MAE 2017

Numerical Analysis of the Influence of Fibre Orientations in a two-layered Biomimetic Flapping Wing

Saiaf Bin Rayhan[1], Yu Chunjin and Shi Jian

Institute of Flight Vehicle Engineering, Nanchang Hangkong University, Nanchang, China

Abstract. A numerical study was carried out to investigate the effects of fibre orientation angles in an adopted biomimetic flapping wing having two-layered Carbon/Epoxy Composite T300/5208. The purpose of this paper is to understand how different orientation angles with different combinations affect the stresses of a flapping-wing. One flapping cycle was divided into twelve segments and both maximum stress and deformation were calculated for all the segments. The results revealed that, the maximum stress was produced in [0/-45] combination, where the least was found for [45/0]. For all the simulated wings, deformation was found less than 1.8 mm. ANSYS DesignModeler and Static Structural was used to design and perform structural analysis. The findings are helpful in answering why insect wings are so impeccable, thus providing a possibility of improving the design of flapping-wing aerial vehicles.

1 Introduction

Significant amount of researches in the design of micro-sized air vehicles (MAVs) have been carried out over the past 15 years [1–6]. Alternatively, tremendous progress was done in the field of aerodynamics and Kinematics [7-16]. Vincent [17] found that the material of insect wing was similar to composite materials, many other researchers also believed that the wing of natural insects are composite structures [18-20]. Composite materials are currently used for UAVs [21,22].

Unfortunately, most of the works mentioned above have been generally limited to CFD and CSD. Still research in deformation and stress analysis continues to lag, especially for composite models. In this paper, total sixteen different orientations and layer combinations were analysed for maximum aerodynamics force and inertial force. Because the structure of insect wings is very complex and their elastic Module and thickness in irregular, at present we can only simulate it to manufacture flapping-wing aerial vehicle.

2 Analytical Model

2.1 Aerodynamics Force and Inertial Force

Three-dimension unsteady vortex lattice method (UVLM) [23] is used to calculate the aerodynamics. The aerodynamics of flapping-wing aerial vehicle

[1] Corresponding author: rayhan.saiaf@mail.ru

$$\Delta F_{ij} = (\Delta P \cdot \Delta S)_{ij} \quad (1)$$

where ΔP is pressure of element & ΔS is area of element.

The flapping motion of the present model wing is simplified to a cosine rotation around a fixed axis located at the root.

$$\beta(t) = \beta_{max} cos\,(\omega t) \quad (2)$$

The periodic inertia force generated from the acceleration and deceleration of flapping wing, is the main acting load on it during every cycle. Its specific load distribution can be predicted as

$$F_i(t) = -m_i \frac{d^2\beta(t)}{dt^2} = \omega^2 \rho m_j r_j \beta_{max}\, cos(\omega t) \quad (3)$$

where $F_i(t)$ is the inertial force of element i, m_i is the mass of element i, r_i is rotational radius of element i, ω is angle velocity, t is time, β_{max} is amplitude of flapping angle.

Because the units of aerodynamic force is N/m^2, the calculation formula of inertial force can express as

$$F_i(t) = \omega^2 \rho_s A_i t_i r_i \beta_{max}\, cos(\omega t) \quad (4)$$

$$f_i(t) = \frac{F_i(t)}{A_i} = \omega^2 \rho_s t_i r_i \beta_{max}\, cos(\omega t) \quad (5)$$

where A_i is area of element i, ρ is material density.

2.2 Composite Material Model

The carbon/epoxy composite T300/5208 is used for the membrane of the flapping wing; its material properties are shown in Table 1. For this analysis, it exhibits viscoelastic behavior, and all the insect wings have the viscoelastic behavior.

Table 1. T300/5208 material characteristics

$E_{11}(GPa)$	$E_{22}(GPa$	$E_{33}(GPa)$	$_{12}\upsilon$	$_{23}\upsilon$	$_{31}\upsilon$	$G_{12}(GPa)$	$G_{23}(GPa)$	$G_{13}(GPa)$
181	10.3	10.3	0.28	0.3	0.28	7.17	3.78	7.17

3 Results analysis

We designed a membrane flapping-wing model, which is shown in Figure 1. A mesh convergence study was performed to obtain the accurate results, which is shown in Figure 2. The parameters used in this model are shown in Table 2.

Table 2. Parameters used in the flapping-wing aerial vehicle model

Semi-span length (mm)	320
Maximum chord length (m)	100
Flapping frequency (Hz)	8
Flapping angle	45°
Flight velocity (m/s)	11
Angle of attack	5°
Outer radius of beam (mm)	3
Inner radius of beam (m)	2
Elastic module of beam (GPa)	40
Poisson's ratio of beam	0.3

DOI: 10.1051/matecconf/201711403005

Thickness of membrane (mm)	1

The boundary condition of the fixed support in the body-fixed coordinate system is adopted. During the flapping flight, the flapping-wing must withstand not only aerodynamic force but also inertial force. Aerodynamic force of every shell element is calculated from Equation (1), and the inertial force of every element is calculated from Equation (5).

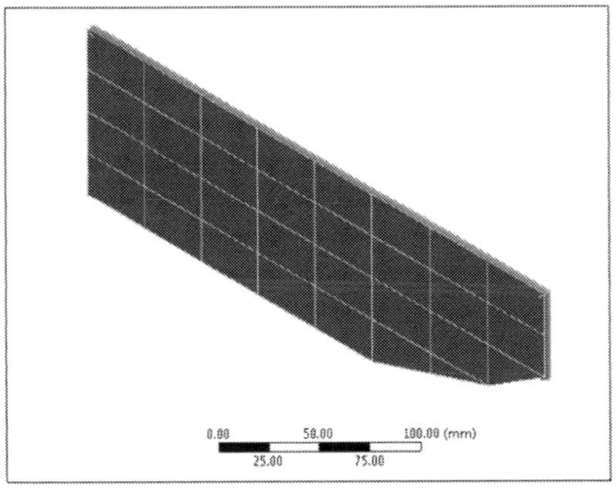

Figure 1. Flapping wing model

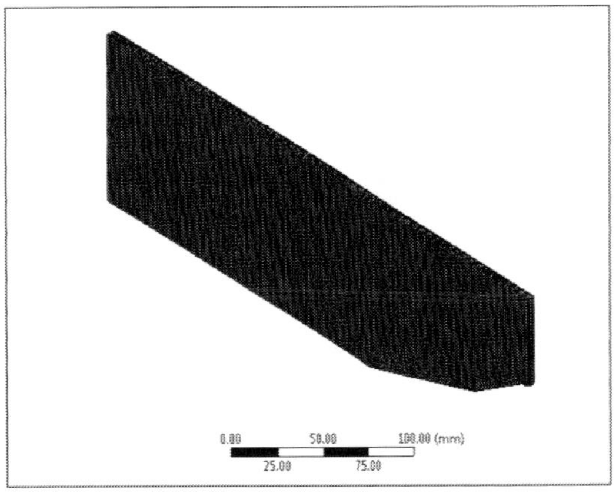

Figure 2. Meshed Body (Element Size=1,5)

For the simulation part, all sixteen possible combinations were calculated for the twelve flapping periods, totaling 192 simulations to find the impact of fiber orientations. The curves of Von-Mises Stress in one flapping cycle are shown in Figure 3. And the curves of maximum deformation are shown in Figure 4. From the figures [3,4] it is clearly understood that [0/-45] will occur maximum stress during a flap, on the other hand, [45/0] will produce the least while deformations are very small, less than 1.8 mm, for all the variants we simulated.

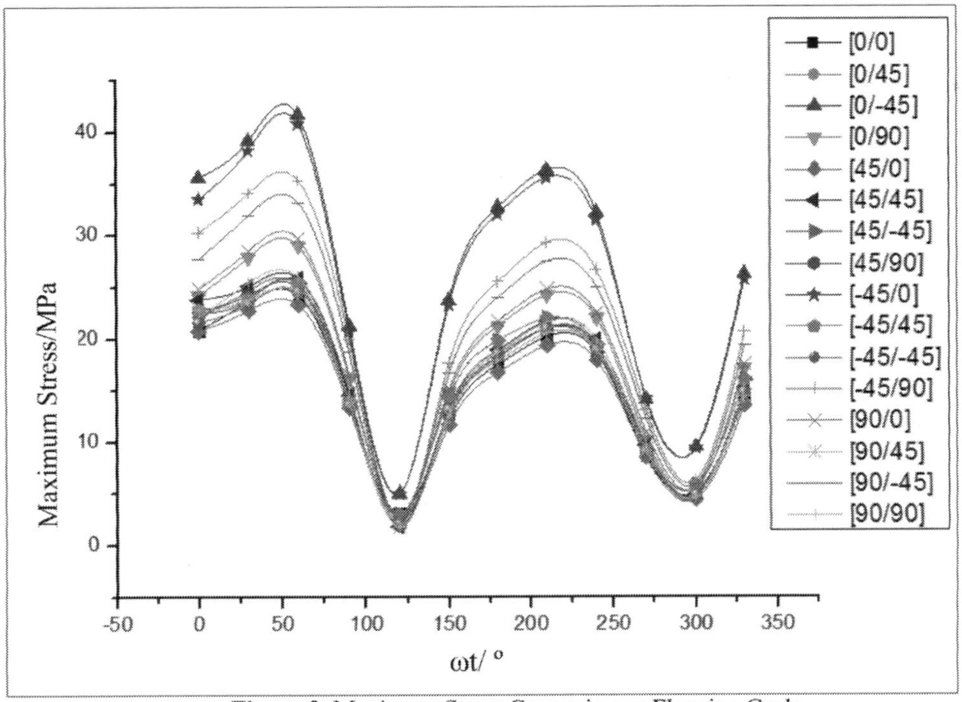

Figure 3. Maximum Stress Curves in one Flapping Cycle

DOI: 10.1051/matecconf/201711403005

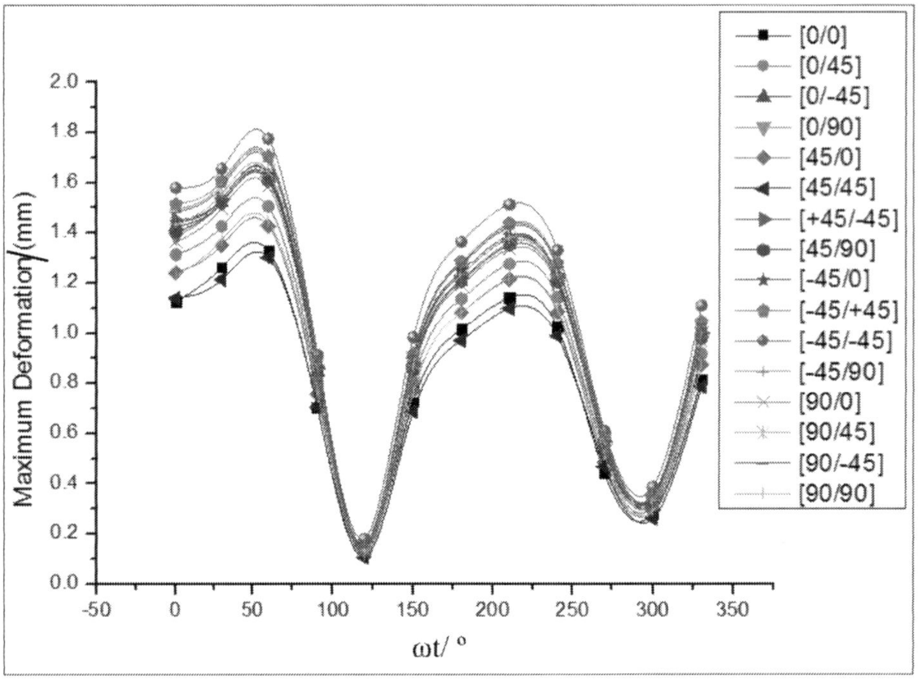

Figure 4. Maximum Deformation Curves in One Flapping Cycle

Figure 5 illustrates the maximum stress zone [45/0], at the root of the flapping wing, on the beam, which is produced during the third segment of the down stroke of a flapping cycle. Figure 6 illustrates the deformation contour [45/45] which is maximum at the tip of the wing.

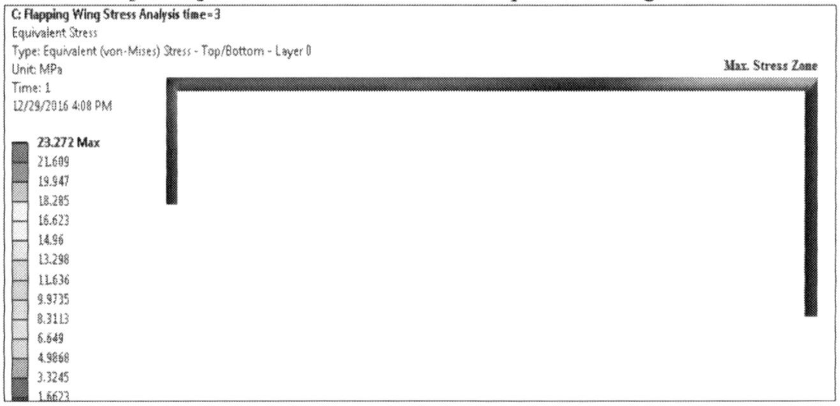

Figure 5. Maximum Stress Zone [45/0]

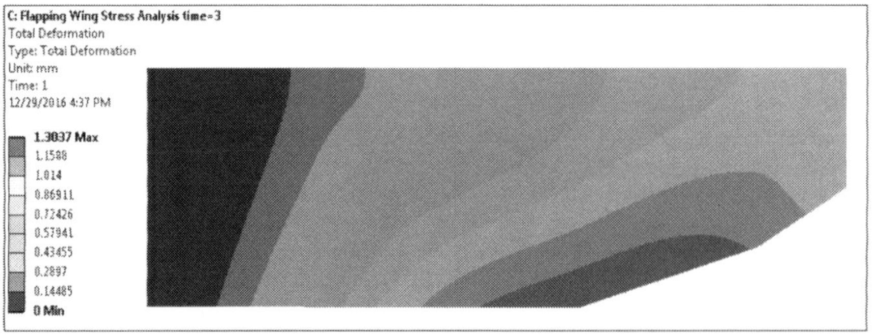

Figure 6. Maximum Deformation Profile [45/45]

Conclusion

Finite element analysis is performed to investigate the membrane stresses of a flapping wing made of Composite [T300/5208]. Our carried out research revealed the structural benefits of two layered flapping wing which experiences both least deformation and stress. In addition to that, we found the best combination of layers and fiber directions [45/0] which produces almost 41% less stress than [0/-45]. This work can contribute in progression of designing two layered flapping wings. In future, we will carry out a research on optimization to find the best performed flapping wing both structurally and aerodynamically.

References

1. P.G. Ifju, S. Ettinger, D.A. Jenkins. Lian, W. Shyy and M.R. Waszak, *Flexible-wing-based micro air vehicles*, Proc. 40th AIAA Aero Sci. Meeting, Reno, NV, AIAA–**0705** (2002)
2. J. P. Whitney and R J. Wood, *Conceptual design of flapping-wing micro air vehicles, Bio inspiration & Biomimetic*, Vol.**7**, No.3 (2012)
3. P. Zdunich, D. Bilyk, M. MacMaster , D. Loewen, J. DeLaurier, R. Kornbluh , T. Low , S. Stanford and D. Holeman , *Development and testing of the mentor flapping-wing micro air vehicle*, Aircraft **44**, 1701–1710 (2007)
4. K. Isogai and Y. Harino, *Optimum aero-elastic design of a mapping wing*, J. Aircraft **44**, 2040–2048 (2007)
5. A.L. Douglas, R. Dudley and C.P. Ellington, *Aerodynamic forces of revolving hummingbird wings and wing models*, J. Zool. Lond. **264**, 327–332 (2004)
6. S.J Furst, G. Bunget, and S. Seelecke, *Design and fabrication of a bat-inspired flapping-flight platform using shape memory alloy muscles and joints, Bioinspiration & Biomimetics*, Vol. **22**, No.1 (2013)
7. H. Cho, J. Kwak and S. Shin, *Computational Analysis for Flapping Wing by Coupling the Geometrically Exact Beam and Preconditioned Navier-Stokes Solution*, The 55th AIAA/ASME/ASCE/AHS/SC Structures, Structural Dynamics, and Materials Conference, National Harbor (2014)
8. J. Guerrero, *Numerical simulation of the unsteady aerodynamics of flapping flight*, PhD. Dissertation, Department of Civil, Environmental and Architectural Engineering, University of Genoa, Genova, Italy, (2009)
9. C. Yu, H. Ang, Q. Chen, *Three-Dimension Unsteady Vortex Lattice Method for Flexible Structure Flapping-Wing Aerial Vehicle*, Journal of Nanjing University of Aeronautics and Astronautics, **40**, 451-455 (2008)
10. C. Yu, H. Ang, K. Ai, *The Study of Deformation for Membrane Flapping-Wing Aerial Vehicles*, Chinese Journal of Computational Mechanics, **26**, 935-941 (2009)
11. D. Silin, *Aerodynamics and flight performance of flapping wing micro air vehicles,* PhD. Dissertation, Aerospace Engineering Dept., University of Arizona, Tucson, Arizona (2010)
12. K. Mazaheri, and A. Ebrahimi, *Experimental investigation of the effect of chordwise flexibility on the aerodynamics of flapping wings in hovering flight*, Journal of Fluids and Structures, Vol. **26**, pp. 544–558 (2010)
13. Y. Lian, W. Shyy, P. Ifju and E. Verron, *A Computational Model for Coupled Membrane-Fluid Dynamics*, AIAA Paper 2002-2972, also AIAA J.,**41**, pp. 2492–2494(2004)
14. Y. Zhang, J. Wu and M. Sun, *Lateral dynamic flight stability of hovering insects: theory vs numerical simulation*, Acta Mech;**28**(1):221–31(2012)
15. JA Grauer Jr, JE. Hubbard, *Inertial measurements from flight data of a flapping-wing ornithopter. J Guide, Control,* Dynamics;**32** (1):326–31 (2009)
16. H. Djojodihardjo, A, Ramli, S. Wiriadidjaja, *Kinematic and aerodynamic modelling of flapping wing ornithopter,* Processing Engineering.;**50(1):**848–63(2012)

17. J.F. Vincent, V. Insect cuticle: *A paradigm for natural composites*. Symposia of the Society for Experimental Biology,183-210(1980)
18. F. Song, K.W. Xiao, K. Bai, Y.L Bai, *Microstructure and nanomechanical properties of the wing membrane of dragonfly*, Materials Science and Engineering: **A, 457**, 254–260 (2007)
19. X.S. Wang, Y. Li, Y.F. Shi, *Effects of sandwich microstructures on the mechanical behaviors of dragonfly wing vein. Composites Science and Technology*, 68, 186–192 (2008)
20. Y. Chen, X. Wang, H. Ren, H. Yin, S. Jia, Hierarchical dragonfly wing: *Microstructure-biomechanical behavior relations, Journal of Bionic Engineering,* **9**, 185–191(2012)
21. J. Callicoat, R. Gaeta, J. Jacob, *Composite Materials Providing Improved Acoustic Transmission Loss for UAVs*. AIAA (2014)
22. C. Yu., D. Kim, Y. Zhao, *Stress Analysis of Membrane Flapping-Wing Aerial Vehicle Based on Different Material Models,* Journal of Applied Mathematics and Physics**2**, 1023-1030 (2014)
23. J. Katz and A. Plotkin, *Low-Speed Aerodynamics*. **2nd Edition**, Cambridge University Press, New York, NY, (2001)

DOI: 10.1051/matecconf/201711403006

A United Allocation Method of Spare Parts and Ground Maintenance Equipment for Civil Aircraft

Li Yongkai[1,a], Feng Yunwen[1], Xue Xiaofeng[1] and Lu Cheng[1]

[1]*School of Aeronautics, Northwestern Polytechnical University, 7100772 Xian, China*

Abstract. Aimed at multi-echelon inventory allocation problem with finite repair capacity for civil aircraft, a united planning method based on queuing theory and Vari-Metric model is presented to deploy the aircraft spare parts and ground maintenance equipment. Through the analysis of actual engineering, the effect of repair time is quantified by the number of maintenance equipment. On this basis, the available of fleet and the cost expense are selected as the objective and constraint respectively, marginal analysis method is set as the optimization, a new allocation scheme which was suit to civil aircraft's maintenance resources is put forward. By applied examples analysis, it is shows that this method is reasonable and feasible.

1 Introduction

Due to the rapid development of high-tech equipment, maintenance resources (spare parts and ground maintenance equipment) to ensure the normal use of the civil aircraft has been attached more and more importance. With complex equipment lead to fast-rising prices of spare parts and repair facilities, the contradiction of outlay and resources shortage become more prominent if we allocate maintenance resources according to the traditional method. Therefore, in order to ensure the timeliness of maintenance tasks and reduce operating costs for civil aircraft, we need to make reasonable allocation for civil aircraft's maintenance resources to seek optimal cost-effectiveness between fleet availability and cost expense of resources.

In foreign countries, study on maintenance resources allocation carried out earlier than China. The classical Metric model is proposed by Sherbrooke[1], which laid a foundation for multi-echelon inventory allocation theory. Muckstadt[2] established an inventory management method for the multi-order multi-echelon case, i.e. the Mod-Metric system model. Graves[3] put forward Vari-Metric model to solve the problem of multi-echelon multi-indenture spare parts inventory control. The existing researches about multi-echelon inventory allocation had been applied to various managements of maintenance resources, however, previous studies were based on the assumption of infinite repair capacity, and the maintenance ability of each location is limited in the practical engineering. These methods failed to consider the interaction between the number of maintenance equipment and maintenance ability.

In this paper, the assumption of infinite repair capacity is relaxed and the effect of repair time is quantified by the number of maintenance equipment based on M/M/c queuing theory. According to

[a] Corresponding author : lyk56820@hotmail.com

the Vari-Metric method, a calculation mode for fleet availability of civil aircraft is established. Based on that, this study put forward a united allocation method of spare parts and ground maintenance equipment for civil aircraft, resources allocation scheme is also obtained. In addition, the influence of the inventory scheme between limited and infinite repair capacity is evaluated.

2 Model descriptions

In general, the maintenance points of aircraft is regarded as two-echelon inventory model, respectively, the Local Warehouse (LW) and the Central Department (CD). When the parts on the aircraft are in trouble, broken parts will be replaced if spare parts are available on the LW, otherwise, order a spare part to the CD with a (S-1,S) replenishment policy. The invalid parts are sent to repair shop at the CD unless there have adequate repair capacity on the LW. Repaired parts will act as new spare parts and replenish stocks.

It should follow the principle of queuing theory in repair: the invalid parts can be repaired immediately if there have free maintenance equipment, or waiting to repair until the free maintenance facilities occurred. Which means that the maintenance process of invalid parts satisfying a first-come, first-serve(FCFS) basis. The replacement and maintenance process of the aircraft parts is shown in figure 1.

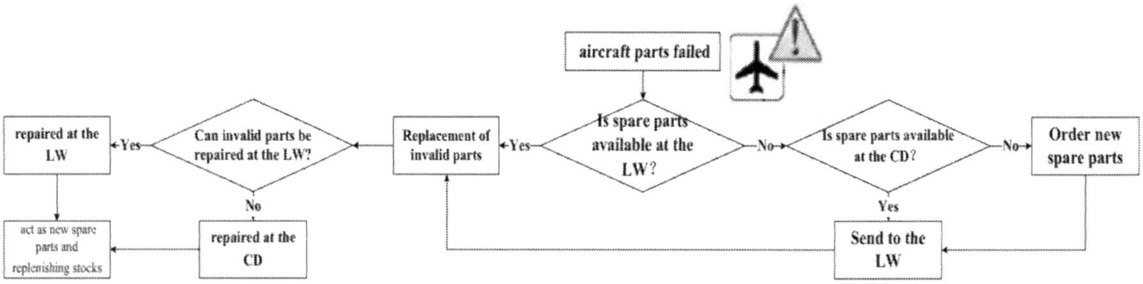

Figure 1 The replacement and maintenance process of the aircraft parts

In order to simplify the modeling process, there are some reasonable assumptions for united allocation method of spare parts and ground maintenance equipment:

- The failure of aircraft parts follows the Poisson process.
- The invalid parts can be repaired perfectly.
- Any aircraft part has a independent mean-time-between-maintenance (MTBM).
- There is no lateral supply.

For a better expression of the model, the variables are listed as follows:

i spare parts number. $i = 1,2,3\cdots,n$

j maintenance points number. CD: $j = 0$, LW: $j = 1,2,3\cdots,m$

k maintenance equipment number. $k = 1,2,3\cdots,1$

S_q the allocation scheme. Spare part: $q = 0$, maintenance equipment: $q = 1$

C_q purchase price. Spare part: $q = 0$, maintenance equipment: $q = 1$

r air route number. $r = 1,2,3\cdots$

D_{rj} flight mileage through maintenance point j at air route r.

M_i the total demand of spare parts i

m_{ij} the demand of spare parts i at maintenance point j.

r_j repair probability at maintenance point j.

T_{ij} the repair time of spare parts i at maintenance point j.

μ_{ij} the average demand of spare parts i getting from maintenance and resupply at maintenance point j.

P_{ij} the steady state probability of spare parts i getting from maintenance and resupply at maintenance point j.

O_{ij} order and ship time of spare parts i from the CD to maintenance point j.

f_{ij} the fraction of all demands of spare parts i at the LW j.

3 Estimating the demand of spare parts

3.1 The demand model of spare parts

The basic formula of spare parts demand is:

$$M=A.I.P+I.N\sqrt{A.I.P} \quad (1)$$

$$A.I.P=S.V.\times C.T./T. \quad (2)$$

Where: M -the demand number of spare parts, $A.I.P$ - the average number of spare parts need to be repaired, $S.V.$-repair rate per year, $C.T.$-repair time, $T.$- annual working days, $I.N$ - index number, represents the level of security.

3.2 The demand for each maintenance point

For the LW maintenance points, the demand of spare parts can be determined according to the percentage of flight mileage through the LW from the total route. Define D_r as the total route of air route r, and $m^r{}_{ij}$ as the demand caused by air route r of spare parts i at maintenance point j:

$$m^r{}_{ij}=\frac{D_{rj}}{D_r}M_i \quad (3)$$

The total demand m_{ij} consist of the sum of demand for all routes that pass through the LW

$$m_{ij}=\sum m^r_{ij} \quad (4)$$

For the CD maintenance points, the demand of spare parts equal to the number of spare parts that cannot be repaired at the LW.

$$m_{i0}=\sum m_{ij}(1-\mathrm{r}_j) \quad (5)$$

4 A united allocation method of maintenance resources for civil aircraft

4.1 The construction of multi-echelon spare parts inventory model

According to the classic METRIC theory, the following relationships exist in inventory level[4]:

$$S=OH+DI-BO$$

S is the inventory level, OH is existing inventory, DI is the number of spare parts getting from maintenance and resupply, BO is the shortage number of spare parts. When using (S-1,S) inventory strategy, the value of S is constant, OH, DI and BO are in a state of dynamic equilibrium. Therefore, the expectation $EBO(\mathrm{S}_0)$ and variance $VBO(\mathrm{S}_0)$ of spare parts shortage can be expressed by the following formulas when the value of OH is S_0:

$$EBO(S_0) = P(DI = S_0+1) + 2P(DI = S_0+2) + ... = \sum_{X=S_0+1}^{\infty} (x - S_0)P(DI = x) \quad (6)$$

$$VBO(S_O) = E[BO^2(S_0)] + [EBO(S_0)]^2 \quad (7)$$

Where, $P(DI = x)$ is the probability that the number of spare parts getting from maintenance and resupply is x. Next, calculating $P(DI = x)$ at the LW and CD respectively.

According to the Palm theorem, the number of spare parts getting from maintenance and resupply followed from Boson distribution with $m_{i0}T_{i0}$ mean. So the average demand μ_{i0} and steady state probability P_{i0} for the number of spare parts getting from maintenance and resupply can be expressed by the following:

$$\mu_{i0} = m_{i0}T_{i0} \quad (8)$$

$$P_{i0}(x) = Poisson(x, \mu_{i0}) \quad (9)$$

The expectation $EBO_{i0}(S)$ and variance $VBO_{i0}(S)$ of spare parts shortage at the CD can be derived from the formulas (6), (7) and (9).

The average demand of spare parts getting from maintenance and resupply at the LW μ_{ij} consists of two parts: one is the fault parts for maintenance at the LW, and the other is the fault parts sent to the CD due to the lack of maintenance capability at the LW:

$$\mu_{ij} = m_{ij}(r_{ij}T_{ij} + (1-r_{ij})\{O_{ij} + EBO_{i0}/m_{i0}\}) \quad (10)$$

The variance for the number of spare parts getting from maintenance and resupply $Var_{ij}(S)$ can be expressed by the following formula:

$$Var_{ij}(S) = m_{ij}\left[r_{ij}T_{i0} + (1-r_{ij})O_j\right] + f_{ij}(1-f_{ij})EBO_{i0}(S) + f_{ij}^2 VBO_{i0}(S) \quad (11)$$

Where

$$f_{ij} = \frac{m_{ij}(1-r_{ij})}{m_{i0}} \quad (12)$$

The steady state probability of $DI = x$ at the LW depends on the ratio of mean and variance $p = \mu_{ij}/Var_{ij}(S)$. If $p > 1$, the number of spare parts getting from maintenance and resupply followed a negative binomial distribution; else if $p \approx 1$, this number according to Poisson distribution; and if $p < 1$, selecting binomial distribution. So we can get the steady state probability P_{ij}, furthermore, we obtain the expectation $EBO_{ij}(S)$ and variance $VBO_{ij}(S)$ of spare parts shortage at the LW.

The total expectation of spare parts shortage consists of the expectation at all maintenance points:

$$EBO_i(S) = \sum_{i=1}^{n} EBO_{i0}(S) + \sum_{i=1}^{n}\sum_{j=1}^{m} EBO_{ij}(S) \quad (13)$$

The relationship between fleet availability A and the total expectation of spare parts shortage $EBO_i(S)$ is:

$$A = 100\prod_{i=1}^{n}\{1 - EBO_i(S)/NZ_i\}^{Z_i} \quad (14)$$

Where, Z_i is the installed number at a plane, N is the number of aircraft in a fleet. Thus, the maximum availability of fleet equal to the minimum value for the expectation of spare parts shortage.

The investment cost of all maintenance resources C is composed of two parts: the cost of purchasing spare parts and the cost of the maintenance equipment:

$$C = \sum_{i=1}^{n} c_{0i} S_{0i} + \sum_{k=1}^{l} c_{1k} S_{1k} \tag{15}$$

Where, c_{0i}, S_{0i} are the purchase price and allocation scheme of $i-th$ spare parts, c_{1k}, S_{1k} are the purchase price and allocation scheme of $k-th$ maintenance equipment.

Through the above analysis, the available of fleet and the cost expense are selected as the objective and constraint respectively, a multi-echelon inventory allocation method is put forward:

$$\begin{aligned} &\min EBO(S) \\ &s.t. \quad C \le C_0 \end{aligned} \tag{16}$$

4.2 The effect of ground maintenance equipment on repair time

Based on the multi-service model of queuing theory, the problem of maintenance resources allocation under the condition of limited maintenance capacity is solved. The following are the assumptions of the repair process

- The arrival interval of fault parts followed with negative exponent distribution (Parameters is λ).
- The repair time followed with negative exponent distribution (Parameters is μ).
- System capacity: infinite.
- Queue length: infinite.
- Serve regulation: first-come-first-served (FCFS).

Define $\rho = \frac{\lambda}{\mu}$, $\rho_s = \frac{\lambda}{k\mu}$, k is the number of maintenance equipment. In order to avoid the queue is lengthening, the repair capacity per unit time should be greater than the number of failures, i.e., $\rho_s < 1$

The probability that a fault part does not have to wait for service when it arrives p_0 is:

$$p_0 = \left[\sum_{n=0}^{k-1} \frac{\rho^n}{n!} + \frac{\rho^k}{k!(1-\rho_s)} \right]^{-1} \tag{17}$$

The number of parts waiting to be repaired in the system L_q is:

$$L_q = \frac{p_0 \rho^k \rho_s}{k!} \frac{\mathrm{d}}{\mathrm{d}\rho_s} \left(\sum_{n=1}^{\infty} \rho_s^n \right) = \frac{p_0 \rho^k \rho_s}{k!(1-\rho_s)^2} \tag{18}$$

According to the Little formula, the average repair time T (including service time and waiting time) of the parts in the maintenance point is obtained:

$$T = \frac{L_q + \rho}{\lambda} \tag{19}$$

Then take repair time T into the inventory allocation model of spare parts, a united planning method based on queuing theory and Vari-Metric model is obtained to deploy the aircraft spare parts and ground maintenance equipment.

4.3 Solving process based on marginal analysis

According to the relevant literature[5], it can be proved that the expectation of spare parts shortage is a monotone increasing, convex function of inventory level, so the multi-echelon inventory allocation model can be solved by the method of marginal analysis.

Step.1 Set the initial allocation of spare parts and ground maintenance equipment, which the number of ground maintenance equipment should be avoid the queue is lengthening.

Step.2 Calculate the marginal increment of spare parts and ground maintenance equipment, i.e., calculate the maximum value of $\Delta i = [EBO(S_i) - EBO(S_i+1)]/c_i$ and $\Delta k = [EBO(S_k) - EBO(S_k+1)]/c_k$, Increase the unit number of spare parts or ground maintenance equipment according to the maximum marginal increment.

Step.3 Calculate the total investment cost C, if $C \le C_0$, go to step.2; otherwise, end of the calculation.

5 Numerical example

A typical numerical example is provided to illuminate the application of our model. The maintenance resources allocation problem about the engine of Air China's aircrafts that used on American route is selected as the research object. The fleet consisted of 5 Boeing 747-8 and 10 Boeing 777-300ER aircraft, the Boeing 747-8 aircraft used on Chongqing - Beijing - San Francisco and Beijing - New York routes, the Boeing 777-300ER aircraft used on Beijing - Los Angeles, Beijing - Houston and Beijing -Washington routes. The two kinds of aircrafts were used GEnx and GE90 engine respectively, the specific parameters are shown in Table 1 and Table 2. Airport maintenance points are regarded as the Location Warehouse (LW), the company called Air China Technics are regarded as the Central Department (CD). A set of GE engine maintenance equipment at the LW worth \$1 million, and the price of maintenance equipment at the CD is \$40 million.

Table 1 GEnx aeroengine parameters

name	The number of aircraft	Installed number	Flight hours/Hour	Mean Time Between Failure/Hour	Cost/USD
GEnx-2B67 engine	5	4	2920	4000	1.7M
Order and ship time(internal)/day	**Order and ship time(abroad)/day**	**Average repair time/day**	**Repair probability at the LW(internal)**	**Repair probability at the LW(abroad)**	
30	50	60	0.2	0.1	

Table 2 GE90 aeroengine parameters

name	The number of aircraft	Installed number	Flight hours/Hour	Mean Time Between Failure/Hour	Cost/USD
GE90-115B engine	10	2	3942	4000	27.5M
Order and ship time(internal)/day	**Order and ship time(abroad)/day**	**Average repair time/day**	**Repair probability at the LW(internal)**	**Repair probability at the LW(abroad)**	
30	50	90	0.2	0.1	

Through the establishment of the joint allocation model proposed by this paper, the method of marginal analysis is adopted to get the fleet availability. Finally, we obtained the allocation scheme with the fleet availability of 0.96 under the investment expenses of \$684 million. The allocation scheme of spare parts and ground maintenance equipment are shown in Table 3 and Table 4. The curves about expectation of spare parts shortage and fleet availability that are changed over investment cost as shown in Figure 2.

Table 3 the number of spare parts

Location	Beijing	Chongqing	San Francisco	New York	Air China Technics
The spare parts number	2	0	1	2	6

of GEnx engine					
location	**Beijing**	**Los Angeles**	**Houston**	**Washington**	**Air China Technics**
The spare parts number of GE90 engine	2	1	1	1	7
Cost	$ 517 million				

Table 4 the number of maintenance equipment

Location	**Beijing**	**Chongqing**	**San Francisco**	**New York**	**Air China Technics**
The number of maintenance equipment	1	1	1	1	4
	Los Angeles	**Houston**	**Washington**		
	1	1	1		
Cost	$ 167 million				

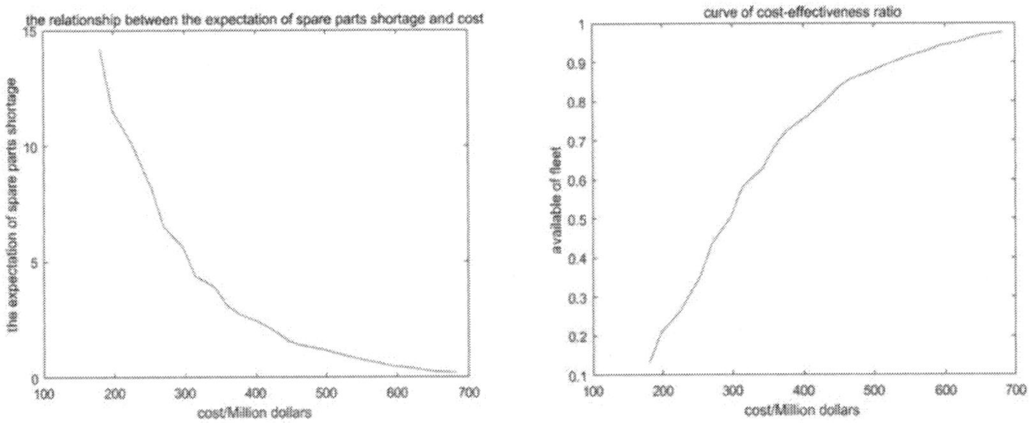

Figure 2 The curves of expectation of spare parts shortage and fleet availability changed over cost

The results show that the following two points: (1) The fleet availability increased with investment cost, but the growth trend is more and more slow. So it's usually to maintain the fleet availability between 0.85-0.95 in engineering applications, some important parts such as the engine can be maintained at 0.95-0.97. (2) The lack of ground maintenance equipment caused significantly impact on spare parts shortage, but when the maintenance equipment achieve a certain amount, the influence to spare parts from the number of maintenance equipment will decrease sharply. So it's impossible to improve the fleet availability through continue to increase the number of maintenance equipment.

If we do not consider the impact of maintenance equipment, and using the assumption of infinite maintenance capability, we can get the engine spare parts inventory scheme under the same availability (see table 5).

Table 5 the number of spare parts

Location	**Beijing**	**Chongqing**	**San Francisco**	**New York**	**Air China Technics**
The spare parts number of GEnx engine	2	0	1	1	4
location	**Beijing**	**Los Angeles**	**Houston**	**Washington**	**Air China Technics**
The spare parts number of GE90 engine	2	1	1	1	4
Cost	$ 397.5 million				

Observing the allocation of engine spare parts under infinite maintenance capacity, we found that the effect of repair capacity for spare parts allocation scheme is mainly reflected in the CD maintenance point. This is because the CD maintenance point has a heavy maintenance task, fault parts repaired quickly and translated into stock of spare parts timely under the infinite maintenance capacity assumption, thereby the demand for the number of spare parts at the CD is reduced.

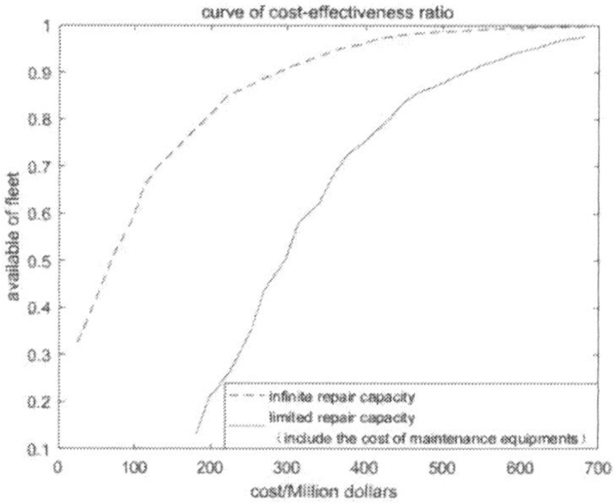

Figure 3 comparison of limited and infinite repair capacity

Due to the limited repair ability and ground equipment costs, there is a big gap of fleet availability in the same cost from whether to consider the ground maintenance equipment (see Figure 3). But with the increase in investment costs, the gap between the two is reduced gradually. The main reason is that after the maintenance equipment reached a certain number, it's influence on fleet availability decreased sharply. At this time, spare parts become the decisive element of fleet availability.

To sum up, the united allocation model proposed in this paper can effectively solve the planning problem for spare parts and ground maintenance resources.

6 Summary

For the current study, the spare parts inventory allocation model has the problem of over estimating the maintenance capability. In this paper, we take the allocation problem of aeroengine maintenance resources as an example, focusing on the use of M/M/c queuing model to analyze the effects of repair time from the number of maintenance equipment. A united planning model, which is used to allocate maintenance resources for civil aircraft is put forward, and the model can be solved by marginal analysis method.

The research of this paper provides an effective method to reasonable allocating the maintenance resources for civil aircraft. The method can not only reduce maintenance cost, improve support efficiency of civil aircraft operation, but also enrich the theory, method and technology of maintenance resources allocation for civil aircraft. It has important application value in engineering for civil aircraft in using and development. Based on the current research, the future will consider the allocation of civil aircraft maintenance resources under the imperfect maintenance case.

Acknowledgements

The work is supported by the National Natural Science Foundation of China (No. 10577015) and Special Scientific Research Project of Civil Aircraft(MJZ-2014-F-06)

References

1. Sherbrooke C C, Operations Research, **16**, 122-141 (1966)
2. Mcclain J O, Maxwell W L, Muckstadt J A, et al, Management Science, **30**, 772-773 (1984)
3. Graves S C, Management Science, **31**, 1247-1256 (1985)
4. J Xu, H Zuo, L Sun, Information Technology Journal, **13** (2014)
5. Costantino F, Gravio G D, Tronci M, Reliability Engineering System Safety, **119**, 95-101 (2013)

Research On Globe-cone Joint Assembly Of Spacecraft Propulsion System

Yanlei Zhang[1,2,a], Feikai Zhang[3], Yuchuan Bian[1,2], Yougao Fan[1,2] and Kai Xu[1,2]

[1]*Beijing Institute of Spacecraft Environment Engineering,100094, Beijing*
[2]*Beijing Engineering Research Center of the Intelligent Assembly Technology and Equipment for Aerospace Product, 100094, Beijing*
[3]*Beijing Institute of Technology, 100081, Beijing*

Abstract. As one of the key step of spacecraft propulsion system assembly, globe-cone joint assembly determines the spacecraft's high reliability and long life on orbit. Through test, simulation analysis and calculation on sealing mechanism of globe-cone joint, in this paper, we make sure the relationship of the tightening torque and globe-cone joint contact stress and contact area, and obtain assembly torque range of different specifications of globe-cone joint, which gives necessary reference for mechanical environment verification.

1 Background

Propulsion system is the most important subsystem of spacecraft, its performance directly influence control precision, reliability and life on orbit. Generally, the propulsion system consists of thruster, propellant tank, latching valve, fill/drain valve, pressure sensor, filter, etc. The propellant is transported by been pushed in pipeline, so, propulsion system has high pressure inside. To insure sealing reliability, the way of welding is adopted in pipeline connecting, while globe-cone joint is adopted between pipeline and thruster, which have cone surface on thruster and globe surface on pipeline, and this way connecting is widely used on DFH(Short for Dongfanghong)-3 satellite platform, DFH-4 satellite platform, remote sensing satellite platform and navigation satellite platform. The tightening torque is one of the factors to affect the globe-cone joint assembly performance. Through test and simulation, this paper gives the minimum and maximum torque value during certain type of globe-cone joint assembly, which provides necessary parameters for further research to satisfy mechanical environment.

2 Globe-cone joint sealing mechanism and research approach

The globe-cone joint seal structure belongs to the type of metal contact and static seal, through giving external force, it forms a compact seal surface after incessant pressure. The force diagram is shown in figure 1.

[a]Corresponding author:yan_hust@163.com

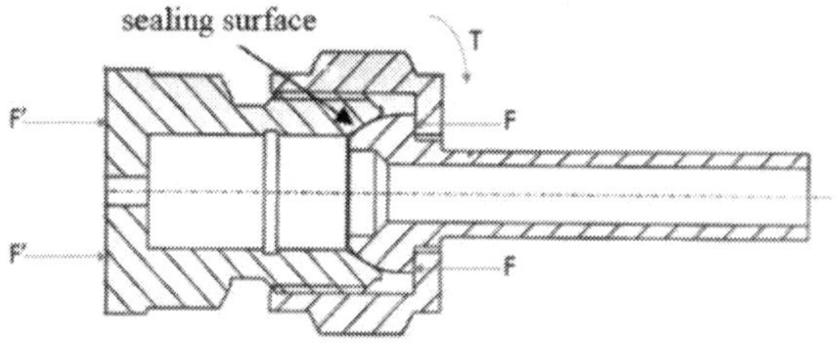

Figure 1. Globe-cone joint structure force diagram

Figure 2 gives the failure tree of the globe-cone joint seal leak rate. It can be seen that the factors to affect seal quality are divided to two kinds by property: seal surface machining factor and assembly process factor. The machining factor refers to the surface of the sealing surface roughness, parts with surface shape error, which is mainly embodied in the globe-cone geometry. Assembly process factor refers to the comprehensive factor including technique, operation experience, operating environment and other factors, mainly reflected in the neutral adjustment, globe-cone relative motion control, tightening torque, tightening stress control and redundancy control etc. The control of tightening torque is the most critical factor in the practice.

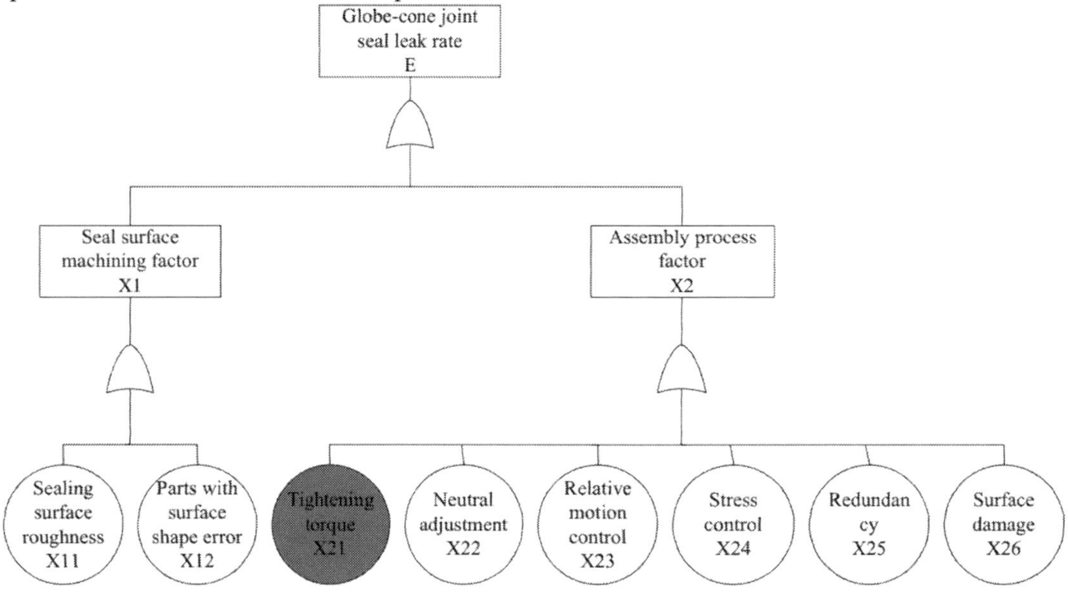

Figure 2. Effect of globe-cone joint structure factors

The research flow of assembling tightening torque shown in Figure 3 is presented.

1. Using experimental measurements to obtain the relationship of globe-cone joint between pre-tightening force and torque assembly method, and the coefficient of friction between the friction pair measured at the same time, to be the parameters of finite element analysis.
2. Using the method of finite element simulation of globe-cone joint structure model and the relationship between the structure of the globe-cone contact state and pre-tightening force.
3. Using the method of judging the relationship between the sealing performance and contact state.

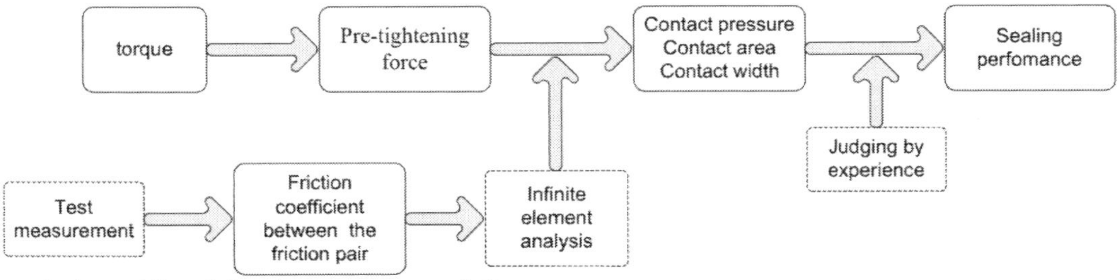

Figure 3. Assembling tightening torque research flow

3 Research on relationship between torque and pre-tightening force

There are two main methods to obtain the globe-cone assembly and other similar cone sealing structure torsion relationship: by theoretical calculation [1] and measured by experiment [2].

Including the ball screw thread structure head sealing structure, in the assembly process, the applied around the axial torque tightening torque, the torque generated by the pre-tightening force in the axial internal thread structure; therefore the pre-tightening force of the screw thread structure will be affected by their assembly tightening torque influence.

The relationship between the axial torque and the pre-tightening force is known as the formula 1.

$$T = KFd \tag{1}$$

In the formula, T is the tightening torque, K is the torque coefficient, F is the pre-tightening force, and d is the nominal diameter of the thread.

The torsion coefficient K can be expressed as a series of calculations:

$$K=\frac{1}{2}\left[1.15\mu_s+\tan\beta+d_W\mu_w/d+D_{MP}\mu_N/(d\cdot\cos\alpha)\right] \tag{2}$$

F is preload, α is cone angle, D_{MP} is globe-cone contact surface equivalent diameter, μ_N is globe-cone friction coefficient, d_W is the support surface equivalent diameter, μ_w is the support surface friction coefficient, μ_s is the thread friction coefficient, β is helix angle.

The friction coefficient of the globe-cone structure is unknown and it changes in a certain range, related to materials, surface contact friction coefficient and substrate surface morphology and other factors. At the same time, equivalent diameter of the supporting surface and equivalent globe and cone contact areas are difficult to be accurately determined. Therefore, the torque coefficient cannot be well matched with the actual value, and the formula method is not practical.

Through the method of testing the relationship of the globe-cone torsion, we can obtain accurate measurement torsion of globe-cone joint structure, and also can measure the coefficient of each friction pair. This method is shown in Figure 4, the out nut is cut from the non-screw thread section, turned the globe-cone inner force to out nut sections outer force, for convenient measurement.

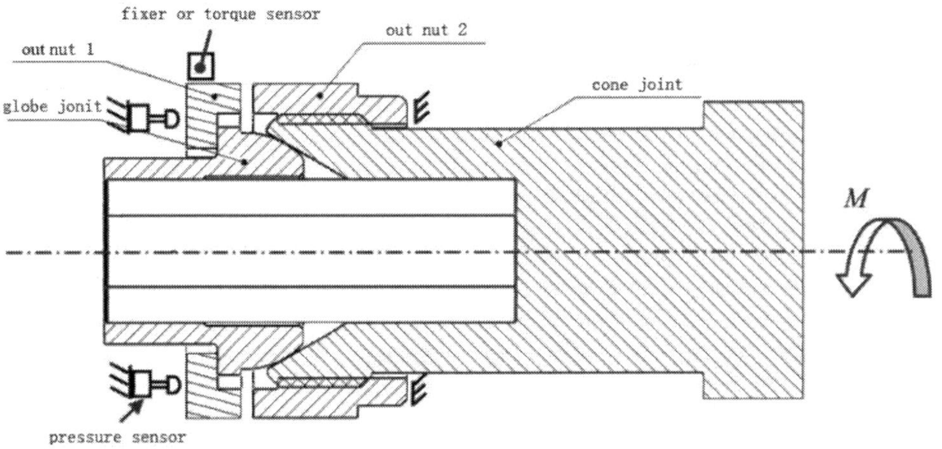

Figure 4. The globe-cone structure torsion relationship test principle

Take the type of 60-SR5.5-4 globe-cone joint sealing structure as an example, the detailed test results are shown in table 1.

Table 1. The type of 60-SR5.5-4 test results

No.	MaxF/kN	MaxT/Nm	T_{tot}/Nm	T_{th}/Nm	T_b/Nm	K	μ_{tot}	μ_{th}	μ_b
1	5.42	30.00	16.14	4.76	11.38	0.38	0.43	0.47	0.35
2	5.52	30.00	15.97	4.46	11.51	0.38	0.42	0.48	0.33
3	5.07	30.00	16.55	6.68	9.87	0.39	0.44	0.41	0.49
Note: *Max T* is the pre-tightening force(the test stop condition, namely 30Nm); T_{tot}, T_{th}, T_b respectively is the total torque, thread torque and support surface torque; *K* is torque coefficient; μ_{tot}, μ_{th}, μ_b, is respectively the total friction coefficient, thread friction coefficient and support surface friction coefficient.									

From table 1, we can see that the torque coefficient K of three groups is basically the same, respectively 0.38, 0.38 and 0.39. There are some differences in thread friction coefficient and support surface friction coefficient. The causes of this phenomenon may be relative to the sliding surface which is more than one.

4 Finite element analysis of pre-tightening force and contact state

The material of different specifications of sealing globe-cone joint is different. Take the type of 60-SR5.5-4 for example, globe-cone seal structure material are titanium alloy TC4 and stainless steel 1Cr18Ni9Ti. In finite element analysis of the globe-cone sealing structure assembly, the bilinear isotropic hardening model is taken to define the properties of the two materials. The materials and their properties are shown in Table 2.

Table 2. Materials of globe-cone joint sealing structure

Specifications/Properties	Cone material	Outer nut material	Globe material
60-SR5.5-4	stainless steel 1Cr18Ni9Ti	titanium alloy TC4	titanium alloy TC4
Young's modulus	206GPa	109GPa	109GPa
Poisson ratio	0.3	0.34	0.34
Yield strength	205MPa	840MPa	840MPa
Tangent modulus	6.8GPa	2GPa	2GPa

In order to get the accurate result, structured grid is used to partition the model. The globe-cone sealing structure of the model is not a common rule model, and the contact surface is related to plastic deformation, therefore the model has nonlinear material and structure. The element type is SOLID185 unit with 3D 8-node, the number of grid model is 338,000, as shown in figure 5.

 DOI: 10.1051/matecconf/201711403007

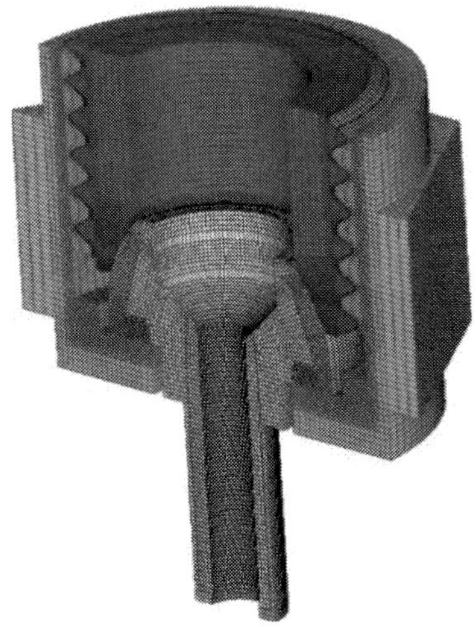

Figure 5. Grid density of globe-cone joint structure

In the model of the project, the three parts are all elastic bodies, and the final state of the contact pair is surface contact. So their contact surfaces are defined as surface contact and elastic-elastic contact.

Based on the grid density model test, the contact interface parameter setting and the boundary conditions imposed, we set up the globe-cone sealing structure assembly simulation of the finite element model as shown in figure 6. Globe-cone joint sealing structure contact friction coefficient can be found in table 1.

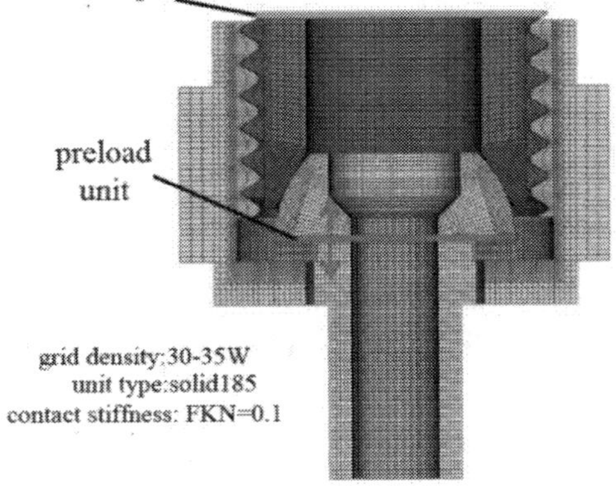

Figure 6. Assembly simulation of the finite element model

On the finite element model simulation analysis, we can get the globe-cone sealing structure of the equivalent stress (Von-mise stress), contact stress, change of base area with pre-tightening force.

Table 3 shows the equivalent stress of the sealing structure under increasing preload. Comparison of four images, we can see that, with the increase of pre-tightening force, equivalent stress increases, and the stress concentration point exits in the maximum diameter of contact surface circle between the out nut and globe.

Table 3. Transformation of contact stress with increasing preload

	Equivalent stress distribution	contact stress distribution

MATEC Web of Conferences **114**, 03007 (2017) DOI: 10.1051/matecconf/201711403007
2MAE 2017

7.5kN	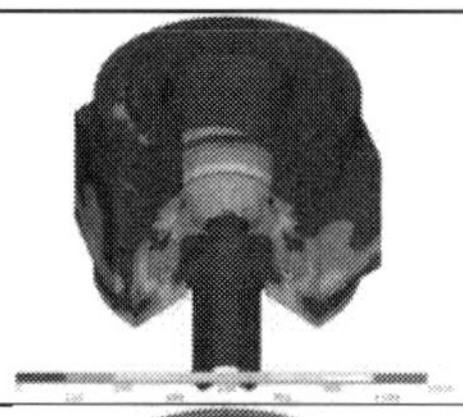	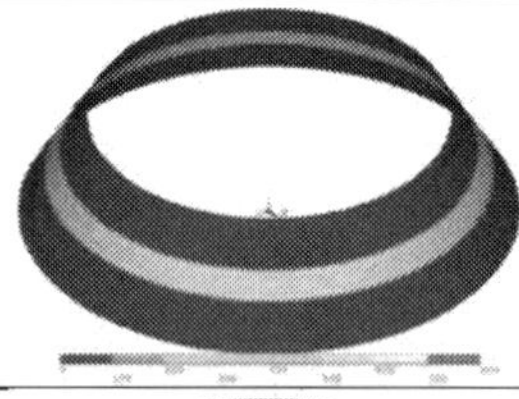
15kN	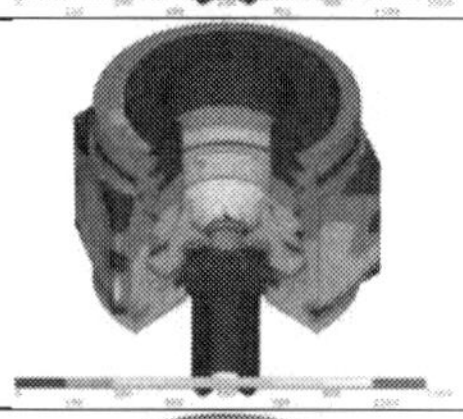	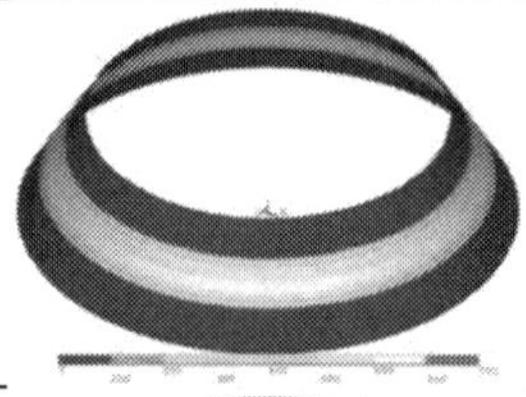
22.5kN	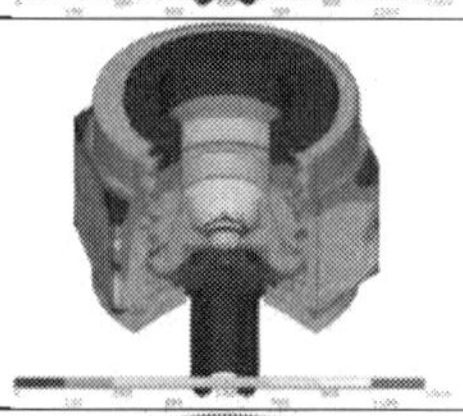	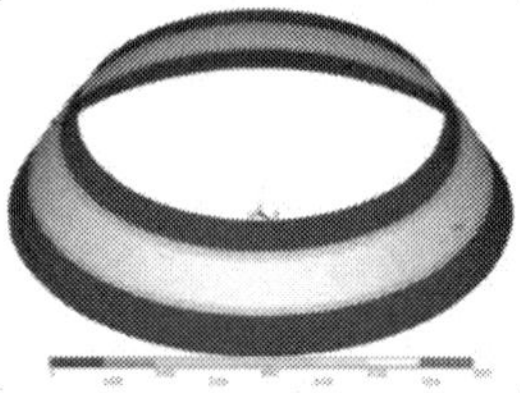
30kN	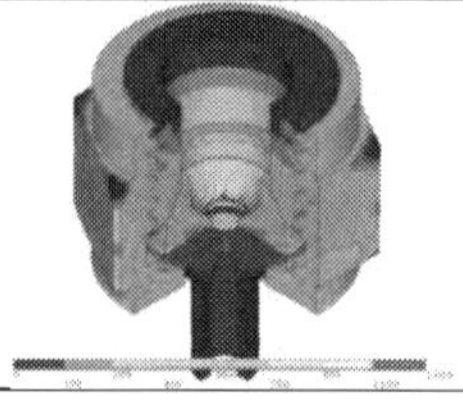	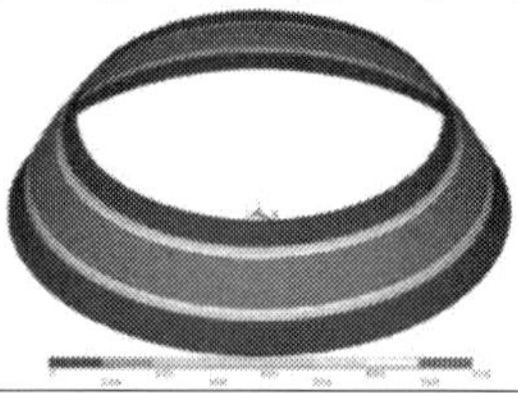

5 Study on the relationship between contact state and sealing

In this project, the average contact stress and the contact width are used to determine the preload range.

5.1 Contact pressure criterion

On average contact pressure, the overall judgment criterion is to ensure that the average contact pressure of the sealing surface is greater than the yield strength of the materials [3].

5.2 Contact area criterion

On the contact area, it needs the contact analysis to judge the contact width required, to ensure the sealing performance of the contact area. Based on the commonly used leakage model A Roth, this project has improved, the cone contour surface is simplified as a plane and the globe surface is simplified as a cylinder [4-5], whose radius is the radius of the globe. The improved model is shown in figure 7.

 DOI: 10.1051/matecconf/201711403007

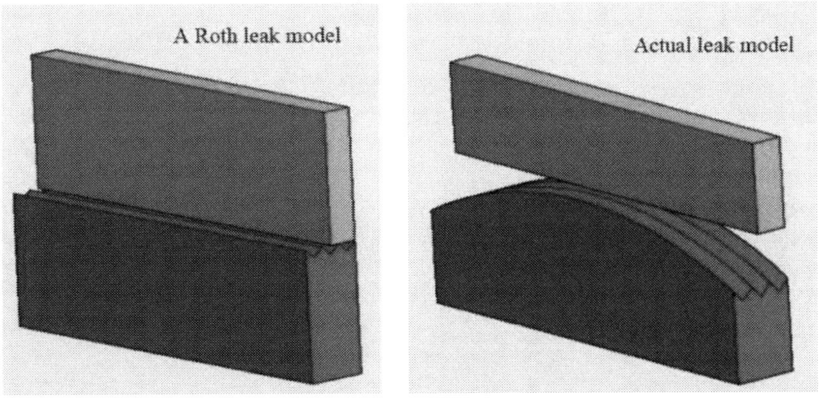

(a) (b)

Figure 7. Microscopic contact model

Because the actual micro contact model is a symmetrical structure in the flow direction, and a triangular wave periodic structure perpendicular to the flow direction [6-7], so in the finite element model, we select a single flow channel perpendicular to the flow direction and half the channel along the flow direction to analysis. The solid45 unit is chosen as the cell type of the model, which is divided into a finite element model with 325,800 structured meshes. The triangle wave surface is defined as the contact surface, and the contact element type is Contact174. The corresponding ideal plane is defined as the target plane, and the target cell type is Target170. Because of the special model, the symmetry constraint is imposed on the plane and the symmetry constraint is applied in front and behind of the plane. At the same time, a uniform pressure load is applied on the top surface, and a static constraint at the bottom, as shown in figure 8.

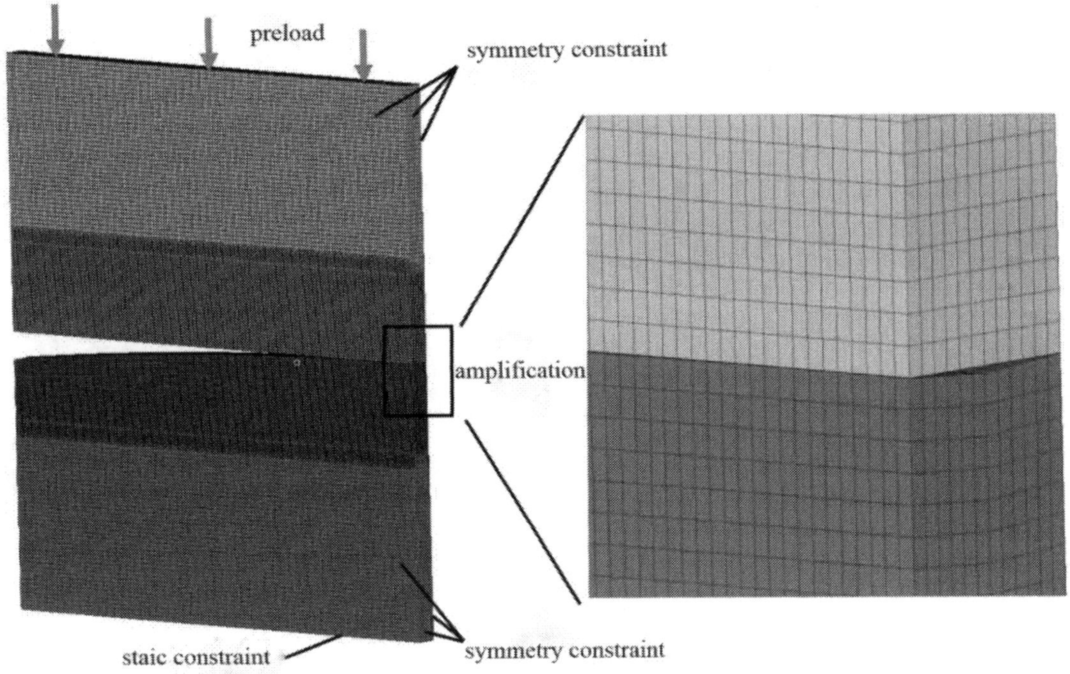

Figure 8. Leakage channel cross section and microscopic contact model

The evolution of the simulation calculation of globe-cone joint sealing structure and micro contact width of the contact area is shown in figure 9. The contact area increases with the gradual increase of the contact width, and the width of the channel leakage begin to decrease when the contact width increases to a certain value, finally, the isolated contact areas are connected to each other and leak channel is completely blocked.

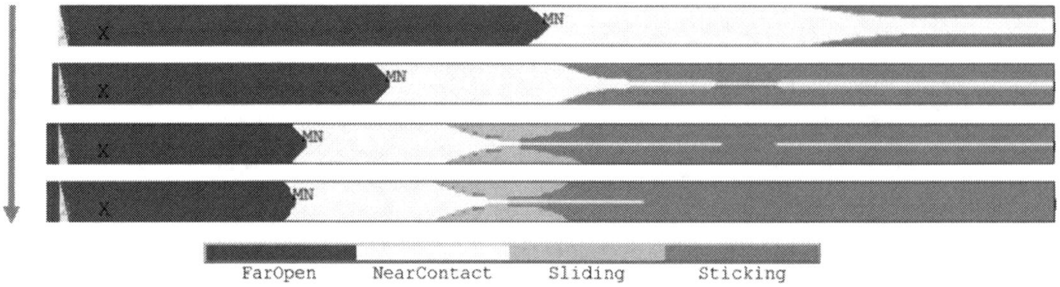

Figure 9. The evolution of micro contact and width the contact area

According to the approximate solution equation 3 of the contact area, we can get the globe-cone contact area standard as shown in table 4.

$$A=\sqrt{3}\pi lr \tag{3}$$

A is the sealing interface contact area, l is contact width, r is radius of the globe.

Table 4. Globe-cone contact area standard

Specification	Contact width/mm	Contact area/mm^2
60-SR5.5-4	0.714	21.36

6 Pre-tightening force judgment and conclusion

6.1 Determination of lower limit of pre-tightening force

In order to satisfy the ideal sealing result, the average contact pressure preload should be up to 205MPa to seal the interface, at the same time, the contact area should be more than 21.36mm^2. As shown in Figure 10, the contact area and contact press change with the pre-tightening force, we choose the larger pre-tightening force as the lower limit of preload, which is 4.34kN.

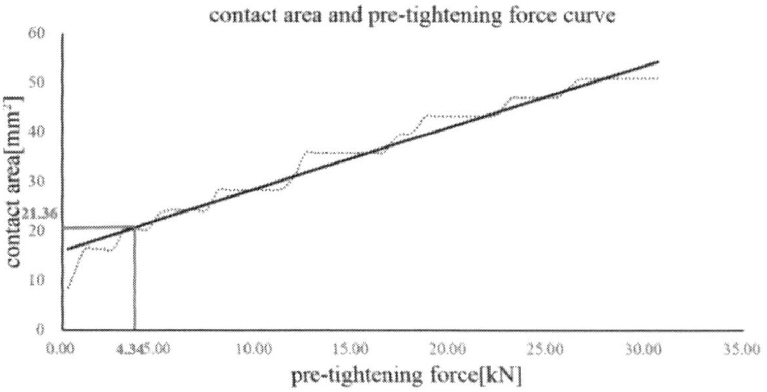

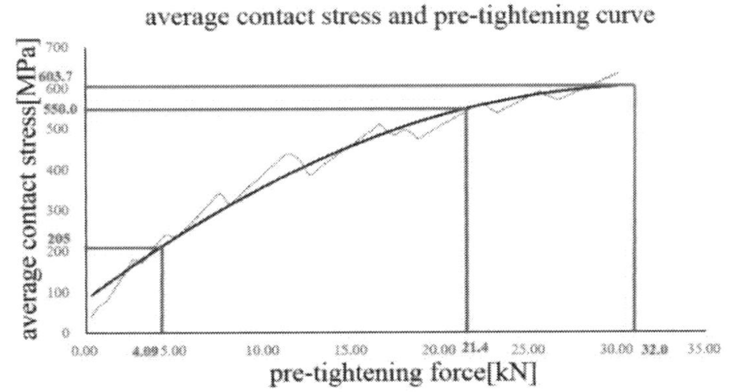

Figure 10. Contact area and press change with pre-tightening force

6.2 Determination of upper limit of pre-tightening force

Also from Figure 10, when the average contact stress is greater than 603.7MPa, the average contact stress decreases with the pre-tightening force. When the pre-tightening force comes to 21.4kN, the average contact stress reaches the ultimate strength of the structure material of 550MPa.

For the thread structure (M14), the maximum tightening torque applied on the part of the screw should ensure that the average stress is not greater than 70% of the yield strength, the average stress on the screw part can be got from formula 4. In 60-SR5.5-4 globe-cone structure, D1 is 12.376mm, d1 is 8mm, and σ_s is 205MPa. It can be calculated that F is no more than 10.0kN.

$$\sigma = F/\left(\frac{\pi}{4}(D_1 - d_1)\right) \leq 0.7\sigma_s \tag{4}$$

Based on the above conditions, it can be concluded that the upper limit of pre-tightening force for sealing condition is 10.0kN.

6.3 Conclusion

According to 3rd section of the globe-cone assembly torsional tension test results, the torque coefficient K is 0.38, and according to the formula 1, we can calculate the tightening torque range of the sealing structure is 23.1-53.2Nm.

With the same analysis and calculation method, we can obtain the tightening torque range for other type of globe-cone joint widely used on spacecraft at present, which is shown in table 5.

Table 5. Tightening force of different joint structure

Thruster specifications	Specifications	Pre-tightening force/kN	Tightening torque/Nm
1N, 5N, 20N	60-SR5.5-6	4.75-11.1	24.6-57.5
10N	60-SR5.5-4	4.34-10.0	23.1-53.2
490N	60-SR7.3-8	6.97-20.1	47.7-137.5
E-propulsion	60-SR2.75-4	2.29-4.94	7.15-15.4

Besides the simulation analysis and calculation of the torque range, it needs further research to verify the torque range under mechanical environment, and to narrow the range.

References

1. Zhou Xin, Pang Hewei, Liu Hongyang, Yan Shaoguang, Influence analysis of assembly error on seal status of spherical seal structure, J. Spacecraft Engineering. 2005, (04) 35-39.
2. Liu Kai, Li Qiang, Xie Zubao, Study on control of pipe joint tightening torque based on torque - test, J. Quality and Reliability. 2014, (5) 16-19.
3. H.H. Buchter. Industrial sealing technology [M] .Beijing: Chemical Industry Press, 1988.
4. Robbe-Valloire, F., and Prat, M., 2008, "A Model for Face-Turned Surface Microgeometry: Application to the Analysis of Metallic Static Seals," Wear, 264(11), pp. 980-989.
5. Zhang F, Liu J, Ding X, Yang Z. An Approach to Calculate Leak Channels and Leak Rates Between Metallic Sealing Surfaces. ASME. J. Tribol. 2016;139(1):011708-011708-11.
6. Zhou Xin, Pang Hewei, Liu Hongyang, Yan Shaoguang. Nonlinear contact analysis of globe-cone connection structure [J]. Spacecraft Environment Engineering,2005,(04):211-214.
7. Wang Xiaogang, Zhang Fangxiao, Huang Peng, Design of cone-head & cone-hole pipeline cone sealing structure based on ANSYS[J]. Machinery,2009,(08):37-40.

Spacecraft Assembly Process Bill of Material Construction Technology For Single-Piece Production

HE Wen-xing[1,a], CHEN Chang-yu[1] and LIU Yu-gang[1]

[1] *Beijing Institute of Spacecraft Environment Engineering, Beijing 100094, China*

Abstract. For the spacecraft assembly Process Bill of Material (PBOM) traditional construction technology, the product assembly relationship, product assembly information are not visual for spacecraft assembly and integration department, leading to the manual PBOM construction by using Product Data Management (PDM) system, which is inefficient and unreliable. This paper sorts all spacecraft products into seven Primary Product and others are Affiliated Product. Then this paper proposes the rules of PBOM automatic construction technology, which are Product Types mapping rule, assembly constraint rule and geometric rule, and introduces the PBOM automatic construction principle and workflow based on Pro/E. Finally, according to a real satellite manufacture, this paper describes the improved efficiency and reliability of the software developed with the technology, realizes the import into the PDM system, and analyzes the fault of the technology.

1 Introduction

Spacecraft is a complex electromechanical thermocouple system, which composed of several subsystems and thousands of products. The spacecraft assembly process is to assemble the products in the main structure of the spacecraft, and completes the relevant detection and testing. The spacecraft development unit can be divided into overall design, subsystem design and manufacturing integration department.

In recent years, based on Model Based Definition (MBD), domestic and foreign universities and enterprises have carried out digital product design and manufacturing technology research. Min Liu et al. studied the transformation between EBOM (Engineering Bill of Material) and maintainability BOM [1]; Chen Zhi et al. used BOM to carry out research on multi-product joint inventory management technology, which involved BOM construction technology [2]; Jang Hyun Lee et al. proposed a suitable BOM construction technology for shipbuilding, which can display the product structure and information [3]; Gonen Singer et al. proposed a BOM conversion technology from plan to production for small batch of complex products , which can add additional information in the BOM [4]; Tsai Chi Kuo built a low-carbon products BOM in PDM system through the standardized data collection forms and search engine system [5]; C. Martinez-Olvera studied the influence of BOM on the process [6].

In the analysis of PBOM construction process and the spacecraft assembly integration, it is not hard to find:

[a] Corresponding author :hewenxing@163.com

(1) Although there is a wide range of spacecraft products (including a total of 24 categories: instruments, pipe fittings, pipe fittings, insulation pads, insulation film, grounding components, fasteners, electrical connectors, straight parts, cables, thermistors, heating circuit, Heating, multi-layer, thermocouple, external heat flow, equipment accessories, piping, ventilation systems, thermal control foam and flame retardant cloth, cabin structure, hose, interior decoration and pipe fittings). In order of simplifying the expression of product assembly relationship ,the 24 categories of products can be simplified to seven categories of objects, consisting of instrument, cable, piping, heating circuit, direct parts, deck and thermal control multi-layer (known as main product object), and the others are affiliated products (known as Affiliated Product Object) .

(2) The PBOM construction of Spacecraft assembly integration is essentially analyzing the product assembly relationship displayed in the form of BOM, which will be instantiated in the form of parameters or database. The process is the visualization of BOM.

(3) Assembly integration PBOM can be automatically analyzed and visual processing in the CAD software, and then imported into PDM, which improves work efficiency and quality.

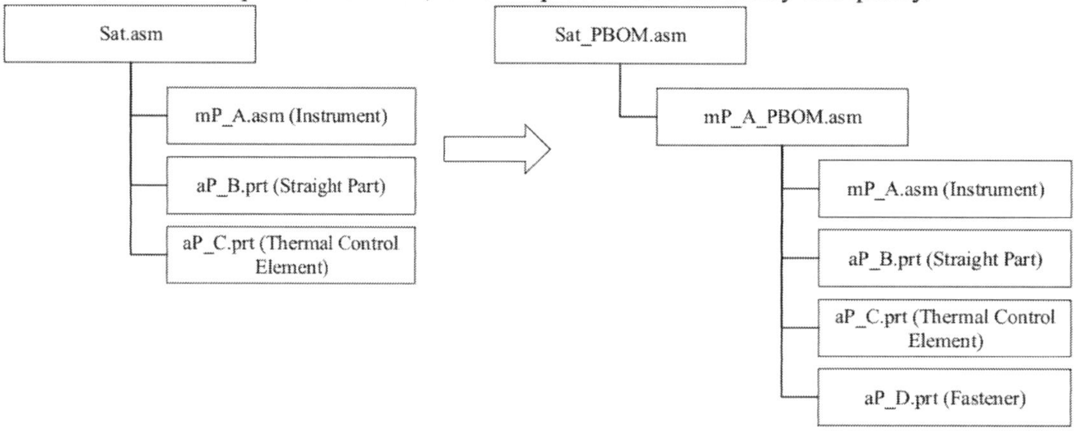

Figure 1. BOM Visualization Processing

According to the overall design division of the spacecraft, the whole satellite model is composed of three types of professional models: the whole mechanical structure models including satellite deck structure, instrument, straight parts and others; the electrical product models including cable, electrical connector and other affiliated product, etc.; the thermal control product models including thermistor, thermocouple, heating circuit and heating film, thermal control multi-layer etc. The spacecraft assembly integration department needs to assemble the models as an integral satellite model in accordance with the whole satellite coordinate system after receiving the design models, as shown in Figure 1. It is gradually constructed out of each corresponding PBOM node to the main object in Pro/E, and the whole satellite PBOM node, Sat_PBOM.asm.

For the assembly integration department, five categories of objects including cable, piping, heating circuit, deck, thermal control multi-layer and their affiliated objects in EBOM have fulfilled the task requirements. In this paper, instruments, straight parts and their affiliated fasteners are taken as examples to illustrate the PBOM visualization process by using the Pro/E software.

2 Fundamental Principles

Set all node elements of the whole satellite PBOM as the set U , and set u_i as a node element of PBOM, then there is:

$$U = \{u_i\} \tag{1}$$

Determining the process of u_i is to inherit, find or create the product component objects that meet certain conditions, such as mP_A.asm, aP_B.prt, aP_C.prt, ap_D.prt in Figure 1 and other components or parts.

In addition to fasteners and other standard parts of the product which need to be instantiated and assembled with the corresponding master products, the other product components have a solid model and establish with the whole satellite coordinate system. The determination of such product components needs to determine whether the object is the required one in accordance with certain criteria, the criteria for this paper are as follows:

(1) Product type matching principle, that is, each node element u_i has 1 component of master product objects, and other product objects are affiliated product object type, such as mP_A.asm for the instrument of master product objects, aP_B.prt, aP_C.prt for the straight parts and thermal control elements of affiliated product objects.

(2) Assembly constraint principle, that is, based on the matching principle of product types, distinguish whether there is assembly constraints (such as face matching) between the master product mP_A.asm and aP_B.prt, if so, determine aP_B.prt as a subsidiary object of mP_A.asm, and B.prt as a component of A_PBOM.asm.

(3) Geometric relations principle, that is, based on the product type matching principle, calculate the normal distance between the master product object mP_A.asm and aP_C.prt, if the distance is the shortest in the product master objects and other affiliated product objects, and the normal distance between aP_C.Prt and other product objects is larger than the value, then determine aP_C.prt as an affiliated product object of mP_A.asm.

Respectively, set a_i, b_i, c_i as the product objects to meet the above criteria, and

$$A = \{a_i\}$$
$$B = \{b_i\}$$
$$C = \{c_i\} \tag{2}$$

According to PBOM construction guidelines, there is

$$U = (A \cap B) \cup (A \cap C) \tag{3}$$

Simplified as

$$U = A \cap (B \cup C) \tag{4}$$

Let d_i be a standard object such as fasteners for instantiation, and

$$D = \{d_i\} \tag{5}$$

To determine the d_i, weneed to obtain product components,such as fasteners, including the number, specifications and standards, through the parameter table or database and then create such product objects in PBOM, including the creation of the corresponding product node (such as aP_D.prt), and assemble it to the corresponding position based on determining its position relative to the whole satellite coordinate system.

Thus we can identify all the constituent elements of the entire satellite PBOM:

$$U = (A \cap (B \cup C)) \cup D \tag{6}$$

3 PBOM Visualization Construction Process

For spacecraft assembly integration, PBOM visualization process is based on the BOM reconstruction to build PBOM without changing the assembly relationship of EBOM products, and instantiates fastener product objects in the form of product parameters, thus enabling the visualization of assembly integration information.

In addition to fasteners and other standard products, the visualization of other products should follow the product type matching, assembly constraints, geometric relations and other three principles, the visualization of these products do not need to be instantiated, but need to complete the product BOM reconstruction and model relocation. Fasteners and other standard products need to read the product assembly with fasteners standards, specifications and other parameters of information to build the corresponding fastener model and the model can be simplified model, and then mounting fastener model in the holes corresponding products.

Satellite assembly visualization PBOM construction process is shown in Figure 2.

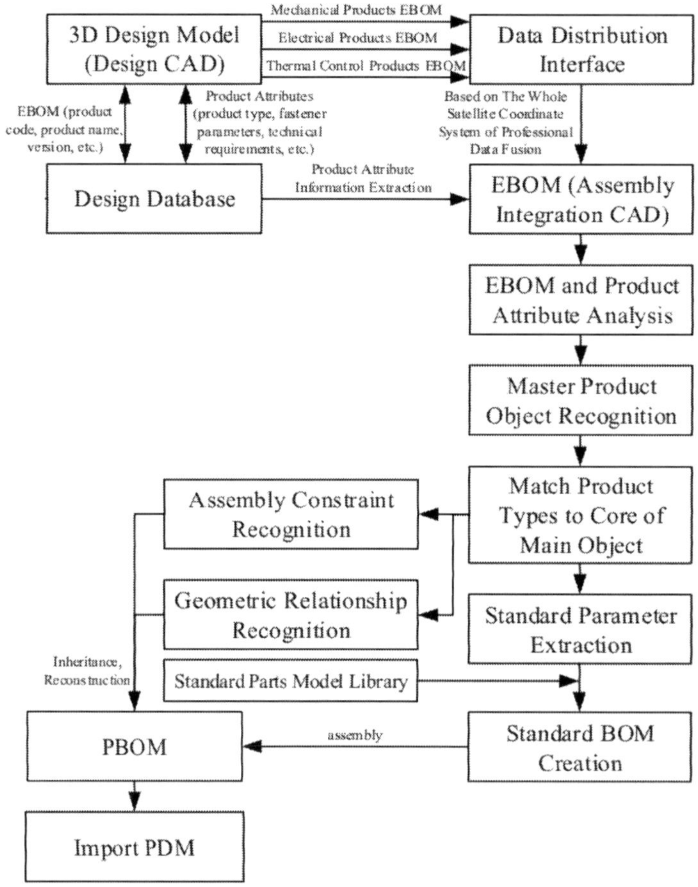

Figure 2. Satellite Assembly Visualization BOM Process

4 System Application

In this paper, Pro/E and ProToolkit are used to establish a set of PBOM visualization construction system for satellite and other spacecraft assembly, which is applied in the practical model development. The system obtains the controlled 3D model and product attribute information from the design data management system of the overall design department (Avidm4.0 system used in this paper), based on EBOM and its related product attributes, creates a visual PBOM that meets the requirements of the assembly integration department through Inheritance, reconstruction and new, etc., and automatically and import PBOM into the PDM system through interfaces with PDM (Teamcenter for this paper). The system not only achieves the automatic transmission of 37 commonly used satellite assembly integration information (more than 30% compared to the traditional model), but also increases the efficiency from design data to assembly integration department more than 1 times (from 20 people down to 8 People days).

5 Conclusion

This paper presents a visualization PBOM processing method for the integration of spacecraft assembly through variety of assembly information generated in the three-dimensional design process of the design departments, realizes the automatic construction and application of the visualization PBOM of satellite assembly integration, succeeds in transmitting to the PDM System, and greatly improves the efficiency and quality of data transmission. The method has good processing power for mechanical products and related thermal control products. However, it is difficult to deal with the problem of products with complicated assembly relationships and many branches such as cables and pipelines, which need to be focused on in the further.

Acknowledgements

This research is supported by Natural Science Foundation of China (No. 51405025), and Beijing Spacecraft Smart Assembly and Equipment Engineering Research Center.

References

1. LIU M, LAI J, SHEN W M. A method for transformation of engineering bill of materials to maintenance bill of materials.[J]. Robotics and Computer Integrated Manufacturing: An International Journal of Manufacturing and Product and Process Development,2014,30(4).
2. CHEN Z, REN C, SHAN M Y. Multi-Product of JMI under BOM.[J]. International journal of advanced pervasive and ubiquitous computing,2014,6(4).
3. JANG H L, SEUNG H K, KYUNGHO L. Integration of evolutional BOMs for design of ship outfitting equipment.[J]. Computer-Aided Design,2012,44(3).
4. GONEN S, MAYA G, YUVAL C. From product documentation to a 'method prototype' and standard times: a new technique for complex manual assembly.[J]. International journal of production research,2014,52(1/2).
5. TSAI C K. The construction of a collaborative framework in support of low carbon product design.[J]. Robotics and Computer Integrated Manufacturing: An International Journal of Manufacturing and Product and Process Development,2013,29(4).
6. C.Martinez-Olvera. An entropy-based approach for assessing a product's BOM blocking effect on a manufacturing process flow.[J]. International journal of production research,2012,50(2/4).

Topology Optimization and Design Guidelines of Sub-Stiffened Panels in Aerospace Applications

Hemza Layachi[a], Yuan-ming Xu and Mohamed Amine Bennaceur

School of Aeronautic Science and Engineering, Beihang University, Beijing 100191, China

Abstract. The present work is a proposed optimization model of grid sub-stiffened panels for aerospace applications, having a better knowledge about the stability analysis of conventional grid sub-stiffened panels against loading and different mode of failure occurring lead us thinking about the topology optimization. In a first stage, we will analyse the stability performance of such structures against specific mechanical loading and conclude some techniques to introduce a topology optimization through PSO algorithm of some design variables. The aim is the better exploitation of the sub-stiffened panel along the plan so that the area less stressed due the charging will be extruded in mass for the enhancement of the concept of looking for better performance with a structure already saved in weight.

1 Introduction

Stiffened panels are primary structures in aerospace engineering as for nuclear, power plate, marine and civil buildings. Stiffening structures is a techno-economic compromise that offer weight saving by using less manufactured material against subjected load conditions, it is more practical to use additional stiffeners than just increasing of thickness of the plate [1, 2]. The common way to optimize stiffened panel's performances in aerospace application has to start with calculating the critical loading, more the structure is needed to carry loads more it has to be better reinforced [3, 4]. It is significant to give a structural design that handle the listed issues. One of the alternative solutions is to introduce a topology optimization to such structures along the plan, as the terminology suggests basing on the study of geometric properties and response of the stiffened panels to mechanical charging by changing the sizes and shapes of figures in question [5, 6].

Topology optimization in design of aircraft structures have been applied in successful industrial case studies [7, 8]. This brand of research develop mathematical and numerical models to optimize material location in design varieties [9], defining the input variables of the proposed design, the basic optimization methodology starts with defining a range of maximum/minimum of parametric variables [6]. Grid-stiffened structure with equilateral rectangular patterns give noticeable strength performance for lightweight against mechanical charging and damage [10]. The repetitive pattern represented by the stiffeners along the panel is also an important factor in minimizing manufacturing cost and complexity of design in one part and, a possibility of such a topology optimization in second part.

[a] Corresponding author : layachi-hamza@hotmail.com

2 Topology optimization procedure

In a first stage, we will focus on the response of a conventional grid sub-stiffened panel to compression load, the first buckling modes will be inspected and the critical load to buckling will be defined. We will look, in such conditions, to the distribution of charging due to the loading. Some zones will be more or less affected so the lesser-charged areas will be enhanced for a topology optimization.

In the table 1we list, the different buckling modes may occur due to charging to the structural panel.

Table 1. Buckling modes and its modality through the stiffened panel.

Buckling mode	Modality	Representative figure
Local buckling	This is happening when the junction between plate and the stiffener does not displace in a vertical way and buckling is located on the skin of the panel.	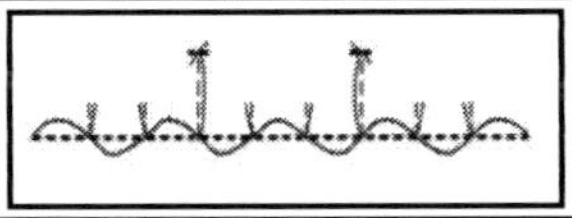
Torsional Buckling	With a local buckling mode, this junction is still stiff and straight.	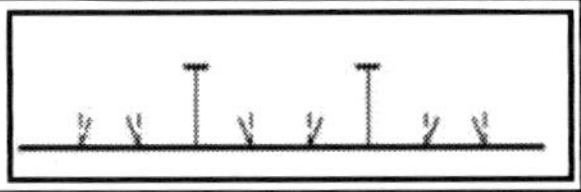
Partial buckling	Stiffeners buckle in torsional mode, partial buckling of a panel is when sub-stiffeners are unbuckled.	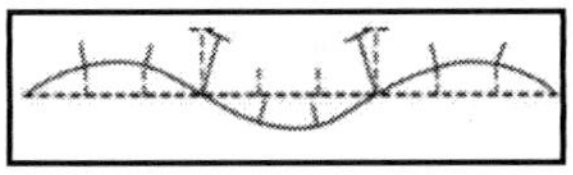
Overall buckling	Catastrophic buckling mode, we observe the buckling of the whole panel as a column including the skin, stiffeners and sub-stiffeners.	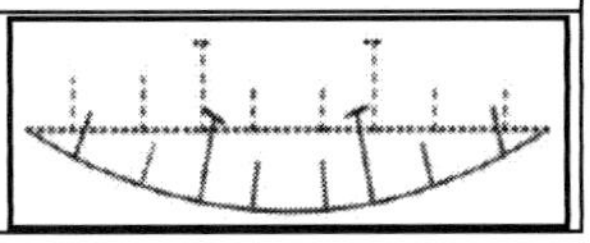

We modify the disposition of the stiffeners and sub-stiffeners along the panel in such a way we extrude in weight, for this we will go through those design guidelines basing on simplifying hypothesis.

Figure 1 give the plan of the targeted topology optimization, we start by the performance of a conventional grid panel in such a way it's given in the best configuration, we look for the possibility of introducing topology optimization procedure through the panel. Dimensioning of selected repetitive patterns will be applied on the stiffeners by comparative study taking care of the conception and manufacturing difficulties

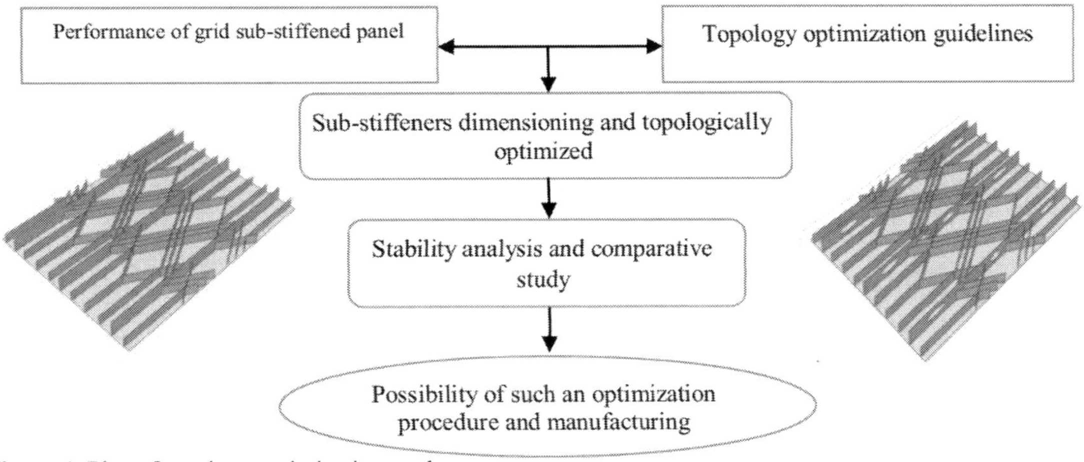

Figure 1. Plan of topology optimization study

- Topology optimization will be applied on the stiffeners due to its capacity to maintain the stability of the sub-stiffened panels in one part and the ability of extruding some material from it, since the plate is conserved in design and sub-stiffeners are typically small for such a topology.
- We opted to introduce elliptic patterns through the stiffeners for topology optimization since it is validated for better performance against crack and fissure propagation and easy in conception, which is more adequate for this study since the sub-stiffened panel will be subjected to compression load.
- The extruded material is a function of the behaviour on the conventional sub-stiffened panel, basing on its response to the imposed loading, we will work on the dimensioning and positioning of the extruded material in a way we guarantee the stability and better performance comparing to the conventional sub-stiffened panel.

3 Case study

In this study and starting from the classic configuration we will investigate the stability performances of sub-stiffened panels, we focus on the buckling and post buckling due to compression load. The critical load and buckling modes will be defined for it in a first stage. The optimization procedure consists to look for the substitute that the structure will be buckled locally which was concluded from previous studies [11] which, give us a possible way to introduce our topology procedure through its stiffeners.

The sub-stiffened panel will be subjected to compression load, figure 2 represent the initial grid panel with its specific dimensions, design variables for the given structure will be dimensioning peer a cross section, the best configuration interpolating stiffeners and sub-stiffeners thickness and high for maximum performance

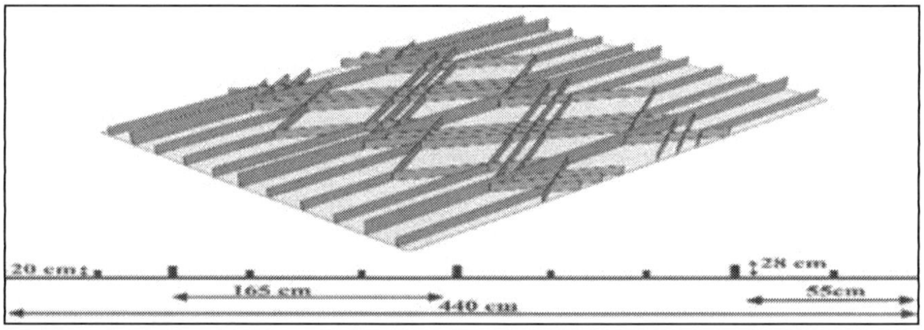

Figure 2. Grid sub-stiffened panel and dimensioning.

Table 2 list the design variable for the grid panel, the respective weight of it, the manufactured material is an Aluminum alloy 2024T3, and here under is the material proprieties of it, as for the topology-optimized panel that will not be changed in material properties and dimensioning.

Table 2. benchmark grid panel design variables and mechanical properties.

Design variables grid sub-stiffened panel		Mechanical proprieties	
sub-stiffened panel dimensioning		Aluminum alloy	2024T3
Plate thickness	1.9mm	density	2770
Stiffeners thickness	2.8mm	Young modulus	73.1GPa
Sub-stiffeners thickness	2.8mm	Poison ratio	0.33
Distance per stiffener/sub stiffener	55cm	Expansion modulus N/m	2.32 10
Conventional panel mass	2.118 Kg	Yield stress	324MPa

Figure 3 give in detail the process of the designed topology optimization for the stiffeners, it involves the extrusion of elliptic patterns $P_i(L_i, W_i)$ which is modeled in the equations below, in aim to get maximum performance comparing to the grid panel an avoid the damage of the stiffeners σ_{max}. The repetitive elliptic extrusion through stiffeners is the same along the plan:

$$P_i(L_i, W_i)\ ;\quad (X^2 / L^2) + (Y^2 / W^2) = R/4 \tag{1}$$

$$\sigma_{max} < \sigma_s \tag{2}$$

$$D_{in} \leq D_{in\text{-}max} \tag{3}$$

$$P_{in}\, L, W \leq P_{in\text{-}max}\, L, W \tag{4}$$

$$P_c\, L, W \leq P_{c\text{-}max}\, L, W \tag{5}$$

$$P_{min} < P_i < x_{max}\, ,\ i = 1,\ldots,n \tag{6}$$

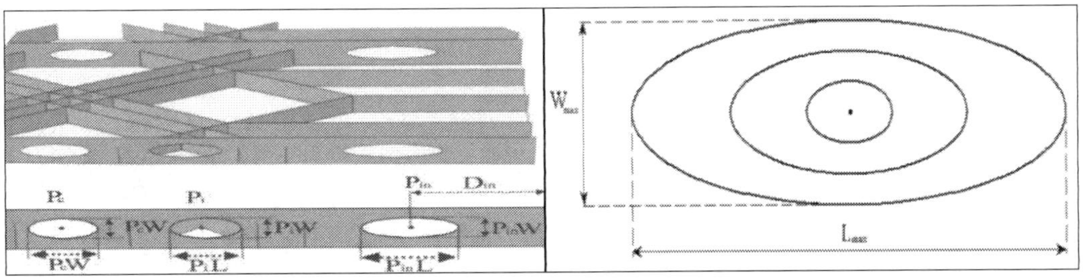

Figure 3. Topology procedure along the stiffeners and dimensioning.

4 Simulation and results

4.1 Buckling analysis

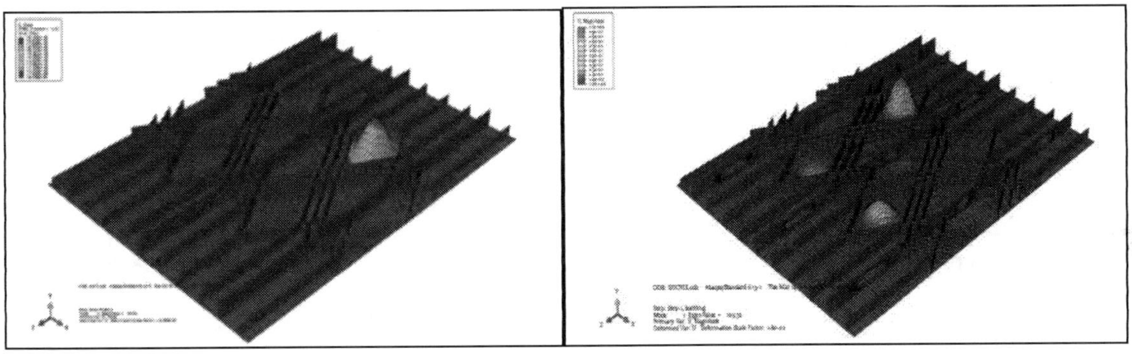

Figure 4. Buckling analysis for structural varieties.

In figure 4, as the results of simulation show, the two proposed sub-stiffened panels buckle in local mode at first stage, which is a great buckling mode due to the capacity of the structure to sustain the loading for more. Even the optimized panel give a greater number of half-wavelet comparing to the conventional one.

Table 3. Buckling analysis results and critical loads.

Structure panel design	Grid sub-stiffened pane	Topology optimized grid panel
Critical load (Kn)	222.4	222.2
Equivalent Mises stress (Mpa)	295.5	312.3
First Buckling mode to occur	local	Local
Half-wavelet appeared in a first buckling mode	1	3

From Table 3, the optimized panel have the capacity to carry more load, the critical load is 222.2Kn that is great similar the conventional panel given at 222.4Kn, the equivalent Mises stress under buckling is set around 295 against 312 Mpa for the two panels. The second part of this study consist to analyze the post buckling range to define the possibility of such a topology optimization.

4.2 Post buckling analysis

The instability process analysis pass by stating the ultimate strength capacity defined by the bearing load [12, 13], we run our numerical study through nonlinear RIKS algorithm by ABAQUS 6.13 for the proposed structures. The structural damage is expected at the bearing critical load, the aim of this part is to define the limit before collapse:

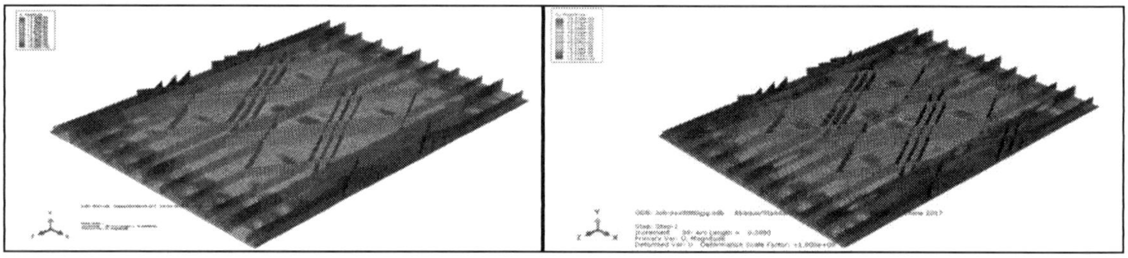

Figure 5. Structural predicted collapse stage.

Figure 5 show the instability process, it is represented by the waveform jumping phenomenon, the buckled surfaces carry more load which affect its stability but it remain local , the two proposed panels have the capacity to sustain the load and keep the structures stable before the bearing load limit.

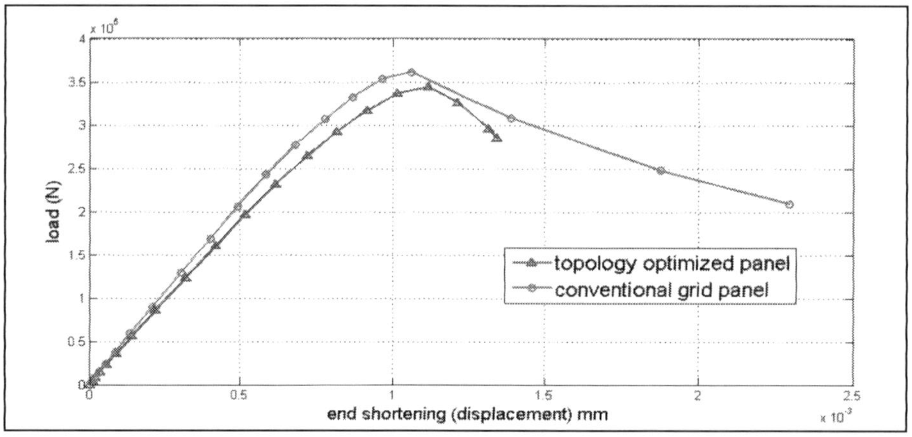

Figure 6. Load Vs end shortening for the structural varieties.

Comparing the results of simulation, we can notice the lesser ratio load/displacement from the second structure to the conventional panel, it is given at 1.4mm as a maximum before the collapse against 2.3mm for the conventional panel. The ultimate bearing load is slightly lower than the conventional panel given at 345Kn and 358Kn respectively for the two panels.

In an overall aspect of analysis and as it is given by figure 5and 6, the stiffeners in both structures remain stable against loading until the failure, it is noticed the multiple stressed zones in the optimized structure than the conventional one but those zones are more located in skin of the sub-stiffened panel. It is more because of the buckling mode in a first stage where the number of half wavelet is greater than conventional panel.

5 Conclusion and perspectives

Topology optimization is a promoting brand of research and engineering in aeronautical science, weight saving for structures remain a necessity looking for better performance and stability. We can conclude from this study the possibility of such a methodology for sub-stiffened panels where topology procedure let us gain 1.5 to 2% of the total weight of a conventional sub-stiffened panel which saving its stability, this will have great reflection to the performance of the aeronautical structure. Regarding the behavior of the topology optimized sub-stiffened grid panel we highlight its great capacity to maintain the stability of the panel against loading.

References

1. K. Doshi and S. Vhanmane, *Ultimate Strength of Stiffened Plates Subjected to Longitudinal Compression and Lateral Pressure, ASME 33rd International Conference on Ocean, Offshore and Arctic Engineering*, V04BT02A051--V04BT02A051 (2014)
2. T. Kubiak, *Nonlinear Plate Theory for Postbuckling Behaviour of Thin-Walled Structures Under Static and Dynamic Load.* INTECH Open Access Publisher, (2012)
3. Y. XU and S. LI, *Approximate calculation method of buckling load on integral sub-stiffened panel, J. Beijing Univ. Aeronaut. Astronaut.*, **3**, 1 (2015)
4. S. B. Mulani, W. C. H. Slemp, and R. K. Kapania, *EBF3PanelOpt: an optimization framework for curvilinear blade-stiffened panels, Thin-Walled Struct.*, **63**, 13–26 (2013)
5. B. Stanford, P. Beran, and M. Bhatia, *Aeroelastic topology optimization of blade-stiffened panels, J. Aircr.*, **51**, **3**, 938–944 (2014)
6. J.-H. Zhu, W.-H. Zhang, and L. Xia, *Topology optimization in aircraft and aerospace structures design, Arch. Comput. Methods Eng.*, **23**, **4**, 595–622 (2016)
7. S. Grihon *et al.*, *Fuselage Structure Optimisation, Adv. Collab. Civ. Aeronaut. Multidiscip. Optim.*, (2010)
8. M. Tomlin and J. Meyer, *Topology optimization of an additive layer manufactured (ALM) aerospace part, Proc. of the 7th Altair CAE technology conference*, 1–9 (2011)
9. M. S. Sohn, *Topics in structural topology optimization, University of Illinois at Urbana-Champaign*, (2010)
10. A. Murphy, D. Quinn, P. Mawhinney, M. Ozakc-a, and S. van der Veen, *Tailoring static strength performance of metallic stiffened panels by selective local sub-stiffening, Proc. of the forty seventh AIAA/ASME/ASCE/AHS/ASC structures, structural dynamics, and materials conference. Newport, Rhode Island*, **1**, **4**, (2006)
11. H. Layachi and Y. Xu, *Performance and Stability Analysis of Sub-stiffening for Mechanical Buckling and Post Buckling: A Selective Study, MATEC Web of Conferences*, **95**, 1004 (2017)
12. T. D. Dang and R. K. Kapania, *Ritz Approach for Buckling Prediction of Cracked-Stiffened Structures, J. Aircr.*, **50**, **3**, 965–974 (2013)
13. D. G. Stamatelos, G. N. Labeas, and K. I. Tserpes, *Analytical calculation of local buckling and post-buckling behavior of isotropic and orthotropic stiffened panels, Thin-Walled Struct.*, **49**, **3**, 422–430 (2011)

MATEC Web of Conferences **114**, 03010 (2017) DOI: 10.1051/matecconf/201711403010
2MAE 2017

Study on Heat Treatment of Aerospace Aluminum Alloy Welded FSW

Luo Chuanhong [1,2,a], Dong Fengbo [3] and Guo Lijie [3]

[1] *Key Laboratory of Hydraulic Machinery Transients (Wuhan University), Ministry of Education, Wuhan, 430072*
[2] *School of Power and Mechanical Engineering, Wuhan University, Wuhan, 430072*
[3] *Shanghai Aerospace Equipments Manufacturer, Shanghai, 200245*

Abstract: The precisely controlled heat treatment was researched to meet the special requirements of the aerospace industry on materials. 2219 aluminum alloy plate was obtained by friction stir welding, and the microstructures and mechanical properties were investigated correspondingly. The causes of the weakened joint strength in friction stir welding were analyzed and summarized. The tensile properties show that the average tensile strength of the samples was increased to 393MPa and the elongation was increased to 12.6% of the base metal. Through the special recrystallization technology with high temperature oscillating heat treatment to hold the fine grain at short time, the ductility can be restored and softening can be eliminated, which will improve the performance in mechanical intensity of aerospace aluminum alloy.

1 Introduction

Aluminum alloy is the main structural material of the equipments in the aerospace industry, which is undergone five generations to promote the development of the aerospace technology. The first was with aging heat treatment for high strength aluminum alloy; The second was with overaging heat treatment for perfect corrosion resistant; the third was based on high purity for high corrosion resistant and high tenacity; The fourth was based on the precise control of multi-scale second phase for super high corrosion resistant and high fatigue resistance of the aluminum alloy; the fifth was with high hardenability for comprehensive performance of aluminum alloy[1]. With the development of the aerospace industry and other materials technology, the aluminum alloy are seriously challenged in the history of performance, such as structure weight loss, high reliability and long life. Moreover, 2XXX and 7XXX aluminum alloy with high strength and high tenacity were mainly applied in aerospace launch vehicle, for example, as shown in figure 1. Friction stir welding (FSW) is a new kind of solid connection technology to solve the traditional problem in welding 2XXX and 7XXX aluminum alloy[2-5], which is now widely used in aerospace structures of the manufacturing process.

[a] Corresponding author: chluo@whu.edu.cn

DOI: 10.1051/matecconf/201711403010

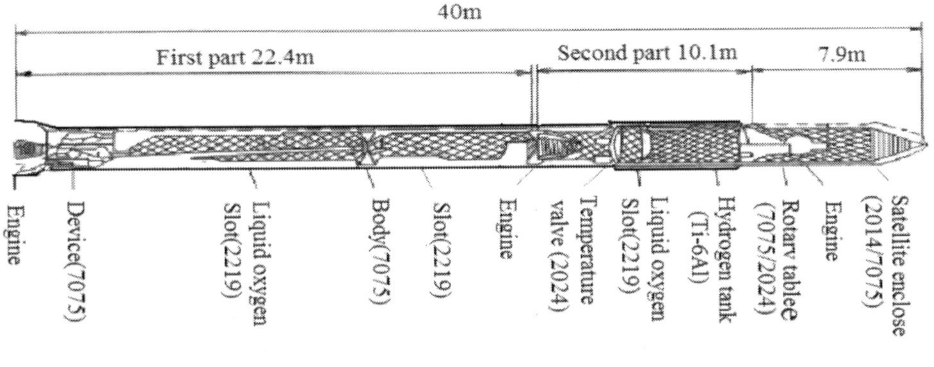

Figure 1 Materials of rocket parts

There are many researches focused on the mechanism and properties of aluminum alloy welded by FSW[6,7]. However the mechanics performance will decline due to material thermal damage and mechanical damage after welding. Generally, the research results and the engineering practices have shown that the ratio of weld metal strength to base metal strength is only about 50%~85%, which apparently makes no full use of the advantages of high strength aluminum alloy[8,9]. It requires precise control by heat treatment to meet the strict requirements of the aerospace industry on materials. After analyzing the joint of 2219 AA weld by FSW, the precipitating of θ phase are argued for the degradation of joint performance, and the heat treatment is put forward that can improve the joint's tensile strength. This research will provide beneficial reference to high strength aluminum alloy used in the aerospace industry.

2 Experimental materials and method

The experimental material is 2219AA plate, whose size is 300×200×6 mm. The stirring needle's diameter is 6 mm, while its shoulder's diameter is 20 mm. The welding method is single butt welding. the post welding heat treatment (PWHT) at high temperature and short time is taken to deal with the weldments, and then an artificial aging at 165℃ × 24h. Two kinds of PWHT for weldments are conventional heat treatment and high temperature oscillation heat treatment. The method of high temperature oscillation heat treatment is as follows: a short stay above the solution temperature about 530℃, and rapidly fall temperature lower than the solution temperature, repeatedly, to realize temperature oscillation around solid solution temperature line of 2219AA. This process can be completed by the two furnaces, it also can be completed by a furnace. After welded or PWHT, the samples are taken along the weld beam and polished, and corroded by mixed acid(1.0%HF +1.5%HCL +2.5%HNO3 +95.0%H2O), and finally observed under OLYMPUS optical microscope, tested tensile by CSS44100 electronic tester, and measured hardness by HXZ 1000 digital tester.

3 Results and discussion

3.1 Morphology and microstructure

The clear boundaries existence on the macroscopic structure of the welded joints, so the welded joints are divided into three zones: the nugget zone (NZ), the thermo mechanical affected zone (TMAZ), and the heat affected zone (HAZ) [10]. The morphology of all regions is shown in fig.2. In the HAZ, the welding thermal cycle causes the original grain coarsening of material, as shown in figure 2a. However, in NZ the grains are broken, and then many small equiaxed structures are formed during grains growth, because the grains undergo a drastic stir and dynamic recrystallization after intense friction. These fine grains have strong strength for whole or the part of compact structure, and have a good metallic luster, when observed under optical microscope, as shown in fig.2b. Moreover, in TMAZ a combined action of strong stirring and welding thermal cycling performs in FSW, resulting in local fragmentations and the adherent growth, where grain size is larger than that in NZ, as shown in Figure 2c.

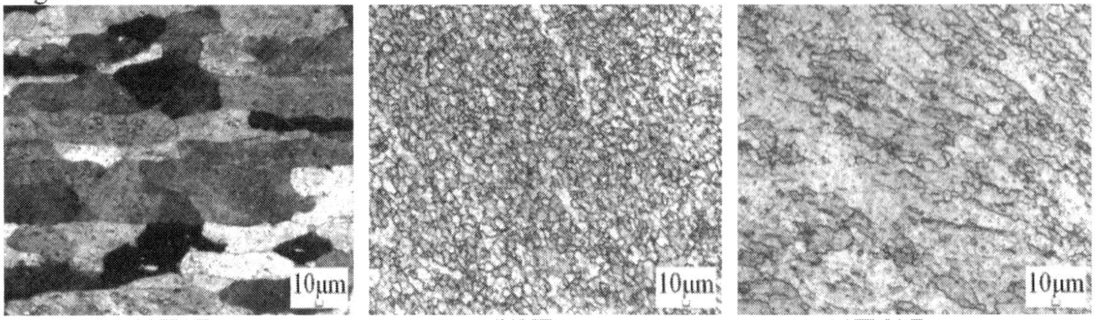

(a)HAZ (b)NZ (c)TMAZ

Figure 2 Microstructures in joint of 2219AA

The normal solid solution heat treatment is that the weldment is maintained at a constant temperature for a long time, but comparatively, a precisely controlling heat treatment of high temperature oscillation is that the weldment is stayed for short-time above the solution temperature, then slightly lower than the solution temperature for short-time to realize heating oscillation near the line of the solid solution temperature. The PWHT has great influence on the microstructure and properties, Microstructures is as shown in Fig.3 and Morphology is as shown in Fig.4.

After normal PWHT, as shown in Fig.3a, it can be seen from the figure that the grains undergo the high temperature recrystallization, the organizations in BM and HAZ remained the same grain size or slightly increased, but those in NZ or TMAZ appear to abnormally grow up. The larger grain size existence at the junction of NZ or TMAZ and the grain boundary is very clear, and the grain growth direction is along the thickness direction of the samples. The research shows that the grain boundary is broken and the organization is formed in NZ under the mechanical action, so no second phase particles on grain boundary, resulting in boundary energy is low. However, much distortion energy and a large number of dislocation is stored in the grains, so the energy threshold of grain growth is reduced. The adjacent grain even with high angle grain boundaries are easy to merge. Therefore, in the process of heat treatment after welding, the abnormal grain growth phenomenon is easy to occur, and the grain size in NZ has a great thermal instability. Traditionally, the abnormal grain growth usually occurs in the conditions that the second phase particles solutes into matrix above solution temperature, such as austenitizing process of the steel, but in aluminum alloy welded by FSW, the abnormal grain growth occurs in the recrystallization process under lower temperature.

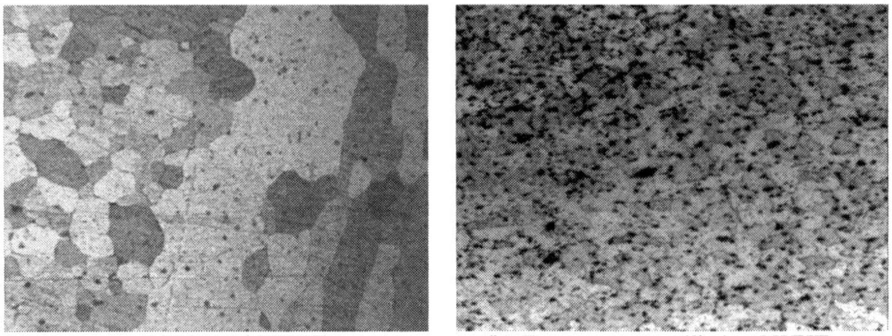
(a) Normal PWHT (b) Temperature oscillation PWHT

Figure 3 Microstructures in joint of 2219AA after PWHT

(a) Welded (b) Normal PWHT (c) Temperature oscillation PWHT

Figure 4 Morphology of FSW joints

3.2 Behavior of θ phase in PWHT

Heat has a great influence on aggregation and growth of the precipitation of θ phase (A12Cu) in the matrix of α (Al) in 2219AA, in which the precipitation is distributed in the matrix to strengthen the material. In the welding process, the precipitation is gradually separated and gathered. It is difficult to form large particles because of the fragmentation of the precipitation. Thus, there are no large particles of the precipitation in NZ. However, in the HAZ, the deformation is small, so the precipitation is easy to gather and grow, and the particle reaches a maximum size of about 15 μm, as shown in the fig.5a. After PWHT, when temperature oscillated around the low limit of solution temperature, the stable precipitates preferentially formed in grain boundary to anchor grain boundary and can restrain the generation of coarse grain effectively. Meanwhile, in HAZ there are some changes of the precipitation in form, but there is no obvious evidence of precipitation in number. Due to the short duration of high temperature, the transferring velocity of copper atoms is slow, so the large precipitation still exists, as shown in the fig.5b. These aggregated particles would reduce the number of precipitation surrounding matrix and weaken the intensity of the heat affected zone.

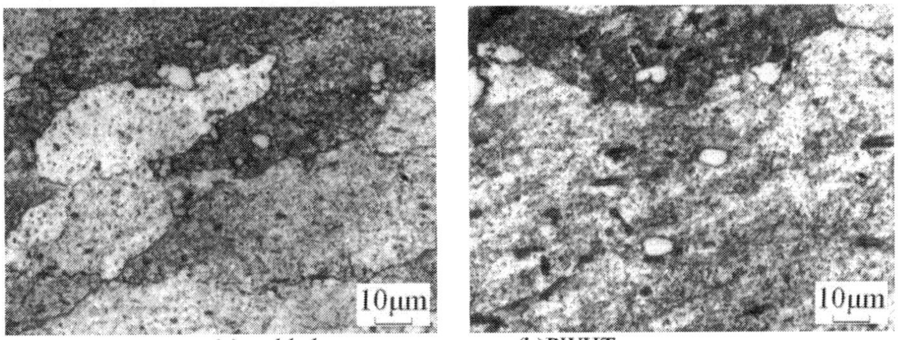

(a) welded (b)PWHT

Figure 5 Distributions of the θ particles in HAZ

3.3 Results of mechanical tensile test

Table 1 shows the test results of mechanical tensile properties of 2219 aluminum alloy. The base material has good plasticity about 23% under the condition of annealing. The plasticity is poorer and the fractures of most samples are located in HAZ when the material is welded by F SW. The plasticity is further reduced but the tensile strength is higher in the condition of solution and aging with normal PWHT. The elongation is about 7.5% and decreased about 50% of welded joints, the average tensile strength is 300MPa, most of the fracture position on the advancing side. After the high temperature solution, the average tensile strength of the samples was increased to 393MPa, the elongation was increased to 12.6%. It means if grain size can be controlled in the heat treatment, the plasticity will be improved. The results show that the grain size is controlled by the high temperature oscillation and the fine grain structure is maintained, thereby the plasticity is improved to a certain extent as well as the tensile strength recovery.

Table 1 Mechanical properties of base metal and joints of 2219AA

Sample number	Ultimate tensile [Mpa]	Elongation [%]	Fracture location	Note
1	150.3	23.1	-	BM
2	156.4	13.4	RS	Welded Joints
3	300.4	7.5	AS	Normal PWHT
4	393.0	12.6	AS	Precisely PWHT

4 Conclusions

This experiment try to improve the tensile strength of joints through heat treatment, which can restore the metal's plasticity, redistribute θ phase, prevent the softening of heat affected zone and eliminate the influence of those inherent weakening factors. It develops the advanced heat treatment technology to improve the comprehensive performance and meet the requirements of the aerospace aluminum alloy.

When temperature oscillated around the low limit of solution temperature, the stable precipitates preferentially formed in grain boundary to anchor grain boundary and can restrain the generation of coarse grain effectively; when the temperature oscillation amplitude is greater than 90°C, the grain size is about 20 μm in diameter. The joint can restore the ductility and eliminate the softening. High temperature oscillatory solution heat treatment can effectively solve the problem of thermal instability in the nugget of friction stir welding.

Acknowledgment

This work was supported by Shanghai Aerospace Science and Technology Innovation Fund under grant no. 2015056.

References

1. C.H. Luo, W.P.Peng, J.Q.Zhang. Transactions of Materials and Heat Treatment.**36**,35(2015)
2. L. E. Murr, G. Liu, J. C. Meclure. J. Mater Sci. Lett., **16**,1801 (1997).
3. F.J.Humphreys, M.Hatherly. Elsevier Science Ltd., Oxford, United Kingdom, (2004).
4. D.K. Lim, T. Shibayanagi,A.P. Gerlich. Materials Science and Engineering A. **507**,194 (2009)
5. I.S.Lee, C.J.Hsu, C.F.Chen, etc. Composites Science and Technology. **77**,693 (2011)
6. L.M.Ke, C.Huang, L.Xing, etc. Journal of Alloys and Compounds. **503**,494 (2010).
7. L.M. Ke, J.L. Pan, L. Xing. Journal of Mechanical Engineering, **45**,89 (2009).

8. C.H. Luo, L.J.Guo, F.B.Dong. Transactions of the China Welding Institution, **37**,90 (2016).
9. D.Y. Wang, J.C. Feng. Transactions of The China Welding Institution, **25**,46 (2004).
10. D.Q. He, H. Deng, P.Z. Zhou. Transactions of the China Welding Institution. **28**,13 (2008).

A New Method on Calibrating PSF of Remote Sensing Space Plane Arraycamera

Wei Liu[1], HuiTing Gao[1], ShiXiang Cao[1],Wei Tan[1] and YunFei Bao[1]
[1]Beijing Institute of Space Mechanics & Electricity, Beijing 100094, China

Abstract: Point spread function (PSF) of an imaging system is an important parameter to a remote sensing imaging system, and it is the important basis for image restoration [1]. The knife-edges method has been widely used to measure PSF because it provides a relatively easy way to get the point spread function. However, the location of the edge points in edge spread function (ESF) is difficult to determine of the knife-edges method, resulting in the inaccuracy of the point spread function. In this paper, we propose a new method to solve the problem of phase influence for calibrating PSF of space plane array camera. This method include: the design of target, the process of calibration and data processing. Then this method was used on calibrating PSF of GF-4 satellite panchromatic camera. The results showed that the new method could determine the PSF correctly.

1 Introduction

PSF, which is derived from physical optics, provides an accurate representation of the real blurs about an imaging system. A variety of methods have been used to calibrate the PSF of a remote sensing imaging system. Some commonly used methods are described as follows:

(1) Impulse input method [2]. Theoretically, the PSF is an impulse input of spatial response of an imaging system. However, in practice this method is very difficult to apply for an imaging system because it is difficult to obtain an impulse input for an imaging system.

(2) Pulse input method [3]. This method is to find a linear object that is a contrast to surrounding objects from the remote sensing image. It also estimates the PSF from its edge response.

(3) Knife-edges method [4, 5, 6]. This method is to search a knife-edge object form the remote sensing image. The profile of the edge is called ESF. The differentiation of the ESF results in a one-dimensional version of the PSF.

(4) Other method [7]. Some researcher brought forward some other methods such as image-derived point spread function method.

The knife-edges method is the most commonly used method to calibrate the PSF of a remote sensing imaging system. It has two major advantages: one is that the target of knife-edge on the ground is easy to design, and the other is that it is easy to understand and compute. But the knife-edges method also possesses some problems. The major problem is it is difficult to determine the location of sampling points in the ESF.

In this paper, we proposed a new method include staggered array target design and process of calibration.

 DOI: 10.1051/matecconf/201711403011

2 Design of staggeret target

To be influenced by many factors such as vibration of the test platform, the phase of single point target is stochastic. To get effective point resource image, we design a staggered array target (Figure 1). This target could assure that there would be one point in a pixel.

There are 100 square points whose widths are equal to pixel width in the target ($e_{i,j}(i = 0,1,\cdots 9,\ j = 0,1,\cdots 9)$). The points are non-light tight, and other area is light-tight in the target. The distance of every point in x and y direction is W.

$$W = (x+0.1)*(\frac{f_{col}}{f_{cam}}*d)$$

$$x > 10 \tag{1}$$

f_{col} is the focus of collimator, the f_{cam} is focus of camera, d is the width of one pixel in camera focal plane device.

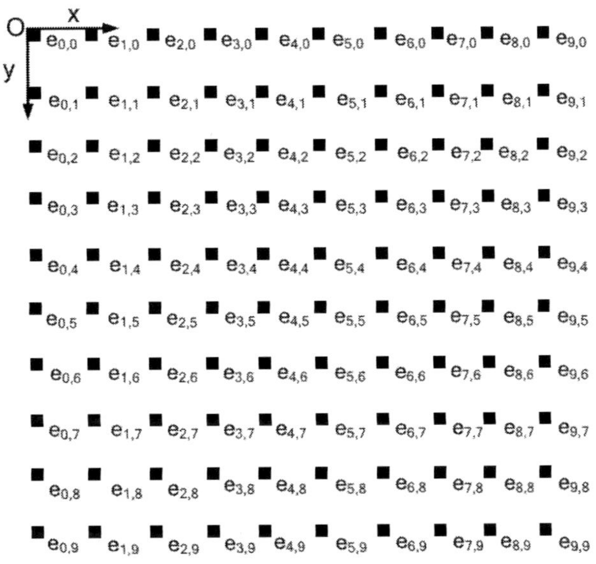

Figure 1. Ponits Target Design

Suppose the phase of point $e_{0,0}$ is $(\varphi x_{0,0}, \varphi y_{0,0})$, then the phase of point $e_{i,j}$ is:

$$\begin{aligned} \varphi x_{i,j} &= \mathrm{mod}(\varphi x_{0,0} + i \cdot 0.1,1) \\ \varphi y_{i,j} &= \mathrm{mod}(\varphi y_{0,0} + j \cdot 0.1,1) \end{aligned} \tag{2}$$

Mod(x,y)is the reminder when x is divided by y. Formula (2) means that the phase difference of neighbouring pontits in x and y direction is 0.1.

This points target design ensures that one point must be in a pixel in every sampling picture [8].

3 Process of calibration data

① Picture sampling

We take the target's picture using the camera which needs to be calibrated.

② Phase template

Suppose the value of DN light is 1, the value of DN dark is zero. When the square point is in one pixel, its DN value is itself. If the square point is not in one point, we can calculate the DN values (Figure 2):

$$\begin{aligned} dn_{i,j} &= DN_{light} * (1 - phase_x)\ (1 - phase_y) \\ dn_{i,j+1} &= DN_{light} * phase_x\ (1 - phase_y) \\ dn_{i+1,j} &= DN_{light} * phase_y\ (1 - phase_x) \\ dn_{i+1,j+1} &= DN_{light} * phase_x * phase \end{aligned} \tag{3}$$

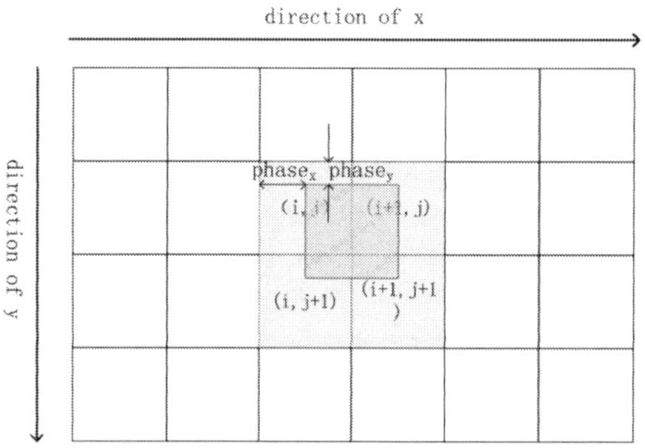

Figure 2. Phase of Point

Form formula (3), we could calculate phase template $D^{l,h}(x, y)$ ($l = 0,1,\cdots 9; h = 0,1\cdots 9$).

③ Phase searching technique and fitting function

After we get the target picture, we cut it to 100 sub-picture $P^{m,n}(x, y)$ $(m = 0,1,\cdots,9; n = 0,1,\cdots,9)$ which is 11*11 pixels and include one point resource. Every brightness point is in location (6, 6).Then we search the phase of the sub-picture:

$$J(m, n) = \sum_x \sum_y P^{m,n}(x, y) \times D^{l,h}(x, y) \tag{4}$$

The value of (m, n) that maximizes $J(m, n)$ is the phase.

$$(m, n) = J^{-1}(\max(J(m, n))) \tag{5}$$

Where $J^{-1}(m, n)$ is the anti-mapping of $J(m, n)$. We could get the phase of every sub-picture according formula (4) and formula (5).

Then, we choose the phase (0, 0) sub-picture, and use the 2 direction Gauss function as the fitting function to get the PSF.

4 Results

GF-4 is the first high resolution satellite of earth synchronous orbit. Its panchromatic camera space resolution of sub-satellite point is 50m. We test GF-4 panchromatic camera image. The test instrumentations include air bearing platform, integrating sphere, designed target, collimator and image acquisition system. We put the target in the focal plane of collimator. The axis of collimator and the optic axis of camera are same. The target is lightened by integrating sphere, then imaging in camera's focal plane device through collimator and optical system of the camera.

Figure 3 is the PSF test picture, the contrast of non-light tight area and light-tight area is 20:1. Figure 4 is different phase sub-picture.

After finding phase (0, 0) sub-picture, we fit 2 direction Gauss function (Figure 5).The PSF test result of GF-4 satellite panchromatic camera is $N(0,0,0.868^2,0.912^2)$. Compared to the laboratory measured MTF of GF-4 satellite panchromatic camera, the error is lower than 5%.

Figure 3. Targetimage of GF-4 panchromatic camera

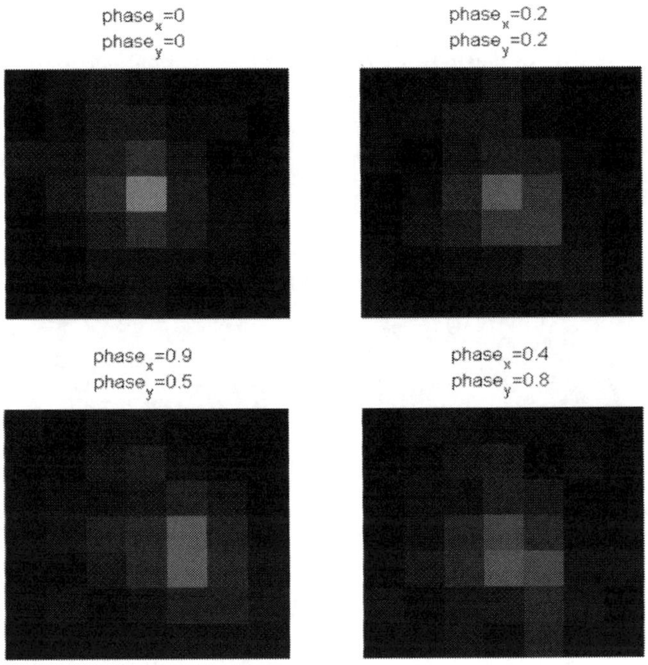

Figure 4. Point image of different phase sub-picture

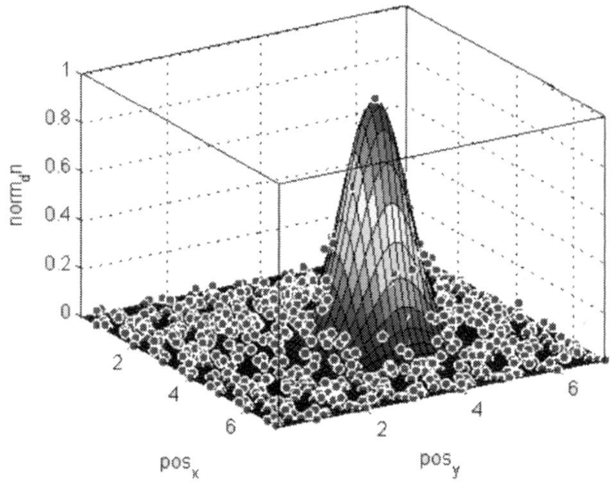

Figure 5. Result of PSF reconstruction

5 Conclution

In this paper, based on the study of the PSF calibration and the analysis of the problem of traditional knife-edges method, a new method on PSF calibration of space plane camera is proposed.

Researchers would evaluate a remote imaging system better and enhance the quality of the reconstruction remote images through the new method they could get PSF/MTF more correctly.

References

1. Liu Wei, Gao Hui-ting. Direct Demodulation Method Applied in Remote Sensing Images of SJ-9A Satellite. Spacecraft recovery & Remote Sensing, VOL. 34, NO. 5 (2013).
2. Zhongxing Zhou, Qingzhen Zhu, Huijuan Zhao,et al. Techniques to Improve the Accuracy of Presampling MTF Measurement in Digital X-ray Imaging Based on Constrained Spline Regression. IEEE Transactions on Biomedical61,4 (2014).
3. Robert Schowengerdt, Charnchai Archwamety. Operational MTF for Landsat Thematic Mapper. SPIE Vol.549 Image Quality: An Overview, 1985.
4. Li Tie-cheng, Tao Xiao-ping, Feng Hua-jun. Modulation Transfer Function Calculation and Image Restoration Based on Slanted-Edge Method. Acta Optica Sinica30,10 (2010).
5. Takacs P. Z., Kotov.I,Frank.J,et al.PSF and MTF Measurement Method for Thick CCD Sensor Characterization.High Energy, Optical,and Infrared Detectors for Astronomy IV,California,USA,7742(2010).
6. Dennis L, Helder, Taeyoung Choi. IKONOS Satellite on Orbit Modulation Transfer Function (MTF)measurement using edge and pulse method. Electrical Engineering Department South Dakota State University(2002).
7. Qiang Guo, Changjun Yang, and Caiying Wei. A New Approach to the On-Orbit Evaluation of Point Spread Function of Thermal Infrared Images With Applications to FY-2 Satellite Products. IEEE TRANSACTIONS ON GEOSCIENCE AND REMOTE SENSING, VOL. 48, NO. 3(2010).
8. Hidekazu Takano, Yoshio Suzuki, Kentaro Uesugi,et al. PSF Measurement of Imaging Detectors with an X-Ray Microbeam4499(2001).

Experimental Study on Thermal Vacuum Environment Sensitivity of Spacecraft Antenna's Typical Failure

Bi Yanqiang[a],WangJing, QieDianfu,Shang Yonghong and Li Xiyuan
Beijing Institute of Spacecraft Environment Engineering, Beijing, China

Abstract: With the development of space applications, spacecraft antenna has become an indispensable part of any space system. The spacecraft antenna affects and constrains the performance and functionality of the entire wireless communication system as well as the entire spacecraft. Spacecraft antenna has to withstand the noise, vibration, shock and acceleration as launched, and weightlessness, high vacuum, radiation, extreme hot and cold alternating space environment on-orbit[1].The influence of different environmental factors on the typical failure modes of spacecraft antenna is different. The environmental adaptability of the spacecraft antenna depends mainly on its structural design, material, process and other factors. In this paper, the influence of different environmental factors on the typical failure modes of the spacecraft antenna is studied. The sensitivity analysis of the typical failure modes of the thermal vacuum environment is verified by experiments, which provides support for the development of the spacecraft antenna.

1 Introduction

Antenna Thermal Vacuum Environment Fault Mode Sensitivity refers to the degree of influence of different thermal environmental factors such as vacuum, high and low temperature, temperature rise and fall rate, and number of thermal cycles on the antenna failure. An environmental factor is likely to cause thermal failure mode of the antenna. This indicates that the factor influence degree is high, and this failure mode is highly sensitive to this environmental factor[2].The research process is based on the results of the failure model study and analyzes the results of changes in environmental factors.

The influence of each environmental factor on each failure mode is analyzed, and the sensitivity of each failure mode to each thermal vacuum environment is divided into three grades: high, medium and low according to the degree of influence. Certain change of environmental factors can induce a fault, and the environmental factors sensitive to the failure is high; If an environmental factor to be combined with other environmental factors induce a failure, the environmental factor is middle sensitivity to the failure; An environmental factor in normal range in any case will not induce a certain type of failure, then determine this environmental factor is low sensitivity to the failure.

Antenna structure in the hot environment is prone to skin delamination, tear, surface accuracy deterioration, gas out, circuit abnormalities and other failures. In this paper, the sensitivity of thermal environment factor to failure mode is analyzed and verified by experiments.

2 Thermal Vacuum Environment Sensitivity Analysis to Typical Fault

[a] Corresponding author: biyq05@mails.tsinghua.edu.cn

DOI: 10.1051/matecconf/201711403012

Thermal vacuum environmental factors include vacuum, high temperature, low temperature, number of cycles, cooling rate and so on. The sensitivity of the degree of vacuum, high and low temperature, and cycle times to typical faults is analyzed separately.

2.1 The vacuum sensitivity

One of the reasons for the occurrence of antenna reflector skin bag is the residual gas inside honeycomb structure and poor ventilation, and there is pressure difference inside and outside the honeycomb under vacuum conditions. Assuming aluminum honeycomb side length L=6mm, high h=15mm as shown in Figure 1, aluminum film thickness d_L=0.03mm, so under low pressure p the pressure difference $\Delta p=p_0-p$ and p_0 is initial pressure inside honeycomb. The stress between the skin and honeycomb is $s=\sqrt{3/4}LDp/d_L$.

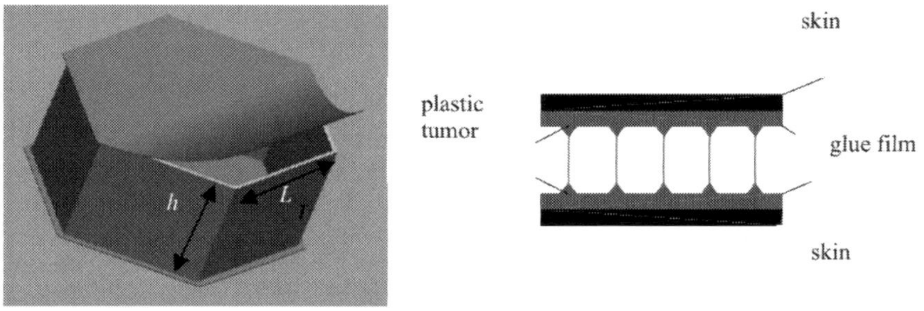

Figure 1. Aluminum honeycomb unit

When the pressure p is 10^{-3}Pa, s=17MPa,close to tensile strength of RTV glue. So vent hole is added in processing technology, the other method is increase the size of tumor appropriately. Figure 2 shows tensile stress of structural adhesive with size changes of tumor under different degree of vacuum, increasing size of tumor can significantly reduce stress inside the structure.

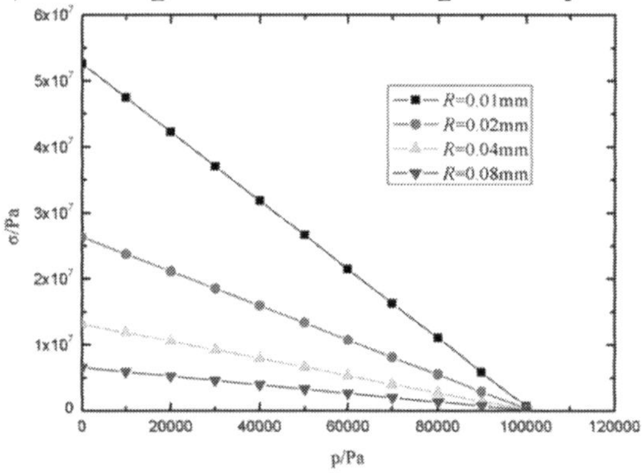

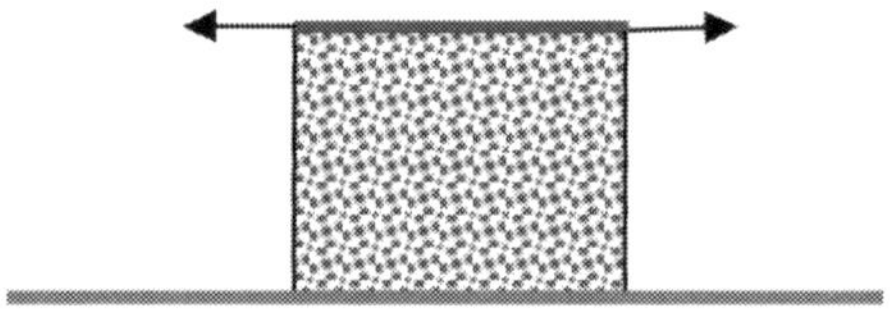

Figure 2. Effect of vacuum on stress under different size **Figure 3.**Vacuum degree on skin

The stress on skin generated by gas in honeycomb is shown in Figure 3. When the maximum vacuum is 0 Pa, the maximum tensile stress of skin is $s = ph/d$, where P is atmospheric pressure, d is thickness of skin and structural adhesive and take 0.5mm. The maximum tensile stress of skin is 3MPa, which is much smaller than the allowable stress of material 295MPa.Therefore, in the case of other environmental conditions unchanged, the vacuum is not enough to cause the antenna skin tear phenomenon, so the sensitivity of vacuum to skin tear is low.

2.2 High, low temperature sensitivity

The cell's temperature is higher, the pressure would be greater. Assuming pressure in honeycomb is

P_0at atmospheric temperature T_0, the pressure in honeycomb would be $P=P_0T/T_0$when temperature rises to T. In sealed honeycomb core, tensile stress of structural adhesive between skin and aluminum foil changes with temperature to $\sigma = \frac{\sqrt{3}L}{2d}\left(\frac{T-T_0}{T_0}\right)P_0$. Figure 4 shows stress changes with temperature under different tumor size.

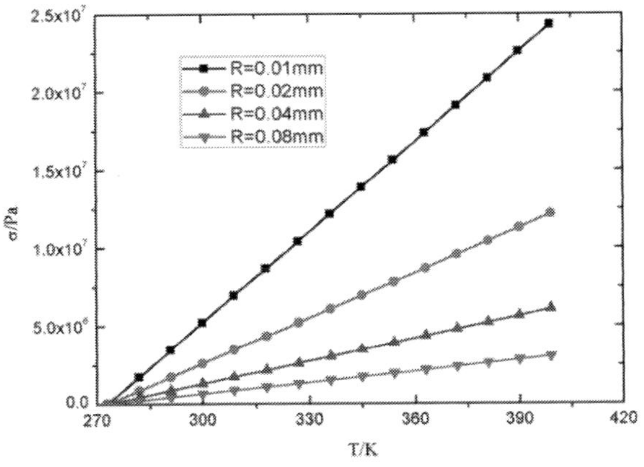

Figure 4. Effects of temperature on stress under different size

In addition after temperature is changed, thermally stress would be exist between inner and outer layers of antenna skin due to difference of thermal expansion. When thermal stress of outer layer is greater than allowable stress, skin will be damaged. The physical model of tear failure mode is shown as Figure 5.In order to analyze rationality and convenience, following assumptions were made for analysis model. 1)Initial temperature is ambient atmospheric temperature, and no internal stress;2)Ignoring the effect of inner layer gas-out of the antenna on tearing of the skin;3)The curvature of the antenna skin is zero; 4) Temperature of each layers is the same.

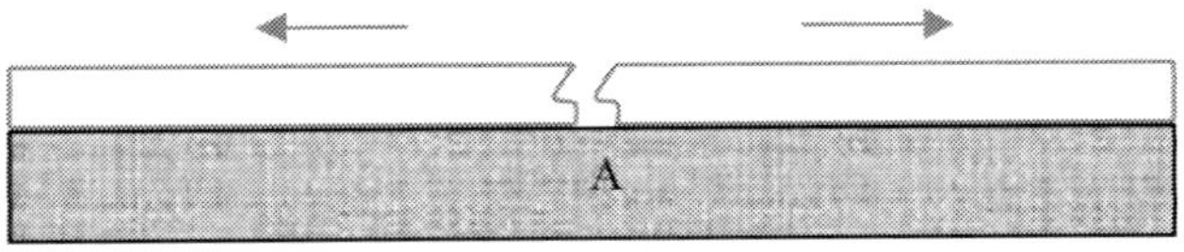

Figure 5. Skin tear analysis physical model

The mathematical model was established based on the previous assumptions and physical model, and simplified as shown below,

$$\begin{cases} \Delta L_A = \beta_A \cdot (T-T_0), \Delta L_B = \beta_B \cdot (T-T_0) \\ \delta = (\Delta L_A - \Delta L_B)/2, \sigma_T = \delta \cdot E \end{cases} \quad (1)$$

So $\sigma_T = [\beta_A(T-T_0) - \beta_B(T-T_0)]E/2$,Where$\beta_A$ and β_B are the thermal expansion coefficients of layer A and B respectively. σ_T is the stress of layer B material. When the stress of B layer is greater than its allowable stress, tearing phenomenon will occur. Through analyze of effect of the temperature on the skin tear, the higher the temperature the higher the tensile stress of the skin, and the temperature continues to rise there is a tendency of tear on skin.

Therefore, high temperature is sensitive and low temperature is insensitive to skin tear failure mode.

2.3 The number of cycles sensitivity

In actual flight of spacecraft, material and structure would be suffered thousands of times alternating stress[3]. Ground thermal cycles can whether expose fatigue failure of antenna skin was analyzed.

If structure is absence of defects, fatigue failure is mainly related to stress amplitude, stress ratio, cycle period and number of cycles. In thermal test corresponding to test conditions are: high and low temperature difference, high and low temperature retention time, number of cycles. The maximum stress at which the infinite stress cycle occurs without breaking occurs is the fatigue limit. The antenna design should meet design requirements, which the stress is lower than fatigue limit. The relationship between the maximum stress and the number of cycles at which the failure occurs can be expressed as the *S-N* curve. Material *S-N* relationship expression is generally $S^mN=C$. Where m and C are the material-related constants[4].General rubber materials' m and Care taken 2 and 10^{12}.Figure 6 shows the relationship between the maximum stress and number of cycles of glue material.

The design will be checked whether fatigue limit is met. That is, stress of material is below the straight line in Figure 6.As can be seen from Figure6, even if material stress is subjected to a 10-fold increase, it is necessary to undergo nearly a thousand thermal cycles to expose fatigue failure. Average number of cycles of the antenna thermal test is only 10 times, and far from the number to exposed fatigue failure. Thus, the number of cycles is less sensitive to the failure.

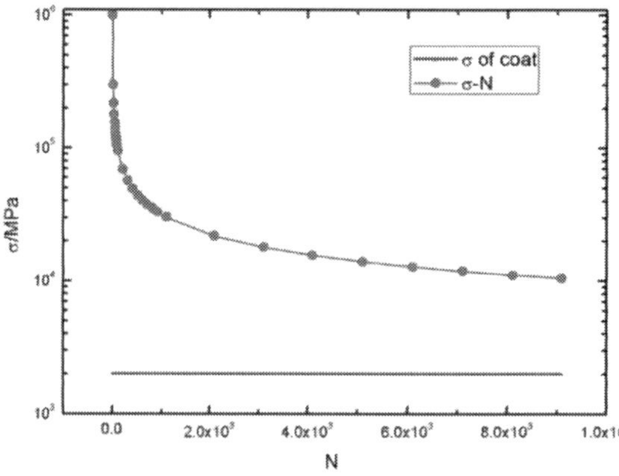

Figure 6. The relationship between the maximum stress value and the number of cycles

3 Experimental verification

3.1 Experimental design

According to the level of sensitivity and experimental conditions, the experimental and experimental environment were designed[5,6]. The experiment was conducted through the following stages:

1) Vacuum section, the specimen to withstand only vacuum effect;
2) Low temperature section, the specimen to withstand vacuum and low temperature;
3) Narrow temperature range, that is, small high temperature and low temperature;
4) Very low temperature, the specimen to withstand very low temperature;
5) High temperature, the temperature from 80℃, heated interval 20℃, up to 160℃.

3.2 Experimental results

The experimental process obtained the antenna temperature, container vacuum degree, respectively is shown in Figure 7 and Figure 8. The antenna specimen before experiment is shown as Figure 9 and after experiment as Figure 10.During experiment the antenna's skin was damaged.

 DOI: 10.1051/matecconf/201711403012

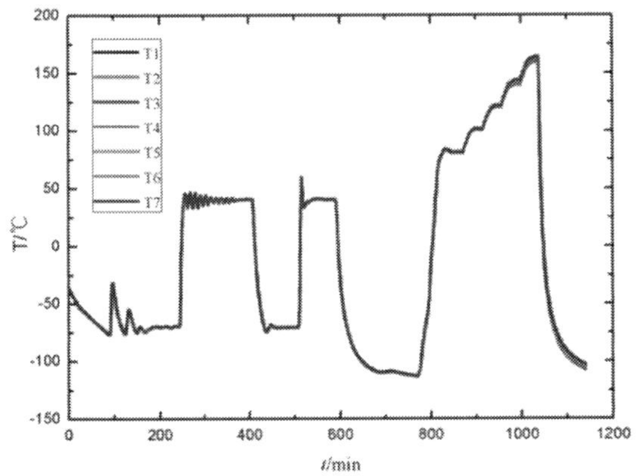

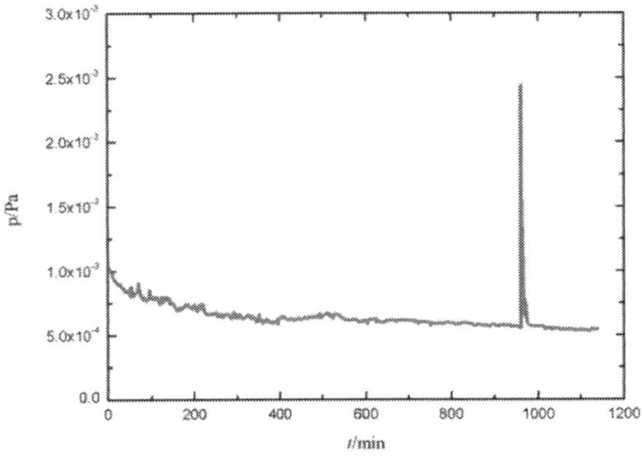

Figure 7. The temperature changes with time **Figure 8.**Vacuum degree changes with time

During the process of experiment, only the temperature reached 133°C, the vacuum degree curve appeared a peak jump, indicating that at this time there was a gas release in container. The specimen was destruction at this time. The results of sensitivity analysis show that high temperature and vacuum have higher influence on the delamination of antenna skin. In the process of verification, the combination of vacuum and high temperature conditions leads to the occurrence of fault, which verifies the correctness of fault analysis and provides support for the following test conditions design.

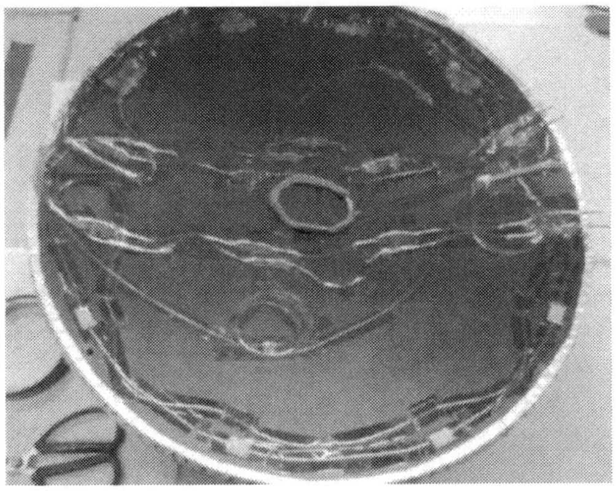

Figure 9. Antenna specimen before experiment **Figure 10.**Antenna specimen after experiment

4 Conclusion

Through the analysis, it is found that the spacecraft antenna has high sensitivity to high temperature and vacuum environment, and has low sensitivity to low temperature and cycle times. Through a reasonable experimental design, low temperature, cycle heating and other stress were put on the antenna firstly, no damage was found. Then high temperature, vacuum and other stress were put on the antenna, vacuum degree inside the test chamber decreased because of the antenna's gas-out. The antenna was damaged. Sensitivity of the antenna to the high temperature vacuum environment is verified by experiments. Based on the research results, it is possible to design the antenna thermal vacuum test reasonably, evaluate the reliability of the antenna design and ensure the performance of the spacecraft.

References

1. Ye Yunshang. Spacecraft Antenna Beijing: China Science and Technology Press, 2008.

2. Ghalaty, N, F.; Aysu,A.; Schaumont, P. Analyzing and eliminating the causes of fault sensitivity analysis. Design, Automation an Test in Europe Conference and Exhibition(DATE), 2014.
3. Paden, R. S. Choosing thermal cyclic screening requirements to achieve desired test effectiveness. Proceedings of the 21st Aerospace Testing Seminar, Manhattan Beach, CA, 2003.
4. Fan Qinshan, Yin Yajun, Tan Jinglin. Mechanics of Materials. Beijing: High Education Press, 2014.
5. Huang Bencheng, Ma Youli. Space Environment Test Technology of Spacecraft. Beijing: National Defense Industry Press, 2002.
6. Welch, J. W. Unit thermal vacuum test consideration. Spacecraft Thermal Control Technology Workshop, The Aerospace Corporation, Feb. 2001.

MATEC Web of Conferences **114**, 03013 (2017) DOI: 10.1051/matecconf/201711403013
2MAE 2017

Aircraft Route Recovery Based on An Improved GRASP Method

Yang He [1,a], Jinfu Zhu [1], Qiang Gao [1], and Bo Zhu[2]

[1] *College of Civil Aviation, Nanjing University of Aeronautics and Astronautics, Nanjing 210016, China*
[2] *Post-doctoral Station, Shenzhen Airlines, Shenzhen 518128, China*

Abstract. Aircrafts maintenance, temporary airport closures are common factors that disrupt normal flight schedule. The aircraft route recovery aims to recover original schedules by some strategies, including flights swaps, and cancellations, which is a NP-hard problem. This paper proposes an improved heuristic procedure based on Greedy Random Adaptive Search Procedure (GRASP) to solve this problem. The effectiveness and high global optimization capability of the heuristic is illustrated through experiments based on large-scale problems. Compared to the original one, it is shown that the improved procedure can find feasible flight recovered schedules with lower cost in a short time.

1 Introduction

Over the course of normal airline operations, some unavoidable disturbances can be caused by unplanned factors like technical aircraft failures, inclement weather conditions, crew absences, etc. These could lead to the delay and/or cancellation of some flights. As a consequence, an aircraft's routing can become disconnected.

The airline controller in the operations control centre can repair disconnected aircraft routings using a variety of options, such as delaying all subsequent flights, cancelling a set of consecutive flights or swapping the affected flights to other aircraft.

However, the scale of disruption can easily grow to the extent that it is impossible for a human to fulfill this task with a viable outcome in a timely manner.

Instead, an intelligent algorithm is required to identify disruptions and create options to help the controller recover from the disruption and bring operations back to normal. The ideal solution should identify a variety of recovery options and choose the best one based on its revenue and cost to the airline in a reasonable period of computational time.

Teodorovic et al. [1] are pioneers to solve irregular flight recovery problem. With one type of fleet consisting of three aircraft executing eight scheduled flights, the branch and bound method is adopted to recover schedules. Teodorovic et al. [2] improve above problems, in that they put forward a dynamic programming greedy heuristic algorithm. To minimize lost passenger revenue and subject to constraints of curfew and crew availability, Clarke et al. [3] proposes an aircraft recovery model that can be promoted to a multi-fleet, with demonstrating calculation with forty-nine aircrafts and more than two hundred schedules. A column generation scheme is developed by Eggenberg et al. [4] to

[a] Corresponding author : heyang@nuaa.edu.cn

solve it. Tang et al. [5] combines GRASP algorithm with simulated annealing, and reduce the possibility of falling into local optimum. Yan et al. [6] also research flight recovery problem under the circumstances of airport closures. On basis of that, Thengvall et al. [7, 8] solves the problem of multi-fleet and long closures of hubs by proposing a model on time-space network, in which strategies of flight delay, cancellation and swap are brought together. Although applied by ILOG CPLEX, it can only provide one solution, while controllers prefer to choose from serval solution in actual scheduling. Petersen et al. [9] are known as the first scholars that put forth the integrated recovery formulation. Similarly, Sinclair et al. [10] design a large neighbourhood search heuristic algorithm to solve the problem above. Additionally, dynamic framework of the aircraft recovery is established by Vos et al. [11]. Optimized by an efficient aircraft selection algorithm, the framework is established on a linear-programming model that is able to trace aircraft status on parallel time-space networks. And to deal with the stochastic aircraft recovery problem, Mou et al. [12] solve it through building an uncertain programming model based on Hungarian algorithm. Furthermore, Arias et al. [13] combine simulation and optimization techniques to cope with it.

Although linear programming is widely applicable, its poor flexibility leads to the run time being exponential. Considering that the schedule recovery problem is NP-hard, existing experiment results demonstrate that some available methods can solve relatively small problems but difficult to promote to practical flight operation and management. In contrast, heuristics can generate feasible solutions of high quality in limited time and reasonable alternatives can also be offered.

Argüello et al. [14] proposes the greedy randomized adaptive search procedure (GRASP) to solve the flight schedule recovery problem when aircraft are in shortage caused by failure. This method proves its superiority in irregular flight recovery problems that involves sixteen aircraft, forty-two flights and thirteen airports. However, in real world data, there is still much scope for improvement in convergence effect. In this paper, we present an improved GRASP for irregular flight recovery with better convergence and more adaptation. Hence, confronted with two types of disruptions, an improved GRASP for aircraft route recovery is proposed in this paper, and various scenarios is considered in computer experiments to verify the reliability.

2 Problem Description

2.1 Problem statement

The recovery problem considers the scheduled flights that are assigned to aircrafts during a fixed time, generally about three days regarded as recovery period. When disruptions occur, the operational control centre needs to collect information about airport status, maintenance schedules, etc. This paper considers two types of disruptions, airport closure and unplanned maintenance.

Due to any number of unforeseen reasons, an airport may be required to close during a given period of time. During this time, flights are not allowed to departure or arrive at this airport. Apart from airport closures mentioned above, in a real scenario, if an aircraft requires unplanned maintenance due to a faulty component, it is required to remain at the airport for a given period of time. Because the maintenance is unplanned, it may overlap with some downline flights for which it has been assigned. Therefore, the aircraft cannot operate any flights during the unplanned maintenance. The aircraft must be at the specified airport at the beginning of the maintenance period, which means the maintenance can begin immediately upon the aircraft's arrival at the airport, and it can only take off again when the maintenance has been completed.

Apart from the assumptions aforementioned, the operational hard constraints below should be taken into consideration when rerouting aircraft: (i) At any time, at most one task (flight or maintenance) can be assigned to an aircraft; (ii) Each aircraft has a start available time and end available time. The start time of the first task that is assigned to this aircraft must be later than or equal to the aircraft start time. Similarly, the end time of the last task must be earlier than or equal to the aircraft end time; (iii)A fixed 30 minutes idle time (also known as turn time) is required for an aircraft between any two consecutive flight tasks; (iv) Each aircraft has a start available airport from the input of the problem. The aircraft start airport must match its first task's start airport. Any two consecutive tasks (flight or maintenance) that are assigned to an aircraft should be connected. That means the airport where the first task is completed should be the same as the airport where the second task begins; (v) If there is any airport closure, flights are not allowed to departure or arrive at this airport.

Apart from that, there are also some soft constraints: (i) The number of aircraft that are ending at a particular airport should equal to that in the original input. For example, if there are 2 aircrafts at PEK at the recovery end time, it also requires 2 aircraft at PEK at the recovery end time in the solution if terminal airport balance is satisfied. However, this constraint is a soft requirement. If there is the terminal airport balance is violated, a cost penalty will be incurred. (ii) The terminal aircraft positioning for an aircraft requires that the aircraft should be at the same airport as its end available airport in the solution. If the terminal aircraft positioning is violated, there will be a slight cost.

2.2 Recovery policy

During the recovery period, different recovery operation can be applied to recover from the disruption as soon as possible and avoid delay propagation. In practice, four recovery options, including flight delay, flight swap, flight cancellation and maintenance cancellation, are allowed. The four options are described in details below. (i) In a recovery situation, this flight can be delayed for a certain time. During the actual operation, there will be a global delay time threshold of 180 minutes, meaning that flights can be delayed any time between 0 to 180 minutes. The shorter the total delay time is, the better the solution quality will be; (ii) Generally, a scheduled flight is assigned to an aircraft, which is identified by its tail number. During the recovery process, the airline can assign the scheduled flight to any aircraft among a given list of aircraft. If the flight is assigned to an aircraft whose tail number is different to the original assigned aircraft, we call this is a flight swap. The lower the number of flight swaps, the better the solution quality will be; (iii) If in no way the flight is able to be assigned an aircraft, it is cancelled. The lower the number of cancelled flights is, the better the solution quality will be; (iv) If in no way the maintenance is able to be assigned the original planned aircraft, this maintenance is cancelled. The lower the number of cancelled maintenance is, the better the solution quality will be. It is noted that a maintenance cannot be swapped to another aircraft.

In GRASP, in every iteration, it constructs neighborhood alternatives by selecting a pair of aircraft randomly. Besides, it does not involve the consideration of maintenance and airport closures. The complexity in original GRASP algorithm may cause a mountain of operation of choosing and cannot get the optimal solution in specific time.

Compared to original GRASP, in the new algorithm, when the pair is neither disrupted, the routes will be retained. In other words, in every iteration, it constructs neighborhood by selecting a pair of route, one of the route comes from disrupted aircraft routes, the other one comes from the reminder of routes which contains normal, disrupted routes and fictitious flight. The main reason is that all disrupted routes could be improved through swapping a segment of route with the rest ones, with original normal routes remaining the same. Additionally, the new algorithm can deal with situation involving temporary airport closures by delaying, swapping or cancelling relevant flights.

3 Improved GRASP for Flight Recovery

To solve the recovery problem, a heuristic algorithm that integrates the superiority of GRASP and simulated annealing algorithm is proposed. The improved GRASP algorithm answers three key issues: how to construct initial feasible solution, how to construct neighboring solution and how to search globally optimal solution.

3.1 Construction of initial feasible solution

When flight schedule is disturbed by some unplanned events, initial feasible solution can be obtained by simply postponing subsequent flights. Moreover, temporally cancelling the most severely affected flights is also a reasonable method to make the schedule executable. It is important to note that subsequent flights should be removed from the route if one aircraft's available time beyond its first flight departure time, until the aircraft is available. The main process of constructing initial feasible solution is represented as follows.

Step 1. Create initial incumbent solution, and define aircraft route according to flight schedules.

Step 2. Check and deal with the disconnection between start available airport and the departure airport of the first flight, as well as the space disconnection on an aircraft routing.

Step 3. Reassign the departure and arrival time of every aircraft route considering the aircraft start available time, minimum connection time and the relative airports closures

Step 4. Generate feasible solution, which deals with the flights that delayed longer than max delay time and beyond the end available time of aircraft.

Step 5. If the new feasible solution generated in Step 4 avoids constraints on aircraft start available time, connection time and airports closures, then proceed to Step 3, otherwise proceed to Step 6.

Step 6. The construction of initial feasible solution stops and output feasible solution.

3.2 Construction of neighbouring solutions

Route pairs are adopted to construct neighborhood solutions through part of disrupted routes being replaced by other disrupted routes or newly generated route, keeping undisrupted pairs unchanged. There are three main types of methods to generate neighbouring solution: route augmentation, route exchange and route cancellation, which can be divided into seven specific operations. Examples of the methods to generating neighborhood solution are shown in Table 1 and Figure 1.

Table 1. Routing pair sample.

Route	Flight	Departure airport	Arrival airport
1	11	PEK	TAO
1	12	TAO	WUH
1	13	WUH	SHA
1	14	SHA	NKG
1	15	NKG	SHA
2	21	SHA	CKG
2	22	CKG	TAO
2	23	TAO	DLC
2	24	DLC	SHA
2	25	SHA	XMN

(1) route augmentation: It removes one flight or a sequence of several flights from one route source and places these flights in the other target route. Thus, the target route is augmented by the flights removed from the route, which can be implemented by three different operations: (i) pre_change: A circuit that originates and terminates at the same airport in the source route can be placed in front of the target route; For example, flight 14 and 15 in Route1 make a circuit, and put them in beginning of Route2 (as shown in Fig.2 Route1/2 a); (ii) middle_change: A circuit mentioned above in the source route can be placed in the middle of the target route when its starting and ending airport intersects a point in the route; An example of middle_change involves removing the circuit from route1 and relocating them in front of flight 24 in route2 (as shown in Fig.2 Route1/2 b); (iii) tail_change: It removes a sequence of flight, which origins the same airport to target route's termination airport, from the source route and then appends it to the end of the target route, and causes exchange of termination stations. An example of tail_change involves removing the flight 15 from route1 and putting them at the end of Route2 (as shown in Fig.2 Route1/2 c).

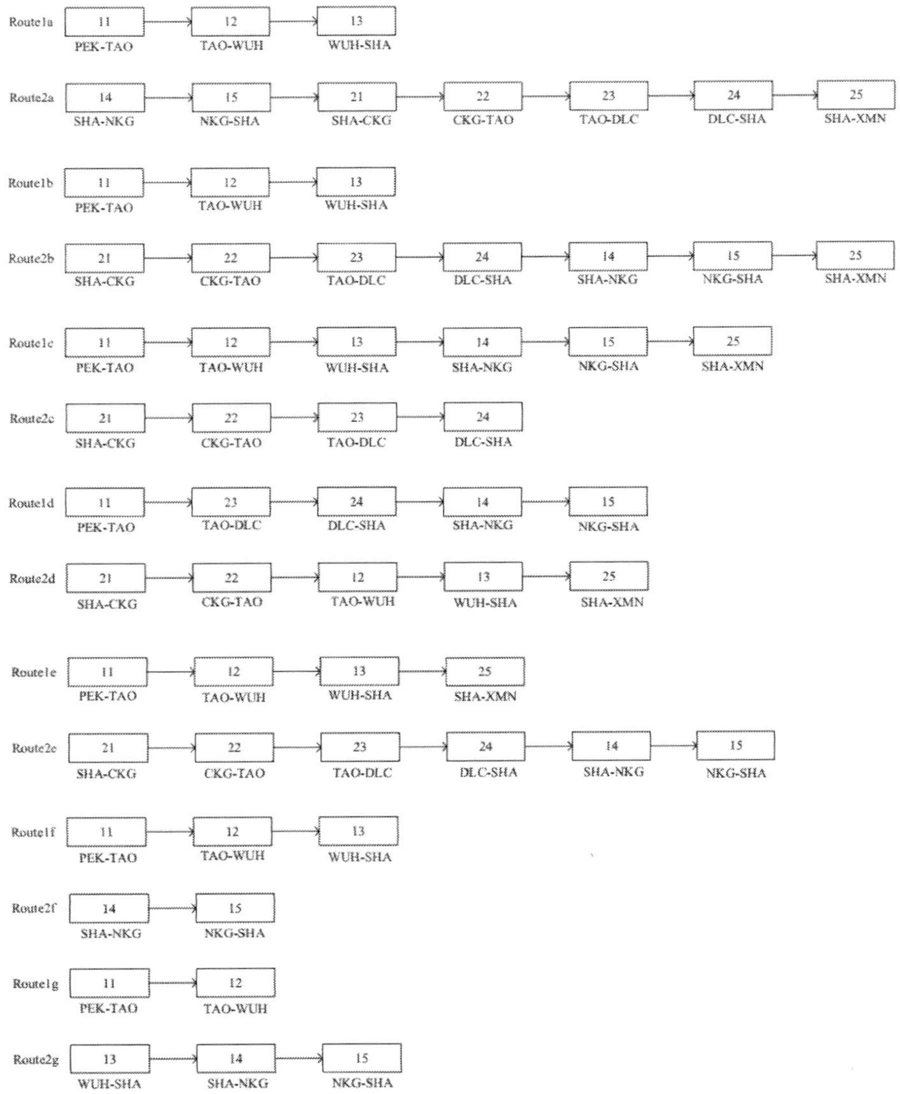

Figure 1. Sample of routing pairs and neighbours.

(2) route exchange: The exchange of a pair of flight sequences is conduct between source route and target route. (i) d_same: The sequence of flights should have the same origination airport, with the termination airports swapped. For example, flight23 and flight24 in Route2d substitute for flight12 and flight13 in Route1d; (ii) d_a_same: The sequence of flights should have the same origination airport, as well as the same endpoint. An example of that is flight14 and flight15 in Route1d substituting for flight25 in Route2d.

(3) route cancellation: It removes a sequence of flights from the source route and then puts it into a newly cancellation route (virtual route). (i) cancel_circuit: It removes a circuit from the route and drops the circuit in a virtual route. As shown in Fig.2 Route1/2 f, it cancels the circuit of flight 14 and flight15, and put them in virtual route (Route2f); (ii) cancel_tail: It removes a sequence of flights from start point to termination airport, as long as the start point differs from the termination airport. For example, flight 13, 14, and 15 do not form a circuit, so they can be cancelled and placed in virtual route (Route2g).

3.3 Searching strategy

After all feasible neighbouring solutions of each route pair are constructed, optimal neighbouring solutions will be chosen and deposited in the restricted candidate list (RCL) and the bad restricted candidate list (BRCL) respectively. The RCL is prepared for cost-reduced neighbours, compare to BRCL which contains cost-added neighbours. However, the original GRASP algorithm only go on along the cost-reduced direction. Therefore, to achieve the global optimal solution, accepting a temporary degraded solution is sustainable in a sense. To some extent, it can get an optimal solution rather than a local optimal solution within a certain range.

In the following, we introduce the procedures of the improved GRASP algorithm, and then the algorithm is explained step by step in flow chart for irregular flight recovery in Figure 2.

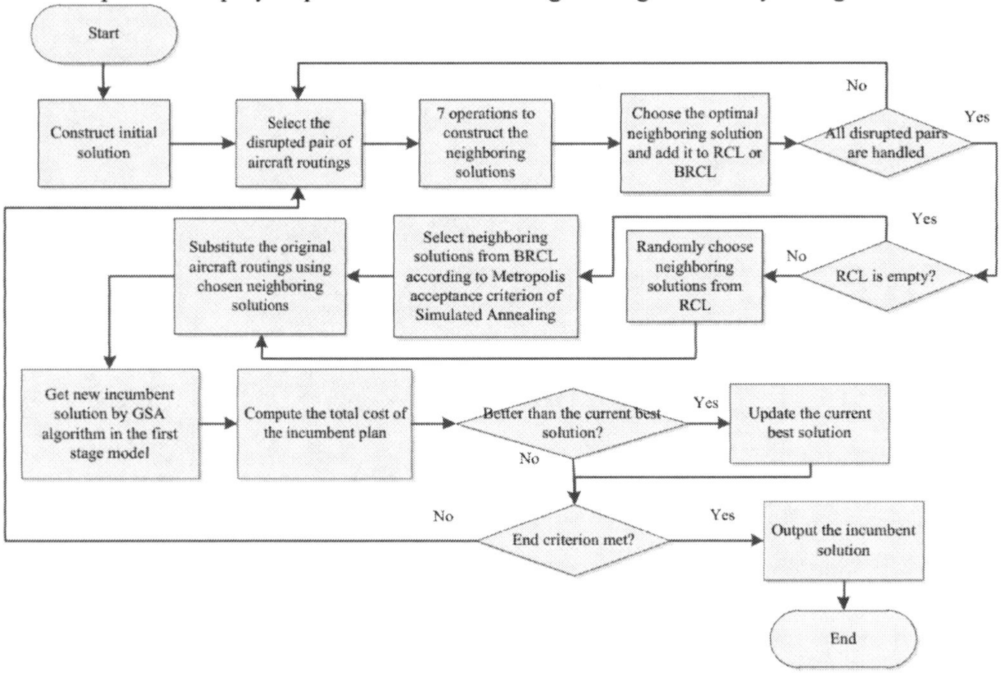

Figure 2. Process of improved GRASP algorithm in recovery problem.

Step 1. Calculate the changed cost incurred by neighbours and determine whether it is cost –saving or cost-adding, and add the chosen neighbours in RCL or BRCL.

Step 2. If RCL is not empty, choose neighbours as much as possible to improve the current solution. If RCL is empty, randomly choose one neighbour from BRCL according to Metropolis acceptance criterion of simulated annealing algorithm.

Step 3. If the end criterion in simulated annealing algorithm is met, output the incumbent solution, otherwise, proceed to Step 4.

Step 4. Substitute the counterparts in the current solution using the chosen neighbours.

Step 5. Calculate the objective of the current solution and proceed to Step 1.

4 Computer Experiment

In order to evaluate this method, several types of scenarios are generated. All computations were performed on a ThinkPad T460s with Intel i7-6600 CPU and 7.4 GB of RAM. The improved GRASP algorithm was coded in C++ language on Microsoft Visual Studio, with calculating times limited to 30min.

The objective of this problem is to minimize the total weighted cost. The total weighted cost can be calculated by the number of cancelled flights, the number of terminal airport balance violated, the number of terminal aircraft positioning violated, the total flight delay time and the number of flight swaps multiplied by the weights, respectively.

The list of weights is a set of input parameters. An example of this list is given in Table 2 below.

Table 2. Weighted cost sample.

	Cancelled maintenance	Cancelled flights	Terminal airport balance violated	Terminal aircraft positioning violated	Total flight delay time
Cost	1000	800	500	10	1

Table 3 shows the solution with 4 scenarios. The first row is the number of aircrafts; the second row is the number of flights, the third row is the number of airport closures, and the last row is the number of maintenance.

Initial feasible solution rows show the delayed and cancelled flights without interference. This is divided into six rows. The first row is the number of cancelled flights, the second row is the number of cancelled maintenance, the third row is the number of terminal airport balance violated, the forth row is number of terminal airport positioning violated, the fifth row is total flight delay in minutes, and the last row is cost calculated by objective function.

The GRASP row is the solution obtained using the improved GRASP algorithm. This row is divided into eight rows. The first row is the number of cancelled flights, the second row is the number of cancelled maintenance, the third row is the number of terminal airport balance violated, the forth row is number of terminal airport positioning violated, the fifth row is total flight delay in minutes, the sixth row is the number of total flight swaps, the seventh row is the consuming CPU time (minute), and the last row is cost calculated by objective function.

Table 3. Solution comparison.

		Scenario1	Scenario2	Scenario3	Scenario4
#AIRCRAFT		12	44	85	16
#FLIGHTS		95	638	417	59
#AIRPORT CLOSURE		1	0	1	0
# MAINTENANCE		0	29	25	23
Initial feasible solution	cancelled flights	0	0	0	5
	cancelled maintenance	0	5	0	4
	terminal airport balance violated	0	0	0	2
	terminal airport positioning violated	0	0	0	1
	total flight delay in minutes	6600	0	7680	0
	cost	6600	5000	7680	22910
Solution by GRASP	cancelled flights	2	2	4	2
	cancelled maintenance	0	0	0	0
	terminal airport balance violated	0	0	0	0
	terminal airport positioning violated	0	0	2	0
	total flight delay in minutes	275	0	345	0
	total flight swaps	14	10	37	1
	consuming CPU time	5	30	30	0.5

	cost	4364	1100	4817	1010

5 Conclusion

In this paper, we propose an improved GRASP method to solve the aircraft routing recovery problem when flight schedules confronted by disruptions. An initial feasible solution is first obtained by postponing or temporally cancelling subsequent flights. Then, neighbourhood solutions are constructed through parts of disrupted routes being substituted by other disrupted routes or newly generated route. Three neighbouring solutions generation operations are applied to aircraft route pairs and single aircraft routes to enumerate feasible neighbour solutions. The optimal neighbour solutions are deposited in a restricted candidate list (RCL), compared to inferior neighbour deposited in a bad restricted candidate list (BRCL). This procedure is continuous before the end criteria is met.

Our contribution is to demonstrate that through the original GRASP algorithm to solve aircraft route recovery problem, combined with simulated annealing algorithm, there are three main points of improvement, the first is its capacity to reserve the inferior neighbourhood to BRCL instead of stopping the search process as original GRASP when no better solution exists currently. Another is limiting aircraft route pairs exchange in itineration between disrupted routes contributes to improvement of efficiency, which could be applied to solve large-scale programing problem. Besides, we take the consideration of aircraft availability, maintenance and airport closures into this method, the weights of which are alterable so that the method can adapt to the different situation and preferences in airlines.

Acknowledgement

This work is supported by "the Fundamental Research Funds for the Central Universities", No.NZ2016109.

References

1. D. Teodorovic, S. Guberinic, E.J.O.R **15**, 178 (1984)
2. D. Teodorovic, G. Stojkovic, TRANSPORT. PLAN. TECHN **14**, 273 (1990)
3. D. Clarke, *The airline schedule recovery problem* (Massachusetts Institute of Technology, 1997)
4. N. Eggenberg, M. Salani, andM. Bierlaire, COMPUT OPER RES **37**, 1014 (2010)
5. Wei. X. T., Q. Gao, J.F. Zhu, Forecasting **29**, 66 (2010)
6. Tu. Y. Yan. S., E.J.O.R **103**,155 (1997)
7. J.F Bard. Thengvall B.G., Yu G., TRANSPORT RES A-POL **35**, 289 (2001)
8. J.F Bard. Thengvall B.G., Yu G., IIE TRANSACTIONS **32**, 181 (2000)
9. J. D. Petersen, G. S lveling, J.P. Clarke, E. L. Johnson, S.Shebalov., TRANSPORT SCI **46**, 482 (2012)
10. K. Sinclair, J.-F. Cordeau, and G. Laporte, E.J.O.R **233**, 234 (2014)
11. H-M. M. Vos, B. F. Santos, T. Omondi, TRPRO **10**, 931 (2015)
12. D. Mou, W. Zhao, J APPL MATH **2**, 485 (2013)
13. P. Arias, D. Guimarans, M. M. Mota, G. Boosten, *Proceedings of the 8th EUROSIM Congress on Modelling and Simulation* 265 (2013)
14. M.F. Argüello, J.F. Bard, Yu G, J COMB OPTIM **5**, 211 (1997)

MATEC Web of Conferences **114**, 03014 (2017) DOI: 10.1051/matecconf/201711403014
2MAE 2017

The Research of Technologies of High Precision & Long Life Space Tracking Mechanism

CAO Qian[a], LI Ming and XU Pengmei

Beijing Institute of Space Mechanics & Electricity, 99 Zhongguancun East Road, Haidian District, Beijing, China

Abstract. In the field of space remote sensing technology, this paper researches on the technologies of high precision and long life tracking mechanism. The especial requirements in space for tracking mechanism are presented firstly, and the technologies are studied. Then a tracking mechanism bearing by two flexural pivots and driven by a voice coil motor is designed. Flexural pivot is ideally suited for limited angular rotation. Voice coil motor has a simple structure, without any structure for reversing, and is easy to be controlled. The mechanism's tracking accuracy can reach to 30″ , and the life of this tracking mechanism can reach to 3.6E+6 which are all proved by some test in the end.

1 Introduction

For space instrument the demands of light mass, compaction, low power cost, high precision, long life and high reliability are all indispensable. In most instruments tracking mirror is usually needed to swing regularly within a certain range of angles. Then the space instruments can track targets stably when it is working in orbit. Also the life requirement of space instrument is usually 5 years, or even 8 years nowadays. So a high precision and long life tracking mechanism is seen as necessary component for most space optical instruments.

2 Technologies of space tracking mechanism

2.1 Key indicator requirements

For application-specific requirements of space optical instruments, angular rang for tracking is usually $\pm 2^\circ \sim \pm 6^\circ$, and tracking accuracy is usually needed to $\pm 30'' \sim \pm 60''$. Besides working angular velocity of space tracking mechanism is very low, usually about 10e-3 rad/s [1].

2.2 Mechanism bearing

[a] Corresponding author : 271673978@qq.com

Whether it is space or ground product, rolling element bearings are used usually in lots of traditional rotary devices. But rolling element bearing is only applied to the condition when the mechanism rotates in complete rotations or at least the rolling elements in bearing rotate in complete rotations.

Firstly, the bearings will get deviation when working in a very low velocity, then lead to a low tracking accuracy for mechanism. Secondly, the bearing's life will be seriously reduced when working long term in a small angle range of swing, because the rolling elements in bearing are pressed unevenly. And thirdly, there is metal-to-metal contact in rolling element bearing, which means some treatments have to be done to protect against space cold welding, then some uncertainty and risk come behind[2].

Flexural pivot is used as tracking mechanism bearing in this study. The flexural pivot consists of flat springs crossed at 90° , and supporting cylindrical counter-rotating sleeves, as shown in Figure 1. Friction, stiction (the force required to cause one body in contact with another to move), fretting corrosion, lubrication, space cold welding and lubricant out-gassing in space vacuum applications are all avoided in the flexural pivot because there is no metal-to-metal contact.

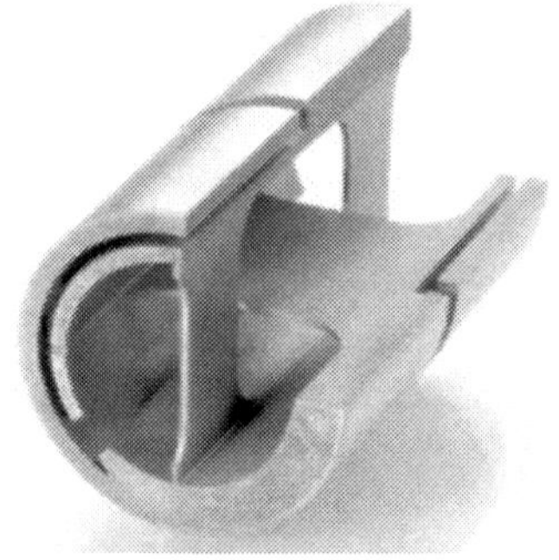

Figure 1. Structure of Flexible Pivot

By matching pivot size and spring rate to the application, cyclic stress on the springs may be maintained at a low enough value to assure infinite cyclic life.

2.3 Design of whole structure

The tracking mechanism consists of a base, a scan mirror, two flexural pivots, a voice coil motor, a resolver and other components, as shown in Figure 2. The scan mirror is supported by two flexible pivots, connected to the base; the motor rotor and the resolver rotor are connected to the scan mirror through two adapter structures. Then the mechanism achieves the functions that the motor drives the mirror directly and the resolver feedbacks the angular position.

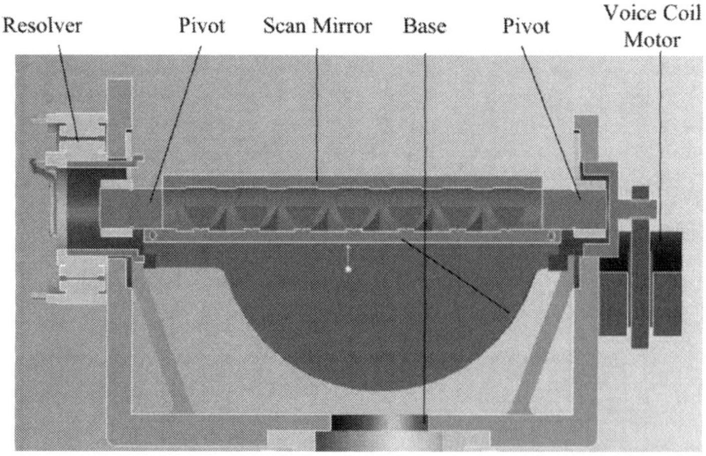

Figure 2. A Design of tracking mechanism

 DOI: 10.1051/matecconf/201711403014

Voice coil motor is a kind of torque motor, and has a special limited angular rotation structure. This can well meet the requirements of this space tracking mechanism.

The structure of rotary voice coil motor is shown in Figure 3. Voice coil motor can drive the mirror directly. Its working principle is Ampere force (that's why it's named like a speaker). An electrified coil (conductor) placed in a magnetic field generates a force, and the size of the force is proportional to the coil current. Voice coil motor have a series of advantages like small size, light weight and fast response. Voice coil motor has been widely used in lots of high precision positioning systems and high acceleration devices, such as disk positioning and optic lens positioning on the ground [3-5].

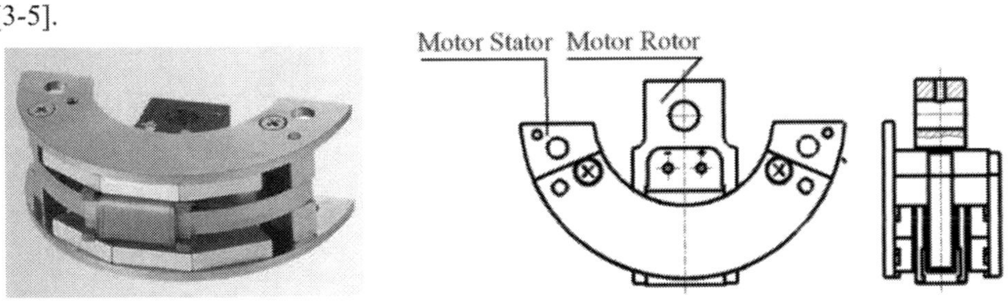

Figure 3. Structure of Rotary Voice Coil Motor

2.4 Control system

The control method is a closed-loop based on position feedback. The high-precision resolver, which electrical error is ±15″, provides the position feedback information. The control system is digitalized based on DSP servo controller. The detail control scheme is: Instructed by upper lever's parameters, DSP generates a position curve command corresponded, using the angle feedback from resolver, concurrently calculates the angle error. Then according to the control algorithm, adjust the voltage on the motor, through allocating the PWM duty cycle, the closed-loop position servo is achieved [6].

3 Measurement and life test

3.1 Tracking accuracy test

Based on all above researches, we designed and realized a tracking mechanism, which is shown in Figure 4. Then the tracking accuracy of is measured, and key parameters and test result is shown in Table 1. Through the test results we can see the tracking accuracy is better than 30″.

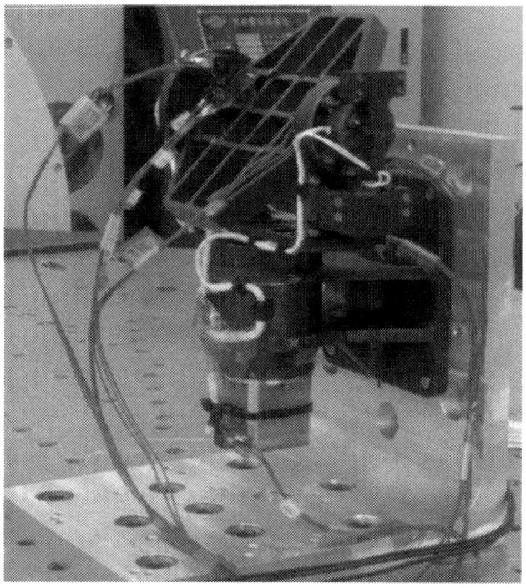

Figure 4. The tracking mechanism

Table 1. Key Parameters and Test Results

Load weight (Kg)	2.2
Load inertia (kgm2)	1.7E-2
Rotation range(°)	±6
Maximum load torque (Nm)	0.08
Torque coefficient of motor(Nm/A)	0.63
Electrical errors of resolver (″)	±15
Tracking accuracy (″)	±26.4

3.2 Lone life test

To demonstrate this tracking mechanism's long life, a life test is prepared. The basic signs of mechanism's long life stability include many features, first it should operate steady through the whole life test, second the tracking accuracy and maximum load torque and maximum output torque of motor should be stable, without any significant changes.

The tracking mechanism above was cycled for over 3.6E+06 cycles over a period of nine months, ±6° for each cycle.

We tested the maximum load torque of this tracking mechanism and maximum output torque of the voice coil motor every 4.0E+05 cycles. The test data is shown in Table 2, the trend of maximum load torque during the entail life test is shown in Figure 5, and the trend of maximum output torque during the entail life test is shown in Figure 6.

Table 2. The test of load and output for life test

Test num.	Total cycles	Operation test	Maximum load torque	Maximum output torque
0.	0	smooth	0.08	0.39
1.	4.0E+05	smooth	0.08	0.38
2.	8.0E+05	smooth	0.075	0.39
3.	1.2E+06	smooth	0.073	0.39
4.	1.6E+06	smooth	0.075	0.38
5.	2.0E+06	smooth	0.073	0.38
6.	2.4E+06	smooth	0.075	0.38
7.	2.80E+06	smooth	0.075	0.39
8.	3.2E+06	smooth	0.075	0.38
9.	3.6E+06	smooth	0.075	0.38

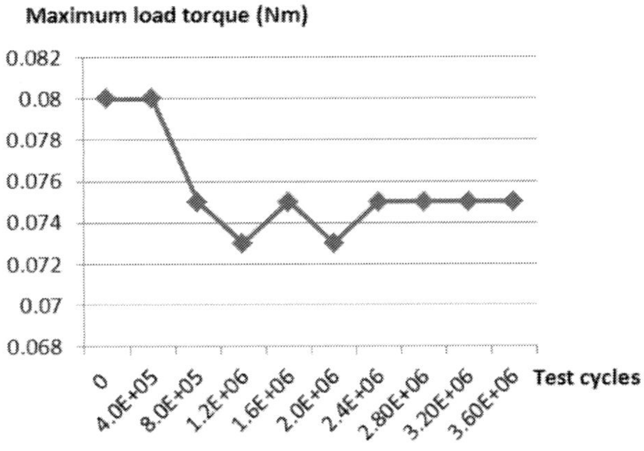

Figure 5. The trend of maximum load torque

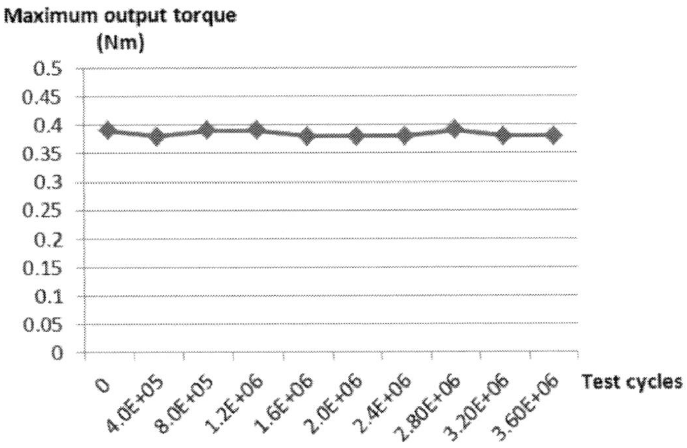

Figure 6. The trend of maximum output torque

Through the test data and trend of maximum load/output torque results we can see that the maximum load torque of mechanism has a slightly fluctuation in the early stages of the life test, after that stabilized. While the maximum output torque of voice coil motor is stabilized during the entail life test.

Also the tracking accuracy is compared before and after the long life test, and test result is shown in Table 3. Due to the error of measurement system, small fluctuation of accuracy results should be partially ignored. Through the test results we can see that the tracking accuracy is stable after the 3.6E+06 cycles life test.

Table 3. The test of tracking accuracy for life test

Num.	Test items	Before life test	After life test
1.	Operation test	smooth	smooth
2.	Tracking accuracy (″)	±26.4″	±28.8″

Through this 3.6E+06 cycles and all the tests, the high precision and long life stability of this tracking mechanism is proved.

4 Conclusion

In summary, the design of high precision and long life tracking mechanism, which bearing by two flexural pivots and driven by a voice coil motor, got certain superiority on space optical instruments. The tracking accuracy can reach to 30″ , and the life of this mechanism can reach to 3.6E+6. This method is suitable for many similar pointing and scanning mechanisms in space products.

References

1. Y. Shouwang. The Research of Image Scanning Control System of Aerial Camera Based on VCM. FMP. D. (2011)
2. X. Xin-hang, W. Bing, H. Xu-dong, W. Heng-Kun, L. Ting-xia. Design of fast-steering mirror with spherical pair supporting structure driven by voice coil actuators. FMP. J. **19** ,6 : 1320-1324. Jun. (2011)
3. Z. Da-wei, F. Xiao-mei. The Technical Principle of Voice-Coil Actuator. SME. J. **27** ,3 : 224-228. Sum. (2006)
4. X. Lian-guo, Z. Hui-xing, H. Shu-lin, C. Rong-min. Research and Application of Voice Coil Motor. Mic. J. **44**, 8 : 83-88. Aug. (2011)
5. L. En-chao, L. Xin, Z. Yan-na, X. Chun-feng, Z. Xiao-bing. Design of Precise Scanning and Positioning Based on Voice-Coil Actuator. J. **9**, 5 : 384-390, Sep. (2014)
6. W. Chun, X. Nihui, L. Zhe. Application of Delta-Sigma Modulator in Space Pointer Mirror Position Servo System. SRRS. J. **35**, 5 : 54-60. (2014)

Research and Simulation on Division Attack Airspace

MAO Houchen[a], ZHAO Guhao, GAO Wenming and Sun Tianchi

ATC and Navigation collage, Air Force Engineering University, Xi'an Shanxi 710051, China.

Abstract. Division attack is the basic tactical action of air strike operation carried out by the Air Force. The aircraft tactical action model is analyzed and established according to the Division attack pattern. Track points are determined on the basis of parameters of the weapons, and the ideal trajectory is determined under the constraint of the flights' ideal parameters and the type of target. The upper winds and pilot' operating error are taken into account on the ideal track, and the total deviation is calculated. We put forward a method of establishing airspace based on the probability deviation of the attack track, and adopt the relevant flight data to carry on the simulation verification. The final results show that the effectiveness and practicability of this method, and it is helpful for the rapid generation of relevant airspace.

1 Introduction

Implementing air raid operations, There may be some weapons such as the Aviation of Air Force, the Long Range Artillery of Army, the ship - to - ground missiles of Navy, the cruise missiles, the Tactical Missiles of Rocket Force require to use a certain airspace in the attack process[1-2]. If we can't plan the airspace before the combat for a reasonable allocation, it will lead to conflict of airspace, and even appears accidental injury.

At present, there are many literatures on airspace planning at home and abroad, but mostly concentrated in the civil aviation airspace, and less on the military. In 2011, Jia Linsheng plans airspace for early warning based on the patrol routes of early warning aircraft, and analysis the early warning effectiveness according to different positions of early warning aircraft in the air line. In 2014, Chen Yunxiang determines the patrol boundary according to the target airspace to be detected and security requirements for early warning aircraft, and then determines the center point and size of patrolling airspace within the patrol boundary. But the airspace on fighter is also basically focused on training airspace, fuel airspace and other non-combat airspace. While the research and planning for air-raid operational airspace is lagging behind.

Due to the particularity of the cooperation, the attack airspace should contain the flight path, attack area of target and the necessary safety separation. This paper references the method which plans airspace based on airline in literature [3]. Aiming at the mode of diving attack in air-raid operation, the attack track model is established. The factors of airborne wind and pilot error are added into the ideal track. The maximum deviation of track is determined. Finally, the safe separation is joined and the airspace module is built [4].

[a] Corresponding author: 347741625@qq.com

2 Diving Attack Path Planning

Implementing dive attacks usually involves adding tactical maneuvers to avoid enemy's missiles or radar threats [5]. The basic track is shown in Figure1.

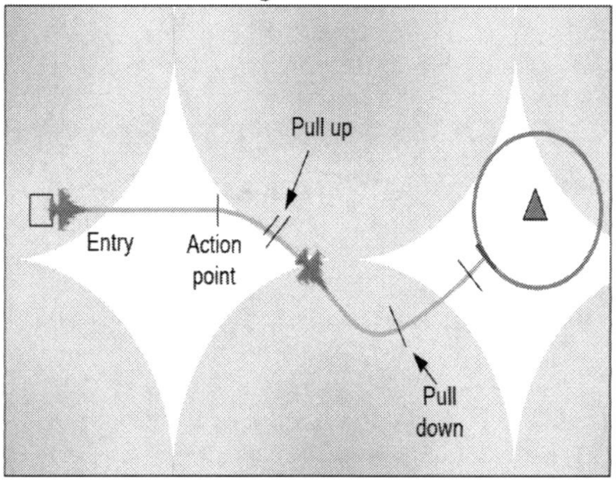

Figure1. Diving Attack Path

Before implementing an air-raid operation, the mission trajectory is generally planned to determine the specific points and parameters in the flight path. The track planning steps are as follows:

Step1: Determine the position of the target point;
Step2: Select the appropriate ammunition to determine the bomb height and bombing distance;
Step3: Select the appropriate dive height;
Step4: Determine the entry point position;
Step5: Determine the safety separation.

3 The motion model of Fighter

In order to simplify the calculation, the geodetic coordinate system as shown in the figure is established. Only the speed, pitch, slope, tangential overload and normal overload are taken into consideration. The aircraft motion model is constructed without considering the aircraft inertia, angle of attack and angle of attack. As shown in Figure.2.

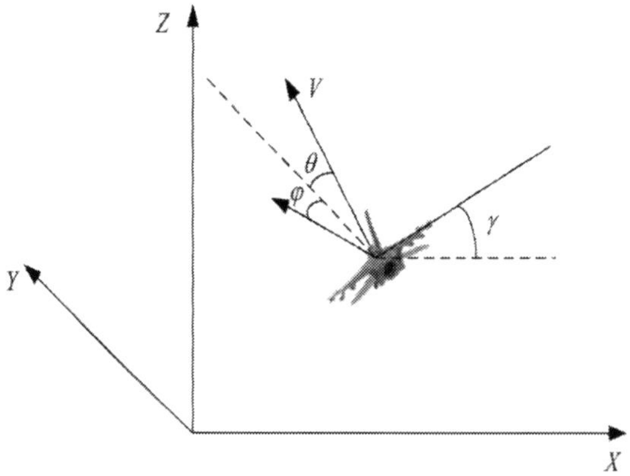

Figure2. The motion model of Fighter

The kinematic equation is as follows:

$$\begin{cases} \dot{x} = v\cos\theta\cos\varphi \\ \dot{y} = v\cos\theta\sin\varphi \\ \dot{z} = v\sin\theta \\ \dot{v} = g(G_x - \sin\theta) \\ \dot{\theta} = g(G_f\cos\gamma - \cos\theta)/v \\ \dot{\varphi} = -gG_f\sin\gamma/(v\cos\theta) \end{cases} \quad (1)$$

v is the true speed, g is gravitational acceleration, θ is pitch angle, γ is slope, φ is course angle, G_x is tangential overload, G_f is normal overload.

4 Analysis flight errors

Before the implementation of air-raid operations, the task category, aircraft type, flight action have been identified, this paper only consider the stand-alone situation, and discuss these uncontrollable factors include the pilots technical level and air wind.
The fighter lateral offset is calculated as follows:

$$X_p = X_c + E_\alpha \quad (2)$$

Among them, X_p is the total offset of the aircraft, X_c is the offset caused by pilot error, E_α is wind deviation.

4.1 Winds

The omnidirectional wind specified by the ICAO is used to describe the error. The wind spiral can quantify the effects of wind on the airplane during the turn and maximize it. Therefore, this paper takes this method into the airspace planning of air-raid operation. The wind spiral is shown in Figure.3.

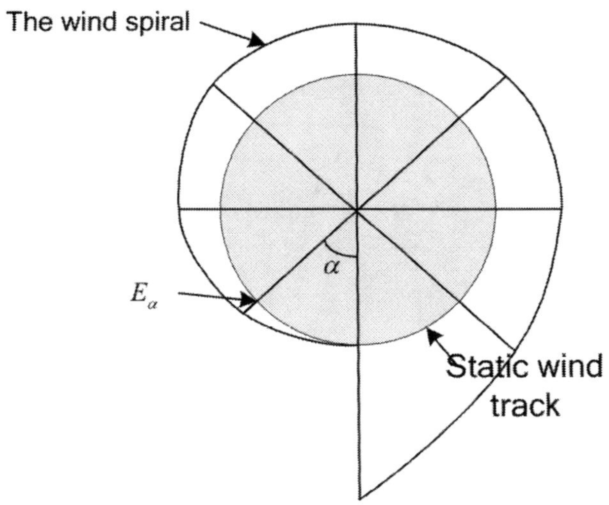

Figure3. The wind spiral

The influence formula of wind:

DOI: 10.1051/matecconf/201711403015

$$E_a=\left(\alpha/\omega\right)\times\left(v_{wind}/3600\right) \tag{3}$$

In the formula: E_α is drift distance, α is turning angle, ω is turning angular velocity.

4.2 Pilot operating error

In this paper, the civil aviation flight error is taken as a reference, the operational error of the fighter pilots is considered as Gaussian distribution [6] based on the ideal operation.

$$W \sim N\left(W_0, \sigma_W^2\right)$$

$$f(W)=\frac{1}{\sqrt{2\pi}\sigma}\exp\left(-\frac{(W-W_0)^2}{2\sigma^2}\right) \tag{4}$$

Among them: W_0 is ideal operation, W is actual operation, σ is standard deviation, the value refer to references [6].

Since the normal distribution value is infinity range from the positive to negative, and when the pilot operates the aircraft, the operation error will not tend to be infinite, in order to reflect the two points, the normal distribution of the distribution function is truncated, the truncation points[7-8] is $(\mu-3\sigma)$ and $(\mu+3\sigma)$, That is, the probability is 1 in this interval.

According to the probability distribution $P=\int_{-x_1}^{x_1} f(W)dx$, determines the effective probability P, and then the maximum value of the error can be calculated.

Total error can be calculated according to the above operational error model:

$$X_C=\sum_{j=1}^{m} X_j, j=1,2,\cdots \tag{5}$$

5 Simulation Study of Example

Due to the particularity of military flight, this paper takes civil flight data [9-10] as reference, simulated in Matlab 8.0.

The simulation results are shown in figure.4.

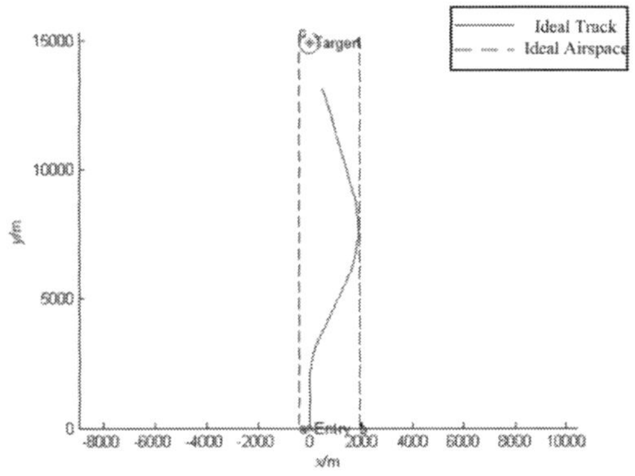

Figure4.(a) Top view of ideal airspace

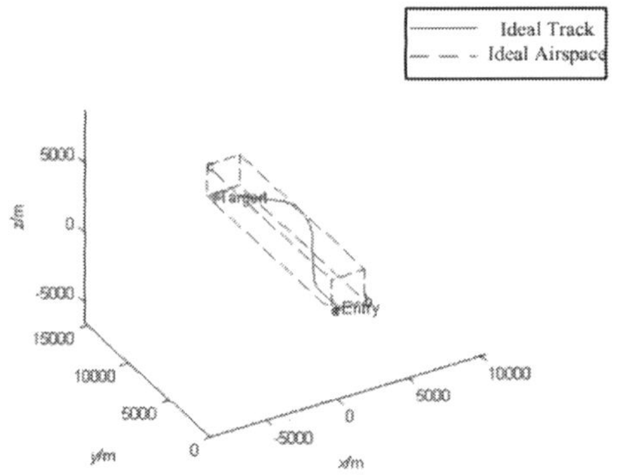

Figure4.(b) Top view of ideal airspace

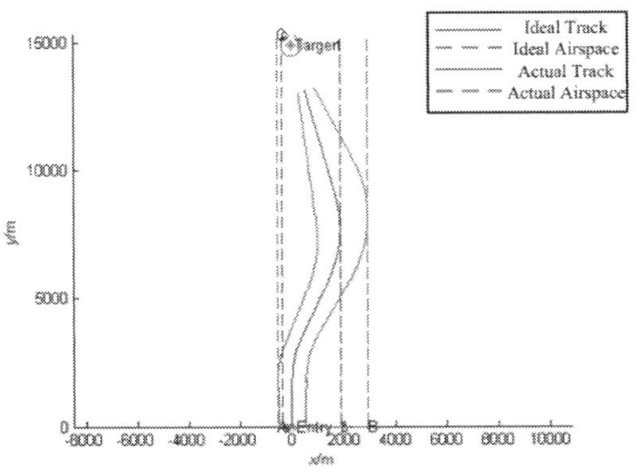

Figure4.(c) Top view of actual airspace

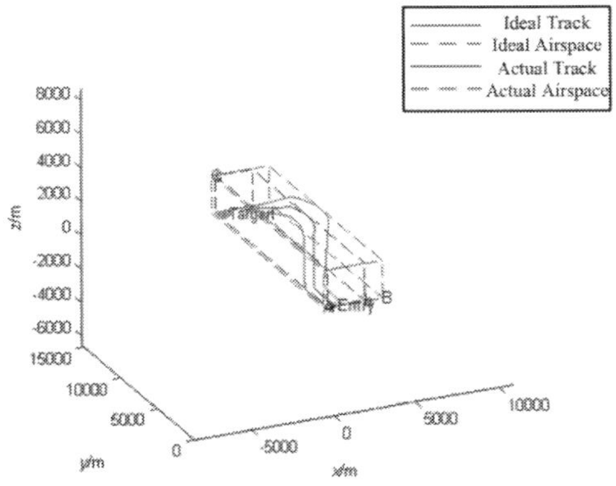

Figure4.(d) 3-D map of actual airspace

In figure 4(a) (b), the blue wireframe indicates the flight path and the required airspace size under ideal conditions. In figure 4(c) (d), the red wireframe represents the maximum deviation trajectory and the actual required airspace size after adding wind and operating errors.

We can obtain the airspace's boundary according to the simulation data as shown in Table1.

Table 1. Airspace Boundary Parameters

Ideal Boundary	a	b	c
Coordinate	(-400,0,0)	(1.8982e+3,0,0)	(-400,1.5295e+4,2500)
Actual Boundary	A	B	C
Coordinate	(-576.01,0,0)	(2.9356e+3,0,0)	(-576.01, 1.5295e+4,2500)

The width of ideal airspace is 2298.2m, the length is 15295m, height is 2500m. But the width of actual airspace is 3511.61m, the length is 15295m, height is 2500m. We can see that the actual airspace is larger than the ideal state.

From the simulation, we can see that this method can plan airspace quickly, which has versatility and can be applied to the planning of other related airspace. At the same time, it can take the concept of modular into the planning and can be combined with other airspaces easily.

6 Summary

This paper aims at the airspace planning problem in air-raid operation, a airspace planning method based on maximum deviation of trajectory is proposed in the case of dive attack model. Based on the modeling of fighter tactical trajectory, the important trajectory point is determined according to the target attribute, the effect of wind and pilot error is added, and the airspace size is determined considering the possible region of fighter attack track. This method can be used in the modular planning of airspace, which is fast and efficient. But in the planning process it does not take the withdrawal route after the completion of the fighter attack into account, and this paper takes the wind and the pilot operating error into consideration only for the lateral yaw by a certain type of aircraft. The next step will consider the enemy radar and air defense missile deployment as well as kinds of aircrafts, and planning withdrawal airspace rationally.

Acknowledgment

This paper is supported by the National Social Sciences Foundation（16GJXXX—XXX）

References

1. JIA Linsheng, WU Wenhai, GAO Wei, WANG Junyan. Analysis on Early Warning Airspace of Airspace Patrolling Route Adopted by Foreign Airborne Early Warning Airplanes[J]. Journal of Naval Aeronautical and Astronautical University. 2011.26(1),36-40.
2. CHEN Yunxiang, ZHANG Yi, ZHUANG Jun2, CAI Zhongyi. Method on Demand Determination of Early Warning Airplane Based on Combat Direction[J]. Fire Control & Command Control. 2014,39(8),94-97.
3. JI Jun, WU Feng, MA Pei-be, SUN Xiaoshu. Research on the Patrol Region Dimension Disposition of AEW in Long-range Air Defense Operation of Carrier Formations[J]. Command Control& Simulation. 2014.36(1),78-83.
4. YU Lung. XING Changfeng. SHI Zhansong. Modeling of Airspace Resource for Cooperative Air-defense Operation[J]. Journal of Naval University of Engineering. 2014.26(1),54-59.
5. HUANG Da, CHEN Jing, REN Min. Research on Trajectory Generation of Air-to-surface Attack Based on Constraint Model [J]. Computer Simulation. 2010.27(6),74-78.
6. XUE Hongtao, SHEN Lincheng. Flight Maneuverability Analyze and Computation of Combat Aircraft Weapon Delivery Tactics[J]. Command Control & Simulation. 2010.27(6),74-78.
7. International Civil Aviation Organization, Doc 8168, Part I — Section 2, Chapter 2, I-2-3-4[S], Montreal: International Civil Aviation Organization, 2008.
8. ZHAO Yifei, WANG Chen, WANG Hongyong. Research on laws of flight path error distribution based on ADS-B[J]. JOURNAL OF CIVIL AVIATION UNIVERSITY OF CHINA. 2012.30(6),48-52.
9. ZHAO Hongsheng, XU Xiaohao, LI Rui, XIONG Zhiyong, LI Dongbin. Overview of research on flight technical error estimation in performance based navigation[J]. Journal of Traffic and Transportation Engineering. 2014.14(5),101-110.
10. Xu Xiaohao. Theoretical Method Research of Safety Separation for Air Traffic[M]. Civil Aviation University of China. 2004.12.

MATEC Web of Conferences **114**, 03016 (2017) DOI: 10.1051/matecconf/201711403016
2MAE 2017

The Review and Development of the Landing Gear Emergency Release System

Xin Li[1,a] and Yu Li[2, b]

[1] Landing Gear System Research Institute, FAI AVIC, Xi'an, CN
[2] System Integration Center, FAI AVIC,Xi'an, CN

Abstract. Landing gear emergency release system is a common design of modern aircraft. In this paper, the typical landing gear emergency release system is analyzed, and the advantages and disadvantages of various emergency release forms are compared. This paper analyzed the research status of here and there electric emergency release technology, and made a preliminary study on the key technologies of electrical actuation. The development direction of the landing gear emergency release system is proposed.

1 Introduction

Landing gear system is one of the important systems that affect the safety of aircraft[1-2]. The retraction system is an important part of the landing gear system, also is a department with high failure probability. In order to improve the safety of landing, China's civil aviation regulations CCAR25.729 (c)[3] stipulate that under the condition of normal appearing in any reasonably likely failure or a single hydraulic source or equivalent energy with landing gear retraction system, the aircraft should have at least one landing gear emergency release method.

2 Landing gear emergency release system present situation analysis

The working principle of landing gear emergency release is making the upper locks of the landing gear and the landing gear doors of unlocked through certain way, so two chambers of the actuating cylinder bypass to relief pressure, then relying on the landing gear its own gravity and aerodynamic loading free down and be locked. Typical emergency release unlocked methods include: mechanical emergency release, electric explosion emergency release, electrical hydraulic actuation emergency release, electric mechanical actuation emergency release and so on. In addition, military aircraft or UAV also uses air as emergency energy or explosive blasting to unlock[4], the application of which is certain limited.

2.1. Mechanical emergency release

[a] 370713464@qq.com

[b] kingliyu@sina.com

Mechanical emergency system is applied commonly in civil aircraft. A typical mechanical emergency system uses cable to transmit, unlock by mechanism, release pressure by bypass valve. As shown in Fig. 1[5], linkage parts of the system include the landing gear doors upper lock, the gear upper lock and bypass valve parts, and the system is independent with normal retraction system using hydraulic actuation. The pilot pulls the emergency release handle, the system operates the landing gear components with the following sequence: (1) the bypass valve and cut off normal pressure supply and normal loop and loop bypass to the return line; (2) landing gear and landing gear door upper lock locked.

The advantages of mechanical emergency system is safe and reliable, so it is widely used in large civil aircraft. Such as Airbus A320 (Fig. 2), the Boeing B373, and China's regional aircraft employ the mechanical cable emergency system.

The disadvantages of mechanical emergency release system are:

- Mechanical structure is complex, take up more structural space.
- A large number of the pulley mechanism and the transmission rod are designed in the systems, so the assembly requirements between the key components are higher.
- The maintenance work is big and difficult. Because of the change of external load, and the change of environment temperature of the plane, body structure and aircraft control system generate different relative deformation, so to cause the cable loosening and elastic mechanism clearance, or to lead to tight and additional steel friction mechanism. At the same time, due to wear, crushing, corrosion and broken wire etc. often happen during cable use, the pulley part due to wear can also lead to the rotation is not flexible. All of these are difficult for normal maintenance work.

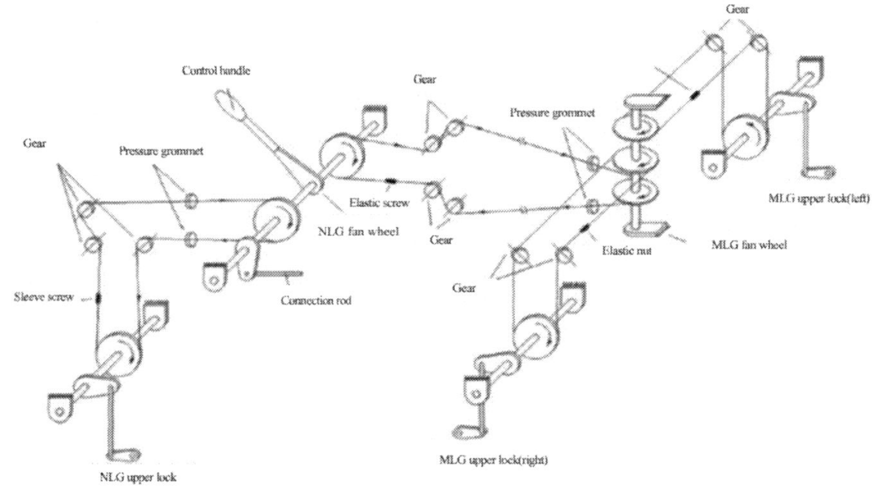

Figure 1. Schematic diagram of mechanical emergency release system

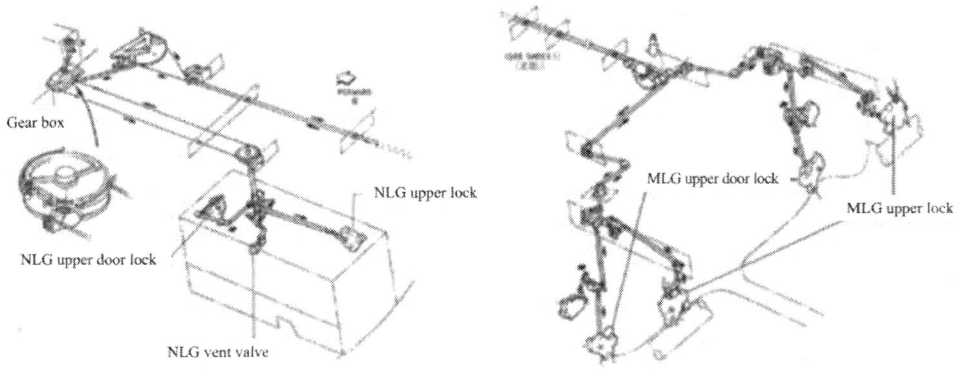

Figure 2. Emergency release system of A320 diagram

2.2. Electric explosion emergency release

The principle of electric explosion emergency unlocking is, when the hydraulic system failure, the electrical explosion impact air flow, the air flow instead of hydraulic pressure to push the piston rod drives the unlocking actuating device, shown as Fig.3.

Electrical explosion emergency unlocking method abandons the complex structure of the electric emergency drive or mechanical cable device, without increasing the complexity of the motor drive system. Electric explosion with the compact structure, light weight, simple and easy to implement, is applied for UAV and small aircraft.

Electrical explosion emergency unlocking carries out the problems of redesigned lock mechanism which is relatively complex, regular maintenance and replacement because of the electric explosion belongs to consumables, and difficult to meet the requirements of the airworthiness.

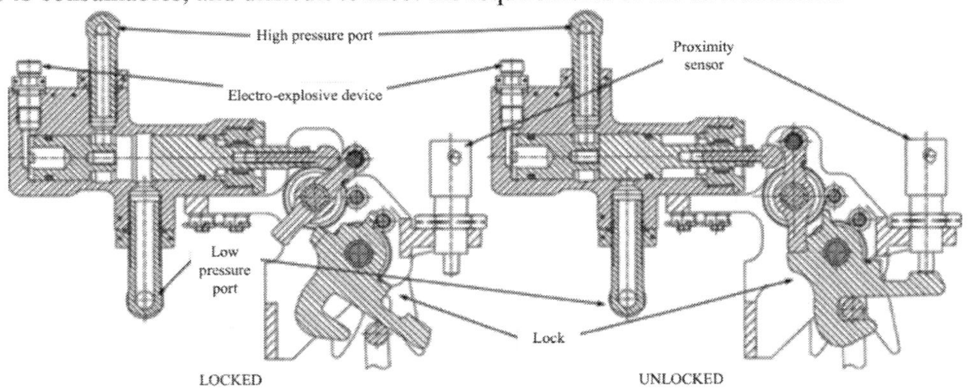

Figure 3. Working principle of the mechanism lock of electric explosion emergency release

2.3. Electrical hydraulic actuation emergency release

Electrical hydraulic actuation is known as Electro hydraulic Actuator(EHA), the system includes DSP controller, power drive unit, motor, servo pump and other components. As an electrical actuator technology, EHA is also used in a large number of modern aircraft technology (Fig. 4), such as A380 turning and braking systems and B787 landing gear emergency release system. For emergency landing gear and the door put itself, electro hydrostatic actuator can be used as an alternate energy emergency unlocking, and without the addition of actuator (use original hydraulic actuating cylinder), but need to have the valve block system corresponding to the original hydraulic pipeline to isolate.

Figure 4. The application of EHA in the nose landing gear of a certain type of aircraft

The advantages of electrical hydraulic actuation emergency release are showed as follow: the actuating cylinder of the lock mechanism is small, so the electric hydraulic actuator can easily meet the requirements without large flow hydraulic pressure, at the same time, the system changes to the lock mechanism is small, no need to add additional operating device, telex control system to facilitate the realization of the monitoring of components, so it is easy to detect, good maintainability, the system is simple and reliable, and the fault tolerance capability is high.

The disadvantages of electrical hydraulic actuation emergency release are also obvious: a complete set of EHA and a separate electric power system will increase the weight of the system itself, larger system changes, need to increase a set of power system and to coordinate with the original hydraulic system to switch working mode, for the original to hydraulic actuating system, adding a set of EHA costs quite highly.

2.4. Electric mechanical actuation emergency release

Electric mechanical actuation is also known as electro Mechanical Actuator EMA[6-8]. The system includes emergency release controller, electro - hydraulic upper lock, electric lock device and so on. In this emergency release method, controlling the hydraulic valve block component in the bypass mode to release pressure, unlocking the upper locks of various components, and transmitting the locked and unlocked signals of upper and lower locks are all operated in the emergency release controller. The electric unlocking device is an independent driving device (shown in Fig.5), which is based on the original hydraulic actuating unlocking device. The device is generally composed of a motor and a worm wheel, working principle is shown in Fig.6. The electric mechanical actuation emergency release system is the development trend, the newly developed A380 landing gear emergency release system is the use of the electric actuator.

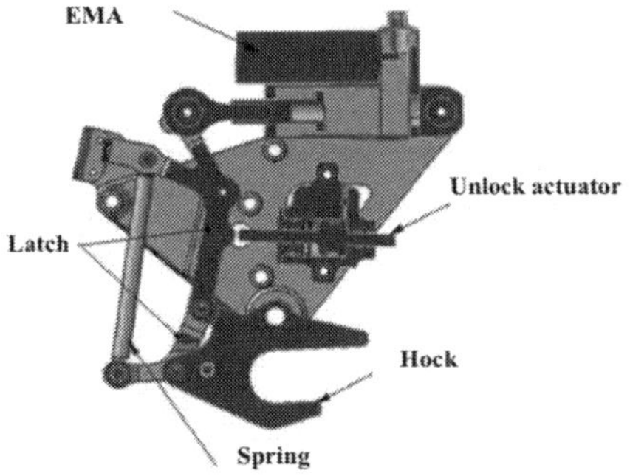

Figure 5. Lock mechanism of electric mechanical actuation emergency release

 DOI: 10.1051/matecconf/201711403016

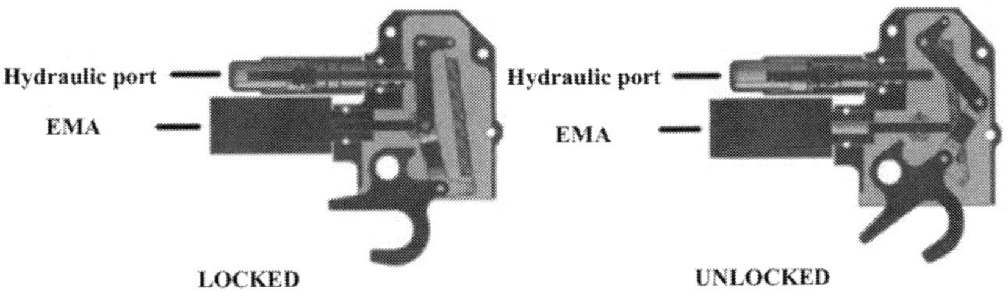

Figure 6. Working principle of lock mechanism

The advantages of electric mechanical actuation emergency release are: compared with the mechanical emergency release system, the weight is obviously reduced, and the volume is small and easy to be arranged, FBW system facilitates the realization of the monitoring of components, so it is easy to detect, good maintainability.

The disadvantages of electric mechanical actuation emergency release are: larger system changes, need to increase a set of power system and the need to coordinate with the original hydraulic system to switch working mode; with the increase of the reliability of electronic components, and the redundancy design will increase the safety and reliability of the system, but still need to work on the full power of the emergency response system for reliability risk assessment; the lock mechanism needs to be redesigned, which is more complicated.

2.5. Comparison

Comparison of the four forms of emergency release is shown in Tab. 1.

Table 1. Comparison of the four forms of emergency release

form	Scope of application	advantage	disadvantage
mechanical emergency release	Widely used in large civil aircraft A320, B737	Safe and reliable	(1) the structure is complex, occupying more structural space; (2) the assembly requirements of key structural parts are higher; (3) the maintenance work is high and difficult.
electric explosion emergency release	Suitable for unmanned aerial vehicles and small aircraft	(1) compact structure, light weight, easy to decorate; (2) the system is simple and easy to implement.	(1) the lock mechanism needs to be redesigned and more complex; (2) electrical explosion type belongs to consumables, need regular maintenance and replacement; (3) reliability needs to be tested and verified.
electrical hydraulic actuation emergency release	wide, such as B787, especially for the situation that the lock mechanism can not be changed	(1) the system has little change to the lock mechanism, no need to add additional operating device; (2) easy detection, good maintainability; (3) simple and reliable, and the fault tolerance capability is high.	(1) the EHA and a separate electric drive system will increase the weight of the system itself; (2) the system changes greatly, need to increase a set of electric drive system and need to coordinate with the original hydraulic system in order to change the working mode; (3) for the original hydraulic drive actuating system, by adding a set of EHA cost quite highly.
electric mechanical actuation emergency release	Widely applicable to all types of models, such as A380	(1) the weight is obviously reduced, and the volume is small and easy to be arranged; (2) FBW system to facilitate the realization of the monitoring of	(1) the system changes greatly, need to increase a set of electric drive system and need to coordinate with the original hydraulic system in order to change the working mode; (2) the reliability risk assessment of all electric

		components, so it is easy to detect, good maintainability;	driven emergency release system; (3) the lock mechanism needs to be redesigned and more complex;

3 Research on the status of electric mechanical actuation emergency release

In 1990s, the U.S. Air Force leaded "more electric aircraft (MEA) " project, , including the army, Navy, and NASA more than and 50 companies to participate in the program, aims to replace the complex actuation system, high maintenance, mixed hydraulic and pneumatic machinery and electric propulsion system with non power consumption power, in order to improve the reliability of aircraft power system to reduce the dependence on ground support equipment, thereby reducing the cost of launching weapons.In the MEA project, more electric aircraft technology has been fully developed in the United States, and is widely and fully tested, and many techniques have been applied in practical application, has made a lot of research.In 1999, Collins set up the model of the landing gear all electric brake system, based on the simulation of the braking system based on MATLAB/Simulink, and some of the key structural strength of the brake was analyzed. In 2000, NASA made EMA technology successfully applied to F18 aileron drive improvement on the original hydraulic actuating device by the electric actuator to replace, designed the electric actuator control system, and the success of the test results showed that the electric actuator performance and hydraulic actuator performance is very close. In 2009, AbdElhafez etc. Compared the difference between the MEA and the traditional hydraulic aircraft, as well as the advantages of MEA. In 2010, Skorupka etc. applied the all electric technology to UAV landing gear retraction, turning and brake system, and the design of the electric actuator control system successfully tested.

In China, the development of all electric technology is slow, but in recent years, many scholars have done a lot of research. In 2009, Sun Jiyong designed a small aircraft electric actuator control system, realized the multi-point coordinated control and multi task coordination control, in order to improve the reliability and stability of the system, used CPU redundancy technology on the control board and driver board were made redundant design. In 2011, Wang Tao etc. designed a landing gear controller of double - double redundant structure, which is in the board setting center ruling redundancy, between boards setting non center ruling redundancy, then combined the two kinds of redundancy. Otherwise, it used a non similar redundancy design and hardware error detection, greatly improved the stability of landing gear controller. In 2012, Li Lin etc. designed a ball-screw type double redundancy electric actuating cylinder, which is mainly composed of a main actuating device (DC servo motor), emergency actuation device (pneumatic motor, pneumatic clutch), a mechanical device (locking mechanism, a piston cylinder, a piston rod and connecting parts etc.), transmission device (reducer, synchronous belt, ball screw), detection device (switch) and other components. They also designed a set of electric actuator systems with feedback (including emergency release), and verified the systems. In 2012, Bao Qunmin and others proposed a double-redundancy controller design for a small aircraft landing gear electric control demand. The controller used DSP as the control core, achieved the four quadrant operation of the motor control, and the system had the function of over-current protection.

4 Key techniques of EMA

4.1. EMA controller

EMA emergency release system is controlled by independent controller. Firstly ensure that the landing gear system valve assembly in bypass mode for pressure relief, the emergency release controller controls each gear upper lock unlocked, the landing gear release free by gravity and locked down, at the same time, independent assessment about controller downlock signals are transmitted to EICAS (the engine indication and crew alerting system).

When the emergency release clearance is placed in the emergency position, the following actions will be carried out in sequence:

(1) the landing gear to put down the discrete signal command sent to the emergency release controller;

(2) landing gear system control valve assembly emergency discharging controller after receiving the instruction, the high and low pressure pipe, hydraulic pressure relief loop control system;

(3) emergency release controller sends commands to drive the corresponding motor, unlock the front, the main landing gear PC Lock, the landing gear will be under the action of gravity and aerodynamic force to put down;

(4) the emergency release controller independently assess the landing gear under the lock signal, and the evaluation results will be transmitted to the EICAS system and display.

4.2. The locked rotor protection

In the locked rotor protection, there are three kinds of common blocking protection. Using relay realizes the overload protection or the low voltage circuit breaker with the relay realize the over-current protection. When the motor current switch starts after a long time and can not be reduced, thermal overload protection relay act, and power supply main circuit contactor disconnect the motor. This method is only suitable for the long time the motor overload caused by the operation of the high temperature protection. Using speed relay realizes the protection. It makes the motor tripping by controlling contactor coil with time relay and the speed switch contact with logic and relations. This kind of traditional blocking protection has certain limitations, and not suitable for low speed operation, shift and angle of operation of motor, stepping motor and servo motor blocking protection. Using PLC realizes blocking protection. By detecting the external signal of photoelectric sensor (displacement or velocity signals) and PLC internal timer, counter device, preparation of blocking protection control program, can realize the operation of the motor stall protection of low-speed operation, shift and angle.

5 Conclusion

This paper introduce four typical emergency release unlocked methods, such as mechanical emergency release, electric explosion emergency release, electrical hydraulic actuation emergency release and electric mechanical actuation emergency release. Compare the four methods from Scope of application, advantage and disadvantage. Analysis the status of EMA emergency release. For the design of EMA emergency release, give some proposals.

Even though the cost will be increased, the EMA landing gear emergency release system is already the development direction. The operation is very convenient, just press the button or switch; because of electric cable replace mechanical cable, layout of the system in the fuselage and cockpit becomes very flexible and convenient, and the weight of the entire system is significantly reduced. In the future, it will gradually replace the mechanical landing gear emergency system.

References

1. Aircraft design manual general editorial board. Aircraft design manual 14: Takeoff and landing system design[M]. Beijing: Aviation Industry Press, 2002.
2. Boitsov. Reliability of aircraft landing gear. Zhen Guo, Peifan Guo, translation. Beijing: national defence industry press, 2002.
3. Civil Aviation Administration of China 2001Part 2-airworthiness standards: transport category airplane(Beijing) pp 181-183.

4. Yao Xiao, Yifei Tang, Hong Cao. R&D of high pressure pneumatic ELGE system[J]. Hydraulics Pneumatics & seals, 2015, 06: 61-63.
5. Cunfang Zhang, Yongsheng Feng. Research on landing gear emergency Extending System[J]. Civil Aircraft Design and Research, 2010(1): 19-21.
6. Beaver aerospace & defence. Electromechanical actuator(EMAS)[EB/OL]. 2013-5-23. http://www.beaver-online.com/64.html.
7. Kai Chang, Shengjun Li. Application and development of more electrical technology in landing gear system[J]. Aeronautical Science & Technology, 2014, 25(07): 01-05.
8. Lin Li. Research for dual redundancy electromechanical actuation system of UAV landing gear[M]. NUAA ,nanjing, 2013.

MATEC Web of Conferences 114, 03017 (2017) DOI: 10.1051/matecconf/201711403017
2MAE 2017

The Application of Advanced Technique of Fan Frame Unit on High Bypass Ratio Aero Engine

HOU Peng[1,a], LUO ai-ran[1] and HAN fang-jun[1]

[1] *Shenyang Aero Engine Design and Research Institute, 2 Department, Shenyang, China*

Abstract. High bypass ratio aero-engine was widely used on military and civil aviation domain, as the power of larger aircraft. Fan frame unit was the main bearing frame of high bypass ratio aero-engine, which composed of strut, HUB MID BOX and external bypass parts. Resin/composite was used on external bypass parts(acoustic liner, containment ring, fan outlet guide vane and fan case skin fillets), which not only reduced the weight and manufacturing cost, but also improved the noise absorption, containment and anti-fatigue ability of engine. The design of composite was becoming a key technique for high bypass ratio aero-engine. In special test of the core engine, nitrogen cooling system was designed to cool the cavity of spool. The nitrogen pipeline passed through the inner cavity of fan frame, then inserted into NO. 3 bearing seal, so nitrogen gas was sent into the cavity of core engine spool. On high bypass ratio aero-engine, the external bypass and fan frame inner cavity were the design platform for advanced technique, such as composite and pipeline system, and also provided guarantee for reliable operation of engine.

1 Introduction

If the bypass ratio of turbofan aero-engine was greater than 4, it was called high bypass ratio aero-engine. The advance of high bypass ratio aero-engine was larger thrust, lower noise, better safety and good reliability. High bypass ratio aero-engine was the power plant of large military and civil aircraft. Fan frame unit was the main bearing frame of high bypass ratio aero engine, which composed of 12 struts, HUB MID BOX and external bypass parts. The strut cavity and HUB MID BOX cavity was communicated with each other, providing channel for oil pipe, air pipe, testing cable and transmission rod. The fan frame cavity of high bypass ratio aero-engine was relatively bigger, so more pipeline could be routed, such as nitrogen cooling system, as shown of figure 1 (7,8). In order to meet the requirement of higher thrust-weight ratio and lower maintenance cost, composite was widely used on high bypass ratio aero-engine [1]. Compared with titanium alloy, composite had the advantage of lighter weight, higher specific strength, higher specific stiffness, and better damping ability. On high bypass ratio aero-engine, the external bypass parts (acoustic liner, containment ring, fan outlet guide vane and fan case skin fillets) were all used composite as shown of figure 1(1~6), which contained resin glass fiber composite, resin carbon fiber composite and resin Kevlar fiber composite. In ground and flight test, the composite showed excellent performance. The design of composite was becoming a key technique for high bypass ratio aero-engine.

[a] HOU Peng : houpeng2749@sina.com

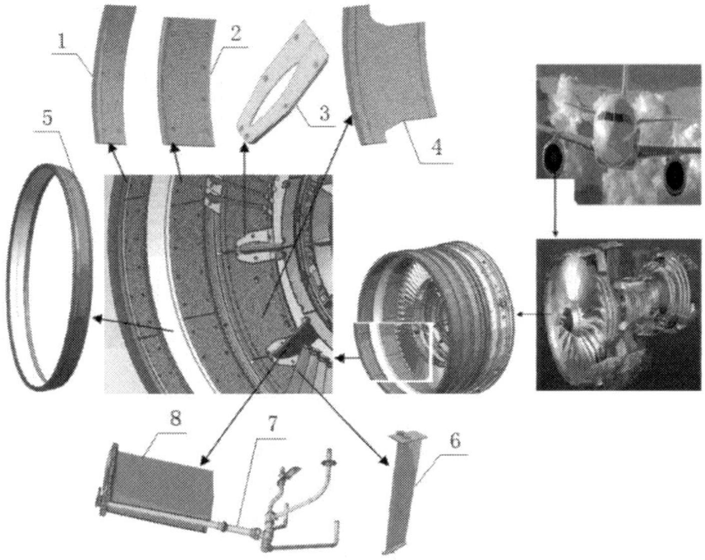

Figure 1. Explosive view of fan frame

1-Acoustic liner, 2-Acoustic liner, 3-Fan case skin fillets, 4-Acoustic liner, 5-Containment ring, 6- Fan outlet guide vane, 7-Pipeline system, 8- Strut

2 Composite Acoustic Liners

Fan was the main noise source of aero engine, and reducing fan noise was the primary task of engine noise reduction [2]. At present, the main method to reduce fan noise was using acoustic liner, and the acoustic liners were assembled on external bypass (secondary airflow bypass), as shown of figure 1(1, 2, 4). A good simple model for an acoustic liner is the Helmholtz resonator, and the acoustic liner model begins with a simplified acoustic system and then is advanced by building it up to the more complicated liner system [3], as shown of figure 2(b).

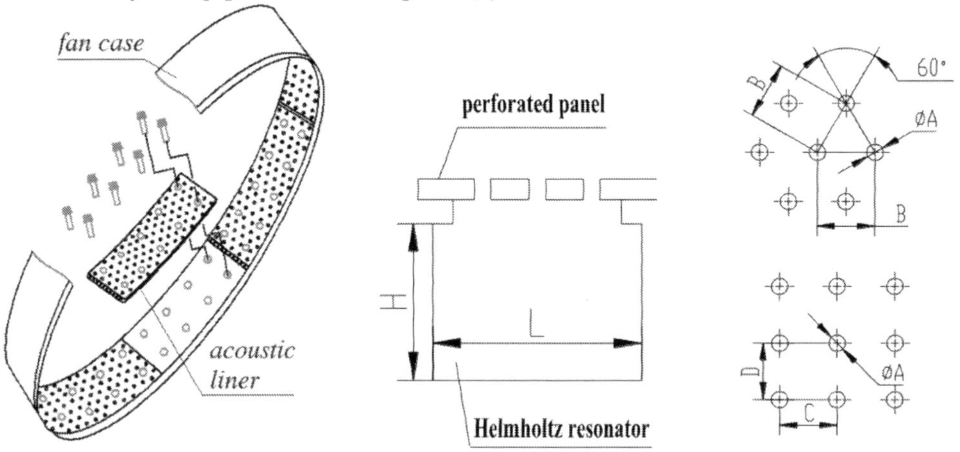

(a)Assembly view (b) Helmholtz resonator (c) Perforation ratio

Figure 2. Assembly and structure of segment acoustic liner

The noise absorption of acoustic liner was determined by the parameter as follow: bore perforation ratio of panel, material of honeycomb, and the dimension of honeycomb core [4, 5]. The rank of bores on perforated panel were regular triangle and rectangular array, the perforation ratio formula were as follow.

DOI: 10.1051/matecconf/201711403017

$$\sigma = \frac{\pi \cdot A^2}{2\sqrt{3}B^2 - \pi \cdot A^2} \tag{1}$$

$$\sigma = \frac{\pi \cdot A^2}{4C \cdot D - \pi \cdot A^2} \tag{2}$$

"A" represented bore diameter, "B", "C", "D" represented bore distance, and "σ"represented the perforation rate of perforated panel. Formula (1) was the perforation rate for regular triangle array bores, and formula (2) was the perforation rate for rectangular array bores, as shown of figure 2(c). As fan mainly produced shock wave noise, the bores of fan case acoustic liner were regular triangle array, and perforation rate was 7.4%. High frequency noise was mainly produced around fan frame shroud, the bores of fan frame acoustic liner were rectangular array, and perforation rate was 14%. For single aero engine, the noise absorption of the whole acoustic liners was 8~10 dB.

For improving the property of noise absorption or adjusting the frequency range of noise reduction, no need to change the structure of acoustic liner and merely change the parameter (perforation rate, perforation mode, dimension and thickness of honeycomb core) concerned. Owning to the excellent noise absorption property, better vibration damping ability and stable structure of composited acoustic liner, it could be popularized on high bypass ratio aero engine.

2.1 Segment acoustic liner

Acoustic liner had a honeycomb panel (aramid fiber or other absorber material), a perforated panel (epoxy resin/aramid fiber), and metal bushings. The perforated panel was a cover body on the surface of honeycomb panel, and metal bushings were inserted in the bolt hole, as shown of figure3. Compared with metal perforated sheet, composite acoustic liner could reduce 10 kg for single engine.

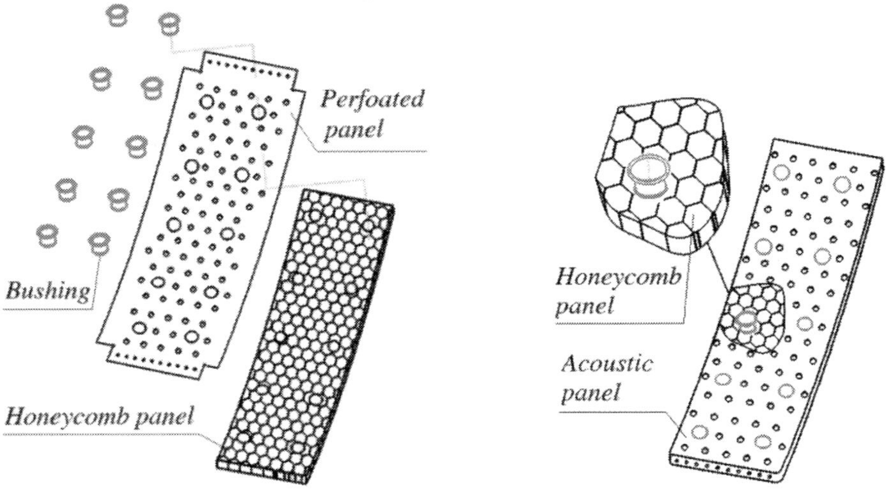

(a) Explosive view (b) Acoustic liner

Figure 3. Structure of segment acoustic liner

The process of assembly is that jointing the segment liners together on fan case and fan frame shroud, fixing the liners with bolts, and sealing the gap of joint with sealant, as shown of figure2(a).

2.2 Integral acoustic liner

The integral acoustic liner had a circle honeycomb panel and a circle perforated panel, which was inserted on the inner wall of fan case, as shown of figure 4. The processing technology, assembly technology and adhesion technology of integral liner is relatively strict, thereby the cost is higher. The inner structure of integral liner was the same to segment liner. Compared with integral liner, the noise reduction of segment liner is lower due to the acoustic scatter from the joint. The disadvantage was overcome by the integral liner, which improves the noise absorption ratio. In order to enhance the noise reduction level of aero engine, segment liner would be gradually replaced by integral liner.

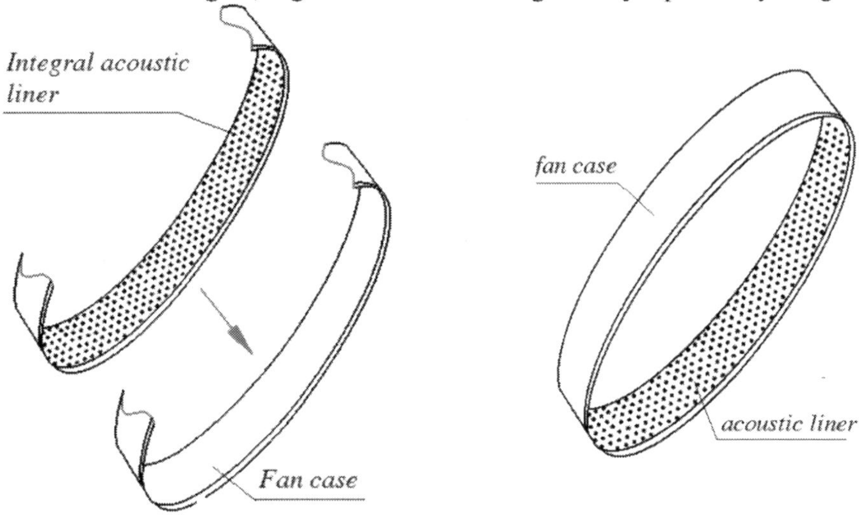

(a) Explosive view of integral liner (b) Assembly view

Figure 4. Structure of integral acoustic liner

3 Composite Skin Fillets

Fan case skin fillets were used for filling the gap between fan frame acoustic liner and strut, and the material was glass fiber reinforced polyether composite. The skin fillet had a lighter weight, better toughness, better anti-fatigue property, and could be repeatedly disassembled and assembled, as shown of figure 1(3). The skin fillet was shell structure and the inner side was reinforced rib and installation base, and the outer side was flow path, as shown of figure 5.

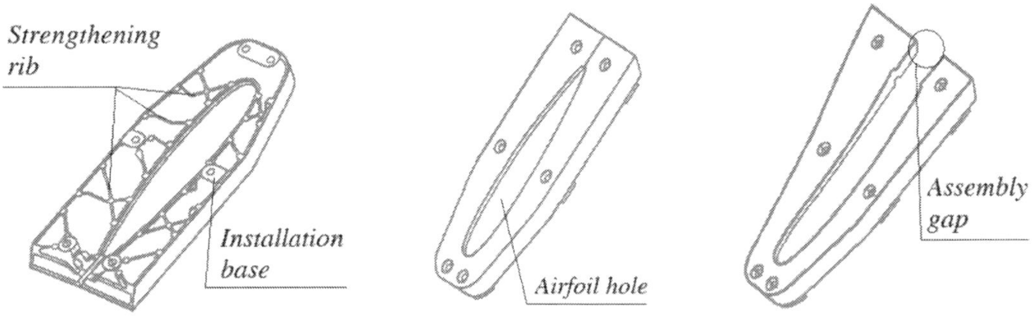

(a) Inner side (b) Outer side (c) Open structure

Figure 5. Structure of fan case skin fillet

Compared with metal material, composite skin fillets could reduce 5~8 kg for single engine.

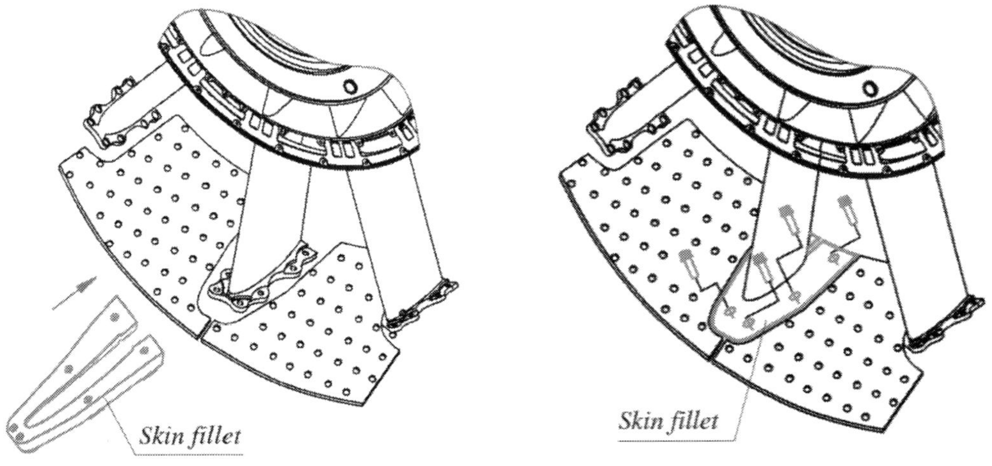

(a) Before assembly (b) After assembly

Figure 6. assembly of fan case skin fillet

Assembly method: opening the skin fillet along airfoil hole, as shown of figure5(c), and then inserting it into strut, finally fixing the skin fillet with bolts, as shown of figure 6.

4 Composite Outlet Guide Vane

Fan outlet guide vanes were the bearing parts of fan frame, adjusting the secondary airflow, as shown of figure 1(6). The vane was made of epoxy resin/carbon fiber composite, and reduced 25~ 40 kg for single engine. The composite vane had higher natural frequency, better damping property, better anti-fatigue and excellent anti-crack ability. For overcoming the weakness of bad anti-impact and easy to delaminate, the nickel overlapping was pasted on leading edge, and the metal bushings were inserted in bolt hole, as shown in figure 7(a).

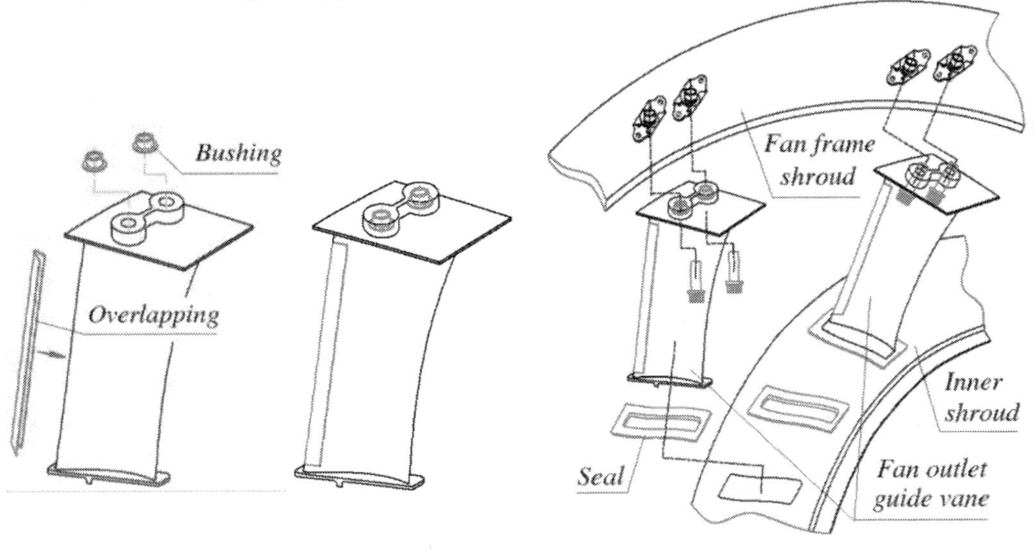

(a) Explosive view (b) Assembly view

Figure 7. Structure and assembly of fan outlet guide vane

The assembly process was as follows: firstly, inserting the inner platform into seal, then, fixing the outer platform to fan frame shroud with bolts, finally, shroud and vanes formed cantilever structure, as shown of figure 7(b).

5 Containment Case

If the titanium alloy fan blade broken off from root, the centrifugal force was more than 40t. In order to prevent the released blade puncturing fan case and ensure the safety of aircraft and passengers, the fan case needed containment structure [6]. Compared with metal containment case, composite containment case could reduce 30~50 kg for single engine.

5.1 Containment ring

The containment ring was formed by winding Kevlar fiber strip, and the inner could be clamped with metal layer to enhance containment ability. The containment liner was inserted into fan case, for preventing radical vibration and deformation, the metal protect ring was fixed on the edge, as shown of figure 1(5). It had a machined groove on inner side, and the groove was coated with talc coating, and the protective coating was coated on both sides of containment liner, after talc coating being completely solidified, as shown in figure 8. Talc coating was blended by carbonate, adhesive and epoxy resin. Through scraping talc coating, fan blade adjusted tip clearance with coating.

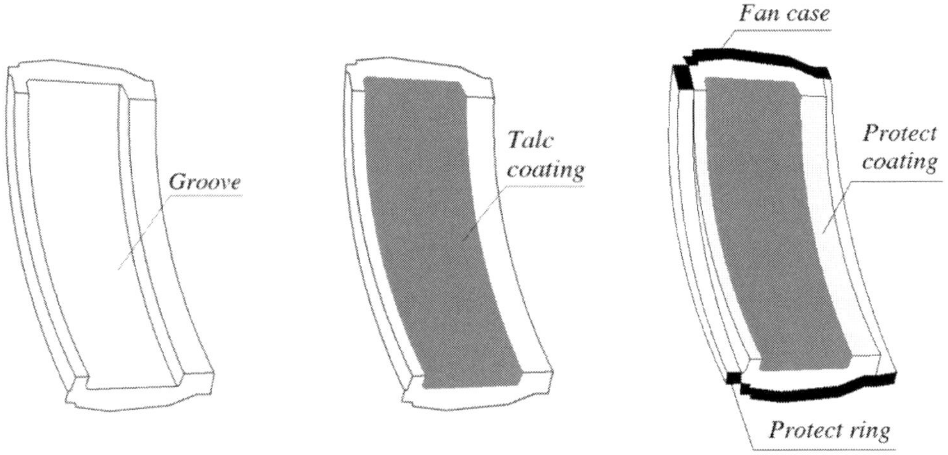

(a) Composited frame (b) Talc coating (c) Assembly

Figure 8. Structure of containment ring

After machining, the outer structure of containment would be destroyed, thereby reducing the containment ability. Three-dimensional weaving technology can avoid mechanical damage to the fiber structure, due to the high cost and technical difficulty, it is too difficult to popularize in a short time.

5.2 Containment layer

The metal fan case is gradually replaced by composite case, and the containment layer was winded around the outer surface of fan case, so as to improve the containment ability. The containment layer was made of carbon fiber or aramid fiber mixed with epoxy resin, and the fiber layer could absorb the impact energy of released blade, as shown in figure 9. Using finite element model with the parameter of fan blade diameter, rotated speed, and the strength of fiber, the amount of fiber layer could be got [7]. Compared with inner containment ring, the containment layer had a simple processing, simplified the structure of fan case, and enhanced the safety and reliability of fan case.

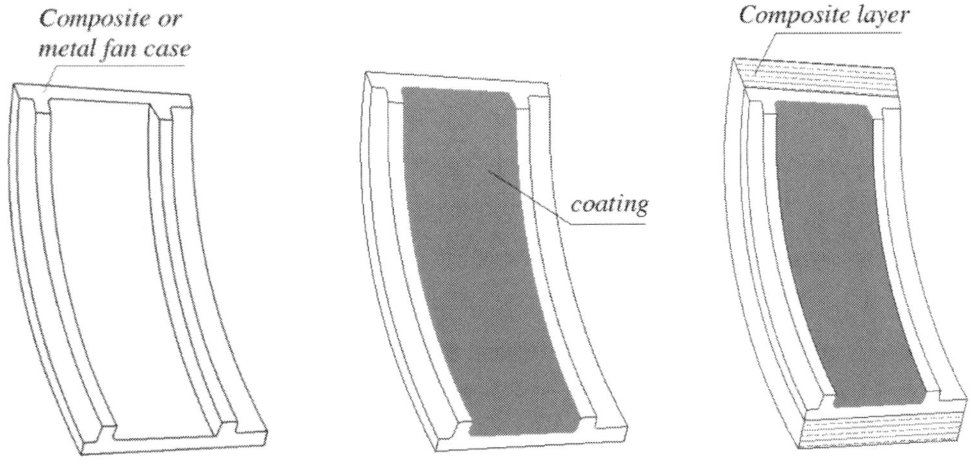

(a) Fan case (b) Coating (c) Containment layer

Figure 9. Structure of containment layer

6 Nitrogen Cooling System

6.1 Pipeline system

Because of the high temperature in spool cavity, the tests for compressor and other components needed cooling nitrogen. The nitrogen cooling pipeline passed through the cavity of fan frame, and then came into the cavity of compressor spool, thereby nitrogen introducing to cool tester. Without changing the structure of fan frame cavity, the nitrogen cooling system was installed in test channel, as shown of Figure 1(7). The main pipe of cooling system was set in strut cavity, and the branch pipe passed through HUB MID BOX cavity, and then inserted into screw holes of NO.3 bearing seal, thereby nitrogen introducing to spool cavity, as shown in figure 10. After the end of the experiment, the cooling system could be disassembled, and the test cable should be assembled again. The assembly and disassembly process of the pipeline system had no influence on the structure of fan frame.

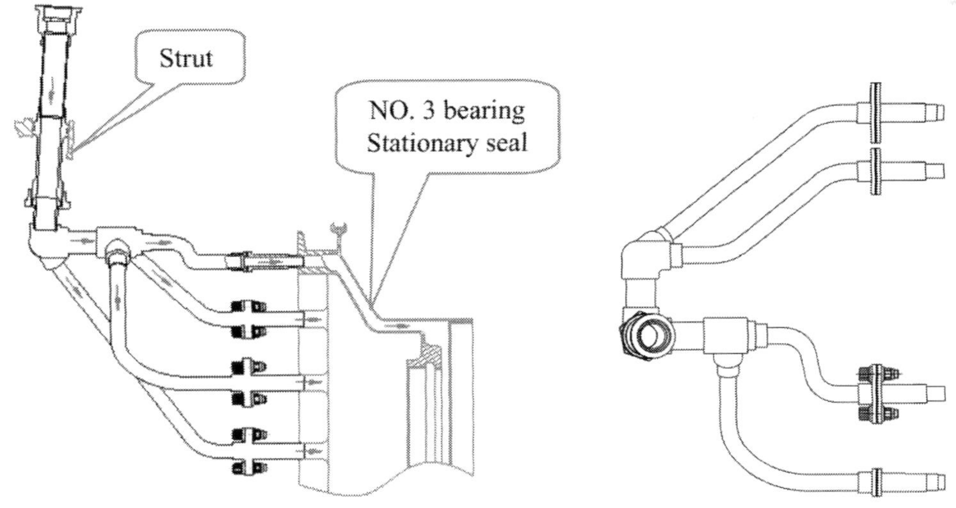

(a) Nitrogen cooling system (b) pipeline system structure

Figure 10. Nitrogen cooling system in fan frame cavity

DOI: 10.1051/matecconf/201711403017

6.2 Seal and fixing structure

Because of the limited space of the cavity in fan frame, there was no available support, so the sealing and fixing of pipeline systems was the difficulty in design. According to the air flow direction, in turn to use the cone seal, rubber ring seal, end face seal and plug seal structure, as shown in Figure 11. The cone seal structure was fixed with a sleeve nut, and solved the sealing and fixing problem simultaneously. The plug seal with small gap and gasket were used on frame interface, and the pipe joint is fixed with pressing broad. The pipe joint is sealed with rubber ring seal and end face seal. For reducing the aerodynamic loss, Pipe line routing is as short and straight as possible [8].

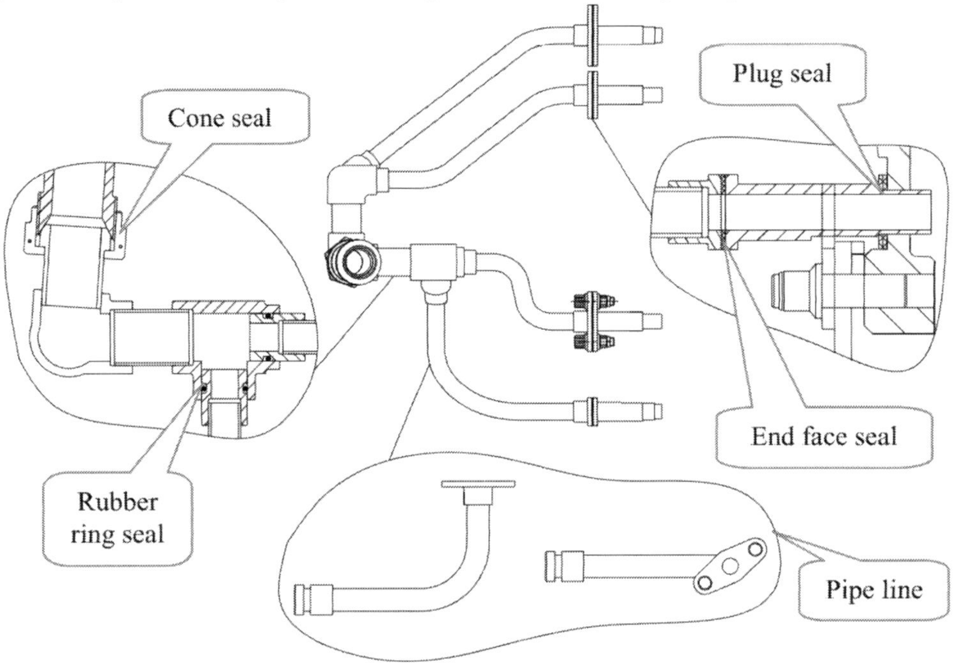

Figure 11. seal structure of pipeline system

7 Conclusion

The fan frame unit of high bypass ratio aero-engine had a larger design space. Using advanced composite could improve thrust-weight ratio, enhance structure performance, and reduce manufacturing cost of engine. The development trend of high bypass ratio aero-engine is super high bypass ratio, and the design technique of composite would become the bottleneck and restrict super high bypass ratio aero-engine. The application range of composite on aero-engine was gradually expanding, such as: fan blade, fan case, front core and thrust reverser.

With the increasing of bypass ratio, the inner cavity of fan frame was gradually enlarging, providing a design platform for more advanced pipeline system. In inner cavity of fan frame, the design of pipeline system could provide more testing projects for the core machine, so as to ensure the engine operating safely and stably.

References

1. DU shan-yi, Advanced Composite Material Sand Aero Space Engineering， Acta Materiae Compositae sinica, 2007,24 （1）
2. HOU Peng, The Application of Acoustic Liner Technology on High Bypass Ratio Aero Engine, 15th Annual Meeting of China Association for Science and Technology, 2013
3. Kinsler,L.E.,Frey,A.R.,Coppens,A.B.,Sanders,J.V., Fundamentals of Acoustics Fourth Edition, John Wiley&Sons,Inc.2000
4. M Perrino, J.E.S.G, Towards Development of an Active Single-layer Acoustic Liner for Jet Engine Noise Reduction, AIAA Aerospace Sciences Meeting Including the New Horizons Forum and Aerospace Exposition, 2013
5. TL Parrott, MG Jones, Status of duct liner technology for application to aircraft engine nacelles, Journal of the Acoustical society of America, 2005,118 (3)
6. Z Fan, D Gao, T Jiang, experimental study and numerical simulation of model casing containment, Journal of Nanjing University of Aeronautics & Astronautics, 2006, 38 (5)
7. YU Ya-Bin, W Chen, Numerical Analysis of the Modeled Blade/Casing Containment, Journal of Aerospace Power, 2005, 20 (3)
8. BAI xiao-lan, Pipe Routing Algorithm for Aero-Engines， Machinery Design & Manufacture， 2013,9 （1）

Experimental Study on Effect of Electrode Location on Ion Current

Hanqing Xu[1], Weijun Fan[1] and Rongchun Zhang[1,a]

[1]*National Key Laboratory on Aero-Engines, School of Energy and Power Engineering, Beihang University, Beijing, China*

Abstract. In an attempt to use ion current method on a suction flame holder for combustion diagnosis, the position of the detection electrode that has the significant effect on ion current, should be investigated. The objective of this study was to experimentally investigate the ion current distribution in different combustion zones by changing the electrode length and streamwise location. An ion current acquisition system was built on basis of the principle of Langmuir probe, and experiments were conducted under different fuel-air ratios ranging from 0.008~0.022. The results suggest that the ion current distribution obeys the same regulation at different fuel-air ratios. With the electrode length increasing, the ion current increases first, then decreases, and finally presents a unimodal curve. As the electrode streamwise location increases, the ion current demonstrates a linear decrease rule. The maximum ion current is obtained in the case that the electrode length and streamwise location equals 60mm and 0.75. Since the larger ion current provides the wider range for setting the misfire threshold, the position where the maximum ion current occurs can be a better choice for arranging the electrode.

1 Introduction

Bluff-body flame holders are commonly used to stabilize flames in high-speed airflows, and one of their engineering applications is turbojet afterburner[1]. Suction flame holder, a variant structure of Vee gutter flame holder, has been successfully applied to a certain turbojet afterburner[2]. Corrugated slots symmetrically arranged both on the tail edge of its sidewalls result in better ignition performance, higher combustion efficiency and improved lean blowout (*LBO*) limit[3]. However, compared with the main combustor, severe inlet conditions for afterburners including lower pressure, lower oxygen and higher velocity have greatly negative effect on *LBO* limit[4]. If blowout occurs, the afterburner will lose thrust augmentation critical to the maneuverability of tactical aircrafts. Whereas if *LBO* is well predicted, the reliability of a fighter will be enhanced.

Many researches have been conducted on predictions of the precursor events prior to incipient *LBO* phenomenon in various combustors. These experimental studies primarily used three methods, which include the acoustic method with a condenser microphone, the optical method with a fused silica optical fiber and the electrical method with a pair of charged electrodes. Moreover, images taken

[a] Corresponding author: zhangrongchun@buaa.edu.cn

by a high-speed ICCD camera and dynamic pressure signals collected by a high-frequency pressure transducer were adopted as auxiliary approaches to understand the transient flame dynamics better.

The first two methods are usually employed together to validate the results by each other. For instance, Muruganandam and Nair used the chemiluminescence and acoustic radiation from a premixed dump combustor to characterize the transient flame behavior approached *LBO* as short duration, localized extinction and reignition events[5]. They also observed the variations in low-frequency acoustic spectrum and increased presence of intermittent events in both acoustic and optical data from a swirl cup combustor[6]. Bompelly and Lieuwen detected the proximity of *LBO* in a premixed swirl combustor using flame chemiluminescence and acoustic signatures, and extended the means for *LBO* limit sensing[7]. Bompelly and Seitzman found that threshold of the low-pass filtered chemiluminescence signal could provide robust precursor identification of *LBO* in a lean direct injection combustor[8]. Yi successfully predicted the proximity to *LBO* based on chemiluminescence signals and global images[9]. Aditya and Muruganandam sensed the precursor events occurred and reoccurred in an interval of several milliseconds before *LBO* using a photodiode sensor and a high-speed imaging camera[10].

The ion current probes, based on chemiionization and thermal ionization, have been applied to combustion diagnosis in some HCCI engines and gas turbine model combustors. As to HCCI engines, Strandh affirmed the existence of ion current sensitive to the fuel-air ratio[11]. Vressner and Andreas examined the effects of fuel type and probe location on the ion current[12,13]. Chen and Panousakis proved that ion current can be a cost effective and adequately informative feedback signal for engine control[14]. Dong constructed the phase relation between ion current signal and combustion phase[15]. With respect to gas turbine model combustors, Chorpening developed an ion current sensor to detect the *LBO* and combustion instabilities[16]. Nair and Rajaram invented a novel ignition system with ion sensing capability used for *LBO* detection[17]. Keshav certified the ion current from a supersonic combustion flow could be used as a flame indicator for feedback combustion control[18].

Optical and acoustic sensors have the advantages of non-direct contact to the high-temperature flames, while are confined to open an observation window. In order to arrange the optical access and devices on a gas turbine combustor, both the combustor casing and liner must be penetrated, which would destroy the structural integrity and strength. Ion current probes have the merits of instant response and simple structure, but are sensitive to their locations mainly due to the heterogeneous diffusion combustion after the flame holder. With a view to the reliability of an aircraft, the ion current probe has more potential for combustion diagnosis than other sensors provided that the best ion sensor location can be determined.

In this paper, an ion current acquisition system was established on basis of the principle of Langmuir probe, and experiments were conducted under different fuel-air ratios. In order to get a better location for the detection probe, ion current distributions in different combustion zones was investigated by changing the electrode length and location. The results could provide guidance for the location arrangement of the detection probes used to diagnose the misfire after a flame holder.

2 Method

2.1 Measurement Principle

As the fuel-air mixture after the flame holder is ignited, a great amount of charged ions stably emerge. The main source of ions is provided by chemiionization. The most important reaction in the process is considered as followed[19]:

$$CH + O \rightarrow CHO^{+} + e^{-} \quad (1)$$

Owing to charge transfer reactions, CHO^{+} is quickly consumed by H_2O via:

$$CHO^{+} + H_{2}O \rightarrow CO + H_{3}O^{+} \quad (2)$$

In SI engines or gas turbine combustors, the high temperature is encountered after ignition, thus thermal ionization may play an important role[20].

After loading an external electric field on the combustion field, these ions will move forth and back, and then a certain type of electric current called ion current flows through the circuit. As shown in Figure 1, a Langmuir probe system, including DC bias power supply (E), sample resistance (R_1), slide rheostat (R_2), positive probe and flame holder, was used to detect the ion current. Table 1 lists the parameter values of the system. By collecting the voltage drop (u) of the sample resistance, the ion current (I) can be expressed as

$$I = u / R_1 \quad (3)$$

During the experiments, a cylindrical metal stick as the central probe was penetrated into the flame region, and another probe was the flame holder connected with the combustor walls. Based on the results of experiments conducted on a constant volume combustion bomb, the combustor wall and central probe respectively served as the negative and positive probe[21].

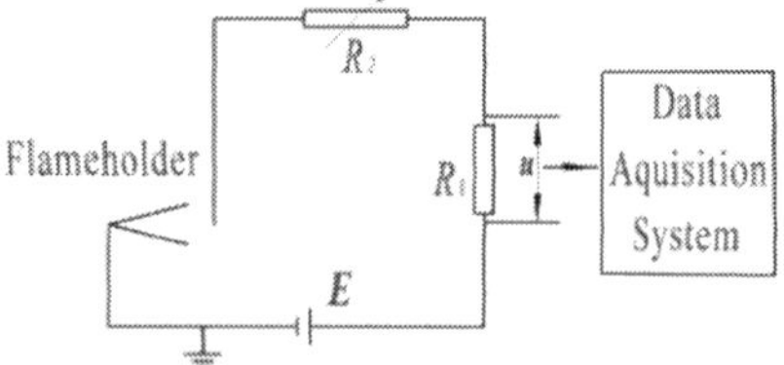

Figure 1. Schematic of the Langmuir probe system

Table 1. Langmuir probe system parameters

Parameter	Value
Bias supply(E)[V]	24
Sample resistance(R_1)[Ω]	110
Slide rheostat(R_2)[Ω]	0-470K
Positive probe(Φ)[mm]	4

2.2 Test Rig Configuration

Test rig mainly consisted of three components: test section, flame holder and oil supply unit.

Figure 2 shows the structure of 2D test section. The spanwise width and vertical height of the outlet section were separately 160mm and 120mm. The positive electrode was mounted vertically on a displacement mechanism.

Figure 3 shows the suction flame holder containing a main flame holder, an outer cover, a fuel injector and a splash plate. Table 2 lists the primary structure parameters of the flame holder. Subscript 0 and 1 severally denote the main flame holder and the outer cover. Three-dimensional Cartesian coordinates were employed in the flow field and the origin of that was placed the middle of the outlet cross section of the outer cover.

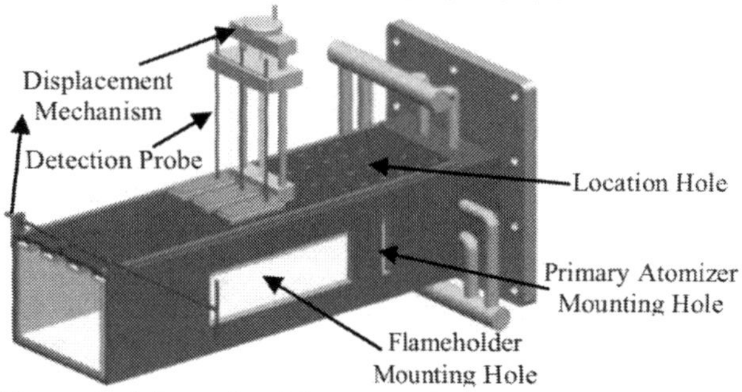

Figure 2. Schematic of the test section

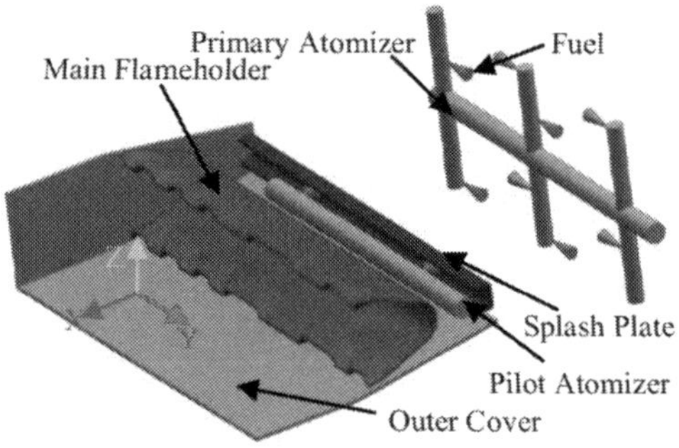

Figure 3. Schematic of the flameholder

Table 2. Flame holder structural parameters

Parameter	Flare angle (α) [°]		Spanwise width (W) [mm]		Groove width (D) [mm]		Streamwise length (L) [mm]	
	α_0	α_1	W_0	W_1	D_0	D_1	L_0	L_1
Value	45	30	156	156	34	40	32	102

Oil supply unit included a pilot atomizer and a primary atomizer. The former was a splash-plate plain orifice injector and the latter was a normal plain orifice injector situated 110mm upstream from the main flame holder lead edge. The orifice diameter was 0.5mm.

2.3 Experimental System

The schematic of experimental system expressed in Figure 4 was comprised of four subsystems.

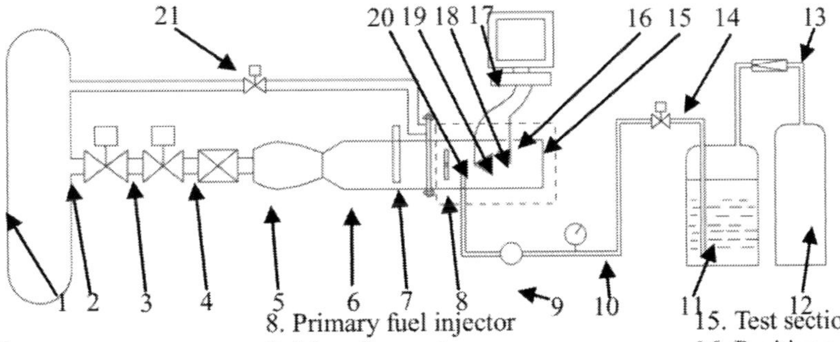

1. Air source
2. Electric stop valve
3. Electric adjustable valve
4. Orifice plate flowmeter
5. Preheat combustor
6. Settling chamber
7. Temperature rake
8. Primary fuel injector
9. Mass flowmeter
10. Pressure gauge
11. Fuel tank
12. Nitrogen bottle
13. Dropping valve
14. Manual needle valve
15. Test section
16. Positive probe
17. IPC
18. Igniter
19. Flame holder
20. Pilot fuel injector
21. Manual stop valve

Figure 4. Schematic of the experimental system

Air Supply System. Air from main air source was divided into the mainstream air and cooling air. The flow rate of mainstream air was controlled by an electric adjustable valve and monitored by an orifice plate flowmeter. The flow rate of cooling air was regulated by a manual stop valve. The mainstream vitiated air was heated by a preheat combustor and the temperature was measured by 22 K-type thermocouples.

Fuel Supply System. It mainly composed of high-pressure nitrogen gas bottle, fuel tank, mass flowmeter and fuel injector. The fuel type was aviation kerosene and the mass flow rate was monitored by a DMF-1A Coriolis mass flowmeter.

Ion Current Transducer System. The key components of this system is depicted in Figure 1.

Data Acquisition System. The voltage induced by ion current was collected by a DH3816 static stress testing system. In consideration of the periodical vortex shedding behind the flame holder, each data point was measured continuously 5 times and the arithmetic mean was taken as the steady state value. Temperature signal was acquired by the ADVANTECH IPC.

2.4 Experimental Conditions

The experimental conditions, including inlet Mach number (Ma=0.2), inlet total temperature (T_{in}=623K), electrode spanwise location ($\tilde{Y}=-1.25$), and fuel-air ratio (f=0.008-0.022), remained the same for all experiments.

The fuel-air ratio (f) was defined as

$$f=\dot{m}_f/\dot{m}_a \tag{4}$$

where, $\dot{m}_f$ was the total mass flow rate of aviation kerosene from the pilot and primary atomizers, $\dot{m}_a$ was the mass flow rate of preheated main airstream. When f=0.008, fuel was injected only by the pilot atomizer. But when f overs 0.008, the mass flow rate of pilot atomizer remained unchanged with f=0.008, and then that of primary atomizer was increased to the preset values.

The electrode streamwise location ($\tilde{X}$) and spanwise location ($\tilde{Y}$) were individually defined as

$$\tilde{X} = X/D \quad (5)$$

$$\tilde{Y} = Y/D \quad (6)$$

where, X and Y were separately the streamwise and spanwise coordinate value of the electrode axis, D was the groove width of the flame holder outer cover.

The electrode length (EL) was defined as

$$EL = H/2 - Z \quad (7)$$

where, H was the vertical height of outlet section for the test section, Z was the vertical coordinate value of the electrode end inside the test section.

3 Results and Discussions

3.1 Relationship between *I* and *EL*

Effect of the EL is shown in Figure 5. Figure (a), (b) and (c) respectively display the data from $\tilde{X}$ =0.75, 1.50 and 2.25.

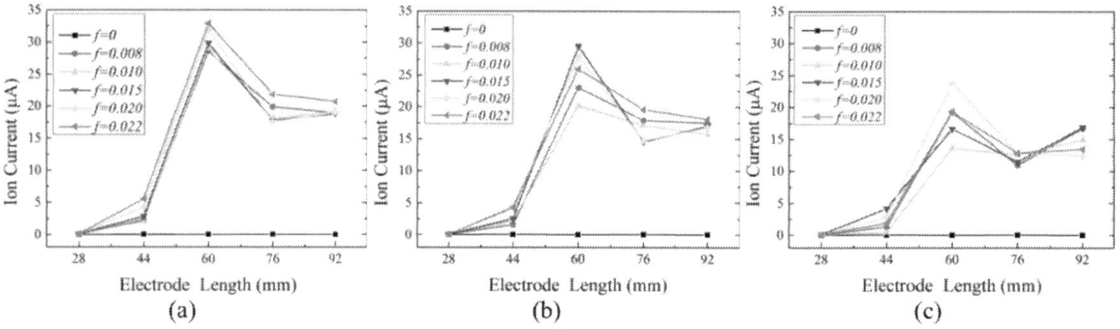

Figure 5. Ion current influenced by electrode length for $\tilde{X}$ =0.75(a), 1.50(b) and 2.25(c)

In an attempt to exclude the possible effect of the vitiated air from the preburner, the I was collected with different EL when f=0. Results shown in Figure 5 suggest that when f=0, the I remains zero throughout with the EL increasing. Consequently, the preheated vitiated air had none effect on the I. One of the necessary conditions for ion current formation is the existence of charged ions. As mentioned above, chemiionization in flame front and thermal ionization under high temperatures are the main sources to provide ions. When only the preburner was operating, the flame couldn't propagate to the flame holder, and simultaneously vitiated air from the preburner couldn't be dissociated at the relatively low temperature of 623K. Therefore, the ion current obtained from the recirculation zone was zero, namely I=0.

According to the any diagram in Figure 5, when the f ranges from 0.008 to 0.022, the shapes of the relation curve between I and EL are approximately the same, which is mainly due to the similar fuel concentration distributions at different f values.

As EL increases, the I increases first, and then suddenly decreases when EL overs 60mm. Finally, it demonstrates a unimodal curve. The peak of ion current appears when EL=60mm.

When EL=28mm, the whole electrode was still not getting into the combustion zone and surrounded by the vitiated air from the preburner. There were not charged ions around the electrode, thus the I remained zero, I_{28}=0. When EL=44mm, the end of electrode had gone across the flame front and part of it was in recirculation zone. Ions produced in flame front and recirculation zone were captured by the electrode and the ion current formed, I_{44}>0. When EL=60mm, the area of surface absorbing the charged particles increased further, which led to the continuous growth of the I, I_{60}>I_{44}. When EL>76mm, adversely, the ion current decreased compared with that when EL=60mm. Reasons for this phenomena can be attributed in two aspects. One is that when EL=60mm, the electrode end was exactly in the middle section along the vertical direction, and thus exposed to the higher temperature burned gas more likely to dissociate into ions compared with the other situations. The other one is the end effect caused by the cylindrical electrode. Experimental results on a simple burner indicate the end effect from a cylindrical electrode resulted in the increased ion current by 75%[22].

3.2 Relationship between *I* and $\tilde{X}$

Effect of $\tilde{X}$ is shown in Figure 6. Figure (a), (b) and (c) respectively display the data from *EL*= 60mm, 76mm and 92mm.

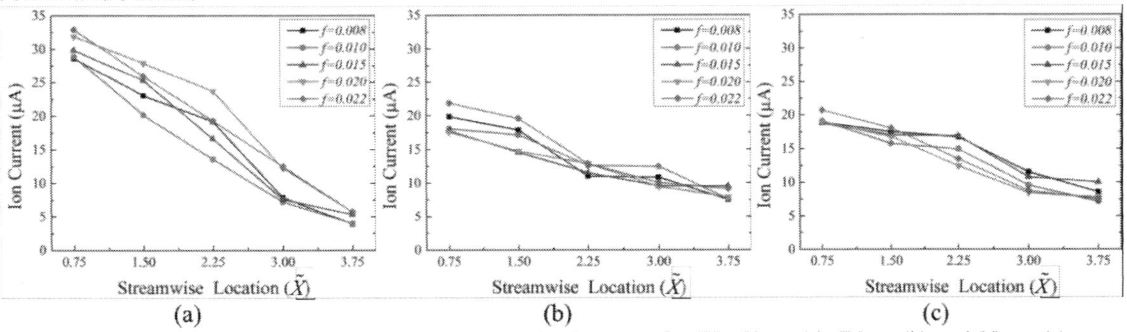

Figure 6. Ion current influenced by electrode streamwise location for *EL*=60mm(a), 76mm(b) and 92mm(c).

As illustrated in any diagram in Figure 6, the relationships between *I* and $\tilde{X}$ under various *f* ranging from 0.008 to 0.022 follow the same rule, which is also attributed to the similar fuel concentration distributions at different *f* values.

With the $\tilde{X}$ increasing, the *I* decreases linearly. When *EL*=60mm, *I* drops from 28~33μA to 4~6μA. Then *EL* grows to 76mm, *I* declines from 17~22μA to 8~10μA. After *EL* reaches to 92mm, *I* reduces from 19~21μA to 7~10μA. The maximum reduction of the *I* appears with *EL*=60mm by an average of 25.5μA.

The *I* declines principally is attributed to the electric field intensity weakening. The flame divergence angle after a V-type flameholder is generally 3~5°and recirculation zone length of that is commonly 3 times of the groove width[23]. That implies the length of the electrode immerged into the high-temperature zones including combustion zone and recirculation zone is nearly the same, even $\tilde{X}$ changes in recirculation zone. Therefore, the increase of $\tilde{X}$ merely denotes the growth of distance between the positive and negative electrodes. The electric field intensity imposed by the bias voltage and the electrode gap are in inverse proportion. Moreover, *I* is in direct proportion to the intensity. Hence, *I* and $\tilde{X}$ are in inverse proportion, which means the *I* declines linearly with the increase of $\tilde{X}$. The mathematical analysis and simulation of the ion current in an SI engine proves that the ion current is inversely proportional to the clearance of the electrodes[24].

4 Conclusions

The experimental study on the effect of electrode location on the ion current was conducted in the suction flame holder. The main conclusions are summarized as followed:

(1) When the fuel-air ratio ranges from 0.008 to 0.022, the relationships between ion current and electrode length are approximately the same. With the electrode length increasing, the ion current demonstrates a unimodal curve. The peak of the ion current appears when electrode length equals 60mm.

(2) When the fuel-air ratio ranges from 0.008 to 0.022, the relationships between ion current and electrode streamwise location are almost the same. With the streamwise location increasing, the ion current demonstrates a linear decreasing curve. The maximum reduction of the ion current appears when electrode length equals 60mm by an average of 25.5μA.

(3) Because the larger ion current provides the wider range for setting the misfire threshold, the electrode length and location for this flame holder can be chosen as *EL*=60mm and $\tilde{X}$ =0.75.

Acknowledgement

This research was financially supported by the National Natural Science Foundation of China (515060 03).

References

1. H. Lefebvre, D.R. Ballal, *Gas Turbine Combustion: Alternative Fuel and Emission*, *Third Edition*, 173 (CRC Press, Boca Raton, 2010)
2. Z.Q. Fu, Areoengine **3**, 28-32 (1998)
3. Q.F. Cheng, J. Aerosp. Power. **3**, 259-262 (1991)
4. J. Lovett, T. Brogan, D. Philippona, B. Kiel, T. Thompson, *40th AIAA/ASME/SAE/ASEE Joint Propulsion Conference and Exhibit*, 4192
5. T. Muruganandam, S. Nair, Y. Neumeier, T. Lieuwen, J. Seitzman, *38th AIAA/ASME/SAE/ASEE Joint Propulsion Conference & Exhibit*, 3732 (2002)
6. S. Nair, T.M. Muruganandam, R. Olsen, A. Meyers, J. Seitzman, B. Zinn, T. Lieuwen, *42nd AIAA Aerospace Sciences Meeting and Exhibit*, 138 (2004)
7. R. Bompelly, T. Lieuwen, J. Seitzman, *47th AIAA Aerospace Sciences Meeting Including The New Horizons Forum and Aerospace Exposition*, 2009-0982 (2009)
8. R.K. Bompelly, J.M. Seitzman, *ASME 2011 Turbo Expo:Turbine Technical Conference and Exposition*, GT2011-46702 (2011)
9. T.X. Yi, E. Gutmark, *42nd AIAA/ASME/SAE/ASEE Joint Propulsion Conference & Exhibit*, 4738 (2006)
10. K. Aditya, T. Muruganandam, *ASME 2012 Gas Turbine India Conference*, GTINDIA2012-9585 (2012)
11. P. Strandh, M. Christensen, J. Bengtsson, R. Johansson, A. Vressner, P. Tunestal, B. Johansson, SAE Technical Paper, 2003-01-3216 (2003)
12. A. Vressner, A. Hultqvist, B. Johansson, SAE Technical Paper, 2005-01-2093 (2005)
13. A. Vressner, P. Strandh, A. Hultqvist, P. Tunestal, B. Johansson, SAE Technical Paper, 2004-01-0934 (2004)
14. D. Panousakis, A. Gazis, J. Patterson, R. Chen, J. Turner, N. Milovanovic, D. Blundel, SAE Technical Paper, 2006-01-0024 (2006)
15. G.Y. Dong, Y.L. Chen, Z.J. Wu, L.G. Li, R. Dibble, Proceedings of the 35th International Symposium on Combustion **3-35**, 3097-3105 (2015)
16. B.T. Chorpening, J.D. Thornton, K.J. Benson, *4rth IEEE Conference on Sensors*, 1597867 (2005)
17. S. Nair, R. Rajaram, A. Meyers, T. Lieuwen, L. Tozzi, K. Benson, *43rd AIAA Aerospace Sciences Meeting and Exhibit*, 932 (2005)
18. S. Keshav, Y. Utkin, I.V. Adamovich, *46th AIAA Aerospace Sciences Meeting and Exhibit*, 2008-1056 (2008)
19. A. Fialkov, PROG. ENERG. COMBUST. **5-6-23**, 399-528 (1997)
20. A. Saitzkoff, R. Reinmann, F. Mauss, M. Glavmo, SAE Technical Paper, 970857 (1997)
21. S. Yoshiyama, E. Tomita, Y. Hamamoto, SAE Technical Paper, 2000-01-2828 (2000)
22. Y.M. Jin, *Probe Technologies in Pyrometry*, 196 (China Metrology Press, Beijing, 1988)
23. S.M. Bush, E.J. Gutmark, AIAA J. **3-45**, 662-672 (2007)
24. J.W. Li, S.L. Liu, T.Y. Wang, W.Q. Liu, S.H. Yang, J. Tianjin Univ.(Sci. Techn.) **S1-39**, 66-70 (2006)

Measurement and Description of Human Hand Movement

Qiang Zhan[1], Chao Zhang[1,a] and Qinhuan Xu[1]

[1] *Robotics Institute, Beihang University, Beijing, China*

Abstract. The actual movement and the coupled relationship of human fingers are very important for the analysis of human hand grasp, but there are very few effective ways to detect it. The paper introduces a method with wearable measuring sensors to realize the real time measurement and description of human fingers' movement during hand grasping. A data glove for real-time measuring the movement of each joint of all the fingers was designed. The finger movements of three typical human hand grasps were measured, and the coupled relationship and the complete motion trajectory of all five fingers were got and described.

1 Introduction

When grasping, human hand usually has two restriction modes: inter-finger-restriction and intra-finger-restriction, due to the complex distribution of biomechanical drive mode and structural constraints between bones and joints inside the hand. The inter-finger-restriction refers to the constraint among finger joints, which makes human hand have a specific range of motion, and the intra-finger-restriction refers to the combined constraint co-produced by tendons and muscles, which reflects the coupled motion produced by the ligaments and tendons' drive among middle finger, ring finger and little finger [1-2]. It's very difficult to accurately observe and describe the true motion of human hand during grasping. Therefore, to accurately measure and describe the movement of each finger and the coupled relationship among fingers during grasping is of great significance for the study of human hand grasp.

Human hand has a very complex structure which is composed of 27 bones, 17 joints and more than 20 DOFs (degree of freedom), and it can complete a variety of complex and sophisticated grasps. Human hand model has great significance for the research of human hand grasp. The complete motion description of human hand needs real-time processing of various data and results in a very large amount of calculation, so it is usual to simplify the human hand model in order to improve the processing speed. To simplify the human hand model, the elements that have little impact on the hand motion are usually ignored. Currently, there are a number of different ways to model the human hand. Bray [3] and Du, Yuan et al. [4] established a hand model with 30 DOFs. Kuch, J. [5] established a human hand model with 23 DOFs. DLR laboratory [6] established a human hand model having 24 DOFs. Van der Smagt P. [7] established a human hand model of 23 DOFs. Chalfon J. [8] established a human hand model with 22 DOFs.

For any human hand model, if the joint movements of the human hand could be obtained, then the motion of the whole hand can be described. There are some researches on the description and

[a] Corresponding author: zhangchao872008@sina.com

 DOI: 10.1051/matecconf/201711404002

measurement of human hand. Nishikawa et al [9] proposed a method based on curvature transition to solver the posture of human hand. Kasprzak et al. [10] developed a computer vision system to detect the static posture change order of human hand by identifying the sequence of color images. N. Miyata et al.[11] used magnetic resonance (MR) instrument to track the hand joint posture. In the previous research of human hand movement, it's very difficult to accurately detect and describe the real joint angles change and the relationship of human hand joints during different hand graspings.

In order to analyze human hand grasp, we developed a hand movement measuring system, proposed a matrix to describe hand posture, and did experiments on three typical human hand grasps with the measuring system and description method. This paper is organized as follows. Section 2 established a simplified model of the human hand and presented a kinematic description method of hand movement. Section 3 introduced the joint angel measuring system. Section 4 introduced the experiments on three typical human hand grasps and presented the real motion trajectories and coupled relationships of five fingers. Section 5 gives the conclusion.

2 Human hand modelling

2.1 Structure model of human hand

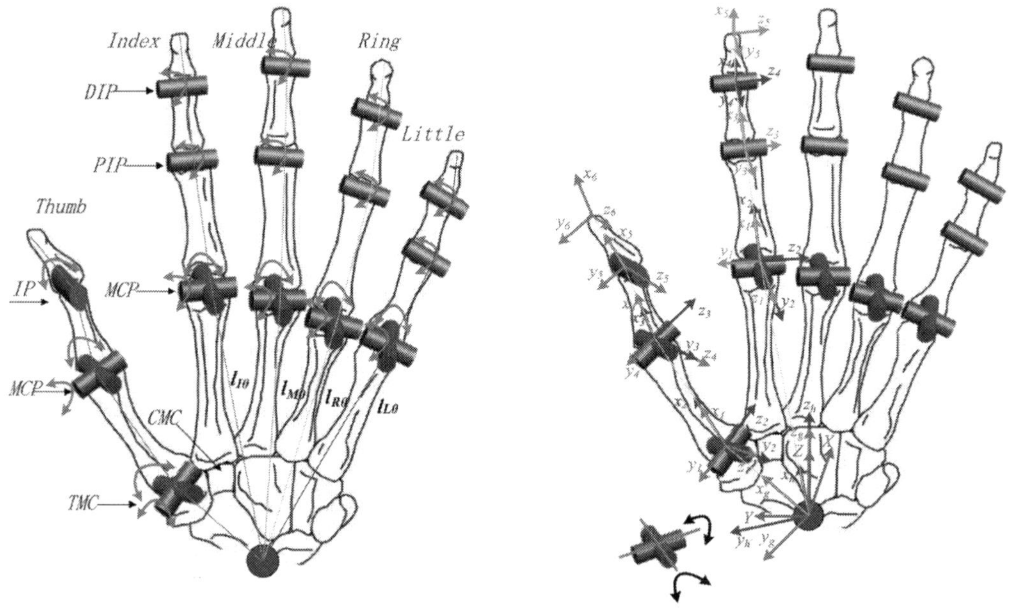

a) Structure model of human hand **b)** Kinematic model of human hand

Figure 1 Model of human hand

By analyzing the anatomical structure of human hand and referring to the medical experience, a simplified structure model ofhuman hand with 21 DOFs was established by following these rules: 1) The thumb is independent of other fingers, 2) The adduction and abduction of each finger are independent, 3) Motion frequency does not affect the relationship of joints, 4) The hand is independent to the wrist and the arm. As shown in Figure 1-a, the thumb has 5 DOFs, of which each of the carpometacarpal joint (TMC) and the metacarpophalangeal joint (MCP) has 2 DOFs (flexion/extension, adduction/abduction), and the interphalangeal joint (IP) has 1 DOF (flexion/extension); each of the other four fingers has 4 DOFs, of which each of the interphalangeal joint (PIP) and the distal interphalangeal joint(DIP) has 1 DOF (flexion/extension), and the metacarpophalangeal joint (MCP) has 2 DOFs (flexion / extension and adduction / abduction).

2.2 Kinematic model of human hand

Referring to the simplified structure model of human hand (Figure 1-a), the kinematic model of human hand was established and shown in Figure 1-b. Suppose the reference frame origin is at the wrist position, and l_{T0}, l_{T1}, l_{T2} and l_{T3} sequentially represent the distance between the carpometacarpal joint and wrist, the metacarpal length, the proximal phalanx length, the distal phalanx length. The other four fingers use the similar description method, l_{F0}, l_{F1}, l_{F2}, l_{F3} ($F=l, M, R, L$) sequentially represent the distance between the wrist joint and MCP, the proximal phalanx length, the middle phalanx length, the distal phalanx length. Referring to the hand measuring method of [12-14] and GB-T16252-1996, we measured the dimension variables of 3 people's hands (right) 10 times, and after the statistical treatment of the measurement data the values of those variables are shown in Table 1.

Table 1 The measured values of the subject hand(mm)

Name	Hand length	l_{T0}	l_{T1}	l_{T2}	l_{T3}	L_{I0}	L_{I1}	L_{I2}	L_{I3}	L_{M0}	L_{M1}
Size	200.5	18.2	57.1	32.1	35.1	11.0	31.3	23.2	23.3	110.	35.0
Name		L_{M2}	L_{M3}	L_{R0}	L_{R1}	L_{R2}	L_{R3}	L_{L0}	L_{L1}	L_{L2}	L_{L3}
Size		25.0	29.1	108	29.2	26.1	27.8	10.8	26.8	18.1	20.1

Referring to Figure 1-b, the Denavit-Hartenberg (D-H) method was used to establish the kinematic models of the thumb and other four fingers respectively. The D-H transformation matrix is:

$$T_{i-1}^{i}=\begin{bmatrix} \cos\theta_i & -\sin\theta_i & 0 & a_{(i-1)} \\ \sin\theta_i\cos\alpha_{(i-1)} & \cos\theta_i\cos\alpha_{(i-1)} & -\sin\alpha_{(i-1)} & -\sin\alpha_{(i-1)}d_i \\ \sin\theta_i\sin\alpha_{(i-1)} & \cos\theta_i\sin\alpha_{(i-1)} & \cos\alpha_{(i-1)} & \cos\alpha_{(i-1)}d_i \\ 0 & 0 & 0 & 1 \end{bmatrix} \quad (1)$$

Substitute the corresponding D-H parameters of each finger into the transformation matrix and multiplyeach joint's transformation matrix in turn as shown in Equation (2), then the motion of the finger can be described.

$$T_0^N = T_0^1 T_1^2 T_2^3 \cdots T_{N-1}^N \quad (2)$$

Transformation matrix T_0^N which represents the position and posture of the finger tip is a function of N variables of the finger joints, where N is the number of finger joints. If we can measure the joint angles of a hand at any moment, then we can calculate the corresponding motion of any finger according to formulas (1) and (2).

3 Human hand movement measuring system

As shown in Figure 2, a human hand movement measuring system composed of a computer and a data glove was developed. The data glove uses fifteen 9-axis magnetic sensor (MEMS) units to measure joint angles of a hand by sticking one sensor unit on the middle of each phalanx of a finger, and each sensor unit consists of an acceleration sensor, a gyro sensor and a magnetic sensor. The computer is responsible for the receiving, processing and displaying of the measured data transmitted from the data glove, and it communicates with the data glove through RS232.

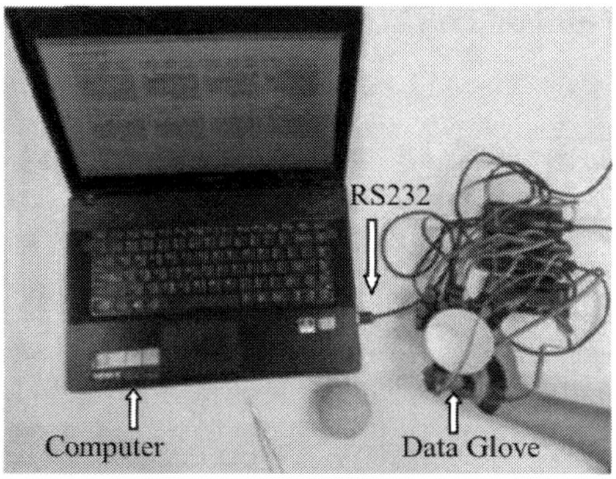

Finger 2 Human hand movement measuring system

Using this data glove, we can obtain each joint's original movement data when a hand is grasping. Then the filtered data with Dynamic Kalman filter was fused with the acceleration data, angular velocity data and geomagnetic data toreduce the constant drift error and the interference of noise. Finally, the obtained joint movement data was substituted into the finger kinematic model so as to get the motion trajectory of each finger. Figure 3 shows the data processing flow ofthe human hand movement measuring system.

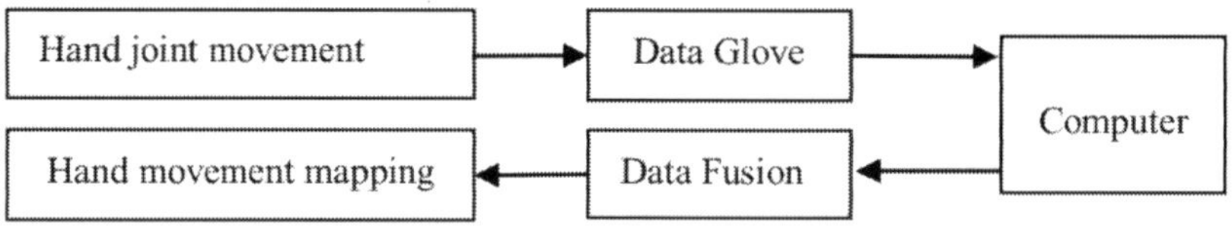

Finger 3 Data processing flow of the human hand movement measuring system

4 Examples of measurement, description and coupled relationship of hand movement

Human hand can do numerous types of grasps. We select the Schlesinger taxonomy as the grasping library to do the measuring experiment which were cylindrical grasp, tip, hook and snap, palmar, spherical grasp, lateral [15].

Human fingers have coupled movements because of the presence of inter-finger-restriction and intra-finger-restriction, which have two main forms as: 1) the coupled movements of distal interphalangeal joint with interphalangeal joint caused by inter-finger-restriction, 2) Multiplefingers' coupled movements caused by intra-finger-restriction (such as the coupled movements of ring finger when the middle finger is in flexion). Such coupled movement relationships of fingers were got from the experimental data as follows.

(1) Coupled movements of inter-finger

The DIP and PIP has coupled flexion when the human hand completes the cylindrical grasp, and the relationship is shown in Figure 4-a and equation (4), where θ_{DIP} is the angle of the DIP flexion and θ_{PIP} is the angle of the PIP flexion.

$$\theta_{DIP} \approx \frac{2}{3}\theta_{PIP} \tag{4}$$

The TMC and MCP have coupled flexion when the human hand completes the spherical grasp and the relationship is shown in Figure 4-b and equation (5), where θ_{FE}^{T-TMC} is the angle of the TMC flexion of the thumb and θ_{FE}^{T-MCP} is the angle of the MCP flexion of the thumb.

$$\theta_{FE}^{T-MCP} \approx \frac{9}{10}\theta_{FE}^{T-TMC} \quad (5)$$

(2) Coupled movements of intra-finger

The MCPs of the middle finger and the ring finger have coupled flexion when the human hand completes the palmar grasp and the relationship is shown in Figure 4-c and equation (6), where θ_{FE}^{M-MCP} is the angle of the MCP flexion of middle finger, θ_{FE}^{R-MCP} is the angle of the MCP flexion of ring finger.

$$\theta_{FE}^{M-MCP} \approx \frac{3}{2}\theta_{FE}^{R-MCP} \quad (6)$$

The MCPs of the ring finger and the little finger has coupled adduction/abduction when the human hand completes the spherical graspand relationship is shown in Figure 4-d and equation (7), where θ_{AA}^{R-MCP} is the angle of the MCP adduction/abduction of little finger.

$$\theta_{AA}^{R-MCP} \approx \theta_{AA}^{L-MCP} \quad (7)$$

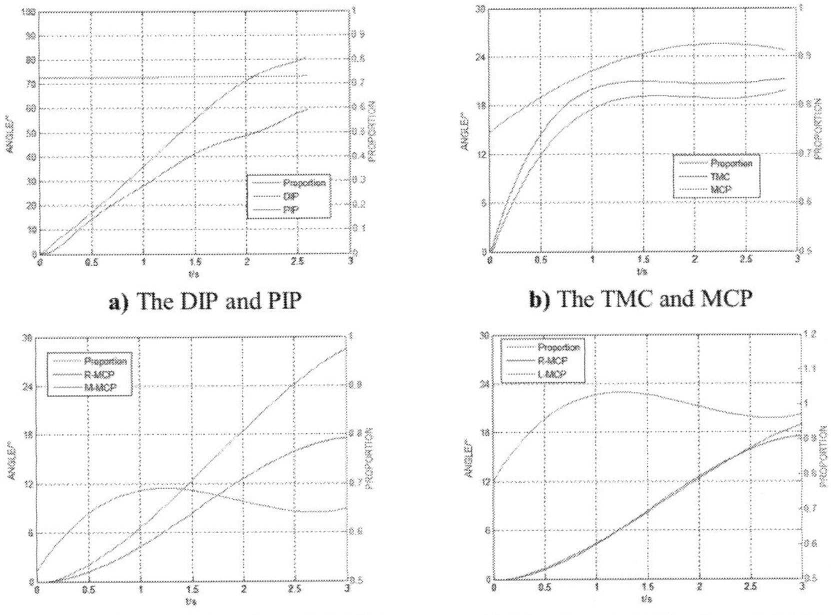

a) The DIP and PIP
b) The TMC and MCP
c)The middle and ring finger's MCP
d) The ring and little finger's MCP

Figure 4 The finger's Coupled flexion

Those above coupled movement relationships got from experiments will be helpful not only for analyzing the human hand grasp as well as the design of bionic hand but also for designing human hand rehabilitation devices.

5 CONCLUSION

In order to survey the actual movements of human fingers, a measuring system and motion description method were presented. Firstly, based on the anatomical structure of the human hand a structure model with 21 DOFs and the kinematic model of each finger with D-H method were established, then a human hand movement measuring system was designed, and last 6 typical human hand grasps were used as grasping library to be measured, and the motion trajectory of each finger and the coupled movement relationships of fingers in different types of grasps were got. The conclusion of coupled movement relationships of fingers has guiding significance for research work such as bionic hand design, human hand rehabilitation and so on. The hand measuring and description method presented in the paper can also be used for the measuring and analysis of other types of human hand grasps.

References

1. Cutkosky, M.R., Wright, P.K.: Modeling Manufacturing grips and correlations withdesign of robotic hands. In: Procedings of the IEEE International Conference onRobotics and Automation, pp. 1533–1539 (1986)
2. Virtual Technologies. Reference Manual. CG081998-2-1 (1998)
3. M Bray, E Koller-Meier, P Muller, NN Schraudolph, L Van Gool: Stochasticoptimization for high-dimensional tracking in dense range maps. In: Proc. IEEE.Vis. Image Signal Process, pp. 501–512 (August 2005)
4. Du, H., Charbon, E.: 3d Hand Model Fitting for Virtual Keyboard System. In:IEEE workshop on Applications of Computer Vision, pp. 31–36 (February 2007)
5. Kuch, J.J., Huang, T.S.: Human Computer Interaction via the Human Hand: A Hand Model. In: Signals, systems and Computers, Conference Record of the Twenty Eighth Asilomar Conference, pp. 1252–1256 (1994)
6. http://www.dlr.de/rm/desktopdefault.aspx/tabid-3820/6244_read-9009/
7. van der Smagt P, Stillfried G (2008) Using MRI data to compute ahand kinematic model. In: 9th conference on motion and vibrationcontrol (MOVIC), München, Germany
8. Chalfon J., Renault M., Younes R., Ouezdou F.B.. Muscle force prediction of the human hand andforearm system in highly realistic simulation [A]. IEEE/RSJ International Conference onIntelligent Robots and Systems Proceedings [C], 2004: 1293-1298
9. A. Nishikawa, A. Ohnishi, F. Miyazaki, "Description and recognition ofhuman gestures based on the transition of curvature from motion images,"Proceedings of IEEE International Conference on Automatic Face and Gesture Recognition, 552-557, 2003.
10. W. Kasprzak, A. Wilkowski, K. Czapnik, "Hand Gesture RecognitionBased on Free-Form Contours and Probabilistic Inference," InternationalJournal of Applied Mathematics and Computer Science, v. 22(2), 437-448, 2012.
11. N. Miyata, M. Louchi, M. Mochimaru, and T. Kurihara, "Finger Joint Kinematics from MR Images," in IEEE/RSJ International Conference on Intelligent Robots and Systems (IROS), 2005, pp. 2750 – 2755
12. J. W. Garrett, "Anthropometry of the hands of male air forceflight personnel," Tech. Rep., Aerospace Medical Research Laboratory, Air Force Systems Command, Wright-Patterson AirForce Base, Ohio, USA, 1970.
13. J. W. Garrett, "Anthropometry of the hands of female air forceflight personnel," Tech. Rep., Aerospace Medical Research Laboratory, Air Force Systems Command, Wright-Patterson AirForce Base, Ohio, USA, 1970.
14. S. R. Habib and N. N. Kamal, "Stature estimation from hand andphalanges lengths of Egyptians," Journal of Forensic and LegalMedicine, vol. 17, no. 3, pp. 156–160, 2010.
15. Taylor C. L., Schwarz R. J.. The anatomy and mechanics of the human hand[J]. Artificial limbs, 1955, 2(2): 22-35

MATEC Web of Conferences **114**, 04003 (2017) DOI: 10.1051/matecconf/201711404003
2MAE 2017

Deformation Test System of Transmission Tower Based on Binocular Vision Method

Haikun Jia[1], Lei Zhang[1,a], Feng Li[2], Dongying Zhang[1], Delu Xu[1], Zhiyong Wang[3], Jianwei Chang[1] and Yusuo Ren[1]

[1]*Shandong Electrical Engineering & Equipment Group Co., Ltd (SDEE), 100071, Beijing, China*
[2] *STATE GRID Corporation of China, 100031, Beijing, China*
[3]*Department of Mechanics, Tianjin University, 300072, Tianjin, China*

Abstract. A test system to measure the deformation of the transmission tower based on binocular vision method is introduced in this work. The test system can achieve the accurate monitoring of the transmission tower deformation, and provide deformation data for the reliability analysis of the transmission tower provides both in static and dynamic measurement. The test system is successfully applied in a tower type test.

1 Introduction

The deformation measurement and the estimate of load capacity for the towers in transmission lines are more and more important. In former years the deformation of transmission towers are usually studied using numerical method, because it can calculate the 3D deformation with low cost. However, researches shown that the numerical results cannot agree with the experiments very well under complex loading conditions[1, 2]. Therefore, real scale tower test for transmission lines under several load cases are more and more performed to estimate the loading capacity in recent years. The measurement of the field deformations of transmission line towers in the real scale tower test is necessary.

The traditional deformation measurement technologies such as the displacement sensor and the strain gauge can only measure deformation in one-dimension within a very limited range at certain points which are selected before test, and it cannot get the full field of the transmission tower deformation [2, 3]. Conventional sensors are usually contact-type which need to be installed on the structure to be tested that must bring certain damage to structure inevitably. In order to avoid these questions, optical measurement method has been reported by a lot of researchers [1]. In addition to its full-field and non-contact measurement capability, the 2D-DIC method (Digital Image Correlation) is well known for its ease of preparation,installation automatic processing, and pervasive applications. 2D-DIC method is mainly used to measure the in-plane displacements and strains. In order to overcome this limitation of DIC in two dimension measurement, researchers developed DIC into

[a] Corresponding author: zhanglei@sdee.sgcc.com.cn

DOI: 10.1051/matecconf/201711404003

three-dimension coordinate (3D-DIC), in which the binocular vision method is introduced into DIC mainly[4].

This work proposes to use measure the deformations of a transmission tower using 3D-DIC, which can measure the position parameters in global coordinate with feature points of the tower. By using the calibration data of binocular cameras and image parallax maps, the 3D coordinates of any points and variations after deformation can be obtained.

2 3D digital image correlation method

2.1 Binocular visions

Fig. 1 (b) illustrates the main principle of binocular vision. O_{cl} and O_{cr} are the left and right optical centers of the two cameras respectively, see In Fig. 1 (b). An object point A is captured as the point A_l and A_r in the image planes of the two cameras respectively. The 3D-DIC is used to reconstitute the 3 dimensional coordinate values at point A in global coordinate with respect to coordinate values of A_l and A_r in the coordinates of left and right cameras respectively. As shown in Fig. 1 (b), it is can be seen that researchers can not determine the 3D coordinates of point A from just one camera. Thus, in 3D-DIC, two cameras are generally used to capture the same points synchronously on the object to be tested from two positions in different directions.

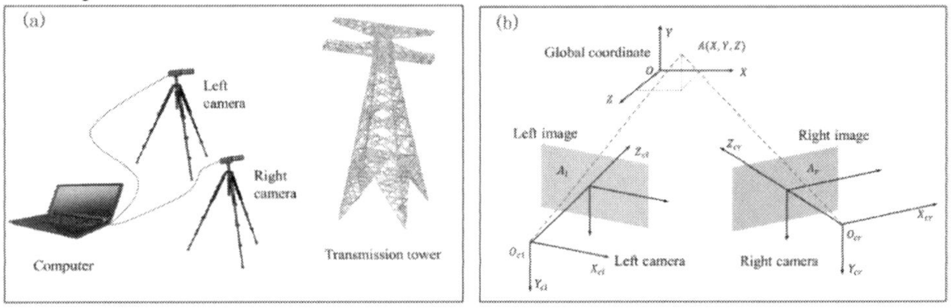

Figure 1. Schematic diagram of the 3D DIC system for measuring the deformation of towers

To measure the 3 dimensional values of the object point A which is a structure node on the tested tower, step 1 of the measurement is to establish a global coordinate. This step can be performed through the calibration of cameras. Step 2 is to calculate the coordinate disparity values of the same structure node of the tested tower between the left and right capures. With respect to the calibrated data of the left and right cameras and the measured coordinate disparity values between the two capure points A_l and A_r, the 3D coordinate values of the object point A will be determined.

2.2 3D matching based on DIC

The 3D matching is used to match the object node on the tower precisely in the left and right images imaged by the two cameras. The matching procedure is commonly considered as one of the most important parts in stereo vision method, can be performed using the subset-based matching technology which is adopted in the 2D-DIC method [5, 6]. The principal concept of the 2D-DIC is to match the same object points with subsets between two images before and after load. In 2D-DIC, only one camera is used, and the position of the camera is fixed during measurement process. It is similar to the 2D-DIC method but with slightly differences. When performing the 3D-DIC, the 3D matching aims to match four subsets on the same object nodes captured by the two cameras with an angle between them before and after deformation of the object. In this work, structure nodes of transmission tower are selected as subsets to determine the displacements after deformation, see Fig. 2.

In the subset-based matching method, the matching can be performed using matching intensity by minimizing or maximizing a function that includes the gray levels of pixels in the subsets in the images. The matching intensity value of a subset are correlated to one another until the intensity value reaches the min/max value. The displacement of the object point, which represents the best possible match, can be determined once the min/max value is obtained. In this work, the Zero-Normalized Sum of Squared Difference, see Eq. (1), was selected as the matching intensity function[7].

$$C = \sum_{i= M}^{M} \sum_{j= M}^{M} \left[\frac{f(x_i, y_j) \quad f_m}{\Delta f} \quad \frac{g(x'_i, y'_j) \quad g_m}{\Delta g} \right] \tag{1}$$

where,

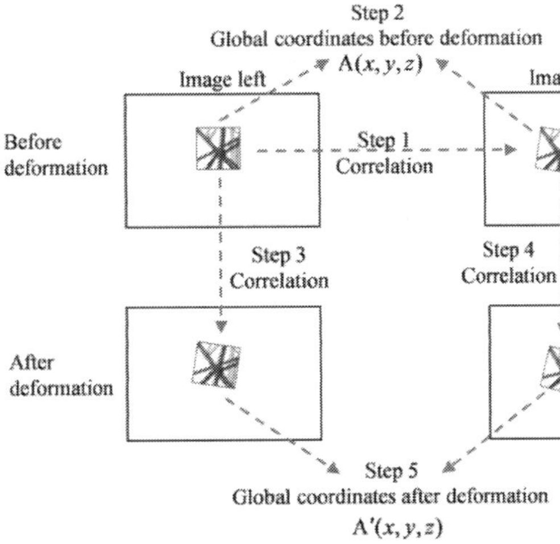

Figure 2. 3D-DIC method for matching a set of images captured by two cameras before and after load.

(a) Left camera (b) Monitoring points of displacements (c) Right camera

Figure 3. Monitoring points, left and right cameras in the real scale model test

Fig. 2 shows the steps used to measure the deformation using 3D-DIC. Once the binocular vision system is calibrated, the space positions and orientations of the two cameras are confirmed and the 3 dimensional positions of the object tower nodes in global coordinate can be calculated by multi-image triangulation.

Using the calibrated data of the two cameras fo the binocular vision system, the triangulation method has been used to calculate the 3 dimensional coordinate values of all the points matched by 3D-DIC. It is important that, prior to the application of triangulation process, all distortions at the object points in images should be corrected using the intrinsic parameters of the two cameras determined in the calibration process. Therefore, a more accurate 3D measurement can be obtained.

2.3 Camera calibration

Calibration of cameras is an important step in the 3D-DIC process, since there is a high dependence of the measurement accuracy on the parameters from camera calibration. The intrinsic parameters of cameras (principle point, effective focal length and lens distortion coefficient) can be determined in the camera calibration procedure. Moreover, in 3D-DIC, determination of the relative location and intersection angle between the left and right cameras are must be performed. In this work, the stereo vision method calibration technique is used followed by a nonlinear least-squares algorithm to optimize the calibration values of the camera parameters[8] .

3 Experiments

The 3D-DIC system consists of 3 parts: two digital single-lens cameras with lens (Basler acA1600 – 20 gm), computer to calculate, shown in Fig. 1 (a). The size of CCD is 7.16 mm × 5.44 mm, the pixel size is 4.5 × 4.5 μm^2, the resolution is 1600 × 1200 pixels, and the optical lens is COMPUTER lens of 16 mm focus.

The binocular vision system is applied in the real scale tower test of a transmission tower. The testing transmission tower is structured to be T-shaped for a single loop ± 500 kV Dc transmission line, see Fig. 3. The design parameters are as follows: practical height 48 m, horizontal span 500 m, vertical span 700 / 300 m, ruling span 300 / 650 m, maximum wind speed 27 m / s and the maximum icing thickness 10 mm. The type test was performed considering two typical working conditions. In the experiment, loads are imposed to the tower step by step on the basis of the designed working conditions. A total station is adopted to measure the deformation of feature points of the tower, which is used as a reference test. The monitoring points of displacement, which are selected at 9 structure nodes, are illustrated in Fig. 3 (b).

Table 1. Calibrated intrinsic parameters of cameras.

	f_x (pixel)	u_0 (pixel)	f_y (pixel)	u_0 (pixel)
Left camera	3634.89	829.4	3635.76	608.11
Right camera	3607.23	791.97	3609.96	633.86

Table 2. Positions and orientations of left and right cameras in global coordinates.

	c_x	c_y	c_z	ψ	θ	φ
Left camera	-6.7 m	-1.7 m	110.7 m	53.2°	24.3°	11.0°
Right camera	-2.5 m	-0.48 m	112.1 m	40.4°	36.8°	17.4°

Table 3. displacements of the monitoring points under working condition 1.

Monitoring point	1	2	3	4	5	6	7	8	9
U_x (mm)	152	115	107	89	25	107	112	144	100
Y_y (mm)	245	234	210	127	58	183	168	248	179
U_z (mm)	-17	-22	-26	-26	-25	-27	-28	-53	-75

Table 4. displacements of the monitoring points under working condition 2.

Monitoring point	1	2	3	4	5	6	7	8	9
U_x (mm)	410	364	343	196	118	305	284	418	312
Y_y (mm)	32	31	30	11	8	14	18	24	17
U_z (mm)	-23	-37	-13	-27	-28	-9	-29	-153	-162

4 Results and discussion

Using this camera calibration method, the intrinsic parameters of the cameras are determined, see Table 1. (u_0, v_0) are the coordinates of the principle points of the cameras respectively; as well as f_x and f_y are the effective focal lengths of each camera along X and Y axes respectively . The intersection angle and relative location of the two cameras can also be obtained using their extrinsic parameters (Table 2).

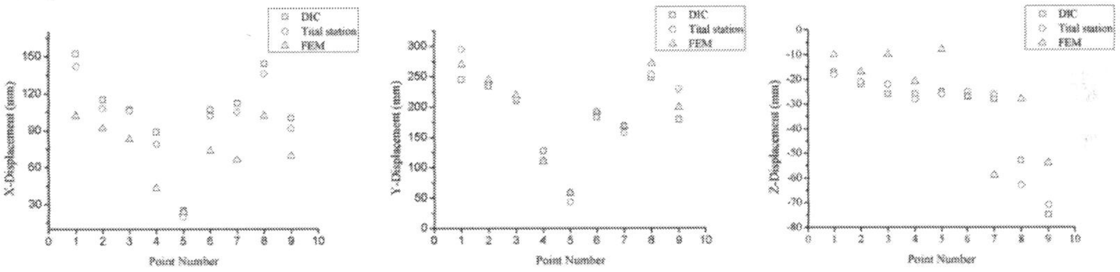

(a) Working condition 1

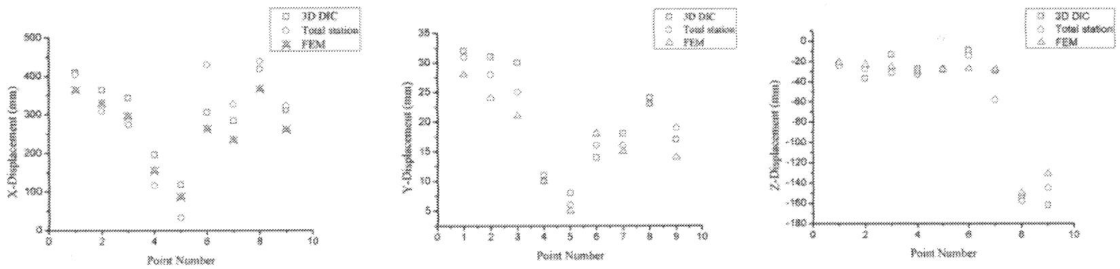

(b) Working condition 2

Figure 4. Comparison of the results from 3D DIC, total station and FEM.

In order to obtain comparable measurement results with the total station, the same monitoring points (see Fig. 3 (b)) on the structure nodes were selected as the points for calculation. To ensure the accuracy of the 3D matching, all pixels within the reference subset, which is located at the current point calculated , should be large enough to cover the image characters around the current structure node. During the procedure of the correlation, the size of the subset was selected to be 31 × 31 pixels. Using the calibrated parameters in Tables 1 and 2, the monitoring node positions of the tower can be determined. The results are shown in Table 3 and Table 4.

To ensure the accuracy of measurement results, the results were compared with that obtained by a commercial 3D coordinate measurement machine (total station). The deformations are also calculated using finite element method (FEM) according to the tower's design report as anther reference. Fig. 4 (a) is the comparison of the displacement results from 3D DIC, total station under working condition 1. Fig. 4 (b) shows results under working condition 2. The measurement results indicate that the displacements of the real scale tower measured using both of 3D DIC and total station are acceptable.

5 Conclusions

In this paper, a 3D DIC measurement system is introduced to measure the deformation of a transmission tower. The results of deformation obtained by the 3D DIC are compared with the results by both of the total station measuring and the finite element method calculation. This work shows the feasibility and effectiveness of the method, thus good prospect can be expected with it. As the trend of smart grid development, development of the intelligent detection technology for transmission line is necessary. As a commonly used technology in the visual field of artificial intelligence, the binocular vision technology proposed a new testing method for monitoring the running state of transmission towers quantitatively.

References

1. Z. Xiao, J. Liang, D. Yu, and J. Liu, Opt. Laser Eng. **48**, 869 (2010).
2. B.-W. Moon, J.-H. Park, S.-K. Lee, J. Kim, T. Kim, and K.-W. Min, J. Constr. Steel Res. **65**, 1 (2009).
3. J. Davis, D. Nehab, R. Ramamoorthi, and S. Rusinkiewicz, IEEE T. Pattern Anal. **27**, 296 (2005).
4. K. Genovese, L. Casaletto, J. A. Rayas, V. Flores, and A. Martinez, Opt. Laser Eng. **51**, 278 (2013).
5. B. Pan, H.-m. Xie, B.-q. Xu, and F.-l. Dai, Meas. Sci. Technol. **17**, 1615 (2006).
6. Z.-F. Zhang, Y.-L. Kang, H.-W. Wang, Q.-H. Qin, Y. Qiu, and X.-Q. Li, Measurement **39**, 710 (2006).
7. H. Shi, H. Ji, G. Yang, and X. He, Opt. Laser Eng. **51**, 47 (2013).
8. T. Xue, B. Wu, J. G. Zhu, and S. H. Ye, Sensor. Actuat. A-Phys. **135**, 185 (2007).

MATEC Web of Conferences **114**, 04004 (2017) DOI: 10.1051/matecconf/201711404004
2MAE 2017

Design and Analysis of Multilayer Solenoid Coil for Faraday Modulator

Muhammad Basharat[1,a], Ming Ding[1], Hongwei Cai[1], Yang Li[1] and Jiancheng Fang[1]

School of Instrumentation Science & Opto-Electronics Engineering, Beihang University, Beijing 100191, PR China

Abstract. A well-designed polarimeter is integral to the realization of the highly sensitive atomic magnetometer. Amongst various detection schemes, optical polarimetry based on Faraday modulator is most commonly used owing to its angular sensitivity at low-frequency range. The multilayered solenoid coil is the key component of Faraday effect based optical detection system. This paper primarily deals with design and analysis of multilayered coil of Faraday modulator. All deterministic parameters that affect the optimum design have been identified and analyzed. Mathematical expressions have been obtained for axial field homogeneity; modulation depth and power dissipation manifesting direct dependence on coil geometry. The design parameters of the coil are optimized with respect to axial magnetic field homogeneity over region of interest and reduced power losses with suitable geometric construction. The influence of different geometrical and electromagnetic parameters on optimum design has been highlighted and guidelines for design procedure are given. Theoretical results have been compared with simulation and experimental results. The mathematical formulation could be implemented in a computer program for recurrence design and to assist the realization of an optimized design of Faraday modulator coil.

1 Introduction

Various scientific and technological applications demand homogeneous alternating magnetic field generated by solenoid coils [1-4]. Typical applications include optical magentometery [5], spectroscopy [6] and biomedical applications [7-9]. Air-core solenoid coils are preferred choice owing to high peak amplitude field and an internal cavity for holding the test element to be magnetically permeated. However, in general, the magnetic field generated by simple solenoid coils is uniform over the limited region. The magnetic field of multilayered solenoid coils is strongly geometry dependent. The optimization of these coils is not merely maximizing the magnetic field over a region of interest rather other associated attributes have to be considered as well. Several coil geometries have been reported for field shaping [10]. The axial field homogeneity can be achieved by increasing the number of turns but it reduces the amplitude of the field for given excitation voltage [11]. On the other hand, poor axial field homogeneity is exhibited by shorter coils having fewer turns. In this case, the span of the peak field is over a limited region and decreases rapidly along the axis of the coil. In general, the dimensions of the coils are larger than the volume of interest. There is always a need for reducing the dimensions without compromising uniformity of field.

[a] Corresponding author: mb@buaa.edu.cn

Sensitive optical magnetometer requires the detection of the extremely small rotation angle. Among various detection schemes, optical polarimetry based on Faraday modulator is most commonly used. The modulation depth of Faraday modulator depends on upon the Verdet constant of the optical material and magnetic field of the multilayered solenoid coil. The optimized performance of modulator depends on upon various parameters. Paramagnetic optical glasses having high Verdet constants are used for large modulation depth. The Verdet constant of paramagnetic glasses is strongly temperature dependent. This dependence dictates the requirement of no heat generated by coil during excitation of the coil. For the optimized performance of the modulator, a tradeoff of different working parameters of coils is inevitable.

The purpose of this work is twofold. First, it aims to identify and analyze all potential geometric and electromagnetic parameters affect the design of the Faraday modulator coil. Second, to develop a mathematical formulation to obtain an optimum design of the coil with the following requirement: (a) Uniform axial magnetic field over the region of interest with suitable geomantic construction and (b) maximize the modulation depth of modulator at the expense of reduced power losses of the coil. The developed formulation could be implemented easily in a computer program for recurrence design. This work is not only limited to the design of Faraday modulator coil but also equally useful for the designing of the coils required for precision instrumentation based on magneto-optics.

2 Design parameters analysis

In this section various geometrical and electromagnetic parameters require for initiation of design are listed and their characteristics effect on the optimum design of multilayered solenoid coil has been analyzed. The interdependence of these parameters has been established. A multi-layered solenoid coil refers to a system many layers of magnetic wire wound on a suitable former. The basic structure of multilayered solenoid coil is shown in Figure 1. Where L is length, r_1 is internal radius; r_2 is the outer radius of the coil, while d is magnetic wire diameter.

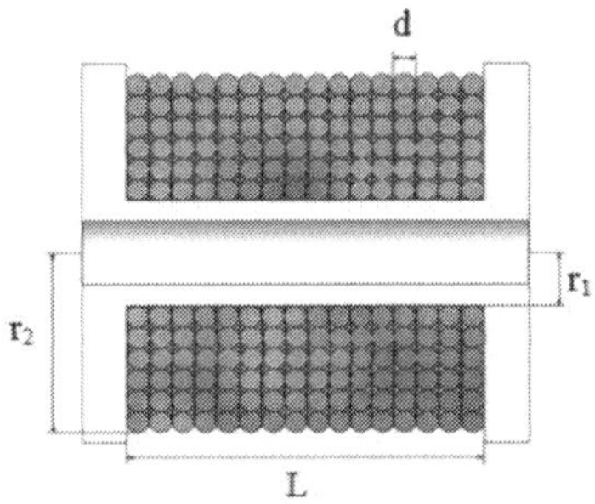

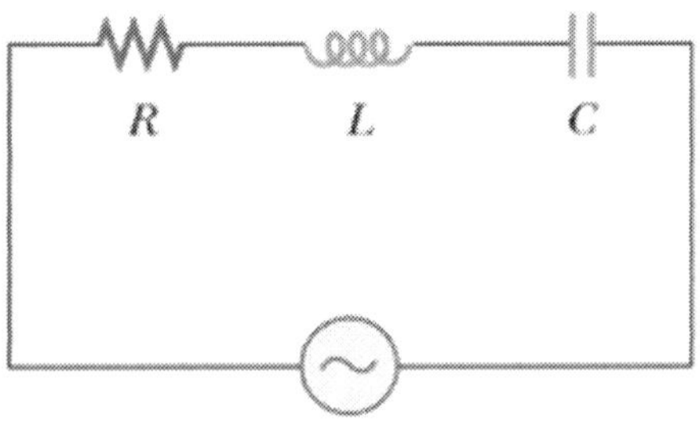

Figure 1. (a) The geometry of the coil, (b) and its equivalent magnetic circuit.

2.1 Geometric and electrical parameters

The number of turns of coil, which depends on diameter d of magnetic wire, coil packing factor k, and dimension of the coil, are given by

$$n = \frac{l(r_2 - r_1)}{kd^2} \tag{1}$$

The coil packing factor k is the ratio of total conduction area of the winding and coil cross section window area. The packing factor may range from 0.2-0.9. With bobbin core, k may be as high as 0.8 for a tight winding.

The approximate length of wire needed for winding can be computed for by the relation

$$l_w = \pi \frac{4(r_2^2 - r_1^2)}{d^2} l \tag{2}$$

The area of the cross-section magnetic wire is given by:

$$A_w = \frac{\pi d^2}{4} \tag{3}$$

The ohmic resistance of the coil is given by

$$R_c = \frac{l_w}{A_w \sigma_{cu}} \tag{4}$$

Where σ_{cu} is resistivity of copper.
The inductance of multilayered solenoid coil can be computed by the relation [12]

$$L = \frac{78.7 n^2 d^2{}_m}{3d_m + 5(d_2 - d_1) + 9l} \tag{5}$$

Where d_m is mean diameter of the coil.
The impedance of coil is given by

$$Z_c = \omega L \tag{6}$$

The coil is resonated by impedance matching; the value of capacitance required is computed by the relation

$$C = \frac{1}{\omega Z_c} \tag{7}$$

If the circuit is properly tuned, the resonant voltage developed across capacitor and coil

$$V_c = Abs[V(\frac{\omega l}{R_{eff}} + 1)] \tag{8}$$

2.2 Copper losses

The DC resistance of the coil depends on dimensions of the winding, the number of turns and the material of magnetic wire. This DC resistance induces the copper loss P_{cu}, it is given by

$$P_{cu} = \rho_{cu} \frac{l_w}{A_w} I_{rms} \tag{9}$$

Where ρ_{cu} is resistivity of the copper l_w is the length of the magnetic wire.

2.3 Eddy losses

In the case of AC, a magnetic flux is produced around the conductor and an eddy current is induced by these flux lines. This eddy current flows in such way that it opposes the induced magnetic flux. The induced eddy current flows in the interior of the conductor to a direction opposite to applied AC current. Consequently, the interior of the conductor is shielded from the applied current. The current density inside the conductor reduces. The depth of the conductor at which internal current density becomes *1/e* of the surface current density is termed as skin depth. The skin depth of magnetic wire at given excitation frequency is computed as:

$$\delta = \sqrt{\frac{2}{\mu_0 \sigma_{cu} \omega}}, \omega = 2\pi f \tag{10}$$

Where, μ_o is permittivity of free space, σ_{cu} electrical conductivity of copper and f is the excitation current frequency. The effective conduction area of conductor reduces which cause an increase in the AC resistance and additional resistive losses.

Another dominant factor which causes an increase in resistance is proximity effect. The current in a first conductor is constrained to smaller region due to the proximity of another conductor. The model developed by Dowell [13] includes both effects and modify the DC resistance accordingly. In this method, round wire is approximated as equivalent foil rectangle wires. For a multilayer coil, the AC resistance is expressed as

$$R_{AC} = (\Delta[G_1(\Delta) + \frac{2}{3}(n_l^2 - 1)(G_1(\Delta) + G_2(\Delta))])R_{DC} \tag{11}$$

Where n_l is number of layer in coil, Δ correction factor due to replacement of rectangular wire

$$\Delta = \frac{d}{\delta}\sqrt{\frac{\pi}{2}} \tag{12}$$

G_1 and G_2 represents the geometry of the infinite coil

$$G_1(\Delta) = \frac{\sinh(2\Delta) + \sin(2\Delta)}{\cosh(2\Delta) - \cos(2\Delta)} \tag{13}$$

$$G_2(\Delta) = \frac{\sinh(\Delta)\cos(\Delta) + \sin(\Delta)\cosh(2\Delta)}{\cosh(2\Delta) - \cos(2\Delta)} \tag{14}$$

The effective resistance $R_{eff}=R_{AC}$ for sinusoidal excitation can be approximated by the relation [14]:

$$R_{eff} = R_{DC}(1 + \frac{\psi}{3}\Delta^4) \tag{15}$$

Where, $\psi = \frac{5n_l - 1}{15}$, n_l is the number of layers in the coil.

2.4 Magnetic field

In order to express the magnetic field of the coil in terms of its geometrical parameters, it instructive to use rectangular coordinates. The magnetic field intensity $B(z)$ due to the finite multilayered coil at the point z on the axis is [15].

$$B(z) = \frac{\mu_0 n i}{2L(r_2 - r_1)}[A \ln \frac{r_2 + \sqrt{r_2^2 + A^2}}{r_1 + \sqrt{r_1^2 + A^2}} + B \ln \frac{r_2 + \sqrt{r_2^2 + B^2}}{r_1 + \sqrt{r_1^2 + B^2}}] \tag{16}$$

Where $A = L/2 - z$, $B = L/2 + z$, while L, r_1 and r_2 are the length, inner radius, and outer radius of the coil respectively. Expressing the axial magnetic field in this form allows the mapping of the field distribution along the axis by knowing the dimensions of the coil.

3 Results and discussion

3.1 Magnetic field homogeneity

The variability of the magnetic field within the volume of interest can be measured and expressed in the form of magnetic field homogeneity (*H*). Normally this variation is expressed as a deviation of magnetic fields from the peak field at the center and it is given by

$$H = \frac{B_0 - B_z}{B_0} \tag{17}$$

The extent of magnetic field homogeneity is application dependent. The dependence of parameter H on the dimensions of the coil can be worked out from Eqs. (16) and (17)

$$H = 1 - \frac{[(L/2-z)\ln\frac{r_2+\sqrt{r_2^2+(L/2-z)^2}}{r_1+\sqrt{r_1^2+(L/2-z)^2}} + (L/2+z)\ln\frac{r_2+\sqrt{r_2^2+(L/2+z)^2}}{r_1+\sqrt{r_1^2+(L/2+z)^2}}]}{L\ln\frac{r_2+\sqrt{r_2^2+(L/2)^2}}{r_1+\sqrt{r_1^2+(L/2)^2}}} \tag{18}$$

The Eq.(18) shows that magnetic field homogeneity depends upon coil geometry only.

Figure 2(a) shows the variation of the axial magnetic field of a coil having inner diameter 15 cm, outer diameter 35 mm and length of 100 mm at 3A current. The homogeneity of axial field is shown in Figure 2(b).Theoretical results have been compared with experimental results in reported in [16]. There is good agreement between experimental and theoretical results over the central region of the curves. At the edges the experimental curve deviates slightly from the theoretical one, it is due to the fact the magnetic field profile of a real coil is prone to errors due to the winding geometry, direction of the windings, errors in wire cross-sections and the insulation layer, filling factor, the geometry of the coil former etc. This result proves the usefulness of the expression given in Eq.(18) for calculating the homogeneity of the magnetic field profile by knowing only the dimensions of the coil with fair accuracy.

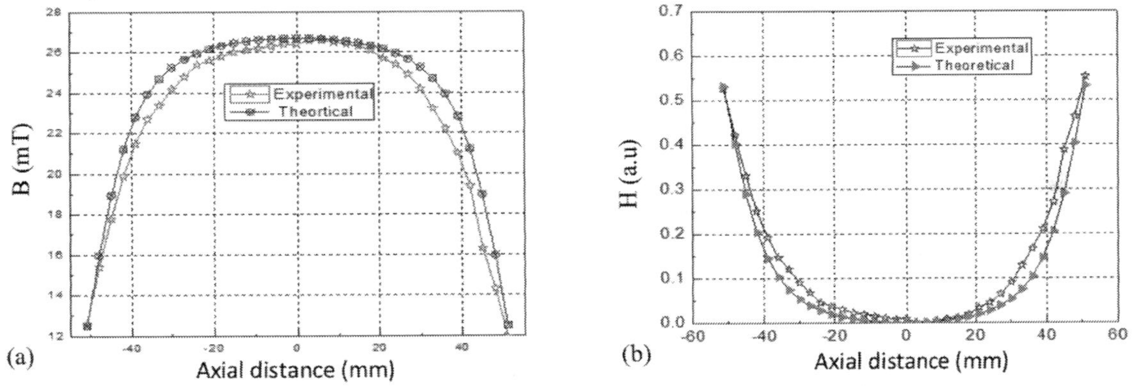

Figure 2. (a) Axial magnetic field variation as a function of distance on the axis of the coil, (b) Homogeneity of the axial magnetic field.

To evaluate the dependence of homogeneity of the coil on its dimensions, the normalized magnetic field distribution curves have been obtained as a function of coil dimension ratios namely length to inner diameter (α) and outer to the inner diameter (β). Figure 3(a) shows regions with a uniform magnetic field for different values of α for fixed value of β. For large values of α, the homogeneity region increases appreciably. As in this case, the inner diameter has been held constant the large value α implies the larger length of the coil. This shows that homogeneity region increases with increasing the length of the coil. For coil dimension $\alpha \geq 10\beta$ the homogeneity region with 10% variation extends to about 80% of the coil length while half of the coil length has homogeneity with variation up to 1%. The coil whose magnetic field profile is given in Figure 2 has dimension α=5.78β and it has a close correspondence with the curve corresponding to α =10 in Figure 3(a).

DOI: 10.1051/matecconf/201711404004

The analysis of axial field homogeneity distribution has been performed using COMSOL Multiphysics. The simulation results of two representative coil configurations are shown in Figure 3(b). Longer coil has parameters L= 100 m, r_1= 7.5 mm, r_2=17.5, and n =730, while shorter has parameters L= 70 m, r_1= 24.5 mm, r_2=30.5, and n =360. Both coils are excited at 3 A current. It is evident that longer coil has uniform magnetic field distribution over extended region compared to the shorter coil.

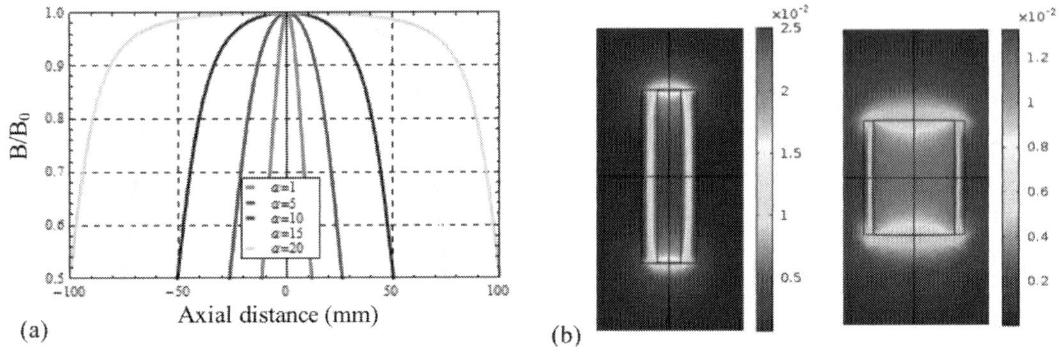

Figure 3. (a) Normalized axial magnetic field for different values of $\alpha(L/r_1)$ at $\beta(r_2/r_1)$=2, (b) Multiphysics simulation of magnetic field distribution (measured in Tesla) of two representative coil conFigureurations.

In Figure 4(a), the normalized magnetic field has been plotted as a function of $\boldsymbol{\alpha}$ for constant coil length and β. It was observed that in this case, inner diameter has characteristics effect on the homogeneity of the magnetic field. For large values of α i.e. smaller inner diameter, the homogeneity regions extend concluding that smallest possible values of inner diameter are desirable. This fact has accentuated by simulation results shown in Figure 3(b). Normalized magnetic field of the square coils ($L=2r_2$) for different values of β is shown in Figure 4(b). In this case length and outer diameter have been kept constant while inner diameter has been varied. It is shown that the square coils have poor axial field homogeneity.

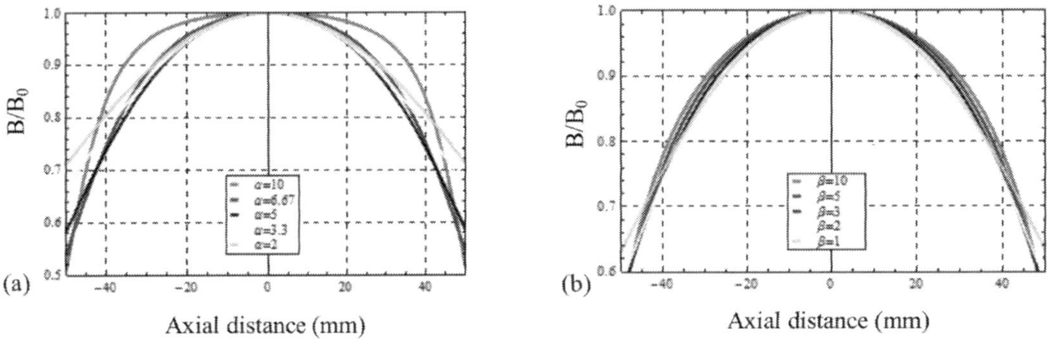

Figure 4. (a) Normalized axial magnetic field for different values of $\alpha(L/r_1)$ at fixed length of the coil., (b) Normalized axial magnetic field for different values of $\alpha(L/r_1)$ for square coils.

3.2 Modulation depth and power dissipation

When solenoid coil is assembled in Faraday modulator, an optical glass is placed inside of bore of air core multilayered solenoid coil. The rotation of the plane of polarization of a plane due to the axial magnetic field of coil is given by

$$\theta = V \int_{-L}^{L} B(z)dz \tag{19}$$

DOI: 10.1051/matecconf/201711404004

Where θ is the angle of rotation (modulation depth), V is Verdet constant of the optical glass of length L, and $B(z)$ is applied axial magnetic field density.

From Eqs. (16) and (19)

$$\theta = \frac{\mu_0 niV}{2L(r_2 - r_1)}[(L^2 \ln \frac{r_2 + \sqrt{r_2^2 + L^2}}{r_1 + \sqrt{r_1^2 + L^2}} + r_2\sqrt{r_2^2 + L^2} - r_1\sqrt{r_1^2 + L^2}) \\ -(\frac{L^2}{4} \ln \frac{r_2 + \sqrt{r_2^2 + L^2/4}}{r_1 + \sqrt{r_1^2 + L^2/4}} + r_2\sqrt{r_2^2 + L^2/4} - r_1\sqrt{r_1^2 + L^2/4})] \quad (20)$$

Eq. (20) can be rearranged

$$\chi = \frac{\theta}{ni} = \frac{K}{\alpha(\beta - 1)}[(\alpha^2 \ln \beta \frac{1 + \sqrt{1 + A^2}}{1 + \sqrt{1 + \alpha^2}} + \beta^2\sqrt{1 + A^2} \\ -\sqrt{1 + \alpha^2}) - (\frac{\alpha^2}{4} \ln \beta \frac{1 + \sqrt{1 + A^2/4}}{1 + \sqrt{1 + \alpha^2/4}} + \beta^2\sqrt{1 + A^2/4} - \sqrt{1 + \alpha^2/4})] \quad (21)$$

Where $\alpha = L/r_1$, $A = L/r_2$, $\beta = r_2/r_1$, $K = \frac{\mu_0 V}{2}$ is a constant, and $\chi = \frac{\theta}{ni}$ is figure merit of the Faraday modulator relating rotation to ampere-turn of the coil

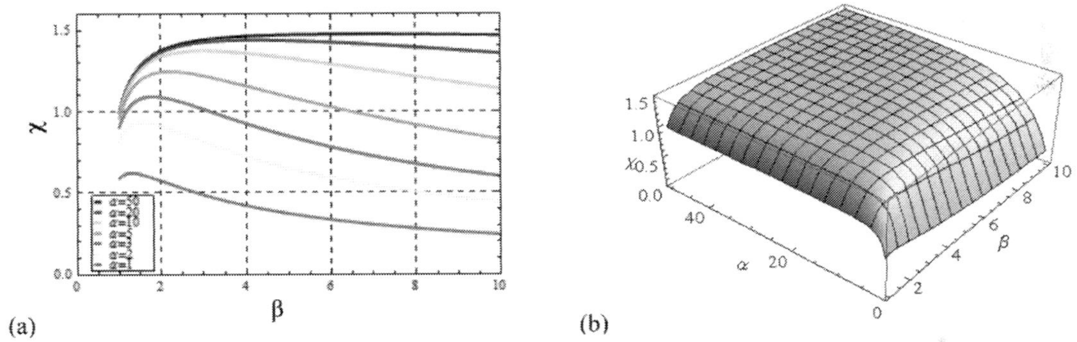

Figure 5. (a) The ratio of rotation to ampere-turn as a function of coil dimension ratios α and β, (b) 3D graph showing the ratio of rotation to ampere-turn as a function of coil dimension ratios α and β for the whole range.

The plot of χ vs coil radii ratio $\beta(r_2/r_1)$ is shown in Figure 5. It can be seen that χ is dependent only on the ratios of coil dimensions and its values are large for the thin coils. However, we have to consider the power needed for required modulation depth.

The power dissipated in coil excited by alternating current is given by

$$P = i^2 R_{eff} \quad (22)$$

If optimal diameter conductor is used for winding, the R_{eff} of multilayer coil using round conductor is two times of R_{DC} [17].

$$P = 2I^2 R_{DC} \quad (23)$$

From Eqs. (1), (2), (4) and using the fact $k = \frac{nA_{cu}}{L(r_2 - r_1)}$

$$P = \frac{8i^2 \rho\pi(r_1 + r_2)n^2}{L(r_2 - r_1)} \quad (24)$$

DOI: 10.1051/matecconf/201711404004

$$\frac{ni}{\sqrt{P}} = \sqrt{\frac{L(\beta-1)}{8\rho\pi(\beta+1)}} \tag{25}$$

Where $\beta = r_2 / r_1$

From Eqs. (21) and (25)

$$\zeta = \frac{\theta}{\sqrt{P}} = k\chi\sqrt{\frac{(\beta-1)}{(\beta+1)}} \tag{26}$$

Where k is constant and $\zeta = \frac{\theta}{\sqrt{P}}$ is figure merit of the Faraday modulator relating rotation to square of power dissipation of the coil

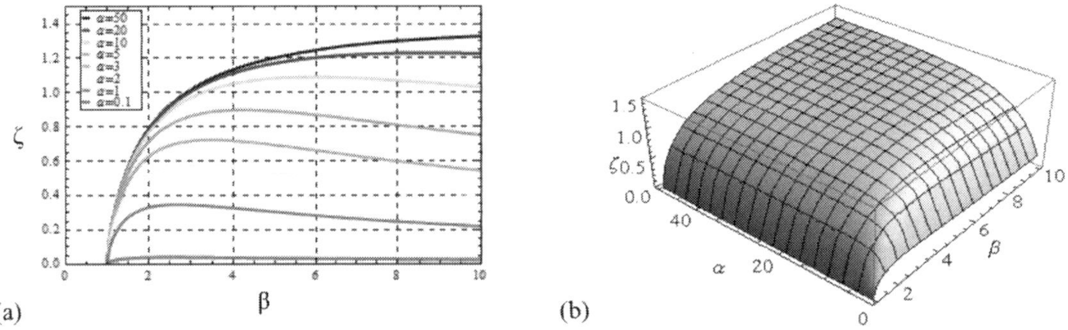

(a) (b)

Figure 6. (a) The ratio of the rotation to $\sqrt{P}$ as a function of coil dimension ratios α and β, (b) 3D graph showing the ratio of rotation to $\sqrt{P}$ as a function of coil dimension ratios α and β for the whole range.

Figure 6 shows that there are well-defined values of the coil radii ratio $\beta(r_2/r_1)$ corresponding to each value of α which maximize ζ. The power needed for a required rotation for a given coil dimensions depend upon the ampere-turns rather than separately on the current or on the coil turns.

The results of this section indicate that the magnetic field homogeneity, modulation depth and power dissipation of the coil are geometry dependent and can be expressed in terms coil dimension ratios. The mathematical formulation given in section 2 and 3 could be implemented in a computer program easily. For instance, we have implemented this into Matlab GUI to assist the design of the coil. Nevertheless, the analysis and findings presented here can be used to design a solenoid coil without any computer program as well. Before proceeding one has to decide what can be changed and what is fixed. In most of the cases, the internal diameter of the coil cannot be changed at will as the optical material along with holding mechanism must be fitted snugly inside it. Based on geometric constraints of the experiment, initial values of length and the inner diameter of the required coil are established. An appropriate gauge of magnetic wire is selected to meet the requirement of frequency and operating voltage and current rating. Depending upon the level of homogeneity of the axial magnetic field and modulation depth for a particular experiment the dimensions of the coil for optimum design are found following the results presented in section 3. Once the design parameters are finalized the working parameters are calculated from the analysis given section 2. By doing impedance matching the resonant rise in voltage may be up to hundreds of Volts, therefore, proper insulation of the coil terminals should be insured.

4 Conclusion

The optimum design of multilayer solenoid coil implies a trade-off between the technical complicacy and best performance parameter such homogeneity, modulation depth and so on. The parameters that affect the optimal design have been identified and analyzed. The interdependence of these parameters has been formulated. The optimum design has been obtained with respect axial magnetic field homogeneity over a region of interest and optimum modulation depth with reduced power losses with suitable geometric construction. The results obtained from this investigation indicate the performance parameters of the coil are geometry dependent and can be manipulated in terms coil dimension ratios. The developed formulation could be implemented in a computer program easily for estimating the optimality of the coil design. This work is not limited only to the design of Faraday modulator coil but also equally useful for the designing of the coils required for precision instrumentation based on magneto-optics.

References

1. N. K. Das, P. Barat, S. Dey, and T. Jayakumar, Prog. Electromagn. Res. M **34,** 99 (2014)
2. S. Sato, S. Sakaguchi, K. Futamata, and K. Katou, IEEE Trans. Magn. **36**, 649 (2000)
3. J. L. Kirschvink, Bioelectromagnetics **13**, 401 (1992).
4. R. Merritt, C. Purcell, and G. Stroink, Rev. Sci. Instrum. **54**, No. 879(1983)
5. S. J. Seltzer, "Developments in alkali-metal atomic magentometery," Ph.D. dissertation (Princeton University, 2008).
6. K. Ehrmann, N. Saillen, F. Vincent, M. Stettler, M. Jordan, F. M. Wurm, P.A. Besse, and R. Popovica, Lab Chip. **7**, 373 (2007)
7. K. R. Minard and R. A. Wind, Concepts Magn.Reson.**13**, 128 (2000)
8. C. McCann and M. D. Sherar, Phys. Med. Biol. **51**, 3851 (2006)
9. D.E. Bordelon, R.C. Goldstein, V. S. Nemkov, A.Kumar, John K. Jackowski, T.L. DeWeese, and R.Ivkov IEEE Trans. Magn. **48**, 47 (2012)
10. P. R. Stauffer, P. K. Sneed, H. Hashemi, and T. L. Phillips, IEEE Trans. Biomed. Eng. **41**, 17 (1994)
11. T. Tominaka, Eur. J. Phys. **27**, 1399 (2006)
12. P.Ripka, Magnetic Sensors and Magnetometer (Artech House Inc, 2001)
13. P. L. Dowell, Proc. Inst. Electr. Eng. **113**, 1387 (1966)
14. W.G. Hurley, E. Gath, J.G. Breslin, IEEE Trans. Power Electron. **15**, 369, (2000)
15. R. M. Duffy and R. P. Netterfield, Rev. Sci. Instrum. **55**, 743, (1984)
16. M. Basharat, Z. Razzaq, S. Javed, and U. Afzal, OAM-RC. **8**, 930 (2014)
17. R.P. Wojda Marian, K. Kazimierczuk, COMPEL - The international journal for computation and mathematics in electrical and electronic engineering, **31**, 1793(2012)

Three-Dimensional Numerical Analysis of an Operating Helical Rotor Pump at High Speeds and High Pressures including Cavitation

Yang Zhou[1,a], Lupeng Wu[2] and Minghui HAO[2]

[1] School of Mechanical Engineering, Zhengzhou University,No.100 Science Avenue, Zhengzhou,China
[2] School of mechatronics Engineering, Harbin Institute of Technology, 92 West Da-zhi Street, Harbin, China

Abstract. High pressures, high speeds, low noise and miniaturization is the direction of development in hydraulic pump. According to the development trend, an operating helical rotor pump (HRP) at high speeds and high pressures has been designed and produced, which rotational speed can reach 12000r/min and outlet pressure is as high as 25MPa. Three-dimensional simulation with and without cavitation inside the HRP is completed by the means of the computational fluid dynamics (CFD) in this paper, which contributes to understand the complex fluid flow inside it. Moreover, the influences of the rotational speeds of the HRP with and without cavitation has been simulated at 25MPa.

1 Introduction

Gear pumps are widely applied in aerospace hydraulic system, chemical industry and civil engineering for the reason of its reliability, simple structure and low cost. Cavitation and trapped-oil are two major limitations to the development of high pressures, high speeds and of gear pumps. Due to trapped-oil, the gear meshing pressure increases suddenly, which dramatically decreases the lifespan of gears. Unfortunately, the vibration and noise are caused by trapped-oil so the stability of pump system drops down[1-2]. In this case, HRP, a kind of special gear pump, becomes relatively popular because it has no trapped-oil and flow ripples, low noise as well[3]. However, cavitation is the other non-ignorable limitation. Besides it can produce pressure fluctuations and noise, it also has effect on volumetric efficiencies[4-7].Along with the cavitation, the problems of two-phase flow appear, which have significant influences on work performance of gear pumps[8]. Numerical simulation of an external gear pumps operating at speeds below 3000rpm including cavitation is studied by D. del Campo[4], who discusses the influences on the volumetric efficiency under different rotational speeds. In order to understand the complex fluid dynamic characterization, Zhou et al[9] simulate 2D gear pumps running at high pressures and high speeds, but cavitation is not considered. Thus,HRP is used to eliminate the effect of trapped-oil in this paper. In order to predict the actual movement of the internal flow inside the HRP, whose three-dimensional simulations including cavitation at high speeds (10000rpm and 12000rpm) and high pressure (25Mpa) are finished. Furthermore, the influences on the meshing pressure and the maximum negative pressure at different rotational speeds are studied.

[a] Corresponding author : rumsfld@126.com

2 Mesh model of HRP

The HRP has no trapped-oil for the reason of its special tooth profile. Two rotors mesh in the HRP, as depicted in Figure1. When the driving rotor spin clockwise, the hydraulic oil trapped in the inter-teeth volumes between rotors and body is moved from the inlet to the outlet by rotors. In this way, the pressure is formed and lifted in high pressure chamber. The transverse contact ratio of the HRP is 0.5 so that rotors keep contacting in one-point. The design of the tooth profile studied by Zhou et al[10] has been referenced to establish the mathematical model of rotors in this paper. As an example, the mathematical model of driven rotor is schematically shown in Fig.1.

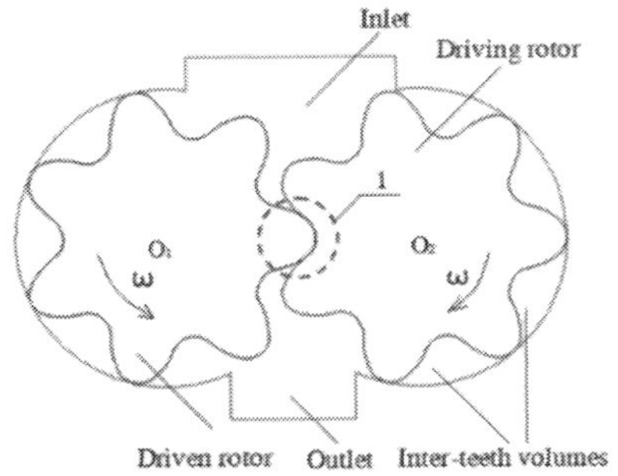

Figure 1. The tooth profiles of the HRP

The tooth profile of the HRP consists of two equal circular arcs and one involute. The mathematical equation for the circular arc and as well as the involute can be obtained with the method presented by Zhou et al[10].The computational domain of the HRP has been divided into three regions, this is schematically shown in Fig. 2. That consists of inlet chamber, outlet chamber and rotoring zone.

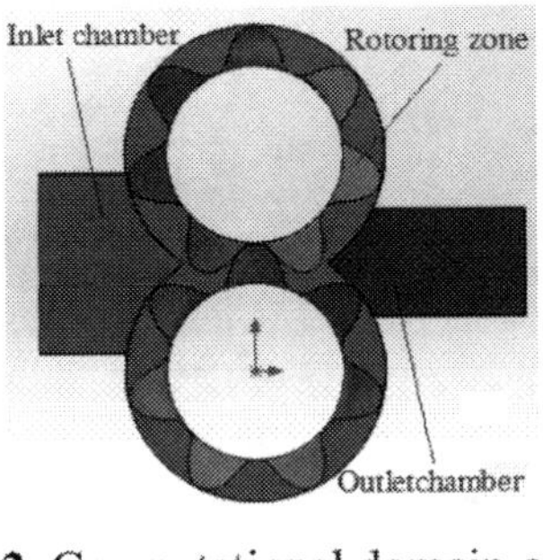

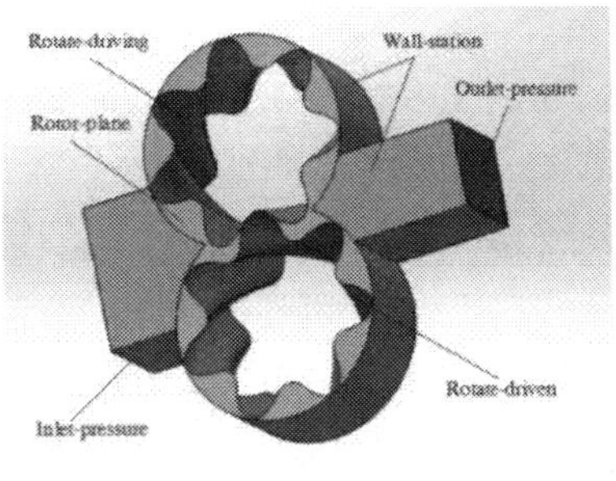

Figure 2. Computational domain of the HRP **Figure 3.** Boundary conditions

Fig.3 shows a detail of boundary conditions ANSYS ICEM CFD is used to create meshes which formed by tetrahedral cells, as shown in Fig. 4. In order to improve the mesh quality, avoiding the errors of negative meshes and non-convergence in the process of simulations, several methods are applied as follows.Triangular prism grid is used to detail the edges of rotoring zone.Meshes whose quality below 0.3 are deleted.The irrelevant grid points are removed.Finally,1085141 cells are obtained from the computational domain, as depicted in Fig. 4.

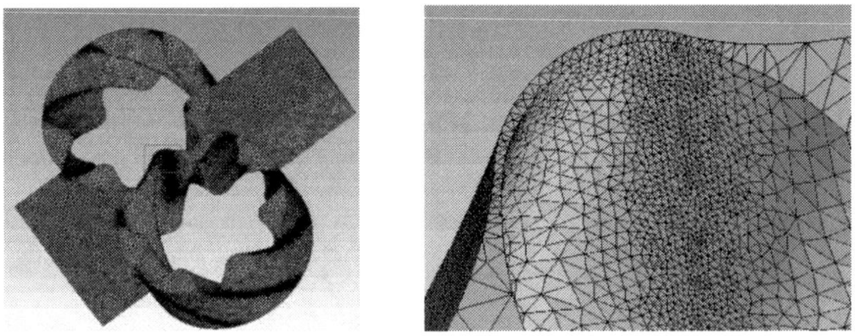

Figure 4. 3D mesh and detail of rotoring zone

Both geometrical parameters of the HRP and properties of mineral oil are presented in table1. The commercial finite-volume-based code ANSYS Fluent 16.0 has been applied in numerical simulations of the HRP.

Table 1. Geometrical parameters of pump and properties of mineral oil

Length of the inlet chamber	22.15mm
Width of the inlet chamber	15mm
Length of the outlet chamber	17.85mm
Width of the outlet chamber	9mm
Height of the rotors	10 mm
Modulus	2 mm
Number of teeth	7
Diameter of rotors	21.334 mm
Distance between centers of rotors	18.08 mm
Pressure angle	28°
Density	960kg/m3
Dynamic viscosity	0.048 $Pa \cdot s$

In order to illustrate that the cells number have no influence on the results of the simulation, the computational domain of the HRP is divided into two groups with different number of cells. The former contains 580329 cells, the later contains 1085141 cells. In the process of simulation, the remeshing algorithm do not stop changing the number of cells in each time step until remeshing is reached. The number of cells of grids in both two groups stabilized around 880 thousands after several meshing cycles. This can be seen in Fig.5. Considering the speed of calculation and quality of grids, the latter is selected in the end.As far as author knows, numerical simulation including cavitation requires more strict convergence criteria than that without it. Singhal et al model has been selected in this paper. This model is attractive because it can solve the continuity and the momentum equation for the mixture. The first phase chooses fuel-oil-liquid while air is used in the second phase. The parameters of the cavitation model are $F_{Vap} = 0.4$ and $F_{cond} = 0.001$.

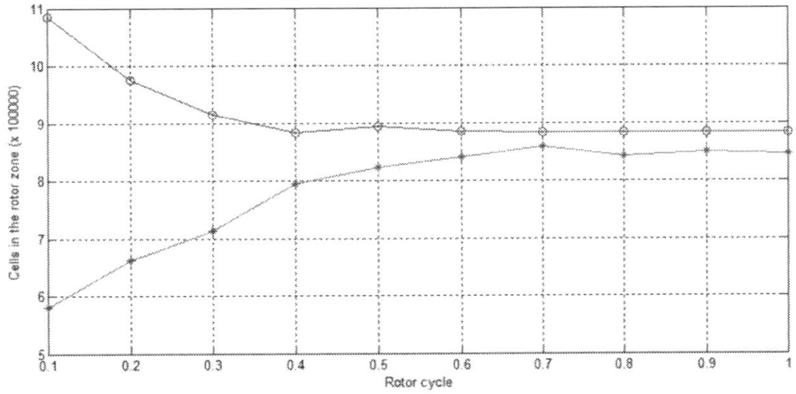

Figure 5. Cells number evolution in computational domain

3 Results of the numerical simulation

Regarding three-dimensional numerical simulations, several methods proposed in this paper allow the author to study the complicated flow inside the HRP. In this case, effects without and with cavitation at different rotational speeds are studied, which can predict the performance of the HRP operating at high pressure and high speeds, although it can not represent the actual values. The pressures of the flow inside the HRP are simulated when it operates at 25MPa and 10000rpm. After 0.0048s, the contours of pressure are shown in Fig. 6.

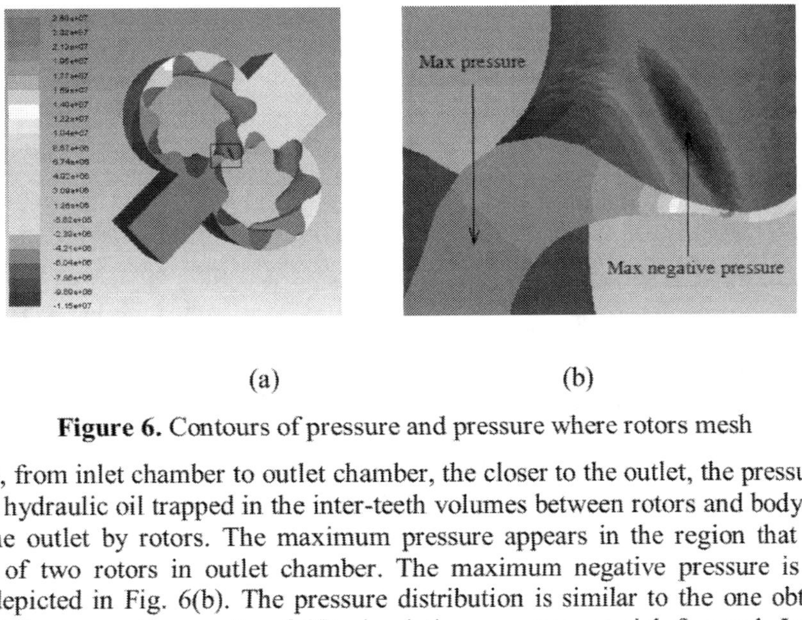

(a) (b)

Figure 6. Contours of pressure and pressure where rotors mesh

In Fig.6(a), from inlet chamber to outlet chamber, the closer to the outlet, the pressure higher. This is because the hydraulic oil trapped in the inter-teeth volumes between rotors and body is moved from the inlet to the outlet by rotors. The maximum pressure appears in the region that is close to the contact point of two rotors in outlet chamber. The maximum negative pressure is found in inlet chamber, as depicted in Fig. 6(b). The pressure distribution is similar to the one obtained from 2D simulation [8], however, the results of 3D simulation are more straightforward. Interestingly, the maximum negative pressure reaches 16MPa in 3D simulation.

Effects without and with cavitation at different rotational speeds are studied when the outlet pressure is assumed at 25MPa. Fig. 7 compares the effects on the maximum pressure without and with cavitation. Meshing pressures gradually increase with the higher rotational speeds. However, the minimum of the maximum pressure appears at 0.8 rotor cycle and this condition is not affected by rotational speeds. Cavitation has little effect on the maximum pressures because it mainly exists in low pressure chamber. Nevertheless, when considering cavitation, the maximum pressures increase nearly 4MPa at 10000rpm. The same figure shows that influence of cavitation on the maximum pressure exists mainly from 0.3 to 0.7 rotor cycle.

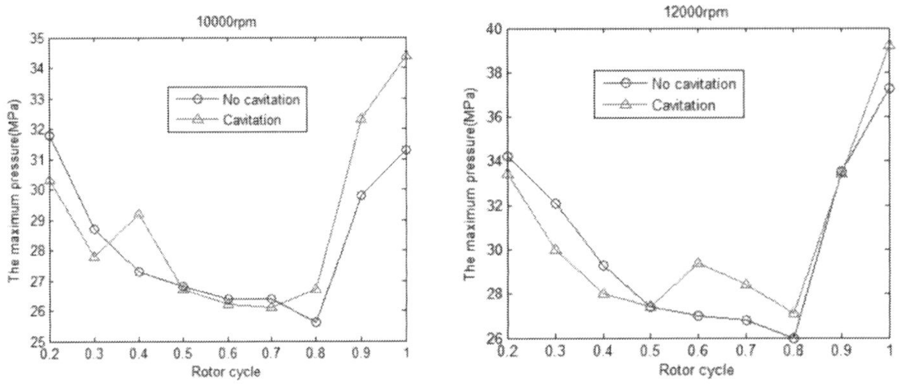

Figure 7. Analysis of the maximum pressure with and without cavitation at 10000rpm and 12000rpm.

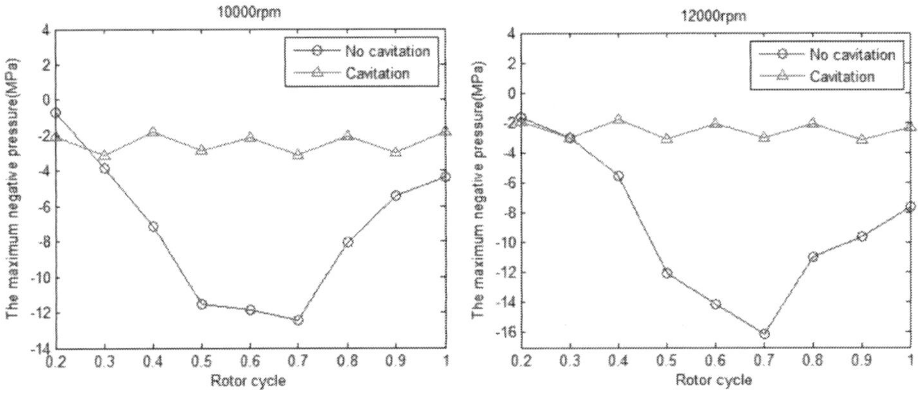

Figure 8. Analysis of the maximum negative pressure with and without cavitation at 10000rpm and 12000rpm

In Fig.8, the effects on the maximum negative pressure without and with cavitation at different rotational speeds are studied. No considering cavitation, the decline of the maximum negative pressures become more urgent with the increase of the rotational speeds and at 0.7 rotor cycle, where the maximum negative pressures reach the lowest points. In this case, the pressure difference between the maximum pressure and the maximum negative pressure becomes larger, which leads to more leakage, and the stability of rotors decreases as well. The maximum negative pressure reaches nearly 16MPa when the HRP operates at 12000rpm. The negative pressure fluctuations are significantly smaller when considering cavitation. Through the whole rotor cycle, the maximum negative pressures maintain around -2MPa, since cavitation mainly exists in low pressure chamber.

Conclusions

Three-dimensional numerical analysis of the HRP has been proposed by means of computational fluid dynamics (CFD). Furthermore, cavitation has been considered, which contributes to obtain more accurate results. The results of simulation do not equal the actual conditions. However, it contributes to study structure optimization and seal of the HRP. Comparisons of effects on the maximum pressures and the maximum negative pressures at 10000rpm and 12000rpm are presented in this paper. The pressure fluctuations where rotors mesh inside the HRP are obvious. The maximum pressure and the maximum negative pressure as the rotational speeds become higher. Cavitation has little effect on the maximum pressure where rotors mesh. However, the low pressure chamber is filled with cavitation, by which a part of leakage is prevented.

References

1. Eaton M, Keogh PS and Edge KA. *The Modelling, prediction, and experimental evaluation of gear pump meshing pressures with particular reference to aero-engine fuel pumps*. Proc IMechE Part I: J Systems and Control Engineering.**220,**396-404,(2006).
2. Shu W, Sakurai H. and Aditya K. *The Optimal Design in External Gear Pumps and Motors*. IEEE/ASME Transactions on Mechatronics **16,** 945-952,(2011).
3. Lee G S, Jung SY and Bae JH. *Design of rotor for internal gear pump using cycloid and circular-arc curves*. Journal of Mechanical Design. 1-12,**134**(1),(2012). .
4. Del campo D, Castilla R, Raush GA, et al. N*umerical analysis of external gear pumps including cavitation.* ASME J Fluids Eng.1-9,**134**,(2012).
5. Neyrat S, Orand N and Jonquet D. *Modeling and analysis of an automatic transmission internal gear oil pump with cavitation*. SAE 2005 Noise and Vibration Conference and Exhibition,16-19 , (2005).
6. Andrea Vacca, Richard Klop and Monika Ivantysynova. *A numerical approach for the evaluation of the effects of air release and vapour cavitation on effective flow rate of axial piston machines*. International journal of fluid power ,33-45, **11**(1),(2010).
7. Abbot, Philip A., Walsh, James. And Halas, Richard. *Cavitation noise investigation of a pump-turbine*. Waterpower '91: A New View of Hydro Resources.2031-2040,(1991).
8. Heath R A. *Theoretical analysis of cavitation conditions in gear pumps*. Internal Report B60, 080, Lucas Aerospace Ltd, Birmingham, 1977.
9. Yang Zhou, Shuanghui Hao and Minghui Hao. *A two-dimensional numerical analysis of a circular-arc gear pump operating at high pressures and high speeds*. Proc IMechE Part E: J Process Mechanical Engineering ,1-12,(2015).(online).
10. Yang Zhou, Shuanghui Hao and Minghui Hao. *Design and performance analysis of a circular-arc gear pump operating at high pressure and high speed*. Proc IMechE Part C: J Mechanical Engineering Science.189-20.**230**,(2016).

Preparation of Waste GFRP Fiber Reinforced Gypsum Block with Water-resistant and Energy Storage Characteristics

Meng-Meng Zhang[1], Xiong-mu Chen[2] and Feng-Qing Zhao[1]

[1]*Department of Chemical Engineering, Hebei University of Science & Technology, Shijiazhuang, 050018, China*
[2]*Hebei Engineering Research Center of Solid Waste Utilization, Shijiazhuang, 050018, China*

Abstract. Gypsum block possesses good performances such as volume stability, lightweight and thermal insulation, is recognized as typical eco-friendly building material. However, its poor water resistance characteristics restrict the application. The semi-hydrated desulphurization gypsum is modified with steel slag, granulated blast-furnace and carbide slag (SGC) composite powder as well as waste glass-reinforced plastic (GFRP) fiber, aiming at producing water-resistant gypsum block. The proper mass proportioning of the modified gypsum block is obtained: semi-hydrated desulphurization gypsum 75%, SGC 25% and waste GFRP fiber 1.0%. The product is of softening coefficient of 0.84 and thermal flexural strength of 8.6 MPa. Phase change energy storage material (PCM) is used to increase the energy saving characteristics of the block. Compared with ordinary gypsum walls, the modified gypsum block with CA-SA exhibits good energy storage property.

Introduction

The development of low-energy wall materials is the main approach to meet the need of energy saving. Gypsum block is a typical eco-type and energy-saving material. Excellent properties, such as volume stability, lightweight and thermal insulation, have made it popular theses years. But its poor water resistance and low mechanical strength restrict the application.

There are various ways to increase the strength and water resistance properties of gypsum block by addition of water-proofing agent, as organic materials, inorganic materials or the both. But the traditional organic water-proofing agents have some negative impacts on mechanical strength of gypsum block. The inorganic materials used are mostly cement or/and industrial wastes, which seems not so satisfactory [1,2]. Much research has done to increase water-resistance and mechanical strength for gypsum block by using super-plasticizer and α-hemihydrate gypsum, and obtained some positive results [3]. The mechanical strength of gypsum block can also be reinforced by fiberglass, plant fibers, synthetic fibers [4,5]. However, high cost of raw materials limits their application.

In this paper, semi-hydrated desulphurization gypsum was modified with SGC powder and waste glass-reinforced plastic fiber, aiming at increasing the water resistance and mechanical strength. In addition, phase change energy storage material is used in gypsum block for being energy saving characteristics.

[a] Corresponding author: zhaofq3366@126.com

DOI: 10.1051/matecconf/201711404006

1 Materials and methods

1.1 Raw materials

The desulfurization gypsum used in this work was provided by China Xibaipo Power Plant; steel slag and granulated blast-furnace slag from Shijiazhuang Iron and Steel Co. Ltd.; carbide slag from Shijiazhuang Chemical fiber Co. Ltd; waste glass-reinforced plastic fiber from Jizhou Mechanical Technology Co. Ltd. The waste mycelium used was from Huabei Pharmaceutical Co. Ltd , which used to prepare gypsum retarder (WPM retarder) in laboratory [6,7]. Phase change energy storage materials were bought from market. The compositions of main raw materials are listed in Table 1.

Table 1. Principal composition of raw materials, wt%

Composition	CaO	SiO_2	FeO	Fe_2O_3	MgO	Al_2O_3	MnO	f-CaO	SO_3
Desulfurization gypsum	37.76	1.38		0.32	0.16	0.27			50.34
Waste glass-reinforced plastic	27.23	55.12		2.202	1.33	8.94			
Steel slag	36.35	15.33	9.27	12.7	5.51	8.49	5.51	4.82	
Granulated blast-furnace slag	34.15	41.20		13.66	8.47	0.84	1.11		
Carbide slag	92.02	3.82		0.50	0.14	2.72			0.63

1.2 Methods

The mixture of steel slag and granulated blast-furnace slag was ground to desired fineness, mixed with carbide slag, forming SGC composite powder, which is used to improve mechanical strength and water-resistant characteristics of modified gypsum block. PCM is added to the mixture to obtain energy storage characteristics. Specimens with dimension of 40 mm×40 mm×160 mm were molded, then de-molded after 2 hours, and cured at desired conditions. The flexural strength value was the average of every three tests, whereas the compressive strength value was the average of every six tests.

For measurement of the fineness of blast-furnace slag powder and steel slag powder, 45μm sieve residue and Blaine values were used. Fig. 1 gives the flow-sheet of the preparation of gypsum block.

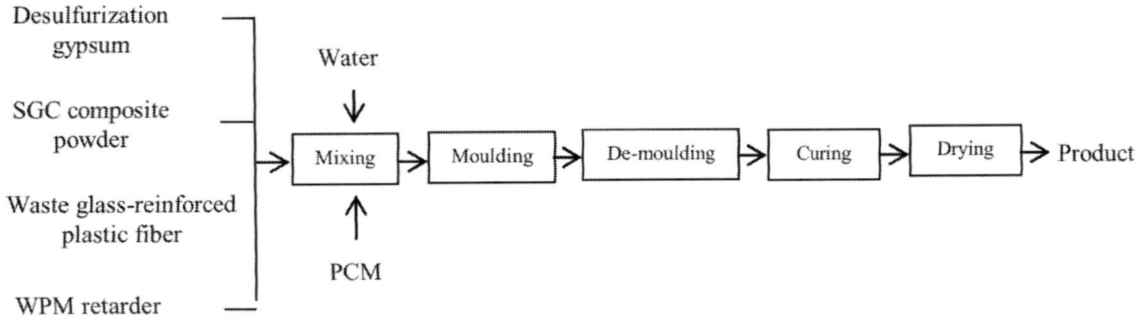

Figure 1. Preparation of water-resistant gypsum block with energy storage characteristics

2 Results and Discussion

2.1 Effect of SGC composite powder on water resistance

The semi-hydrated desulphurization gypsum is modified with SGC composite powder which is used to improve water resistant property. Here, the optimal mass ratio of steel slag: granulated blast-furnace slag: carbide slag is 50:50:0.08, according to initial experiments.

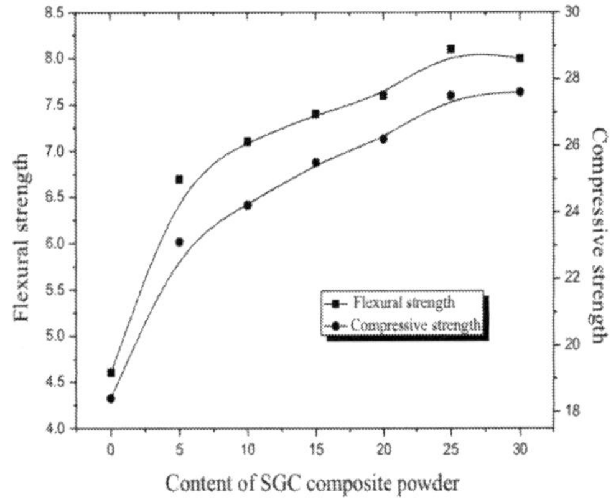

Figure 2. Effect of SGC powder on the mechanical strength

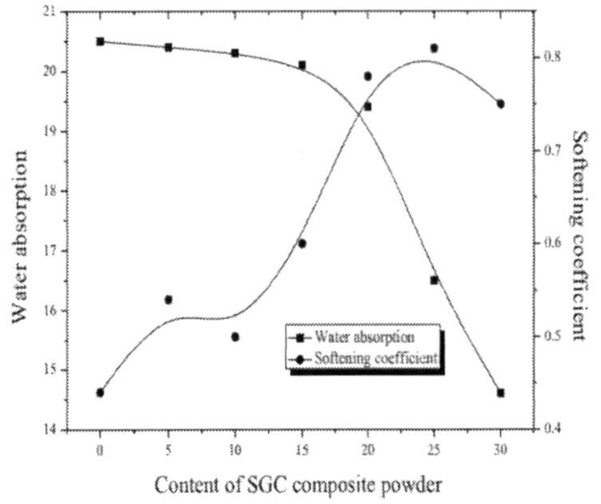

Figure 3. Effect of SGC powder on water resistance

We can see from Fig. 2 that the mechanical strength increases with the addition of SGC, and reaches peak vale when the content of SGC composite powder is 25%. Fig. 3 shows that with the increase of the content of SGC composite powder, the water absorption decreases sharply. However, the softening coefficient of gypsum block with SGC composite powder is improved, and reaches highest value of 0.81 when the content of SGC powder is around 25%. The results show SGC powder have significant effect on the mechanical strength and water resistance of the gypsum block. Steel slag and granulated blast-furnace slag have high content of glass phase, thus possess high potential hydration activity. Carbide slag as alkali activator can enhance remarkably the activity of steel slag and granulated blast-furnace slag. The hydration products are formed such as ettringite and hydrated calcium silicate, and the strength and water-resistance of the gypsum block are increased accordingly.

2.2 Effect of waste glass-reinforced plastic fiber on water resistance

The influence of waste glass-reinforced plastic fiber with different length and addition on flexural strength and water-resistance of the gypsum block were investigated

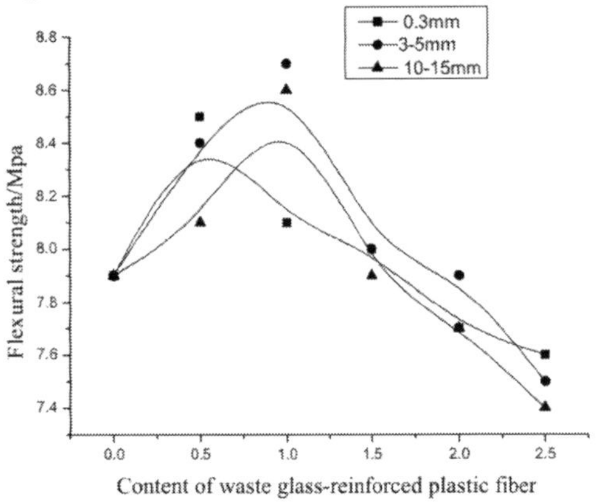

Figure 4. Effect of WGFP fiber on flexural strength of test block

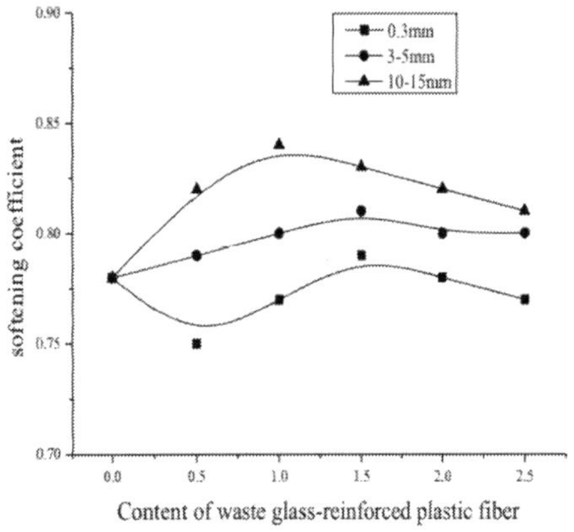

Figure 5. Effect of WGFP fiber on softening coefficient of test block

From Fig. 4 we can see that when the waste glass-reinforced plastic fiber is in the range from 0% to 1%, the flexural strength of gypsum block increases. Fig. 4 and Fig. 5 show that 10-15mm fiber was more suitable for the flexural strength and softening coefficient of gypsum block than shorter fibers. Excessive fiber results in the decrease in mechanical strength of the block.

2.3 Overall modification effect of WGFP fiber on gypsum block

Based on above study, we come to the conclusion that both SGC powder and WGFP fiber have positive effect on the mechanical strength and water-resistance property of gypsum block. The results were obtained for the modifier: SGC composite powder 25%, WGFP fiber 1.0% (fiber length is 10-15mm). On this basis, the performance of gypsum block is tested: softening coefficient of 0.84, compressive strength of 27.5 MPa and flexural strength of 8.6 MPa. Fig. 6 gives the fracture and cross section of the modified gypsum block. It was shown that the gypsum samples containing WGFP fiber can weaken or eliminate the tensile stress of crack, restraining the extension, which increase the flexural strength of the block.

Figure 6. Fracture and cross section of the modified gypsum block

2.4 Preparation and application of phase change energy storage gypsum block.

The phase change energy storage material (PCM) was prepared from capric acid (CA) and stearic acid (SA). Through systematic experiment, the PCM (the mass ratio of CA to SA= 9:1) was obtained with melting point of 26.8℃ and latent heat of 126.4 J/g. The energy storage test were carried out for experiment cubic box (simulating real house) with the addition of PCM in gypsum block 8 % (based on total solid mass). See Fig.7.

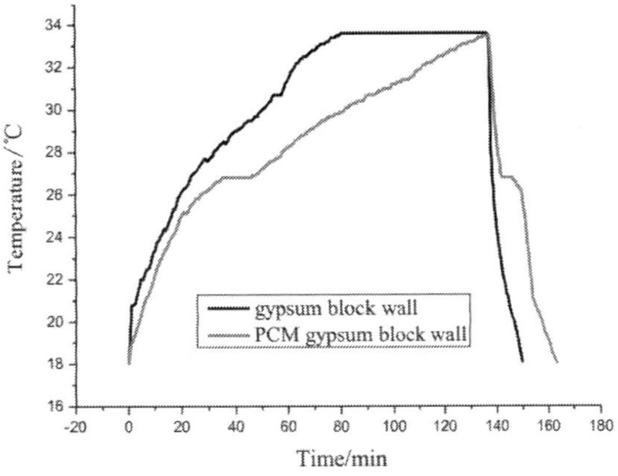

Figure 7. Performance test of PCM gypsum block

As is shown in Fig.7, the inner surface temperature of PCM gypsum block walls has lagged obviously compared with ordinary gypsum walls, indicating that the phase change energy storage gypsum blocks have good energy storage effect.

Summary

When SGC powder is used together with gypsum at wet conditions, ettringite and hydrated calcium silicate formed, which is the source of water-resistance of the modified gypsum block. It is proved that WGFP fiber has positive effect on the flexural strength of the block, and the fiber with the length of 10~15 mm behaves best.

A kind of waste glass fiber-reinforced plastic gypsum block with high strength and excellent water resistance was prepared. The optimal modification formulation is: the content of SGC 25%, WGFP fiber, with the length of 10~15 mm, 1.0%. The product prepared is of softening coefficient of 0.84, compressive strength of 27.5 MPa and flexural strength of 8.6 MPa. It is also shows, compared with ordinary gypsum walls, the modified gypsum block with CA-SA exhibits good energy storage effect.

The block uses large amount of industrial waste material including desulfurization gypsum, steel slag, granulated blast-furnace slag, carbide slag and WGFP fiber. Through this approach, the co-processing of different industrial waste and effective utilization of solid waste are realized.

References

1. F. Q. Zhao, H. J. Liu, L. X. Hao etc, Water resistant block from desulfurization gypsum, Construction and Building Materials. **27** 531-533 (2012).
2. M. M. Zhang, M. M. Chen, et al, Improving Water proof Property of Gypsum Block with Organic-Inorganic Compound Materials, Proceedings of 2nd International Conference on Material Engineering and Application. 707-711 (2015).
3. L. X. Wu, D. Li, Y. X. Chen, Influence of water reducer on performance of cast-in-place wall material with hemi-hydrate desulfurization gypsum, Raliway Engineering. **8** 123-126 (2015).
4. K. Deng, J. X. Liu, K. Tu, et al, Analysis of physical and chemical properties of steel slag cementitious materials with inorganic fibers, Multipurpose Utilization of Mineral Resources. **01** 73-80 (2015).
5. Y. Cao, G. Z. Li, J. Q. Li, Research on water resistance of glass fiber/gypsum composites, Journal of Wuhan University of Technology. **7** 42-46 (2007).
6. X. Q. Ren, F. Q. Zhao, Waste Mycelium Processing and Resource Utilization, Advanced Materials Research. **746** 58-61 (2013).
7. M. M. Chen, M. M. Zhang, F. Q. Zhao, Preparation of Gypsum Retarding Material and Activated Carbon from Waste Mycelium, Key Engineering Materials. **703** 297-300 (2016).

MATEC Web of Conferences **114**, 04007 (2017) DOI: 10.1051/matecconf/201711404007
2MAE 2017

A Reinforcement Learning Approach to Call Admission Control in HAPS Communication System

Shu Yan Ni[1], Pan Feng He[1,2,a] and Nai Ping Cheng[1]

[1]*Equipment Academy, 101416 Beijing, China*
[2]*NanjingTelecommunicationTechnologyInstitute, 210007 Nanjing, China*

Abstract. The large changing of link capacity and number of users caused by the movement of both platform and users in communication system based on high altitude platform station (HAPS) will resulting in high dropping rate of handover and reduce resource utilization. In order to solve these problems, this paper proposes an adaptive call admission control strategy based on reinforcement learning approach. The goal of this strategy is to maximize long-term gains of system, with the introduction of cross-layer interaction and the service downgraded. In order to access different traffics adaptively, the access utility of handover traffics and new call traffics is designed in different state of communication system. Numerical simulation result shows that the proposed call admission control strategy can enhance bandwidth resource utilization and the performances of handover traffics.

1 Introduction

Wireless communication system based on HAPS is a new type of wireless communication system which is being studied in the world at present. Due to the influence of the stratospheric winds and the limitation of the position and attitude keeping technique, HAPS is usually in a quasi-stationary state, also known as perturbation. The perturbation of HAPS will cause the dynamic change of cells, resulting to the cell edge users to handover frequently, which brings more challenges to the call admission control strategy. The purpose of studying call admission control strategy is to ensure the lower new call blocking probability and handover dropping probability without degrading the performance of the system. Since the cellular coverage area can improve the system performance and increase the flexibility of the system[1], Li use location information of HAPS and users to make the overlapping area to ensure the handover strategy, which helps determine and block new calls that may cause a handover failure, resulting in a near-zero handover dropping rate[2]. These methods can adapt to dynamic change of platform position, but cannot deal with the situation of attitude change.

An call admission control for adaptive resource allocation is introduce in [3], which reduces the blocking rate of new call and the handover dropping rate by service degradation. However, since the fixed admission parameters are set for new and handover calls, it has poor adaptability for changes of system state. In [4], the system resource allocation is optimized to maximize the number of users to improve the system performance. Du to this method only deal with the current call request and system state, the adaptability of dynamic state is poor.

[a] Corresponding author : hepanfeng01@126.com

The problem of call admission control can be modelled as a Markov decision processes (MDPs) or Semi-Markov decision processes (SMDPs), which may adapt to the dynamic change of system state[5]. The methods such as policy iteration and dynamic programming can solve this kind of problem [6], but the current research assumes that the link capacity is constant. In fact, due to the presence of signal fading, interference, user mobility and using adaptive modulation and coding(AMC), the link capacity is time-varying in actual communication networks (such as World Interoperability for Microwave Access(WiMAM), Long Term Evolution(LTE) etc.)[7].The quasi-stationary state of HAPS will also cause the change of link capacity in HAPS communication system.

Although the above research can ensure the performance of handover in a certain extent, the dynamic changes of users in cell and link capacity caused by perturbation and mobility of users are considered inadequate, as well as different business requirements. In addition, it is difficult to obtain the exact model parameters of the MDPs. To solve the above problems, we introduced the reinforcement learning which does not need the exact model parameters to maximize the long-term benefits of system, aiming at deal with different traffics. By cross layer interaction and business downgrade ideas, the adaptive call admission control strategy based on reinforcement learning was put forward.

2 System model

Thornton pointed out that the change of carrier interference ratio from cell centre to edge is large, because of the power rolling down quickly at the edge zone[8]. So the adaptive modulation coding can get very good spectrum efficiency. Meanwhile call admission control strategy should not only consider the change of link capacity, but also guarantee the QoS of different traffics.

The multi-beam antennas from HAPS produce an elliptical beam that is projected onto the ground as a circular. For each deviation θ of the beam direction from HAPS, the beam directivity gain $D(\theta)$ can be approximated as

$$D(\theta) = G_{\max} \cdot \max[\cos^n \theta, S_f] \tag{1}$$

Where $G_{\max}$ is directivity gain at the centre of the cell and n is the main lobe decay rate. S_f is sidelobes parameter. Assuming that the center coordinates of cell i is (x_i, y_i), the coordinates of user q in cell i is expressed by (x, y), as shown in Figure 1. The deviation $\theta_{i,q}$ from user q to the beam boresight of cell i can expressed as

$$\theta_{i,q} = \mathrm{a}\cos\left(\left(xx_i + yy_i + h^2\right) / \sqrt{x_i^2 + y_i^2 + h^2} \,/\, \sqrt{x^2 + y^2 + h^2}\right) \tag{2}$$

Where h is the height of HAPS. Due to the quasi-stationary state of the HAPS, the directivity gain of the user changes dynamically in the cell even under the condition of user's still, which will cause the change of signal to interference plus noise ratio($SINR$) . In addition, the wireless channel fading characteristics also cause changes of $SINR$.

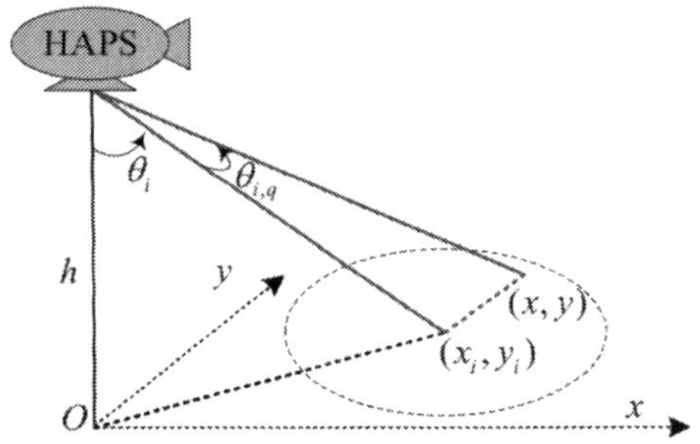

Figure 1. the Relationship between Users Position and the Beam Boresight.

DOI: 10.1051/matecconf/201711404007

Assuming that the radio channel of the multi-user OFDMA communication system from HAPS is flat fading and the channel conditions are stable and independent in each call admission control period. The subcarriers used in the cell are allocated equal power. The instantaneous *SINR* for the subcarriers n in cell i can be expressed as

$$SINR_{i,q}^{n} = \frac{P_i^n D(\theta_{i,q}) | H_{i,q}^n |^2}{N_q^n + \sum_{j=1, j \neq i}^{n_{cc}} P_j^n D(\theta_{j,q}) | H_{j,q}^n |^2} \tag{3}$$

Where P_i^n is the allocated transmit power of the subcarrier n in the cell i, N_q^n and n_{cc} are the Gaussian white noise and the number of all cells in same frequency. $H_{i,q}^n$ is channel response of the user q on the subcarrier n of the cell i, which generally include path loss, small scale fading and large scale fading. $D(\theta_{i,q})$ is the beam gain of the user q in the cell i, and $D(\theta_{j,q})$ is interference gain of the user q from the interference cell j. $\theta_{j,q}$ and $\theta_{i,q}$ can be calculated by (2).

For the AMC mechanism based on the Multiple Quadrature Amplitude Modulation(MQAM), the instantaneous bit rate of the carrier n in the cell i can be expressed as

$$R_{i,q}^n = B / N \cdot \log_2 (1 + SINR_{i,q}^n / \Gamma) \tag{4}$$

Where $\Gamma = -2/3 \cdot \ln(5\text{BER})$ is the signal to noise ratio difference between the spectral efficiency of the AMC mechanism based on MQAM and the Shannon capacity[9]. B and N are the channel bandwidth and the number of subcarrier of the cell.

According to the changing state of the channel environment, the instantaneous bit rate of OFDM subcarriers can be dynamically changed. So the link capacity can be updated based on the capacity of the subcarrier instantaneous state. It is assumed that a single subcarrier can carry the 1bit information at its minimum, its basic transmission capability is R_B bps. If the subcarrier can carry M bit information at a certain time, then the transmission capacity is MR_B bps. For the subcarrier which is not allocated, it has the basic capacity. Therefore, the total reachable link capacity ξ of the cell i can be expressed by

$$\xi = \sum_{N_{occ}+1}^{N} R_{\mathrm{B}} + \sum_{n=1+1}^{N_{occ}} R_{\mathrm{B}} \sum_{q=1}^{Q} \left\lfloor R_{i,q}^n / R_{\mathrm{B}} \right\rfloor \cdot R_{\mathrm{B}} \tag{5}$$

where N_{occ} is the number of subcarriers occupied. Assuming that the link reachable capacity has a total of L states and the link capacity is represented by ξ_l , then $\xi_l < \xi_{l+1}, l = 1, ..., L-1$.

3 Call Admission Control scheme based on reinforcement learning

Reinforcement learning (RL) interacts with the environment to obtain optimal strategy by learning agents. Q learning a kind of model free reinforcement learning algorithm , mainly through MDPs modelling and iterative method to approach the optimal solution. The call admission control(CAC) problem is modelled as a discrete time MDPs, the following are the state space and action set, reward function design, realization process of Q learning and adaptive CAC.

3.1 system state space and action sets

The vector $\boldsymbol{x} = (x_1, x_2, ..., x_k, ..., x_K)$ represents the number of different traffic types in a cell. The number of occupied subcarriers in the cell at the time t is $n_c(t)$, and the link capacity of the cell at the time t is represented by $\xi(t)$. The state space $\boldsymbol{S}(t)$ of the cell at the time t can be expressed as $\left(\boldsymbol{x}(t), n_c(t), \xi(t)\right)$, and can be divided into three kinds according to the link capacity and the number of occupied subcarriers.

$$s=\{(x(t),n_c(t),\xi(t)\}=\begin{cases}S_1,n_c(t)<N,\sum_{k=1}^{K}x_k rb_k\le\xi(t)\\ S_2,n_c(t)=N,\sum_{k=1}^{K}x_k rb_k\le\xi(t)\\ S_3,n_c(t)=N,\sum_{k=1}^{K}x_k rb_k\ge\xi(t)\end{cases}\quad(6)$$

Where rb_k and x_k are the average rate and number of the traffic type k. There are available subcarriers can allocate for the new call request or handover request in state S_1. Although there is no available subcarrier in state S_2, the sum of link rate of all traffics is less than link capacity of the system, which means that it is possible to allow call request by redistributing subcarriers to release some subcarriers. Since the sum of link rate exceeds the link capacity in state S_3, some traffic must be degraded or interrupted. No matter what state the cell is in, when call request of the traffic k arrives, the system must choose to reject or accept the request. Then the action sets can be expressed by

$$A(s)=\{a(s)=(a(s,1),...,a(s,K))\mid a(s,k)\in\{0,1\}\},\forall s\in S_1\cup S_2\cup S_3\quad(7)$$

3.2 Reward Function

The goal of maximizing the reward of the system can be expressed as the sum of the utility functions of traffics in service, we consider four types of traffics, voice, fixed data, multi-media and best effort traffic. The priority of these four traffics are $\beta_k(k=1,...,4)$, and the priority of handover traffic is β_0.

The voice and fixed rate data traffic are belong to constant bit rate(CBR) with strict guaranteed bit rate(GBR) limits, and the quivalent bandwidth rc_k must reach a certain value. While the requirement of multimedia traffic for bit error rate and delay is not high, which belongs to the GBR service with minimum rate guarantee. So the equivalent bandwidth is in a certain range[$rc_{\min}, rc_{\max}$]. The best effort traffic generally does not have constant bandwidth requirements, and the equivalent bandwidth is represented by rc_e .Since QoS requirements for different are measured by the equivalent bandwidth requirements of the traffics, the utility function of new call request with different traffics can be

$$U_r=\begin{cases}\beta_k\cdot rc_k\varepsilon(r-rc_k) & ,k=1,2\\ \beta_k\cdot\sin\left(\dfrac{\pi}{2}\cdot\dfrac{r-rc_{\min}}{rc_{\max}-rc_{\min}}\right)[\varepsilon(r-rc_{\min})-\varepsilon(r-rc_{\max})]+rc_e\cdot\delta(r-rc_{\max}) & ,k=3\\ \beta_k\cdot r/rc_e[\varepsilon(r)-\varepsilon(r-rc_e)]+rc_e\cdot\delta(r-rc_e) & ,k=4\end{cases}\quad(8)$$

Where $\varepsilon(\cdot)$ is the utility function of handover request with different traffics can be $U_{rh}=\beta_0\cdot U_r$. When the state is S_1, there are idle subcarriers. So allowing access to the traffic can increase the overall utility, and the reward $r(s,a)$ equals to $a(s,k)\cdot U_r(k)$ for the traffic k request. If the average blocking ratio $\bar{f}_d$ is zero or blocking ratio for the traffic k request $f_d(k)$ is zero, the reward is also $a(s,k)\cdot U_r(k)$ .If the value of $\bar{f}_d\cdot f_d(k)$ is greater than 0 in state S_2 ,the reward value is $a(s,k)\cdot\gamma_k f_d(k)/\bar{f}_d\cdot U_r(k)$ to access more traffics k than others (γ_k is the blocking probability weighting factor of traffic k). While the state is S_3, the reward value includes both the access revenue and the inherent loss, which can be expressed by $a(s,k)\cdot(\gamma_k f_d(k)/\bar{f}_d-1)\cdot U_r(k)-\bar{f}_d\bar{U}_r$. Therefore the designed reward function $r(s,a)$ is

$$r(\boldsymbol{s},\boldsymbol{a})=\begin{cases} a(\boldsymbol{s},k)\cdot U_r(k)\,, & \forall \boldsymbol{s}\in \boldsymbol{S}_1 or\left(\overline{f}_d\cdot f_d(k)=0, \forall \boldsymbol{s}\in \boldsymbol{S}_2\right), k=1,...,K \\ a(\boldsymbol{s},k)\cdot \gamma_k f_d(k)/\overline{f}_d\cdot U_r(k), & \overline{f}_d\cdot f_d(k)>0, \forall \boldsymbol{s}\in \boldsymbol{S}_2, k=1,...,K \\ a(\boldsymbol{s},k)\cdot(\gamma_k f_d(k)/\overline{f}_d-1)\cdot U_r(k)-\overline{f}_d\overline{U}_r, & \forall \boldsymbol{s}\in \boldsymbol{S}_3, k=1,...,K \end{cases} \tag{9}$$

3.3 implementation process of reinforcement learning

We use the Q value iteration method to achieve reinforcement learning in this paper. The iterative formula is[10]

$$Q_{t+1}(\boldsymbol{s},\boldsymbol{a})=(1-\alpha)Q_t(\boldsymbol{s},\boldsymbol{a})+\alpha\left[r_t(\boldsymbol{s},\boldsymbol{a})+\gamma\max_{a'\in A}Q_t(\boldsymbol{s}',\boldsymbol{a}')\right] \tag{10}$$

where $\alpha\in[0,1)$ is the learning rate and $\gamma\in[0,1)$ is the discount factor. α can be expressed by $\alpha=1/(1+T(\boldsymbol{s},\boldsymbol{a}))$ ($T(\boldsymbol{s},\boldsymbol{a})$ represents the number of visits on state-action). Obviously, when α tends to 0, $Q_t(\boldsymbol{s},\boldsymbol{a})$ converges to $Q_t^*(\boldsymbol{s},\boldsymbol{a})$. The optimal action set can be obtained by repeated learning and decision making. In order to balance the exploration and exploitation, the standard algorithm ε-greey is used to select the action of the controller.

(1) Parameter initialization: make $Q_0(\boldsymbol{s},\boldsymbol{a})$ equal to $r(\boldsymbol{s},\boldsymbol{a})$ and initialize α , γ and the exploration rate ε_0 ;

(2) Obtain the current state: When a new call arrives or a call leaves the cell, the controller collects the number of traffic in the current cell, the number of subcarriers occupied, and the current cell link capacity;

(3) Select action: the random decision is made to access or reject the arrival request for the exploration stage. While the action of the current state maximum rewards value is selected to decide whether to access or reject the arrival request for exploitation stage.

(4) Update the value $Q(\boldsymbol{s},\boldsymbol{a})$: Caculate $Q(\boldsymbol{s},\boldsymbol{a})$ according to the reward function $r(\boldsymbol{s},\boldsymbol{a})$ of the current state-action-pair ;

(5) Update parameters: the parameter α and ε are gradually reduced by anti-scaling function after each iteration, and returned (2).

3.4 adaptive subcarrier allocation strategy

To ensure the long-term gains of the system, the effective subcarrier resource allocation strategy can ensure the QoS requirements and efficient resource utilization. According to the most stringent QoS requirements for CBR traffics, the service of these traffics should not downgrade and the AMC scheme is always successful implementation. While the AMC scheme of VBR and BE traffic is controlled by the adaptive subcarrier allocation strategy to accept more users and reduce the call blocking probability.

Suppose that the equivalent bandwidth of the user q for the traffic k corresponds to the number of subcarriers required is $n_{k,q}^r$, the number of allocated subcarriers is $n_{k,q}^a$. We use N_{occ} to express the total number of occupied subcarrier.

(1) $N_{\text{occ}}<\eta N$: Since the cell is in a state of light load, all users does not distinguish the traffic types. The allocated subcarriers equals to the need of this traffic, namely $n_{k,q}^a=n_{k,q}^r$.

(2) $N_{\text{occ}}\geq\eta N$: The cell should distinguish traffic type in the heavy load state. For the CBR traffics, the number of allocated subcarriers equals to the requirements ($n_{k,q}^a=n_{k,q}^r, k\in\{CBR\}$). While the number of the allocated subcarriers for VBR or BE traffic is discounted on the number of requests

($n^a_{k,q}=\gamma_{k,q}n^r_{k,q}, k\in\{VBR, BE\}$). The discount factor $\gamma_{k,q}$ is associated with the current load state, the link capacity, the new call blocking rate and the handover dropping rate in the cell. Assuming the minimum number of subcarriers allocated is $n^{\min}_{k,q}$ for VBR and BE traffic, the number of subcarriers that can be released by the service degradation is:

$$N_{\text{release}}=\sum_{k=3}^{K}\sum_{q=1}^{x_k}(n^a_{k,q}-n^{\min}_{k,q}) \tag{11}$$

According to the relationship between $N-N_{\text{occ}}-N_{rel}$ and n^r_k, there are two cases:

1) $n^r_k\le N-N_{\text{occ}}-N_{rel}$: the traffic k can be accessed through service degradation in the cell. In order to In order to access serve more traffics, the cell will allow access to the traffic k.

2) $n^r_k > N-N_{\text{occ}}-N_{rel}$: Although service degradation can release some subcarriers, it is not enough to accept traffic k. For the purpose of reducing handover dropping rate of the GBR (including CBR and VBR), the system can force call termination of the BE traffic, which has the largest number of occupied subcarriers currently to release subcarriers. If the arrival traffic is belong to the GBR and the required subcarriers n^r_k is satisfied with the formula

$$n^r_k\ge\max_q\{n^a_{4,q}\},\ 1\le q\le x_4 \tag{12}$$

the system allows this handover access to the cell. Otherwise the access is refused.

4 simulation analysis

The simulation parameters of HAPS communication system are consistent with [11]. The mapping relationship between AMC and SINR used in this system can refer the air interface of IEEE802.16, which can be found in [12]. The simulation traffic parameters are shown in Table 1. It is assumed that the relationship between handover call arrival rate λ_h and new call arrival rate λ_n is $\lambda_h=\rho\lambda_n$, in which ρ is a random variable with mean and variance of 0.5.

Table 1. Traffic Parameters

Traffic types	Voice	Fixed bit rate data	Multimedia	Best effort
Equivalent bandwidth/Kbps	64	56	1128-512	0-120
Call duration/s	100	120	400	250
Traffic priority weight	9	87	2	1
Percentage of all new calls	0.25	0.15	0.4	0.2

In order to verify the performance of the proposed algorithm, the strategy based on guard channel (guard channel with 10%), adaptive bandwidth allocation with handover priority strategy in [3] and the proposed algorithm are compared under the same conditions. We use five metrics to test the performance of the algorithms, which are handover call dropping probability, new call blocking probability, bandwidth utilization, forced call termination probability and Average system revenue. Forced call termination probability include the current dropped call resulting from Link capacity reduction or arrival GBR handover request.

The performances of this three strategies are shown in figure 2. Compared with the guard channel strategy, the proposed method have better performance in handover dropping probability, new call blocking probability and bandwidth channel utilization, the price is slightly higher rate of call termination probability. Although adaptive bandwidth allocation strategy has lowest new call blocking probability and highest bandwidth utilization, the handover dropping probability rate and forced call termination probability are worst. In comparison, the performance of the proposed method is superior, especially handover dropping probability and bandwidth utilization.

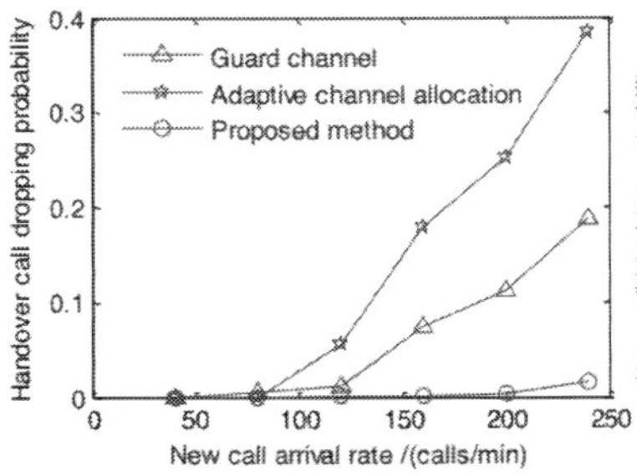

(a) Handover dropping probability

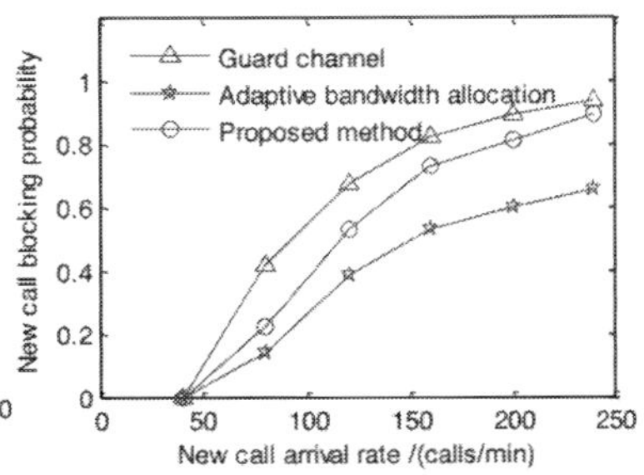

(b) New call blocking probability

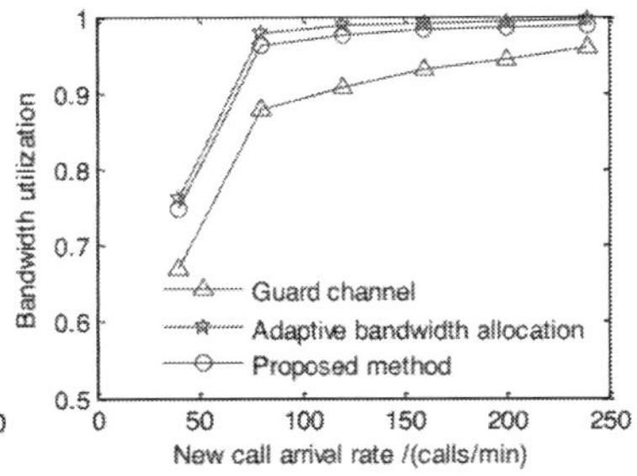

(c) Bandwidth utilization

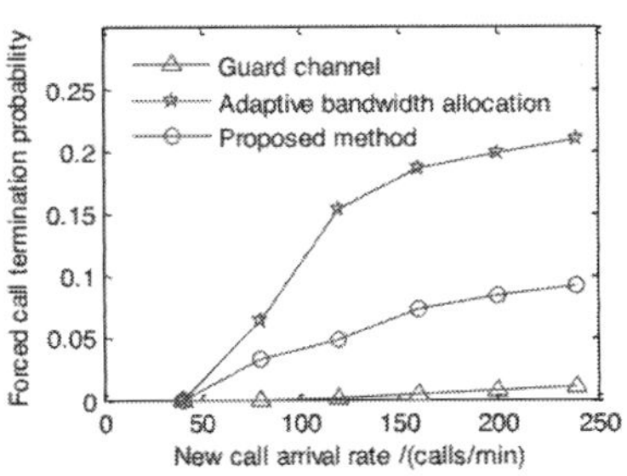

(d) Forced call termination probability

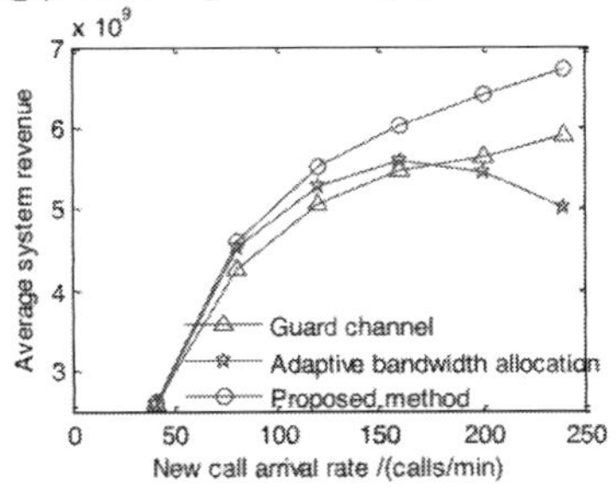

(e) Average system revenue

Figure 2. The performances of the three strategies

Figure 2 (e) shows that the proposed method always has the most average system revenue, The reason is that this method takes full account of the changing system state , the performance of the user and the system, in order to ensure long-term revenue.

5 conclusions

It has always been a hot issue in the research of wireless communication system that how to effectively access various types of traffics, and ensure the QoS requirements of various traffics services and improve the utilization of system resources. By introducing cross layer interaction and service downgrade, the proposed strategy considers the utility and the blocking rate of different traffics and maximizes long-term gains of system to balance the handover performance and system resource utilization. In the future work, we also need to consider the dynamic interference between cells to further improve the system performance.

References

1. D. Grace, C. Spillard, T.C. Tozer, *Wireless Personal Multimedia Communications Conference* (IEEE, New York, 2003)
2. S. FLi ,J. Wei, D. T. Ma, L. Wang, Journal on Communications. **32**, 131 (2011)
3. M. Z. Chowdhury, Y. M Jang, Z. J.Haas, Journal of Communications and Networks, **15**, 15 (2013)
4. A. Ibrahim, A. S. Alfa, IEEE Trans.Wireless Comm., **14,** 5823 (2015)
5. M. Panfili, A. Pietrabissa, International Journal of Control, **89**, 1428 (2016)
6. M. K. Luka ,A. A. Atayero, O. I. Oshin,*Sai Computing Conference* (IEEE, New York, 2016)
7. A. Pietrabissa. . European Journal of Control, **17** 89(2011)
8. J. Thornton, D. Grace, M. H. Capstick ,IEEE Trans.Wireless Comm., **2,** 484 (2003)
9. A. J. Goldsmith , S. G. Chua,IEEE Trans. Comm., **45**, 1218(1997)
10. D. Xiong, Y. Li, Journal on Communications, **36**, 252 (2015)
11. P. He, N. Cheng, S. Ni, *Wireless Communications & Signal Processing* (IEEE, New York, 2016)
12. P802.16Rev3/D5, Air Interface for Broadband Wireless Access System (IEEE, New York, 2012)

An Improved INS Factor Graph Formulation and Its Application in Integrated Navigation System

Gao Junqiang[1,a], Tang Xiaqing[1] and Zhang Huan[1]

[1]*Department of Control Engineer, Academy of Armored Force Engineering, Beijing, China*

Abstract. A new method of establishing INS factor graph is proposed to deal with the serious waste of INS measurements in current integrated navigation factor graph algorithm. The INS observations pre-integrated to construct a smart factor during a time period, are used to solution simultaneously, to get the navigation states estimation. Then the estimated values can be used to construct an INS factor, adding into the INS factor graph. As a result, the new model is larger than the original, but the incremental inference technology makes sure the real-time ability of the factor graph algorithm. To analysis the performance of the proposed algorithm, INS and GNSS data are simulated, and the navigation position error of either the INS factors are added or not are compared. The results show that, the precision and stability of the factor graph algorithm are greatly improved by adding the INS factors. The proposed algorithm makes full use of the INS accuracy, and reflects the advantage of INS high stability.

1 Introduction

Navigation and positioning technology which plays an important role in precision strike and cooperative combat is an indispensable key technology in modern warfare. Presently, navigation means has been developed from the use of a single navigation method to the stage of integrated navigation system. With the increasingly harsh environment and the higher demand on the performance of navigation and positioning, the type and number of sensors integrated together are increasing day and day. Adaptive multi-sensor integrated navigation system [1] and All Source Positioning and Navigation System [2] have become an important tendency of the future navigation system.

The integrated navigation algorithm based on factor graph can flexibly integrate information of a variety of asynchronous sensors. It is very convenient to deal with the dynamic changes of the sensors combination and has great advantages in the application of multi-sensor integrated navigation system [3, 4].

As a very important sub-system of integrated navigation system, INS is a processing difficulty of the integrated navigation algorithm based on factor graph, because its performance depends on the inertial sensors' high frequency measurement of the carrier's movement. However, if the factor nodes of INS are added to the factor graph by such a high frequency, the factor graph will be very large and leads to an unacceptable computational cost in practical application. What's more the navigation information does not need to be output by the INS measurement frequency. All above makes it

[a] Corresponding author : gjqxkl8990@163.com

DOI: 10.1051/matecconf/201711404008

meaningful and acceptable to reduce the INS factor graph through a reasonable method. In the current factor graph algorithm, INS measurements during a time period are pre-integrated to construct an equivalent INS factor [5]. But the pre-integration process leads to a serious waste of INS precision, because it is too simple comparing with the very mature INS solution algorithm.

In view of this, in order to fully utilize INS measurement information and ensure the real-time performance of factor graph algorithm, a new method of establishing INS factor graph is proposed. The pre-integrated INS observations are used to solution simultaneously, to obtain navigation states estimation. Then the estimated values can be used to construct an INS factor and added into the factor graph to realize a stably and accurately fusion with GNSS information.

The remaining part of this paper is organized as follows. The integrated navigation algorithm based on factor graph is introduced in Section 2. Section 3 presents the factor graph of INS/GNSS integrated navigation system. Simulation and results are provided in Section 4, and concluding remarks are suggested in Section 5.

2 Integrated navigation algorithm based on factor graph

The problem solved by integrated navigation algorithm is computing navigation states based on the observations. Essentially, the joint probability distribution function (pdf) is constructed by the navigation states and observations, then the maximum a posteriori (MAP) estimate is calculated and obtaining the optimal estimation of the navigation states. In factor graph algorithm, the joint pdf is described as a factor graph, and the MAP estimate is calculated in the established factor graph.

2.1 Factor graph

In integrated navigation system, we define the sets of all navigation states and calibration states up to the current time as V_k, and letting Z_k represents all the measurements received up to the current time. Then, the joint pdf can be written as:

$$p(V_k \mid Z_k) \tag{1}$$

The optimal estimation of the navigation states in the form of MAP estimate is given by:

$$V_k^* \triangleq \arg\max_{V_k} p(V_k \mid Z_k) \tag{2}$$

The joint pdf (1) can always be factorized in terms of measurement models, and expressed as a factor graph. Formally, a factor graph is a bipartite graph with two node types: variable nodes v_j and factor nodes f_i, where v_j is navigation states or calibration states, and f_i is an error function corresponding to the measurement model, as:

$$f_i(V_k^i) = d(\text{err}_i(V_k^i, Z_k^i)) \tag{3}$$

where the operator $d(\bullet)$ denotes a certain cost function, $V_k^i \subseteq V_k$ and $Z_k^i \subseteq Z_k$ are the state variables and the measurements, respectively, involved in the measurement model. f_i and v_j will be connected with an edge, if and only if V_k^i existences in v_j, which determines the topology of the factor graph.

The factor graph encodes the relationship between the navigation states and the measurements of navigation systems. Defining the measurements of navigation systems as factors, the integration of navigation systems becomes to connecting these factors to the corresponding variable nodes in the factor graph. The defined factor will be added into the factor graph, when the corresponding sub-system is valid, and denied as soon as it's invalid. Such a representation provides a plug and play capability, since no special adjustment will be needed when the sub-systems changed, which greatly improves the stability of the integrated navigation system.

2.2 Incremental inference of factor graph

The basic inference method is firstly obtaining Bayes net by eliminating the factor graph. Then the chordal Bayes net can be converted into a tree-structured graphical model, namely Bayes tree. As a result, optimization and marginalization will be done in the Bayes tree.

The approach using Bayes tree to realize incremental inference [6] of the factor graph is shown in Fig. 1. When new factors are added into the graph, the affected part of the Bayes tree and the involved variables are identified, firstly. Then the affected part of the Bayes tree is converted back into a factor graph, the new factors are added, and the resulting (small) factor graph is re-eliminated to get a new (small) Bayes tree. The new (small) Bayes tree is then merged with the unchanged part of the original Bayes tree, and the incremental update of the Bayes tree is accomplished finally.

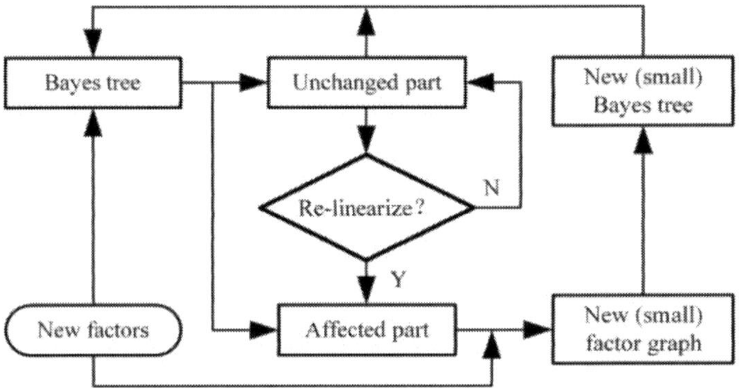

Figure 1. Increment implementation method.

Additionally, it is necessary to consider the linearization problem of the factor graph in this process. A re-linearization strategy is employed to reduce the linearization computation. When the deviation of its current estimate from the linearization point becoming larger than a threshold, the given variable will need to be re-linearized. Then all the relevant information related to this variable has to be replaced by re-linearizing the corresponding original factors. The variables need to be re-linearized are also included in the affected part in Fig. 1.

New factors and re-linearization variables only update the top of the Bayes tree, however, estimating of the variables in all the sub-trees will be changed. But the farther away from the top of the tree, the less the influence, because of the effect of the new measurements is usually limited. Therefore, in order to reduce computational cost, only partial variables being affected significantly will be updated. The solve starts from the top of the tree, and will be stopped when comes to a clique, in which all the variables changed smaller than a threshold.

3 Factor graph of INS/GNSS integrated navigation system

The integration of INS/GNSS is a typical representative of integrated navigation system and also the core of multi-sensor integrated navigation system. So it is significant to study the application of factor graph algorithm in INS/GNSS integrated navigation system.

3.1 INS factor graph formulation

As a premise, an INS factor graph is established based on the state equations, and then an equivalent INS factor graph is proposed to reduce its size through a reasonable method.

3.1.1 INS factor graph

The discrete representation of INS state equations can be written as:

$$\begin{aligned} x_{k+1} &= h(x_k, \alpha_k, z_k) \\ \alpha_{k+1} &= g(\alpha_k) \end{aligned} \tag{4}$$

where x represents the navigation state variables, which include position, velocity and attitude angles, α represents the INS calibration parameters, and z is the measurements of inertial sensors.

The state equations can be used to predict x_{k+1} and α_{k+1}. Then error functions represented by the factors can be given by the differences between these predictions and the current estimates:

$$\begin{aligned} f^{\text{IMU}}(x_{k+1}, x_k, \boldsymbol{\alpha}_k) &\triangleq d(x_{k+1} - h(x_k, \boldsymbol{\alpha}_k, z_k)) \\ f^{\text{bias}}(\boldsymbol{\alpha}_{k+1}, \boldsymbol{\alpha}_k) &\triangleq d(\boldsymbol{\alpha}_{k+1} - g(\boldsymbol{\alpha}_k)) \end{aligned} \tag{5}$$

where f^{IMU} is an IMU factor connecting x_k, x_{k+1} and $\boldsymbol{\alpha}_k$; f^{bias} is a bias factor connecting $\boldsymbol{\alpha}_k$ and $\boldsymbol{\alpha}_{k+1}$.

Fig. 2(a) illustrates the factor graph of INS, where f^{Prior} is a unary factor, defined by the available prior information of the navigation systems. The frequency of adding the variable nodes and the factor nodes in this model is the same as the INS measurement frequency.

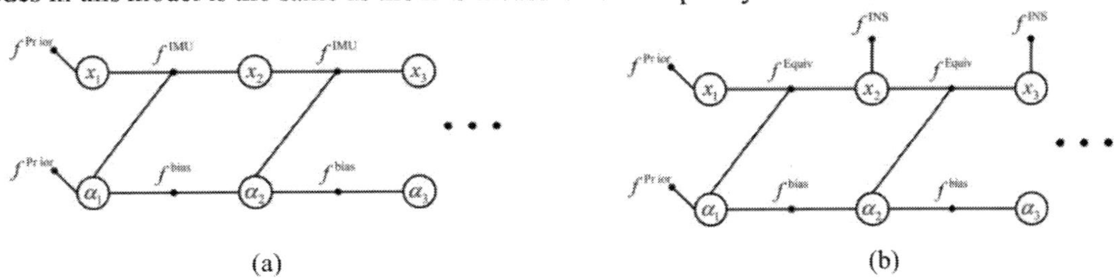

Figure 2. (a) Factor graph of INS. (b) Equivalent factor graph of INS.

3.1.2 Equivalent INS factor graph

The navigation states only need to be output by a low frequency, while the high frequency measurements are necessary to realize the accurate solution of INS. The variables need to be output are defined as target variables, and must be added to the factor graph as variable nodes. The remaining variables, that we call support variables, are not strictly required as output of the estimation problem. But these support variables are essential for the formulation of the optimization problem, and need to be considered when constructing factor nodes of the target variables. Then, these factor nodes can be called smart factors [7]. The idea of smart factor applies to all types of navigation sensors. For INS, the smart factor is obtained by pre-integrating all the INS measurements between two target variables.

However, compared to the state-of-the-art solution method of INS [8], the pre-integration process presents rough and ready, which results in a serious waste of INS precision. Therefore, The INS observations are used to calculate navigation states estimation by INS solution method, simultaneously with the pre-integration process. Then, the difference between these estimates and the current estimates defines an INS factor:

$$f^{\text{INS}}(x_k) \triangleq d\left(x_k - x_k^{\text{INS}}\right) \tag{6}$$

Fig. 2(b) illustrates the equivalent factor graph of INS. f^{Equiv} is the smart factor obtained by pre-integration [3], and the frequency of adding variable nodes and factor nodes in the model depends on

DOI: 10.1051/matecconf/201711404008

the time period of pre-integration. What's more, the equivalent INS factor graph constructed by pre-integration only, can be acquired by removing the INS factors.

3.2 GNSS factor graph formulation

The GNSS measurement equation can be written as:

$$z_k^{\text{GNSS}} = h^{\text{GNSS}}\left(x_k\right) + n^{\text{GNSS}} \tag{7}$$

where h^{GNSS} is the measurement function, relating between the measurement z_k^{GNSS} to the carrier's position, n^{GNSS} is the measurement noise. Eq. (7) defines a unary factor f^{GNSS}:

$$f^{\text{GNSS}}\left(x_k\right) \triangleq d\left(z_k^{\text{GNSS}} - h^{\text{GNSS}}\left(x_k\right)\right) \tag{8}$$

The factor graph of INS/GNSS integrated navigation system shown in Fig. 3 can be established by adding the GNSS factors to the equivalent factor graph of INS.

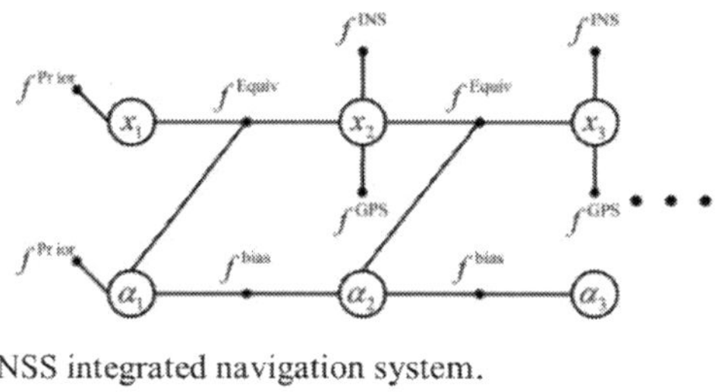

Figure 3. Factor graph of INS/GNSS integrated navigation system.

4 Simulation and analysis

The prosed method is simulated in MATLAB to compare navigation results of the model with INS factors and without INS factors.

Fig. 4 illustrates the simulation trajectory. The initial position is latitude 39°, longitude 116°, height 35 m, and the initial velocity is 0 and heading north. Arriving the terminal of latitude 39.01°, longitude 115.98° height 35 m, after acceleration, turning, climbing, descent, deceleration, etc. The maximum velocity and height are respectively 20 m/s and 236 m, and the total time is 1800 s. The gyro constant drifts along three axes of body frame are 0.02 (°)/h with white noise 0.01 (°)/$\sqrt{\text{h}}$. The accelerometer biases along three axes of body frame are 100 μg with white noise 50 $\mu\text{g}\cdot\sqrt{\text{s}}$. The GNSS noise in north and east are both 1 m, and 3 m in height. The INS sampling period and GNSS output period are 0.01 s and 1 s, respectively.

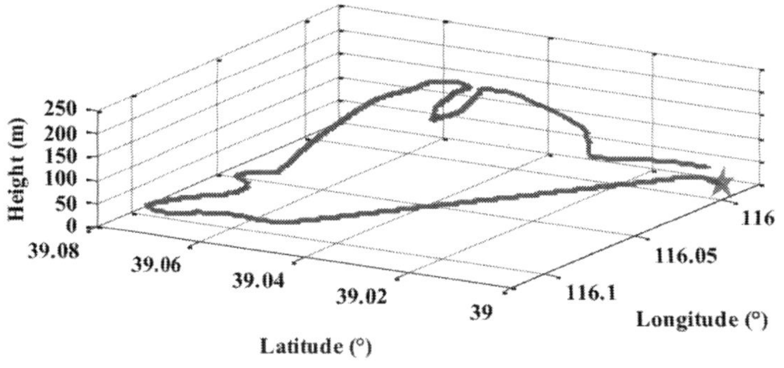

Figure 4. Simulation trajectory.

For the factor graph model shown in Fig. 3, the periods of f^{Equiv} and f^{INS} are both 1 s, and the period of f^{GNSS} is 10 s. Then the incremental inference technology which makes sure the real-time ability of the algorithm is used to calculated navigation results. The comparison of position errors is shown in Fig. 5, and the Max and RMS of position errors are shown in Table 1. Obviously, the position precision of the factor graph integrated navigation algorithm is significantly improved in north and east by adding INS factor. The maximum position error of north and east are reduced by 62%, and 43%, respectively. And the position error RMS of north and east are reduced by 17% and 27%, respectively. The precision and stability of the integrated navigation algorithm based on factor graph can be improved significantly by adding the INS factors.

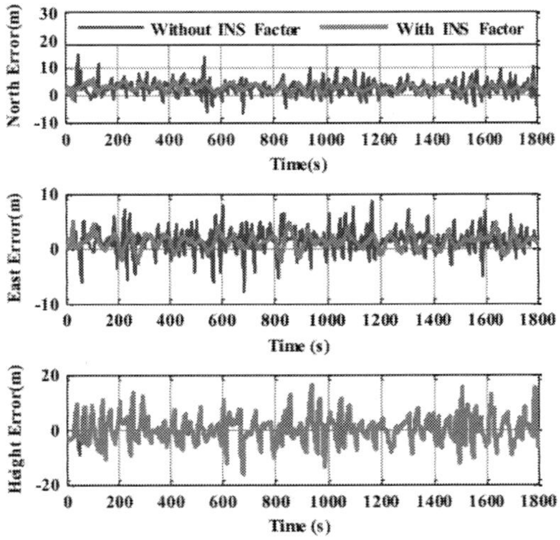

Figure 5. Position errors.

Table 1. Max and RMS of position errors.

		North	East
Max	Without f^{INS}	14.7691	8.6363
	With f^{INS}	5.6731	4.9614
RMS	Without f^{INS}	3.3378	2.2972
	With f^{INS}	2.7650	1.6822

5 Conclusion

Studying the integrated navigation algorithm based on factor graph, and a new method of establishing INS factor graph is proposed. Although the model size is increased, the incremental inference technology of factor graph makes sure the real-time ability of the algorithm. The simulation results show that the proposed INS factor graph presents superior performance when integrated with GNSS. Compared with the original model without INS factors, the new model makes full use of the INS accuracy, and reflects the advantage of INS's high stability. As a result, the precision and stability of the factor graph algorithm are both greatly improved.

References

1. P. D. Groves, J Navigation **67**, 311 (2014)
2. Y. Zhou, J. Lai, X. Guo, J. Yang, *China Statellite Navigation Conference (CSNC) 2015,* 801 (2015)
3. V. Indelman, S. Williams, M. Kaess, F. Dellaert, Robot Auton Syst **61**, 721 (2013)
4. N. Wu, B. Li, H. Wang, C. Xing, J. Kuang, Sci China Inf Sci **58**, 042305 (2015)
5. T. Lupton, S. Sukkarieh, IEEE T Robot **28**, 61 (2012)
6. M. Kaess, H. Johannsson, R. Roberts, V. Ila, J. J. Leonard, F. Dellaert, Int J Robot Res **31**, 216 (2012)
7. L. Carlone, Z. Kira, C. Beall, V. Indelman, F. Dellaert, *2014 IEEE International Conference on Robotics and Automation (ICRA)*, 4290 (2014)
8. Y. Qin, *Inertial Navigation (Second Edition)*, (2016)

Design of Online Washing Process for an Engine Based on Washing Experiments

Zhang Yong[1], Wu Heng[1,a], Li Benwei[1] and Zhao Shufan[1]

[1]Department of Airborne Vehicle Engineering, Naval Aeronautical and Astronautical University, Yantai Shandong 264001, China

Abstract. In order to develop online washing process of a carrier-based aircraft engine and improve washing efficiency, washing equipments were employed in cold operation state to carry out experimental study on structural strength of spray frame, selection of nozzle and determination of injection parameters of online washing spray system, determination of washing injection timing, determination of drying procedure, liquid-washing experiments of different concentrations and water-washing experiments and so on. The results show that if engine is washed in cold operation state, mid-nozzle with upward offset angel of 7° 30′ can be selected and the feed liquid pressure provided by washing vehicle is 0.9Mpa. The start time is 20s after an engine cold start-up, and sustains 30s for once washing. After a liquid-washing with the use of washing fluid, the liquid concentration of which is no more than 20%, twice water-washing can be carried out to rinse the almost all cleaning agent remaining in the engine runner(the residue is less than 1%). The engine can be thoroughly dried by power-on operation, the procedure of which is engine operating 2min in ideal state, then 2min in state of n2=85% and 2min of ideal state lastly. Washing will not affect the engine function.

1 Introduction

As a carrier-based power plant, there are two main purposes of a type of engine in situ washing: one is, considering the sea flight conditions and environment, anti-corrosion washing of engine leaves is carried out to remove the salt spray of the fan, high pressure compressor and other components to prevent corrosion of mechanical parts. The other is, when flow capacity and efficiency of components declined due to fouling and other factors of engine airflow channel resulting in reduction of engine thrust and increase of fuel consumption, the engine airflow channel washing is carried out depending on the situation or combined with the timing of inspection. The main technical contents of engine washing include: special washing agent, washing equipment, washing implementation specifications and washing efficiency assessment [1]. After the development of engine washing equipment and the usage of special washing agent, designing the washing process and developing a washing implementation specification are essential to improve washing efficiency and increase aircraft attendance, as well as provide a basis for engine maintenance procedures.

At present, the United States and Europe and other advanced aviation technology countries have a wide range of research on gas turbine, aero engine washing technology. It mainly includes:

[a] Corresponding author : wuheng89810@sina.com

compressor corrosion and blade washing [2-3], modelling and optimization of multiphase flow in online washing [4] as well as the analysis of the effect of washing on engine performance [5-6] and so on. Domestic research on washing technology has just started and mainly focuses on shore-based aero engines. Major studies include: washing technologies [7-9] as well as simulation analysis of washing effectiveness [10-11] and so on. As we can see, washing technology-related researches are carried out at home and abroad. But it could not find related literature about the establishment of engine online washing process based on a large number of washing experiments.

In this paper, cooperated with industrial sector, a large number of washing experiments are designed to carry out based on previous research, it mainly includes experimental study on structural strength of spray frame, selection of nozzle and determination of injection parameters of online washing spray system, determination of washing injection timing, determination of drying process, liquid-washing experiments of different concentrations and water-washing experiments and so on. Finally, according to the experimental results, a certain type of engine online washing process is determined.

2 Engine washing experiments in cold operation state

2.1 Experimental equipments and preparation work

As the washing parameters in the cold operation state can be adjusted, the washing nozzle flow is small, and the control of corresponding particle size is relatively easy, therefore, cold run immersion and flushing and hot test drying are adopted as an online washing mode by a certain type of carrier-based engine. The test engine used for research on technical state after overhaul qualified is removed from a certain type of carrier-based aircraft. A mechanical valve with relatively high reliability is used by washing car, there are two solution tanks, that is, the same washing solution or rinse solution can be loaded into the solution tank to work together and also the washing solution and pure water can be loaded into the solution tank to work separately. Washing fluid transport power use pneumatic for the system comes with two small high-pressure cylinders. Structural installation diagram is shown in Figure 1. Washing spray frame is shown in Figure 2, finite element is used to obtain the structural static strength and modal analysis of the spray frame and the verification tests of simulated environmental vibration are designed to find that the structural strength of the spray frame could meet the requirements. Cloud diagram of static force calculation (The amount of deformation) and Block diagram of simulated environmental vibration test system are shown in Figure 3 and Figure 4.

A carrier-based aircraft inlet with full state size is adopted and the installation on the gantry simulates the installation status on the airplane. Other experimental equipments mainly include: porcelain probe, high-definition camera, nitrogen bottle (pressure not less than 12MPa), GT-1 washing agent and drinking water.

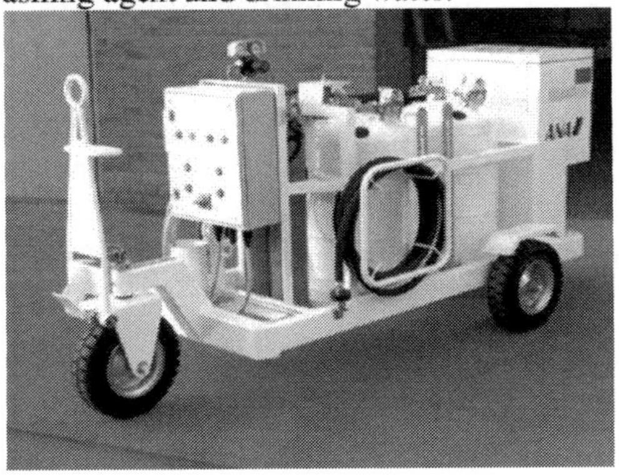

Figure 1. （2×120 litters） Structural appearance of multi-purpose engine washing equipment

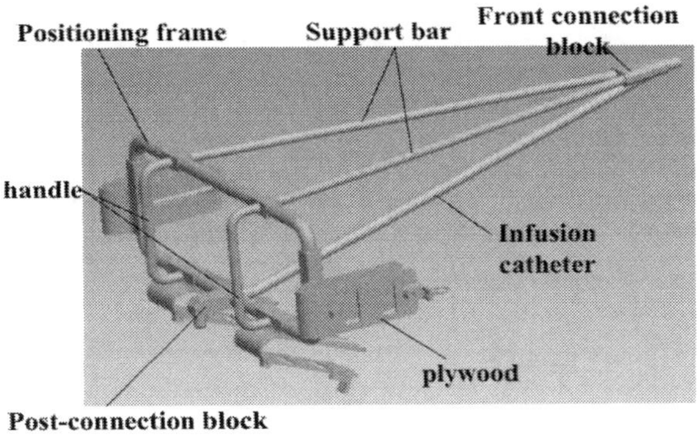

Figure 2. Washing spray frame

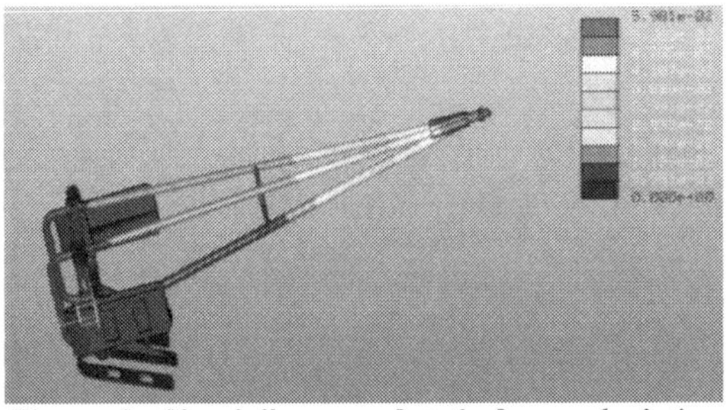

Figure 3. Cloud diagram of static force calculation (The amount of deformation)

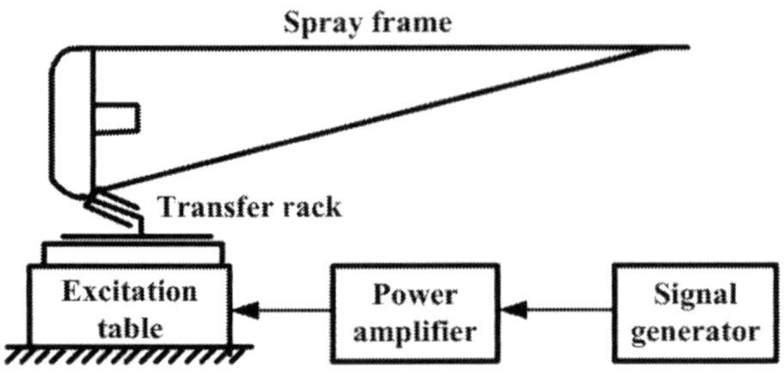

Figure 4. Block diagram of simulated environmental vibration test system

The experiments' preparation mainly includes the following aspects:

(1) install the craft cover only at the outside ducted air-entraining position (aircraft fuel tank pressurization) and seven-stage air-sucking position of high pressure compressor (gas supply from aircraft control system), the remaining interface status is consistent with the installed state. Before the experiments, check the engine function according to the outline requirements of the engine bench washing experiments to ensure that all kinds of parameters and functions are normal;

(2) complete installation of the inlet for test and the transfer section to ensure the same installation state with a carrier-based aircraft inlet;

(3) make sure that the washing vehicle is in good condition;

(4) install the washing spray frame and connect with the washing vehicle, check the connection of the liquid and gas pipeline of the washing vehicle to make sure that it is intact, watertight, no leakage and meets the sealing requirements;

(5) put the monitoring camera in the position aimed at the engine tailpipe to record the engine drain during the whole washing process; Place the camera in front of the test inlet to record water jet during the washing process; record videos and take pictures about the engine water leakage under the engine.

2.2 Washing experimental methods, process and results

2.2.1 Washing nozzle selection and parameter determination

As nozzle of the injection system is far away from the engine entrance, in order to ensure that the washing fluid will not be sprayed out to inlet wall, the inlet side of engine has a certain water flow coverage and the water flow has the capability of multi-stage fan and compressor cascade penetration, it is necessary to optimize the selection of the numerical range of parameters such as flow rate, pressure and installation angle of the nozzle [12]. Compared with the traditional trial method of determining the engine washing injection parameters, the method of flow field simulation and pre-selection can greatly reduce the number of tests and also reduce the test cost. Therefore, commercial software FLUENT is used for flow simulation and solution. The droplet size and motion trajectory are taken as the optimization goal, and the droplet on the airflow inlet face of the engine and the penetrating ability are taken as the effect target to carry out numerical simulation by using Linearized Instability Sheet Atomization（LISA）and Spalart-Allmaras（S-A）models. The specific simulation theory, model and results are described in detail in literature 12. Optimize the selection of the design parameters of nozzle according to a large number of simulation results, that is, when the engine washed in cold operation state, the inlet has certain water flow coverage under the conditions that the spray cone angle is 15～16° and the upward offset angel of nozzle is 5～6° . In addition, the diameter of most atomized droplets of the injection fluid is less than 2mm, the number of droplets exceeding 2mm in diameter is minimal and no more than 3mm. while engineering practice and research analysis show that: the larger the particle size of the washing droplets formed by the nozzles and the greater the particle speed, the better the washing effect. But oversized and too fast droplets will cause damage to the blade surface, so the appropriate washing fluid particle size is not more than

3mm(generally, the number of droplets exceeding 2mm in diameter should not be too much), the appropriate droplet velocity is close to or slightly larger than the velocity of the flow field. So it concludes that in the range of the spray parameters, the distribution of the spray particle size is reasonable so that it will not cause the fan and compressor blade erosion. The droplet's circumferential and radial distributions are shown in Figure 5, we can see that in this condition, the droplets have the capability of multi-stage fan and compressor cascade penetration so that it is conducive to the realization of engine washing.

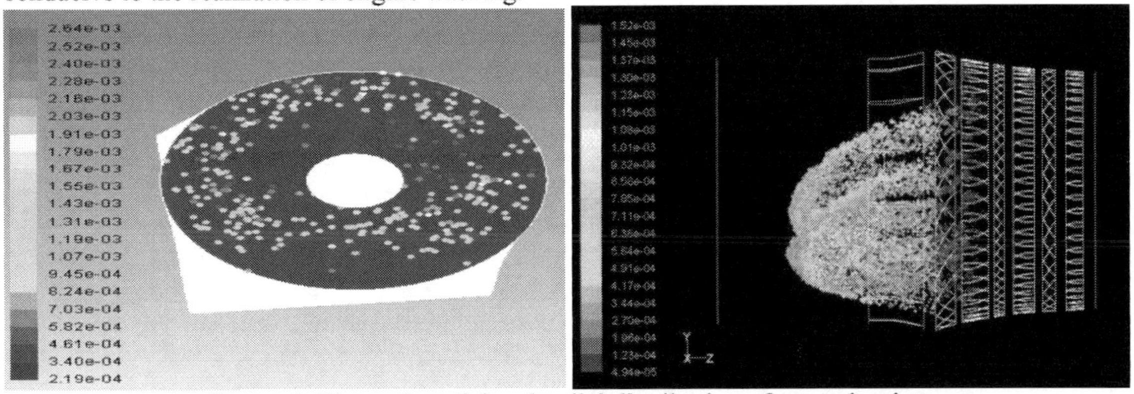

Figure 5. Circumferential and radial distribution of spray droplets
(Cone angle of 16°, upward offset angle of 6°, pressure of 0.8MPa, SMD of 1000)

According to the optimally selected nozzle design parameter data, combined with the nozzle design manual [13], several sets of built-in swirl nozzle are developed. Nozzle test system is adopted in this paper to test the developed nozzle and also make an adjustment of the necessary performance parameters. The schematic diagram is shown in Figure 6. The test system includes several sub-systems, such as flow test system in pipe (mainly water flowmeter), optical goniometer, droplet particle size and velocity test system (laser Doppler test system), et al. Among them, the particle test instrument employs PDI provided by Artium of America. The data transform software packet MATLABTM is implanted into information processing system with multi-user data operation and system control functions. The test system achieves the overall measurement of the injection flow field by the automatic control of axes. Through the tests, it is indicated that the nozzles can meet the design requirements. What's more, multi kinds of nozzles and nozzle parameters with adjustment tolerance have been determined, which lays the foundation for further selection of working parameters and nozzle optimization.

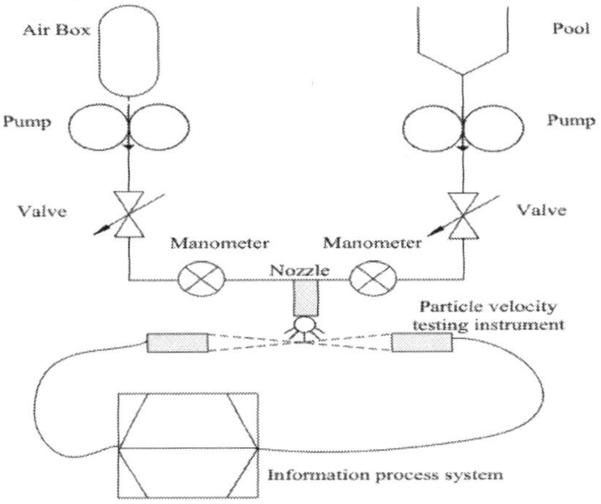

Figure 6. Schematic diagram of experiment system

Figure 7. Medium nozzle with upward offset angel of 7° 30′

An online washing ground simulation test system of engine flow passage is designed for ensuring the reliable washing under the condition that the inlet and injection equipment are equipped into test bench and meanwhile reducing the total test time. Different models of nozzles and their spray parameters are pre-selected in test bench trail. The engine bench tests are carried out according to the results of the pre-selection and verification. The spray effect experiments are carried out in three medium nozzles at different angles under different pressure, respectively there are mid-nozzle with upward offset angel of 8° 30′ and the corresponding feed liquid pressure provided by washing vehicle are 0.6MPa, 0.7MPa, 0.9Mpa and 1Mpa, mid-nozzle with upward offset angel of 8° and 7° 30′ and the corresponding feed liquid pressure provided by washing vehicle is 0.9Mpa. what's more, The spray effect experiments are carried out in small nozzles at different angles under different pressure, respectively there are nozzle with upward offset angel of 8° 30′ ,10° and 4° and the corresponding feed liquid pressure provided by washing vehicle is 1Mpa. According to the visual water injection situation from the front of inlet and the transition section near the observation window, it is found that the forward impact force of small nozzle water flow is weak so that the intake air flow is not sufficient to draw the water at high speed into the engine, therefore, there is no need to do the spray effect validation tests under the higher water pressure of the small nozzle in cold operation state. When the two kinds of nozzles washed in cold operation state, check the engine flow passage through the hole detector, it is found that there are water traces in low pressure and high pressure compressor blades at all levels, there is a small amount of water remaining in the test inlet and engine connections, there is more water discharged from the combustion chamber leakage port, and there are a large number of water discharged from the outer duct leakage port, what's more, there is no difference of the remaining water leakage points basically. In view of the combustion chamber leakage port and the outer duct leakage port have a leaky oil pipeline connected to the engine compartment outside so that washing will not affect the engine compartment and aircraft system.

The nozzle installation angle is adjusted to ensure the nozzle around the center, observe the water flow from the front of inlet to find that: mid-nozzle with upward offset angel of 7° 30′ can be selected and the feed liquid pressure provided by washing vehicle is 0.9Mpa. It can meet the requirements of the injection range at the engine intake box, the ejection effect is shown in Figure 7.

2.2.2 Determination of washing injection timing

The timing of washing injection is determined by reference to the requirements of a certain type of engine channel washing instructions in the outline of the engine bench washing experiments, that is, 10s after an engine cold start-up, and sustains 30s for spraying. During the experiments, it is found that the engine remained at a high speed after stopping the injection. Check the n2 speed of multiple engines in the factory in cold operation state to find that the high pressure speed can reach about 20% after 20s of the engine's start-up, and also drop to around 20% after 50s of the engine's start-up. To ensure that the engine has a high speed when the outfield washing began to spray, the washing injection timing is adjusted and four repetitive verification tests are performed to find that the washing injection timing in cold operation state is finally determined as: 20s after an engine's cold start-up, and sustains 30s for spraying.

2.2.3 Determination of engine drying procedure

Respectively, dry the engine according to the two kinds of drying procedure developed in the outline of the engine bench washing experiments, eight repetitive verification tests are performed. According to the first drying procedure, that is, dry the engine after engine operating at idle state for 5min, a total of three trials are conducted to find that after drying the engine piping is waterless, but there is still a small amount of water remaining in the connection between the inlet and engine; according to the second drying procedure, that is, engine operates 2min in ideal state, then 2min in state of n2=85% and 2min of ideal state lastly. A total of five trials are conducted to find that the engine piping and the connection between the inlet and engine is waterless after drying, that is to say, when the engine operates in state of n2=85%, the residual water in the connection between the inlet and engine will be sucked into the engine. Before and after drying, the oil sample is obtained and tested for a total of 13 times to find that there is no moisture content in the oil, and also the oil viscosity, acid value, flash point and other physical and chemical indicators are in line with the requirements.

In summary, to ensure that the residual water in the connection between the inlet and engine can be sucked into the engine when the engine washed in the outfield, the drying procedure which is engine operating 2min in ideal state, then 2min in state of n2=85% and 2min of ideal state lastly is developed. Two trials are conducted to find that the engine piping is waterless after drying. According to the test result of the seven-stage air-sucking cover between the engine and airplane after each washing and drying, it is found that there are varying degrees of water residue so that when the engine washed, disconnect the aircraft control system and tank pressurized air-entraining piping with the engine, and install the craft cover in the corresponding location.

2.2.4 Cooperative experiments of liquid washing and water washing

After determination of the technical state of the engine washing jet system in cold operation state, the washing injection timing and the post-washing drying procedure, washing experiments of three washing fluid ratio 5%, 10% and 20% are carried out respectively, after spraying washing fluid to the engine airflow channel for 30s at the first time, another four times drink water washing are implemented in cold operation state, and then pick up water samples from outer duct leakage port and analyze residual amount of the washing agent in the water sample after each water rinse. Dry the engine after washing, and analyze the obtained oil samples before and after drying. Liquid washing experiment of 10% concentration in cold operation state is adopted as an example to illustrate the specific experimental process, the experimental process of 5% and 20% concentration liquid washing are basically the same.

The specific experimental process of 10% concentration liquid washing in cold operation state: there is a small amount of washing solution remaining in the connection between the inlet and engine after spraying washing solution for 30s in cold operation state, as is shown in Figure 8; At the first water washing, there is still a bubble leak in the connection between the boost converter valve and the cited trachea road and there are more bubbles in the water sample obtained from outer duct leakage port; At the second water washing, there is no visible bubble in the water outflowed at the connection between the boost converter valve and the cited trachea road and the same as the water sample obtained from outer duct leakage port; At the third and fourth water washing, there is no bubble in the water sample at all. The first and fourth water samples are shown in Figure 9. Dry the engine after washing, check the airtight piping and the test inlet, there is no water residue.

Figure 8. The residue washing fluid in transfer section of inlet and engine junction

Figure 9. The first and fourth water sample

According to the test results of water sample after each water washing, it is found that after a 10% concentration liquid washing, twice water washing can be carried out to rinse the almost all cleaning agent remaining in the engine runner, and also after a 5% concentration liquid washing, twice water washing can be carried out, in this condition the amount of washing agent residue has been very low. Due to the timing of getting water sample is not exactly the same, it is likely that the amount of washing agent residue of a 10% concentration liquid washing and twice water washing is less than that of a 5% concentration liquid washing and twice water washing. But according to the test results, it is still concluded that after a different concentration liquid washing, twice water washing can be carried out to rinse the almost all cleaning agent remaining in the engine runner (the residue is less than 1%). The schematic diagram of residual amount of washing fluid in water sample after a different number of water rinses is shown in Figure 10.

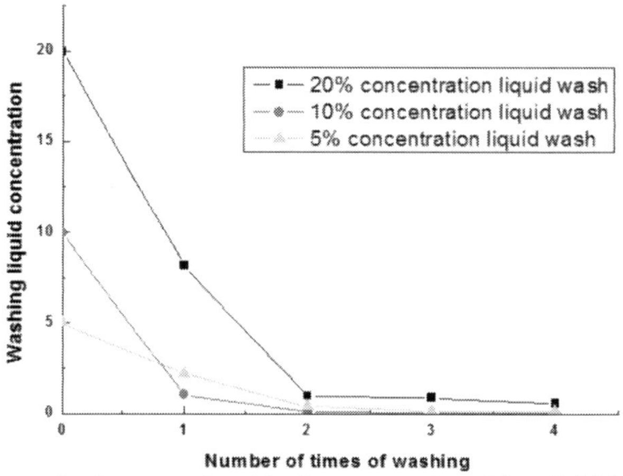

Figure 10. Schematic diagram of residual amount of washing fluid in water sample after a different number of water rinse

After a 20% concentration liquid washing and four times drink water washing, another four times drink water washing are carried out in cold operation state without drying. After a total of nine times washing in cold operation state, dry the engine according to the drying procedure of outfield engine washing, That is, engine runs for 2min in ideal state, then 2min in state of n2=85% and 2min of ideal state lastly, after drying, check the airtight piping and the test inlet, there is no water residue.

In summary, according to the results of engine washing experiments with three different concentrations of washing fluid and the test results of the residual amount of the washing agent in the water sample after each water rinse, it is found that after a liquid-washing with the use of washing fluid, the liquid concentration of which is no more than 20%, twice water-washing can be carried out to rinse the almost all cleaning agent remaining in the engine runner (the residue is less than 1%).

3 Development of the washing process based on experimental results

According to the results of liquid-washing experiments of three different concentrations and water-washing experiments, the washing process in cold operation state can be determined as follows:

(1) disconnect the aircraft control system and tank pressurized air-entraining piping with the engine, and install the craft cover in the corresponding location;

(2) 20s after an engine cold start-up, and sustains 30s for spraying the washing solution;

(3) 20s after an engine cold start-up, and sustains 30s for spraying the drinking water;

(4) after twice water-washing, the engine can be thoroughly dried by power-on operation, the procedure of which is engine operating 2min in ideal state, then 2min in state of n2=85% and 2min of ideal state lastly;

(5) restore connection of the aircraft control system and tank pressurized air-entraining piping with the engine.

After the engine washing is finished, check the engine function according to the outline requirements of the engine bench washing experiments, it is find that all kinds of parameters change reasonably and the engine is functioning properly, that is to say, Washing will not affect the normal operation of the engine.

4 Conclusions

Through the study of this paper, the following conclusions are drawn:

(1) through the engine bench washing experiments, it is possible to determine the technical state of the engine washing jet system, the washing injection timing and the post-washing drying procedure, that is, if engine is washed in cold operation state, mid-nozzle with upward offset angel of 7° 30′ can be selected and the feed liquid pressure provided by washing vehicle is 0.9Mpa. The start time is 20s after an engine cold start-up, and sustains 30s for once washing. The engine can be thoroughly dried by power-on operation, the procedure of which is engine operating 2min in ideal state, then 2min in state of n2=85% and 2min of ideal state lastly.

(2) according to the results of engine washing experiments with three different concentrations of washing fluid and the test results of the residual amount of the washing agent in the water sample after each water rinse, it is found that after a liquid-washing with the use of washing fluid, the liquid concentration of which is no more than 20%, twice water-washing can be carried out to rinse the almost all cleaning agent remaining in the engine runner(the residue is less than 1%). Washing will not affect the engine function.

References

1. E Syverud, L. E. Bakken, Online Water Wash Tests of GE J85-13, JOT, **129**: 136-142(2007).
2. R Kurz, K Brun, Fouling Mechanisms in Axial Compressors, JOEFGTAP, **134**(3): 1-9(2012).
3. F. C. Mund, P. Pilidis, Online Compressor Washing: A Numerical Survey of Influencing Parameters, POTIOME, PA: JOPAE, **219**: 13-21(2005).
4. E. M. Wahba, H. Nawar, Multiphase Flow Modeling and Optimization for Online Wash Systems of Gas Turbines, AMM, **37**: 7549-7560(2013).
5. E. A. Ogbonnaya, Gas Turbine Performance Optimization Using Compressor Online Water Washing Technique, E, **3**: 500-507(2011).
6. S Demircioglu, Detail Analysis of Compressor Online Wash Impact in the Gas Turbine Performance(Master thesis, Germany: FH Offenburg, 2008).
7. B. W. Li, X. X. Wang, G. C. Hu, Research on cleaning technology of turbojet engine, A, **36**(1): 12-16(2000).
8. C. B. Yuan, H. Z. Miao, X. Huang, Washing Technology of Turbo-shaft Engine for Military Helicopter, AMAE, **5**: 36-38(2010).
9. Y. Y. Yang, Y. Zhao, Influence analysis of a certain engine shipboard washing schedule of based on environment test, JONAAAU, **29**(2): 168-172(2014).
10. S. W. Chen, C. Zhang, H. Shi, Numerical Study on the Impact of Fouling on Axial Compressor Stages, JOPT, **33**(3): 377-383(2012).
11. C. H. Wang, Z. P. Xu, X. Y. Yang, Numerical simulation study on two phase flow field of cleaning aero engine, SST, **11**(2):101-105(2015).
12. K. Y. Jiang, W. Shen, B. W. Li, Parameters optimizing for a turbofan engine cleaning system using flow field simulation, JOAP, **26**(3):481-488(2011).
13. L. Y. Hou, X. C. Hou, *Technical manual of nozzle*(China Petrochemical Press, Beijing, 2002).

Storage Life Prediction of Space Relay Based on Elman Neural Network

Wenhua Li [1], Lulu Zhou [1,a] and Binglong Wang[1]

[1]*Hebei University of Technology, No 8 GuangRong Street HongQiao District Tianjin, China*

Abstract. When the linear equation is used to fit the life data for predicting the storage life at room temperature, there is a large error. From this point of view, the parameter degradation is considered and the contact pressure drop parameter is fused according to the comprehensive correlation degree at 60°C, 73°C, 92°C and 125°C. The accelerating factors of four temperature gradients are obtained by using the Arrhenius equation. The parameters and storage life in high temperature state are equivalent to normal temperature. The Elman neural network is used to study and train the fusion parameters, then the storage life of the relay at room temperature is predicted. The prediction error of this method is 2.44%, which is smaller than expected.

1 Introduction

As a commonly used component in aerospace systems, space relay has the characteristics of 'long-term storage but one use'. The relay will age over time during the storage time, leading to its performance decline even failure, which is directly related to whether the product can be put into use. Therefore, the storage life, as an important technical indicator, is a main parameter to characterize the performance [1].

If there is no test data of the product components during the long-term storage time, under some assumptions, the accelerated life test method is usually used to estimate its life [2]. In some normal circumstances, scholars are used to establish model equations between the acceleration stress and characteristic parameters. Then the storage life under natural conditions can be calculated by fitting the life data. In the paper [3], the aging rate of the product at the storage temperature calculated by the Arrhenius equation is substituted into the fitting equation to predict the storage life at room temperature.

The equation for fitting the life data under each stress does not reflect the process of product change well, and does not take into account the relationship between the stresses. In view of this deficiency, the accelerating factors of four temperature gradients are obtained by the Arrhenius equation according to the fusion value of the contact pressure drop obtained in the previous research. Then the fusion parameters and the storage life at high temperature are equivalent to normal temperature. The Elman neural network is used to study and train the fusion data to predict the storage life of the relay at room temperature.

---

[a] Corresponding author : 15222200570@163.com

2 Parameter fusion

2.1 Accelerated storage life test

Oxygen, ultraviolet, vibration, humidity, salt mist, temperature and other environmental factors will cause performance degradation of aerospace relay materials. The accelerated life test is difficult to fully simulate the actual storage profile involved in the environmental factors. In this paper, the relay samples are subjected to an accelerated life test with constant temperature and humidity, and the temperature factor that mainly causes the aging of the sealing materials is selected to be analyzed. The temperature stress is respectively set to 4 stress levels at 60°C, 73°C, 92°C and 125°C.

Two thermostat humidity control boxes are used to simulate two different temperatures. The DC28V voltage is applied to the coils of 25 relays and the DC 6V voltage / 0.1A current is applied to the resistive loads. The parameters of all samples under the two temperature levels are monitored regularly at the same time. The monitored parameters in the accelerated life storage test are contact pressure drop, pull-in voltage, release voltage, pull-in time, release time and turn-off voltage. In this paper, the first four parameters of the 100 contacts are recorded continuously for 16 weeks.

2.2 Parameter fusion

On the basis of the previous period, the fusion parameter value of the contact pressure drop is obtained by using the comprehensive correlation analysis method to analyze the correlation between the pull-in voltage, the release voltage, the pull-in time parameter and the contact pressure drop.

Assuming that the reference sequence $X_i = \{x_i(k)|k = 1,2,...,n\}$ is the i -th standard mode matrix, $Y_j = \{y_j(k)|k = 1,2,...,n\}$ is the j -th mode to be examined. Then the similarity coefficient of the two sequences at the k -th time is defined as:

$$r_{ij} = \frac{1}{n}\sum_{k=1}^{n}\frac{\Delta_{\min} + \rho\Delta_{\max}}{\Delta_{ij}(k) + \rho\Delta_{\max}} \tag{1}$$

$$\Delta_{ij}(k) = |X_i(k) - Y_j(k)|$$

Where, $\Delta_{\max}$ is the maximum value of $\Delta_{ij}(k)$. $\Delta_{\min}$ is the minimum value of $\Delta_{ij}(k)$. ρ is the resolution coefficient, generally ρ=0.5 。

The relationship between the parameters in the paper is blurred. In the fuzzy clustering system, in order to determine the relationship between the detection data better, a similarity coefficient μ_{ij} is used to describe the two groups of data [4]. In this paper, the correlation coefficient method is used to calculate μ_{ij}. The formula is as follow:

$$\mu_{ij} = \frac{\sum_{k=1}^{n}|X_i(k) - \bar{X}_i||Y_j(k) - \bar{Y}_j|}{\sqrt{\sum_{k=1}^{n}\left(X_i(k) - \bar{X}_i\right)^2 \sum_{k=1}^{n}\left(Y_j(k) - \bar{Y}_j\right)^2}} \tag{2}$$

$$\bar{X}_i = \frac{1}{n}\sum_{k=1}^{n} X_i(k), \bar{Y}_j = \frac{1}{n}\sum_{k=1}^{n} Y_j(k)$$

The degree of relevance reflects the similarity between patterns, which describes the degree of association between the function curves of things over time. Similarity coefficient reflects the similarity between patterns, that is, the difference between the distances of things. The characteristic parameters of each fault mode are usually both random and fuzzy. In order to accurately describe the degree of correlation between patterns, the similarity coefficient and the correlation degree are combined to form a comprehensive correlation coefficient in the paper, which can describe the connection between the two series more closely.

The comprehensive correlation coefficient z_{ij} is defined as:

$$z_{ij} = \alpha r_{ij} + \beta \mu_{ij} \tag{3}$$

Where, α and β are weight coefficients and $\alpha + \beta = 1$, reflecting the importance of relevance degree and similarity. The values of α and β can also change the influence degree of $\Delta_{\min}$, $\Delta_{\max}$ and ρ.

The correlation degree of the three parameters relative to the contact pressure drop at four temperature gradients is analyzed according to the comprehensive correlation analysis. The weight of converted parameter is shown in Table 1. On this basis, multiply the value of each parameter according to the corresponding weight and then sum them, as a result, the fusion contact pressure drop can be obtained. The fusion parameter is used to analyze the overall process of the whole process of change, and is substituted into the Arrhenius equation to convert the temperature.

Table 1. The weight factors of three factors.

Parameter	Comprehensive correlation coefficients			
	60°C	73°C	92°C	125°C
Pull-up Time[ms]	0.333	0.381	0.342	0.353
Release Voltage[V]	0.322	0.28	0.375	0.337
Pull-in Voltage[V]	0.345	0.339	0.283	0.31

3 Conversion factor of temperature

Assuming that the normal working stress level of the product is S_0, and S_k is the k-th accelerated stress level. In order to obtain the storage life of the product at room temperature, it is necessary to convert the accelerated stress of the accelerated life test. The Arrhenius model is widely used in the evaluation of accelerated life test when the temperature is used as an accelerated stress [5]. The Arrhenius model linearly unified can be described by the following formula:

$$\ln R(S_k) = a + b \cdot \psi(S_k) \tag{4}$$

Where, $R(S_k)$ is the function of reaction rate. $\psi(S_k)$ is the function of the accelerated stress. a and b are estimated constants. The acceleration factor is calculated by (4).

$$K_{ij} = \exp(b \cdot [\psi(S_i) - \psi(S_j)]) \tag{5}$$

The acceleration factor expression for product in the paper is:

$$K_{ij} = \exp(b \cdot [\frac{1}{T_j} - \frac{1}{T_i}]) \qquad T = t + 273 \tag{6}$$

The value of parameter b is estimated from the fusion parameter.

The Arrhenius equation does not take into account the effect of temperature when calculating the acceleration stress, so the conversion factor is introduced in this paper. The room temperature is selected as 25°C. The parameters at 125°C are converted to 92°C, and the correlation degree between the converted value and the parameter value at 92°C is calculated according to the formula (3). Then the conversion factor of 125°C relative to 92°C is obtained after multiplying the correlation degree and the accelerating factor resulted from the equation (6). And so on, the gradient conversion factors

K_{ij} of the high temperature to low temperature are 0.8465, 0.9572 and 0.9301. The accelerating factor K of 60°C relative to 25°C is 0.92 according to equation (6). Finally, the fusion data at each temperature are converted to the fusion degradation parameters of the relay at 25°C.

4 Prediction of Elman neural network

4.1 Elman neural network

Elman neural network is based on the Jordan network [6]. In contrast to the static feed-forward neural network, in addition to the common input layer, hidden layer and output layer structures, it has a special undertake layer. The undertake layer is similar to the role of the hidden layer, which is the feedback connection within the neural network. The structure of Elman neural network is shown in Figure 1.

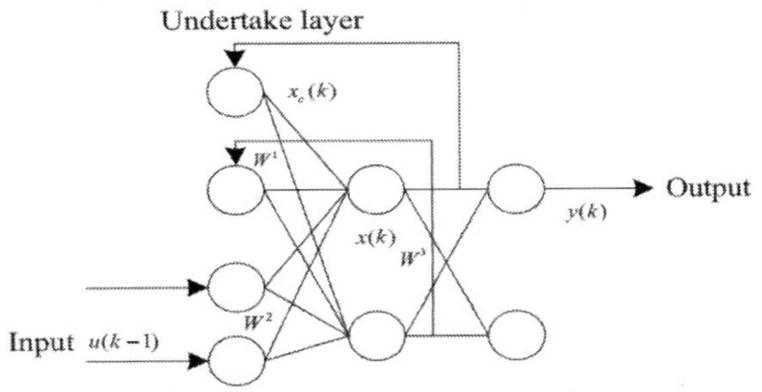

Figure 1. Structure of Elman neural network model.

The nonlinear state space of Elman neural network can be expressed as:

$$R_L(t) = f[\sum_{i=1}^{M} W_{1ij}(t)u_i(t) + \sum_{r=1}^{N} W_{2rj}(t)x_{cr}(t)] \tag{7}$$

Where, $x_{cr}(t)=x_j(t-1)$. $x_{cr}(t)$ is the feedback output of the undertake layer to the hidden layer at time t .

The output of the hidden layer to the output layer is:

$$y_p(t) = g[\sum_{j=1}^{N} W_{3jq}(t)x_j(t)] \tag{8}$$

Where, $W_{1ij}(t)$, $W_{2rj}(t)$ and $W_{3jq}(t)$ are respectively the weight matrices of the input layer to the hidden layer, the undertake layer to the hidden layer and the hidden layer to the output layer.

The functions of the input layer, undertake layer and output layer are all linear function. In order to speed up the convergence speed of the Elman neural network model, the function of the hidden layer is used modified sigmoid function [7]:

$$f(x,a,b,c,k) = a + \frac{c}{1+e^{-k(x+b)}} \tag{9}$$

Where, a is a constant term. b is adjustable bias parameter. c is expanded function gain. The value of the independent variable factor k depends on the input of the neural network

4.2 Prediction of storage life

Elman neural network is sensitive to historical data, which increases the ability of network to deal with dynamic information and network stability [8]. The amount of input equivalent data is 1600. The

DOI: 10.1051/matecconf/201711404010

monitoring period of the degraded data is 16 weeks. Therefore, the cycle period of the neural network is 16, and the network input data is normalized according to Eq. (10).

$$X_i = \frac{x_i - x_{\min}}{x_{\max} - x_{\min}} \tag{10}$$

In order to prevent the extrapolation of the sample data, the training sample is set to 1200 and the test sample is 400. According to the usual method of selecting the number of hidden neurons, the number of hidden neurons is set to 25 and the time of feedback is 4. The Elman neural network prediction model is established. The training and test results are shown in Figure 2.

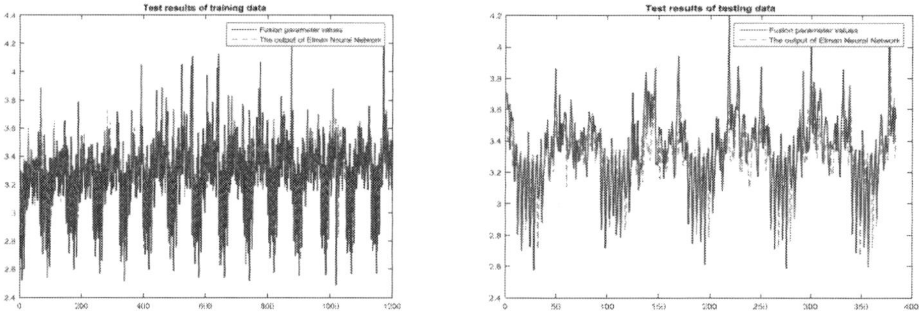

(a) Test results of training data (b) Test results of testing data

Figure 2. The training and test results.

The training mean square error of the Elman neural network 2.44% and the testing error is 3.25%, which satisfies the expected accuracy. Then the trained network is used to predict the data. When the fused contact pressure drop reaches the equivalent failure threshold of 8.9351, the storage life of the relay at room temperature is 1062 weeks, which is in accordance with the practical engineering application.

5 Conclusion

The irrationality of estimating the storage life by linear fitting is taken into account in the paper. The degradation process of the product is studied from the parameter point of view, rather than the simple calculation. The accelerating factors of four temperature gradients are obtained by using the Arrhenius equation, and the parameters and storage life in high temperature state are equivalent to normal temperature. The Elman neural network is used to study and train the fusion parameters, and then the storage life of the relay at room temperature is predicted. The prediction error of this method is 2.44%, which is smaller than expected.

Acknowledgement

This research was financially supported by Hebei Higher School Science and Technology Research Key Project. (ZD2015051).

References

1. J. WANG, J. FENG, JY. YANG, et al. Structure& Environment Engineering, **42**, 54-58 (2015)
2. L. JINSUK, P, RONG. IIE Transactions, **42**,589-598 (2010)
3. X. Zhou, XJ. Ding, W. Wei. Spacecraft Environment Engineering, **31**,287-291 (2014)
4. KT. Chian, FP. Chang, TC. Tsai. International Communications in Heat and Mass Transfer, **33**,744-752 (2006)

5. J. ZHOU, J. YAO, Y. SONG. Journal of Beijing University of Aeronautics and Astronautics, 744-750 (2015)
6. MI. Jordan. Englewood Cliffs: Erlbaum, 531−546 (1986)
7. R. LIU, GF. FANG. Power System Protection and Control, **40**, 113-117 (2012)
8. JM. Zhou, ZX. Zhu, YJ. Liu, et al. Journal of Central South University (Science and Technology), **42**, 3871-3875 (2011)

MATEC Web of Conferences **114**, 04011 (2017) DOI: 10.1051/matecconf/201711404011
2MAE 2017

Analysis on Missile-borne Bistatic Forward-looking SAR Imaging Perfomance

Qi Chen[1,2,a], Xiangping Li[1], Mingbo Zhu[1] and Xiaohai Zou[1,2]

[1]*Naval Aeronautical and Astronautical University, Department of Electronic and Information Engineering, 264001 Yantai, China*
[2]*CEMEE State Key Laboratory, 471003 Luoyang, China*

Abstract. Synthetic aperture radar (SAR) has been gradually applied to missile guidance for its advantages of all-weather, all-time and long-distance capabilities. Nevertheless, missile-borne SAR can not image on the area in front of the missile track restricted by SAR working characteristics. Missile-borne forward-looking SAR will be realized through separated transmitter and receiver. Iso-range contours, iso-Doppler contours, ground resolution and azimuth resolution are used to analyze the performance of missile-borne bistatic forward-looking SAR, as to provide the theoretical basis for its application.

1 Introduction

In modern war, missile-based precision guided weapons have been the key restrictor that determines the process and the result of the war, making lots of countries compete for the development on them[1]. Infrared guidance, television guidance and radar guidance are the traditional ways for missile guidance, among which radar guidance has a good application prospect for its all-weather and all-time capabilities[2-4]. Missile-borne SAR guidance can provide two-dimensional images with high resolution, which has a number of advantages over one-dimensional imaging guidance. However, SAR can not image on the area in front of the platform track limited by the inherent working characteristics. Thus missile-borne SAR can only be used in matching guidance through side-looking or squint-looking imaging rather than homing guidance, restricting its application scenarios. The emergence of bistatic SAR offers a new way to solve this problem[5-8]. Bistatic SAR realizes forward-looking through separated transmitter and receiver, making up the inherent disadvantages of monostatic SAR. Iso-range contours, Iso-Doppler contours, ground resolution and azimuth resolution are used in this paper to research the feasibility of missile-borne bistatic forward-looking SAR.

2 Geometric model

The geometric model of missile-borne bistatic forward-looking SAR is shown in Figure 1.

[a] Corresponding author : chenqiyt@163.com

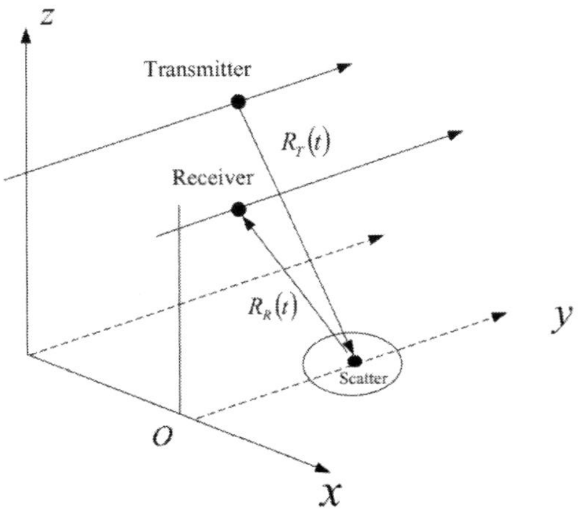

Figure 1. Geometric model of bistatic forward-looking SAR

The transmitter and receiver point together to the front area of receiver in the figure. The signals are transmitted to the target area in a certain PRT, then received and recorded by the receiver. $R_T(t)$ is the squint distance between the transmitter and the scatter, $R_R(t)$ is the squint distance between the receiver and the scatter.

3 Iso-range contours and iso-Doppler contours

Different scatters are distinguished through their distance to bistatic system in range direction. The range distance determines the time delay of the echo signal. The expression of range distance is

$$R(t)=R_T(t)+R_R(t) \tag{1}$$

Iso-range contours are composed of scatters with the same range distances, which are always in the form of ellipse.

The Doppler frequency of bistatic SAR is

$$f=-\frac{1}{\lambda}\left(V_T I_T(t)+V_R I_R(t)\right) \tag{2}$$

Where V_T and V_R are the velocities of the transmitter and receiver, respectively. $I_T(t)$ and $I_R(t)$ are the unit vector from the transmitter and receiver antenna phase centers to the scatter, respectively. Iso-Doppler contours are composed of scatters with the same Doppler frequency.

SAR system can obtain two-dimensional image of the area where the iso-range contours and iso-Doppler contours make rectangle or approximate rectangular partition.

When $V_T=V_R$, $R_T(t)=R_R(t)$, (3) and (4) refer to the calculation equation of monostatic SAR. The iso-range contours and iso-Doppler contours of an ordinary monostatic SAR system are shown in Figure 2. The arrow in the figure symbolizes the location and move direction of the SAR system. As is shown in the picture, the areas in side-looking and squint-looking direction have orthorhombic iso-range contours and iso-Doppler contours, indicating the nice imaging performance in these areas. Meanwhile, the The approximatively parallel iso-range contours and iso-Doppler contours in forward-looking direction account for the poor imaging performance in that areas.

 DOI: 10.1051/matecconf/201711404011

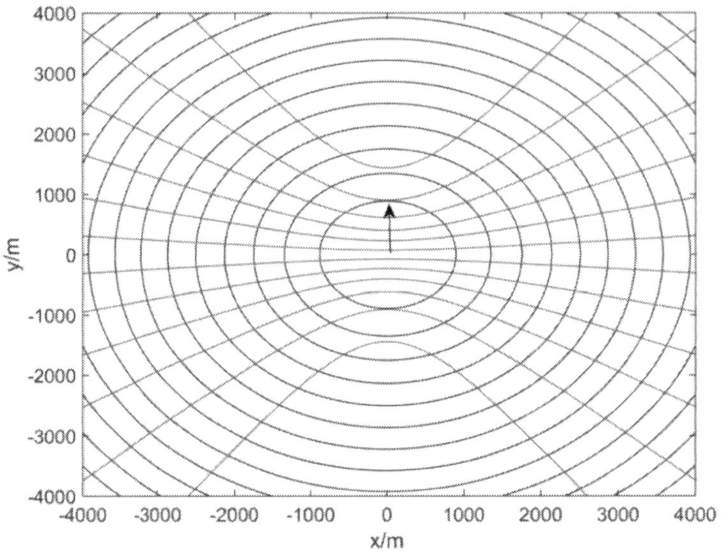

Figure 2. Monostatic SAR iso-range contours (circles) and iso-Doppler contours (lines)

4 Ground resolution and azimuth resolution

Derivated from the gradient method, the expression of bistatic SAR ground resolution is

$$\rho_{rG} = \frac{c}{B\sqrt{2+2\cos\beta-\left(\cos\phi_T+\cos\phi_R\right)^2}} \tag{3}$$

Where c is light velocity, B is signal bandwidth, β is bistatic angle, ϕ_T , ϕ_R are the incidence angles of transmitter and receiver, respectively.

The expression of bistatic SAR azimuth resolution is

$$\rho_a = \frac{D_R}{\sin^2\theta_R + K_V^2 / K_R \sin^2\theta_T} \tag{4}$$

Where K_V is the transmitter to receiver velocity ratio, K_R is the transmitter to receiver range distance ratio at center moment of synthetic aperture, D_R is the azimuth size of receiver antenna, θ_T and θ_R are coangles of transmitter and receiver squint angle, respectively.

5 Simulation and analysis

Assume a bistatic SAR system in which the missile-borne transmitter is working on side-looking mode, the missile-borne receiver is working in forward-looking mode. Parameter Settings are shown in Table 1, iso-range contours and iso-Doppler contours are shown in Figure 3, distribution curves of ground resolution are shown in Figure 4 and distribution curves of azimuth resolution are shown in Figure 5.

Table 1. Simulation parameters

Parameters	Values
Wavelength	0.03 m
Pulse bandwidth	200 MHz
Pulse width	2 μs
Azimuth size of transmitter and receiver antennas	1 m
PRF	2500 Hz
Transmitter velocity	(0,500,0) m/s
Receiver velocity	(0,500,0) m/s
Initial position of transmitter	(-6000,0,1000) m
Initial position of receiver	(0,-6000,1000) m

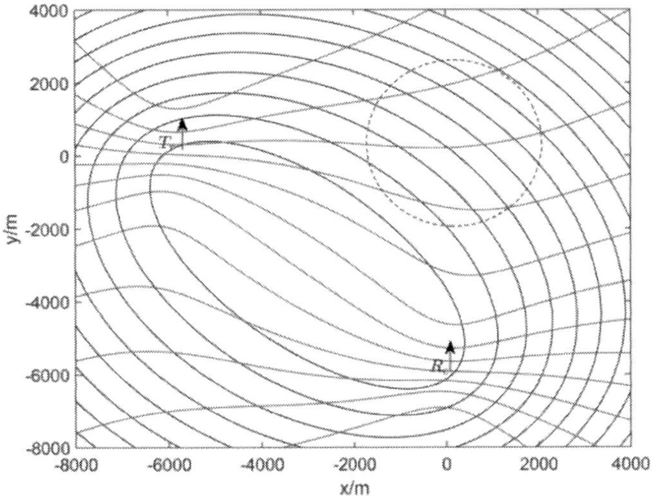

Figure 3. Bistatic SAR Iso-range contours (circles) and iso-Doppler contours (lines)

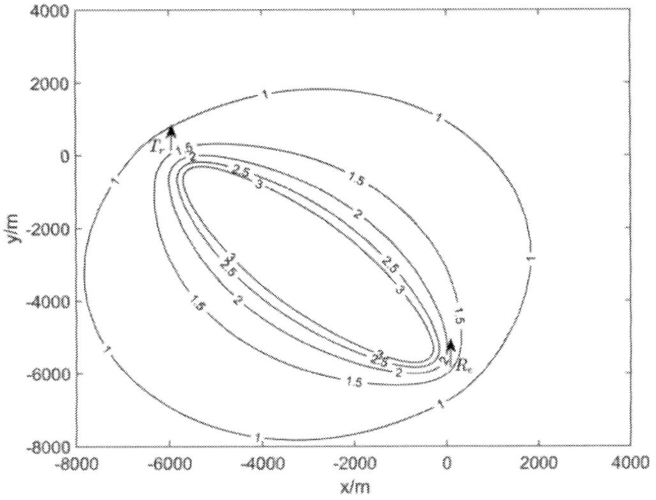

Figure 4. Ground resolution distribution curves

DOI: 10.1051/matecconf/201711404011

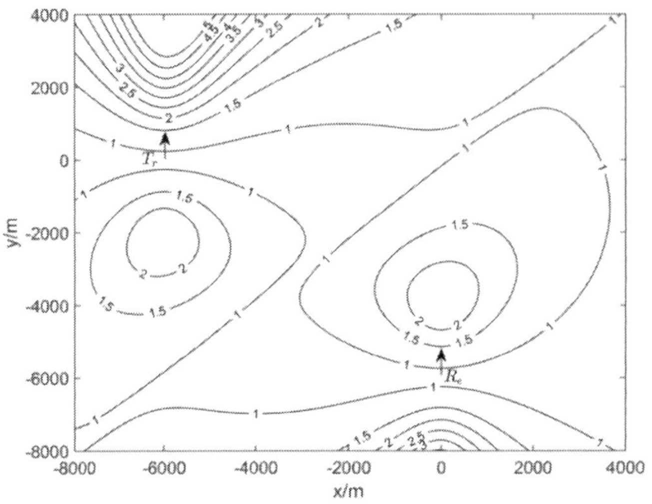

Figure 5. Azimuth resolution distribution curves

T_r and R_e are transmitter and receiver, respectively. The dashed part in Figure 3 is the imaging area. It is observed that the iso-range contours and iso-Doppler contours form rectangular segmentations in the imaging area, which indicates the forward-looking ability of the receiver. Therefore, the bistatic SAR system has high ground and azimuth solutions in the imaging area, as is shown in Figure 4 and Figure 5.

Echo simulation and imaging process are conducted for the point target at (0,0,0). Point target simulation result is shown in Figure 6. The result tells the well focus in range and azimuth directions, which validates the correctness of the above conclusion.

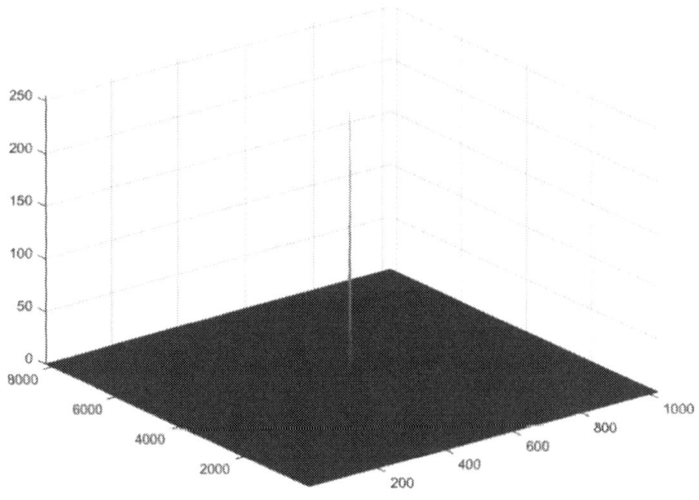

Figure 6. Simulation result of point target

6 Conclusions

To verify the imaging performance of missile-borne bistatic forward-looking SAR, the iso-range contours and iso-Doppler contours of the system in which the transmitter is side-looking and the receiver is forward-looking are depicted. Then the ground and azimuth resolution calculation and point target simulation reveals the imaging ability of missile-borne bistatic forward-looking SAR. The results show that bistatic forward-looking SAR can be applied to missile platform. In practice, the transmitter can be placed on an UAV, a reconnaissance aircraft or a missile to illuminate along with the forward-looking missile-borne receiver to image the target area and realize forward-looking SAR guidance, which has a high application value.

Acknowledgments

This work is sponsored by open research foundation of State Key Laboratory (CEMEE2016K0201B).

References

1. H Wei, G Min. AMJ, **8**, 56 (2016).
2. ZH Zhaorong, G He, ZH Xi, L Tao, K Yuhang. AMJ, **3**, 23 (2016).
3. L Qi, Q Luyan. IT, **1**, 146 (2016).
4. ZH Gang, ZH Mingbo, ZH Zhenbo, L Xiangping. AMJ, **9**, 67 (2011).
5. Y Jianyu, H Yulin, Y Haiguang, W Junjie, L Wenchao, L Zhongyu, Y Xiaobo. IGARSS, 4202 (2013).
6. W Ingo, E Thomas, K Jens, R. B. Andreas, H. G. E. Joachim. IGARSS, 216 (2010).
7. E Thomas, W Ingo, K Jens, R. B. Andreas, H. G. E. Joachim. GRSL, **8**, 765 (2011).
8. W Ingo, R. B. Andreas, K Jens. IGARSS, 327 (2012).

An Analysis of Time-delay for Remote Piloted Vehicle

Feng Wang, Shengjun Qi[a] and Li Jing
Chinese Flight Test Establishment, 710089 Xi'an, China

Abstract. Time-delay is one key factor affecting handling qualities of remote piloted vehicle. In this paper, the composition elements of time-delay were analyzed, in order to determine the relationship between latency and handling qualities. Then, comparing with the definition of time-delay in manned aircraft operation, the expression of time-delay was proposed for remote piloted vehicle and the measuring method of time-delay in piloted airplane was used as a reference. Moreover, considering different types of time-delay in RPV operation, typical open-loop and closed-loop experiments were designed and conducted based on ground testing environment. According to these results, the relation between time-delay and handling qualities was obtained. Finally, validity of this relationship was demonstrated through simulated approach and landing during flight testing. Time-delay is a critical element of RPV handling qualities, the research of which will be helpful and meaningful in unmanned aerial vehicle design and flight testing.

1 Introduction

For remote piloted unmanned aerial vehicle, pilot in ground control station could not get real time feedback as their counterpart, manned aircraft pilot, such as status of aircraft, visual weather and vibration of fuselage. Consequently, vehicle could not receive real time control order via uplink from ground control station. Therefore, time-delay in unlink and downlink has been becoming one primary issue affecting operator who control UAV to complete assigned mission. Nowadays, a large amount of research are carried out to decrease influence of time-delay on handling qualities in remote piloted vehicle and many methods are applied on RPV operation to alleviate operator's workload and to improve handling qualities. However, there are still two unsolved questions: how to define time-delay in unmanned aerial vehicle clearly and how to determine relation between time-delay and handling qualities. In this paper, study both in theoretical analysis and flight testing is started to solve these two questions mentioned above.

2 Definition of time-delay in remote piloted vehicle

2.1 Time-delay in manned aircraft

In the standard of manned aircraft flying qualities, there are two kinds of time-delay definition: equivalent time-delay and effective time-delay [1-2]. Equivalent time-delay is mainly aiming at low

[a] Corresponding author:saintknight372@hotmail.com

level equivalent system. The effective time-delay refers to the time that equals to number of the maximum rate of slope intersecting with time axis, which input of control force/displacement order produce.

Table 1 shows requirements in handling qualities of piloted aircraft. Time-delay in handling qualities of manned aircraft is measured by interval from input of con to response of control surface in either equivalent or effective time-delay.

Table 1. Time-delay requirement in handling qualities

Scale	Equivalent time-delay (s)	Effective Time-delay (s)
1	0.1	0.12
2	0.2	0.17
3	0.25	0.21

2.2 Unmanned aerial vehicle time-delay based on handing qualities

2.2.1 Composition of UAV time-delay

Time-delay of unmanned aircraft system consists of many items, not only the latency from input of manipulator system to response of control surface, but also latency of data link, digital processing in computers, and so on [3-5]. Comparing with latency of aircraft response, magnitude of data link time-delay is greater. Figure 1 is the composition diagram of time-delay in remote piloted vehicle. The total time delay is the sum of time-delay in ground control station, time delay of data link and time delay in UAV. The time delay on ground contains pilot input, computer calculation in ground control station and interaction between ground control station and data link, while time delay in UAV covers calculation in flight control computer, response of actuator and servo, interaction between aircraft and data link. And time delay of data link includes uplink latency and down link latency. It can be seen that the compositions of UAV time delay is rather complex. There are so many factors affecting the analysis of time delay of UAV.

2.2.2 Definition of Unmanned Aerial Aircraft Time-delay

Like definition of manned aircraft time-delay, time-delay of RPV reflects the relation in time-difference between input of order sent by pilot from GCS and display in screen of GCS down link from aircraft [6]. For the remote control, this expression reflexes pilot's feeling of latency in aircraft motion after manipulation, which has approximation for large time-delay RPV control as well as less accurate. It is difficult to measure equivalent time-delay and the main method for measurement is depends on equivalent matching algorithm to calculating latency. For the RPV with large magnitude of time-delay, adaptive condition happens easily. Hence, a pre-knowledge hypothesis is proposed for time-delay of data link.

Taking longitudinal movement of manned aircraft as an example, pilot got feedback of pitch angular acceleration firstly during manipulation, but it is difficult to measure acceleration in flight testing. However, as the integral of angular acceleration, angular velocity refers to variation of angular acceleration directly, and angular speed is easy to measure and has high accuracy. Therefore, that is why using angular speed as main parameter to evaluate time-delay.

DOI: 10.1051/matecconf/201711404012

For remote piloted vehicle, pilot in GCS could not get feedback on change of angular speed, but get information about attitude instead. Hence, it is suggested to choose attitude as main evaluation indicator for time-delay. Meanwhile, from perspective of measurement and calculation, the most part of latency is led by data link, so it is not to emphasize precision of latency in aircraft. Considering all points mentioned above, the definition of time-delay in RPV is proposed as follows. The time-delay in RPV refers to time span from sending input of manipulating order by remote pilot in GCS to display of status of system in GCS via down linking after response of aircraft to input.

2.2.3 Measurement of UAV time-delay

There are two kinds of measurement for time-delay in manned aircraft. Equivalent time-delay is the equivalent matching of angular speed, while effective time-delay is calculated on basis of maximum slope of angular speed and starting time of motion. The time-delay of RPV is to measure time-relation between sending input of manipulation order and receiving status of aircraft shown in screen of GCS. Thus, variation of attitude angle is becoming the main parameter in time-delay measurement. Operator in GCS could monitor changing speed of RPV attitude. The method of equivalent time-delay in manned aircraft could be used in RPV time-delay measurement to determine slope of attitude angle. As shown in figure 1, Step input starts in first second. The time-delay of manned aircraft is the difference value between starting time of order and crossover point of maximum slope and time axis, which is $\Delta t1$. As shown in figure 1, according to measurement of manned aircraft, time-delay of RPV is the interval between start time of step input and the time, which value equals crossover point of maximum slope on attitude angle and time axis.

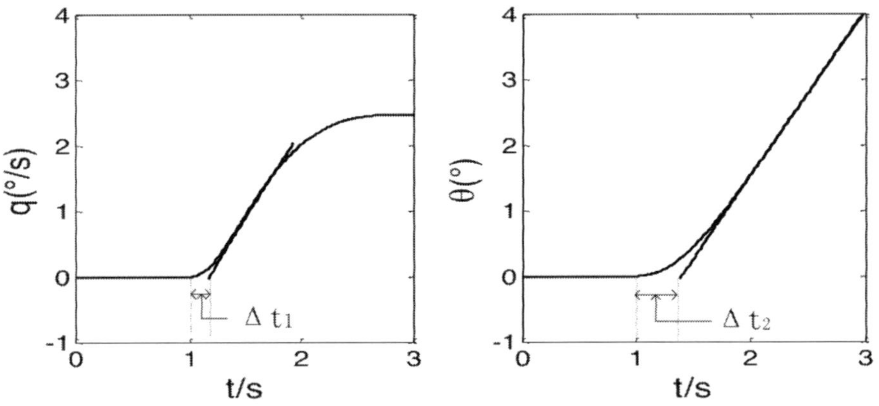

Figure 1. The method of measuring time-delay

3 Testing environment for time-delay

Closed-loop ground test and flight test are conducted in small unmanned aerial aircraft in order to verify the relationship between time-delay and handling qualities.

3.1 Closed-loop ground testing introduction

Closed-loop ground testing system is built using simulator and aircraft, which is shown in figure 2. Firstly, input of manipulation input was sent to dual-redundancy flight control computer through data link. Then, Flight control computer will send orders to actuator to deflect control surface. On the other side, analog computer collects deflection of surface and calculates dynamic response of aircraft, which will be transmitted to ground control station via flight control computer and data link. All of process forms a closed-loop.

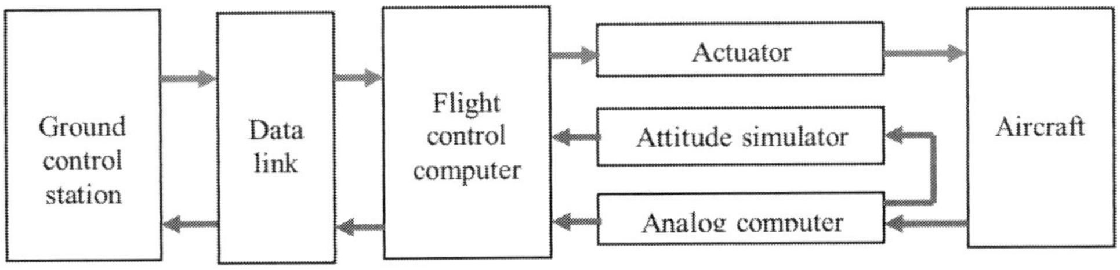

Figure 2. Closed-loop ground test composition

3.2 Flight testing description

The experiment is executed in one remote piloted aircraft that was refitted from trainer aircraft. In order to achieve remote control, fly-by-wire system was added on the basis of mechanical manipulation system in front of cockpit, so that pilot in front could control vehicle through fly-by-wire. The pilot in rear of cockpit still uses mechanical manipulation system as safe guarantee. In the ground control station, side stick was installed and layout of human-machine interface was optimized to make sure operation safety and effectiveness. On the other hand, data transmission system with high-performance was employed in order to decrease latency in data link, which means input of order and status of system could interact in short interval.

During testing, remote pilot control the RPV in GCS to measure the period from sending input of control to receiving status on screen.

4 Results analysis

4.1 Analysis of closed-loop test results

There are 5 sophisticated test pilots having more than 3000 flight hours in fixed wing manned-aircraft, who participate in testing. In the beginning phase of testing, pilots did not adapt to large time-delay manipulation. They can complete open-loop mission with easy input like impulse and step, while it is hard to accomplish closed-loop task using simple input. With much practice, pilots understand the characteristics of RPV time-delay and influence on aircraft motion from different magnitude of latency, and they could finish tracking task in closed-loop test.

The comments from pilots are shown in table 2. The open-loop task includes impulse and step input maneuvers, while closed-loop mission refers to approaching with inspecting disturbance and correcting. Since time-delay is measured on the basis of attitude, the magnitude of time-delay in RPV is greater than manned aircraft. From comments, it is seen that well-trained pilot could accomplish remote control in open-loop task and there is higher probability of successful landing with simulated correcting ability when magnitude of time-delay is around 300 milliseconds. When magnitude locates in 410-460 milliseconds, the operative difficulty becomes greater and coupling oscillation occurs when correcting during landing. Over 510 milliseconds, it is hard to execute normal landing, but there is less probability of occurring coupling oscillation.

With analysis of testing data, time-delay for 400-500 milliseconds is a sensitive range. The magnitude of time-delay coincides with responding time, which may produce coupling phenomenon.

Table 2. Comments about closed-loop test from pilots

No.	Time-delay(ms)	Closed-loop operation

1	310	Task can be accomplished
2	360	Task can be accomplished, but need more mental effort
3	410	Hard to accomplish precise landing under large Disturbance, more probability for oscillation
4	460	Disturbance causing Coupling oscillation, difficult accomplish to landing
5	510	Coupling oscillation, pilot cannot accomplish normal landing。
6	560	Hard to accomplish precise heading
7	610	More difficult to accomplish landing

4.2 Analysis of flight test results

In order to explore further relationship between time-delay and handling qualities, remote pilot control vehicle carried out simulated approach at 80m height, and then go around, shown in figure 3. During testing, remote pilot adjust heading to targeting at runway. It is found that heading is difficult for remote pilot. And it is hard to touchdown. Therefore, Go-around is the only choice for remote pilot due to safety consideration.

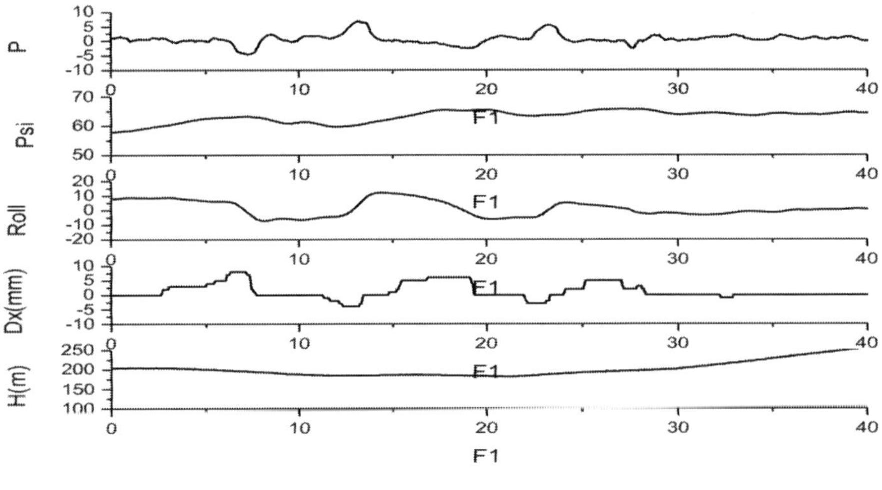

Figure 3. Simulated approach and go around in remote piloted control mode

4.3 Comparison between tests

During approaching phase, step input of heading operation was repeated for several times to determine magnitude of time-delay. The results illustrate that time-delay in heading direction is around 400 milliseconds, as shown in figure 4. This magnitude of latency is almost the same with comments from remote pilots. Hence, the scope of remote operation time-delay locates between 400~500 milliseconds, and manipulation, especially closed-loop operation is hard to implement in this range, which would cause piloted induced oscillation .

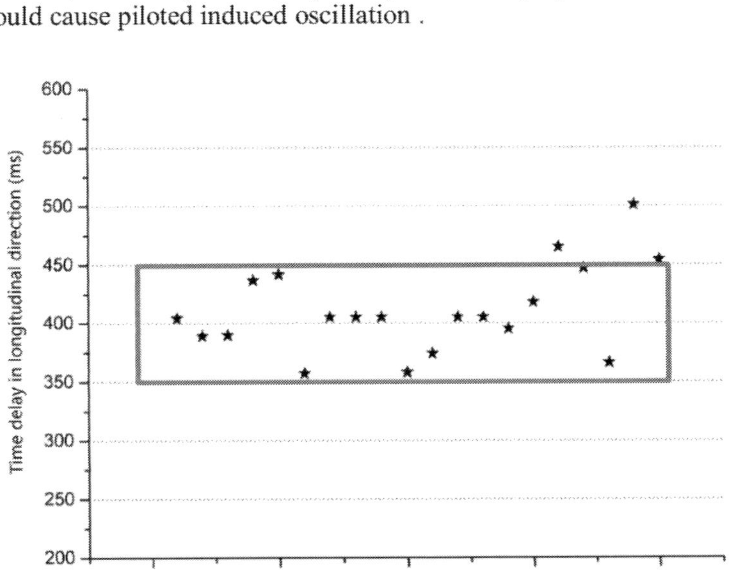

Figure 4. Statistics of heading time-delay

Finally, it is seen that the results between closed-loop ground test and flight test coincide badly based on analysis. From definition and results mentioned above, it is hard to control remote piloted vehicle if magnitude of time-delay is greater than 400 milliseconds.

5 Conclusions

As development of UAV and expanding in applications, latency in operation has been becoming one of primary issues in operation. From perspective of handling qualities, definition and composition of remote piloted vehicle are presented at first. To demonstrate relationship between time-delay and handling qualities, closed-loop ground test and flight test are executed. Results show that scope of latency is found and suggestions are proposed, which is helpful for study of remote piloted vehicle handling qualities in future.

References

1. Department of Defense,MIL-HDBK-1797, 20(1997)
2. C.Mcfalan, T.Richardson, C.Jones, AIAA,7156(2008)
3. T.Yu Jin, W. Jian Pei. F **28**,13(2010)
4. W.Fang, F.Wang,T.Ding Jie. F **28**, 68 (2010)
5. T.Ding Jie, W.Fang, F.Wang. F **29**, 17 (2011)
6. S.C.de Vries. UAVs and Control delays.15 (US,2005)

MATEC Web of Conferences **114**, 04013 (2017) DOI: 10.1051/matecconf/201711404013
2MAE 2017

Optimization Design Method of Optical Remote Sensor Based on Imaging Chain Simulation

Xing Kun[a], Cao Shi-xiang, Yue Chun-yu, Zhou Nan and He hong-yan
Beijing Institute of Space Mechanics & Electricity, Beijing 100094, China

Abstract. Space optical remote sensor of high spatial resolution need achieve both light and small design and high imaging quality. An optimization method of optical remote sensor based on imaging chain simulation is presented. All factors in imaging chain are analyzed according to MTF models and noise models. Additionally, image restoration is considered as one part of imaging chain. The simulation model are built to optimize remote sensor overall performance by the evaluation of remote sensor and simulation image. This method is used to enhance image quality and reduce the development difficulties of optical remote sensor. The test validates the feasibility and predominance of the optimization design method.

1 Introduction

Many countries in the development of remote sensing imaging system, attach great importance to the simulation analysis. With the development of military, civil and commercial remote sensing satellites, the United States has put forward a series of CONOP (Concept of Operation), SSAD (Space System Analyses and Design) and performance simulation and application simulation systems, which promotes the design, manufacture and application of remote sensing satellite system[1].1995, LaRC (Landley Research Center) funded by NASA announced the PATCOD integrated design software platform [2]. It combines Pro/Engineer structural design, Pro/Manufacturing, PATRAN, NASTRAN, SINDA-85, P/Thermal, CODE V and other common software for spacecraft simulation design and analysis process. The system-level simulation software as developed by Alcatel, France, has been used to simulate the optical system of the earth observation, and has been widely applied in MERIS (ENVISAT), SEVIRI (MSG), HRG (SPOT) and HR (Pleiades). In addition, Kodak has developed Physique software. In the image quality assessment and estimation, the U.S. government's IRARS and its collaborators jointly developed the NIIRS standard [3,4]. The GIQE equation provides a forecast for the NIIRS grade [5,6], performance prediction of the remote sensing system can be given in remote sensing system designing and parameters, the quality assessment of remote sensing image can be given before system development [7], so as to optimize the design of remote sensor. In the process of Pleiades system demonstration, France has made a comprehensive balance of several important parameters (MTF, SNR, compression ratio, deconvolution, de noise, etc.)

[a] Corresponding author: xingkunfeixiang@aliyun.com
Fund Project: China 973 Program (61321001), High Resolution Earth Observation Special Fund (GFZX40136-03-02), National natural science foundation(41401411)

which affect the final image quality, through the system optimization and the ground comprehensive processing, the image quality can be guaranteed [8]. It can be seen from the optimization methods, MTF and SNR are important factors affecting image quality. An optimization method of optical remote sensor based on imaging chain simulation is presented.

2 Optical remote sensing simulation model

Imaging simulation are realized through researching the influence of 3d scenes, imaging conditions (The atmosphere and background light, etc.), remote sensing (The optical system, detector, circuit), satellite platform, data transmission, image restoration, etc. Path of Remote sensor imaging simulation technology is shown in Figure 1.

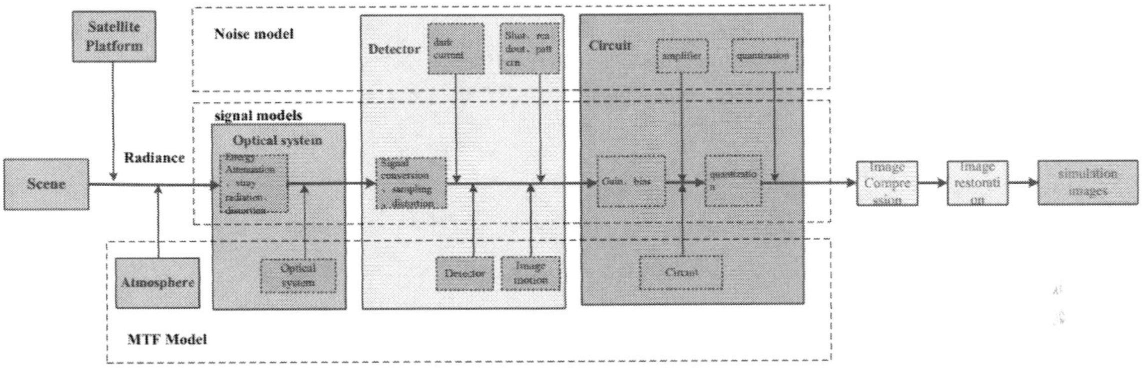

Figure 1. Optical remote sensing imaging simulation technology roadmap

2.1 Scene

Scene is consists of 3 d scene and the target source geometry characteristics, physical properties and thermal characteristics and so on. 3 d scenes target source geometry is based on the geometric module development.

2.2 Atmosphere

The MTF model established in this paper, considers atmospheric degradation characteristics of imaging system. The main consideration is turbulence MTF and aerosol MTF. For space optical remote sensing applications, the turbulent MTF in the far field and short exposure condition can refer to literature [9,10]. Aerosol impact on image comes mainly from low altitude atmosphere part. MTF of degeneration of imaging system cause by turbulence and aerosols is:

$$MTF_{atm} = MTF_{turb-se} \times MTF_{aerosol} \quad (1)$$

In the last formula, $MTF_{turb-se}$ is atmospheric turbulence MTF, $MTF_{aerosol}$ is aerosol MTF.

2.3 Remote Sensor

Remote sensor is mainly composed of optical system, detector and circuit. The optical system MTF model can be represented by the type:

$$MTF_{opt} = MTF_{diff} \times MTF_{aberration} \times MTF_{defocus} \times MTF_{other} \quad (2)$$

MTF_{diff} is the optical system MTF, divided into the central obscuration and no central obscuration of the circular aperture diffraction. $MTF_{aberration}$ is the optical system aberration MTF. $MTF_{defocus}$ is the focal shift MTF. In addition to the above factors, there are parts processing error, installation error, etc.

Line-Array CCD two directions of MTF are as follows:

$$MTF_{DETECTOR\_CCD}(f_x) = MTF_{SL} \times MTF_{other} \tag{3}$$

$$MTF_{DETECTOR\_CCD}(f_y) = MTF_{SL}(f_y) \times MTF_{DIFFUSION}(f_y) \times MTF_{TRANSFER}(f_y) \times MTF_{other} \tag{4}$$

X direction is the direction of satellite flight. Y direction is the direction of detector array. $MTF_{DETECTOR\_CCD}$ is detector MTF. MTF_{SL} is detector geometry MTF. $MTF_{DIFFUSION}$ is photoelectron diffusion MTF. $MTF_{TRANSFER}$ is charge transfer MTF. MTF degradation induced by other factors satisfies Gauss's function.

There are four main types of detector noise: shot noise, dark current noise readout noise and mode noise. The noise size is represented by the RMS noise equivalent electron number. The detector noise is calculated as:

$$N_{\mathrm{detector\_rms}} = \sqrt{(N_{shot\_rms})^2 + (N_{dark\_rms})^2 + (N_{readout\_rms})^2 + (N_{mode\_rms})^2} \tag{5}$$

The circuit model is mainly represented by the electronic filter MTF. The noise model of the circuit needs to consider the amplifier noise and quantization noise.

2.4 Satellite Platform

Single integral time is T, normal push scan speed is v, pixel center spacing is $p = vT$. The initial integration time at level n of line L is $(L+n-2)T$. Suppose the abnormal image motion in the TDI direction is $X(t)$, then the image motion at level n integration time of line L in the TDI direction is $\xi x(t) = v[t-(L-2+n)T] + X(t)$. Suppose the abnormal image motion in vertical TDI direction is $Y(t)$, then $\xi y(t) = Y(t)$. The image motion of different lines is inconsistent. The resulting image quality decline is spatially varying, which cannot be directly obtained by convolution with a PSF and the image. The image needs to be calculated in the sub-region by convolution. The line spread function in the TDI direction is ($\delta(x)$ is Dirac function)

$$\begin{aligned} \mathrm{LSF}_x &= \frac{1}{NT}\sum_{n=1}^{N}\int_{(L+n-2)T}^{(L+n-1)T} \delta\left(x - v\left(t-(L+n-2)T\right) - X(t)\right)\mathrm{d}t \\ &= \frac{1}{NT}\sum_{k=L-1}^{N+L-2}\int_{kT}^{(k+1)T} \delta\left(x - v(t-kT) - X(t)\right)\mathrm{d}t \end{aligned} \tag{6}$$

The line spread function in the vertical TDI direction is

$$\begin{aligned} \mathrm{LSF}_y &= \frac{1}{NT}\sum_{n=1}^{N}\int_{(L+n-2)T}^{(L+n-1)T} \delta\left(y - Y(t)\right)\mathrm{d}t \\ &= \frac{1}{NT}\int_{(L-1)T}^{(L+N-1)T} \delta\left(y - Y(t)\right)\,\mathrm{d}t \end{aligned} \tag{7}$$

The MTF in two directions is obtained by Fourier transform of the diffusion function.

$$\mathrm{MTF}_x = \frac{1}{NT}\left|\sum_{k=L-1}^{N+L-2}\int_{kT}^{(k+1)T} \exp\left\{-\mathrm{j}2\pi f\left[v(t-kT) + X(t)\right]\right\}\mathrm{d}t\right| \tag{8}$$

$$\mathrm{MTF}_y = \frac{1}{NT}\left|\int_{(L-1)T}^{(L+N-1)T} \exp\left(-j2\pi f Y(t)\right)\mathrm{d}t\right| \tag{9}$$

For more complex forms of MTF, numerical analysis can be used to calculate.

2.5 Image Compression

As the new generation of still image compression standard, JPEG2000 with its excellent compression features and higher resistance have been widely used in the fields of astronautics and aerospace [11].

2.6 Image Restoration

In the process of remote sensing image generation, storage and transmission, often because of hardware equipment, weather conditions, light intensity, motion blur, noise. camera mis-focus and a number of factors such as image quality degradation. Image restoration is performed by reversing the process that blurred the image and such is performed by imaging a point source and use the point source image, which is called the Point Spread Function (PSF) to restore the image information lost to the blurring process. In order to recover these corrupt images, a number of mature image restoration methods, including inverse filtering, Wiener filtering [12], constrained least squares filtering and Lucy-Richardson(LR) filtering [13,14].

This paper proposes an adaptive modulation transfer function compensation (MTFC) restoration algorithm of remote sensing images based on spatial domain [15]. The method is based on the modulation transfer function values laboratory-measured, which considers the impact of optical system, detector and electronic components, and construct finite impulse response to calculate deconvolution nuclear. According to the signal to noise ratio (SNR) of different gray values given by the sensor parameters (including integral series, gain, etc), we handle image deconvolution, which combining scene content for adaptive noise suppression processing. The method can improve both the image clarity and the SNR to increase the image quality.

3 Remote sensor optimization design

Considering the influence of each link, such as the scene, atmosphere, remote sensor, satellite platform and image compression on the imaging quality, at the same time, the restoration process is added, the parameters which affect the imaging quality are analyzed synthetically, and the design parameters is iterated for several times. Finally, the optimal design of remote sensing system is achieved. Figure 2 is Optimization design flow chart.

Specific steps are as follows.

(1) According to the needs of users, the five main factors affecting the capability of remote sensor, such as GSD, Swath width, MTF, SNR, Aliasing, are used as the input parameters of the simulation system.

(2) According to the imaging conditions and system parameters of remote sensing system, the scene models with ground objects are used to simulate the radiation images.

(3) On the basis of the mathematical modeling of each segment of the whole link, the remote sensing imaging simulation is realized, and the system designed parameters are determined.

(4) The simulation image is restored.

(5) The restored image is evaluated to determine whether the image quality meets the requirements, and if the requirements are not met, the results are fed back to the system to design new parameters.

(6) The system parameters are evaluated to determine whether they meet the requirements of engineering development, and if the requirements are not met, the results are fed back to the system to design new parameters.

(7) When the image quality and system performance are all meet the requirements, the design parameters is obtained.

The image quality evaluation parameters in step (5) are measurable, including image mean, standard deviation, contrast, definition, detail signal energy, SNR, MTF, etc., which belongs to the objective evaluation criteria. There are some other evaluation parameters that need to be developed according to the requirements of the optical spectrum and application field.

The parameters in step (6) belongs to the remote sensor, including focal length, F-number, aperture, spectral bands, transmittivity, field of view, quantization bits, circuit noise, etc.

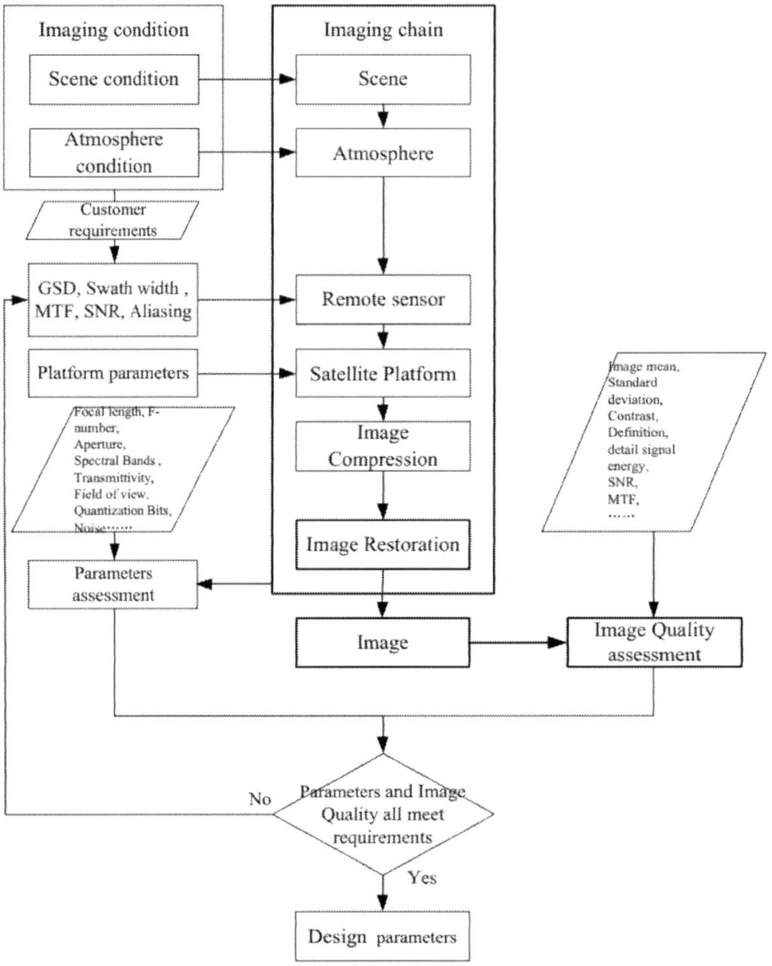

Figure 2. Optimization design flow chart

4 Experimental analysis

4.1 Traditional Design

Suppose the user needs for earth observation: the ability is 0.7m resolution and 12km swath width at 645km height.

According to the traditional design process, in order to shorten the focal length, the pixel size of TDICCD is chosen to be 10μm, and then the focal length of the system is designed to be 9.2m. According to the requirements of MTF at the Nyquist frequency and SNR, the camera is determined as MTF>0.2, $SNR_{max} > 46\text{dB}$, $SNR_{min} > 20\text{dB}$, the obstruction ratio of optical system is 2.25%, wave-front error is 0.12 λ (rms). In order to meet the requirements mentioned above, the aperture of optical system is 1.2m, F-number is 7.6, $\lambda F/p$ is 0.5, resulting in large aperture, image aliasing serious, developing difficulty, long cycle and high cost.

4.2 Optimization Design

User requirements are the same as traditional design.

According to the optimization ideas, we can know the aperture is the key design parameter, which directly affect the cost, size and weight of the camera. Considering the five main factors, such as GSD, Swath width, MTF, SNR and Aliasing, the aperture is determined by optimization design.

The TDICCD is the same as traditional design, focal length is 9.2m. The initial aperture is determined by the requirements of $SNR_{max} > 46\text{dB}$, $SNR_{min} > 20\text{dB}$. The MTF value of each segment is assigned by the requirements of MTF=0.20. The system uses 10bit quantization.

On the basis of the mathematical modeling, the remote sensing imaging simulation is realized.With the solar elevation of 20 degrees, and the ground reflectivity of 0.1, the radiance of 6.23W/(m^2·sr) is used as the minimum input. With the solar elevation of 70 degrees, and the ground reflectivity of 0.6, the radiance of 71.9W/(m^2·sr) is used as the maximum input. The SNR of different apertures and TDI stages is shown as Table 1 and Table 2.

Table 1. the SNR_{min} （dB） of different apertures and TDI stages

Aperture (m)	TDI stages					
	12	24	36	48	72	96
0.72	27.5	30.5	32.3	33.5	35.3	36.5
0.65	25.8	28.8	30.6	32.0	33.6	34.9
0.61	24.9	27.9	29.7	30.9	32.7	34

Table 2. the SNR_{max} （dB） of different apertures and TDI series

Aperture (m)	TDI stages					
	12	24	36	48	72	96
0.72	40.0	43.0	44.7	46.0	47.8	49.0
0.65	38.4	41.4	43.2	44.4	46.2	47.4
0.61	37.2	40.2	42.0	43.2	45.0	46.2

In order to reduce the difficulty of engineering development, Static MTF was reduced to 0.1, after the restoration process, MTF increased to 0.2. It is also necessary to consider whether the aliasing of the image is within the allowable range. This process requires compromise and optimization. The simulation results show that the aliasing parameters $\lambda F / P$ are between 0.8 and 1.2.

With the image mean, standard deviation and the detail signal energy as evaluation parameters, the simulation image quality of different F-Number is shown in Table 3. The mean represents the brightness of the image. The standard deviation represents the richness of the gray level of the image, and the detail signal energy represents the richness of the detail signal in the region.

Table 3. The image quality evaluation of different F-Number

F-Number	image mean	standard deviation	detail signal energy
10	63.3	22.3	124.8
12	59.5	19.8	104.4
14	55.1	17.7	81.1
15	54.0	16.1	76.8

20	38.9	10.0	34.8

As can be seen from Table 3, the larger the F-number, the smaller the image mean, standard deviation and the signal energy, and the worse the image quality.

After the image quality and system performance evaluation, considering the requirements of miniaturization and stabilization, the parameters of remote sensor is designed as: aperture is 0.65m, F-number is 14, $\lambda F/p$ is 0.9. The simulation results are shown in Figure 3. Figure 3 a) is the result of traditional design simulation. TDI stages is 12, aperture is 1.2m, and F-number is 7.6, Static MTF is 0.2. Figure 3 b) is the result of optimization Design simulation. TDI stages is 72, aperture is 0.65m, and F-number is 14, static MTF is 0.1, MTF after restored is 0.2.

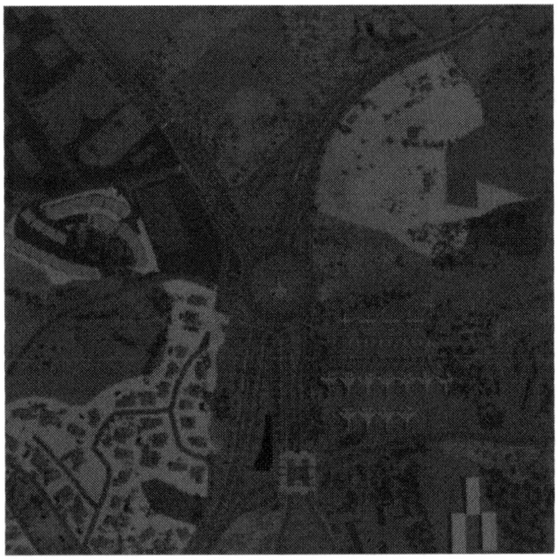

a) Traditional Design

b) Optimization Design

Figure 3. Simulation image

The final design parameters are shown in Table 4.

Table 4. design parameters

Orbit Height（km）	645
GRD（m）	0.7
Swath（km）	12
Pixel Size（μm）	10
Focal length（m）	9.2
Aperture（m）	0.65
F-Number	14
$\lambda F/p$	0.9
MTF Static MTF	Before compensation 0.10 After compensation 0.20
SNR_{max}（dB）	>46(TDI stages is 72)
SNR_{min}（dB）	>20

According to the requirements of the swath with, the optical system field of view is determined. The design scheme of optical system is proposed based on the aperture, focal length, pixel Size, etc. The next step is the detailed design of remote sensor.

5 CONCLUSION

In this paper, based on the end to end simulation, the optimization design method for remote sensor is studied. The parameters which affect the imaging quality are analyzed synthetically, and the design parameters are iterated by the evaluation of image quality and system performance. Finally, the optimal design of remote sensing system is achieved. The test validates the feasibility and predominance of the optimization design method. Using this method, the remote sensor can meet the requirements of the engineering development and image quality. The aperture of the remote sensor is reduced by about 1/2, and the development difficulty is greatly reduced.

References

1. Edson Robert, Frittman Jaime. Illustrating the Concept of Operations (CONOPs) Continuum and Its Relationship to the Acquisition Lifecycle[C]. 7th Annual Acquisition Research Symposium to be held May 12-13, 2010 in Monterey, California. U.S.Government or Federal Rights License.
2. Amundsen R M; Feldhaus W S; Little A D; Mitchum, M. V. Integration of design, structural, thermal and optical analysis and user's guide for structural-to-optical translator (PATCOD)[C]. Mar 1, 1995 Langley Research Center. NAS 1.15110153; NASA-TM-110153 95N24559.
3. John M Irvine. National imagery interpretability rating scales (NIIRS): overview and methodology[C]. Proceedings of SPIE, 1997, Vol.3128, 93-103.
4. Taejung Kim, Hyunsuk Kim, HeeSeob Kim. Image-based estimation and validation of NIIRS for high-resolution satellite images[C]. The International Archives of the Photogrammetry, Remote Sensing and Spatial Information Sciences(ISPRS), Vol. XXXV II, Part B1. Beijing 2008.
5. Leachtenauer J C, Malila W, Irvine J, Colburn L. General Image-Quailty Equation: GIQE[J]. Applied Optics, 1997, 36(32): 8322-8328.
6. Leachtenauer J C. Image Quality Equation and NIIRS[C]. Encyclopedia of Optical Engineering, 2003.
7. Ryan R, Baldridge B, Schowengerdt R A, Choi T, Helider D L, Blonski S. IKONOS spatial resolution and image interpretability characterization[J]. Remote Sensing of Environment, 2003, 88(01):37-52.
8. Philippe Kubik, Veronique, Paseal, Francoise de Lussy. Pleiades-HR image quality design[C], ACSM-ASPRS 2002 annual conference proceedings.
9. Krapels K, Driggers R G, Vollmerhausen R H, et al. Atmospheric turbulence modulation transfer function for infrared target acquisition modeling[J]. Optical Engineering(S0091-3286), 2001,40: 1906-1913.
10. Kobi Buskila, Shay Towito, Elad Shmuel, et al. Atmospheric modulation transfer function in the infrared[J]. Applied Optics (S0003-6935), 2004, 43(2): 471-482.
11. ISO/IEC. JPEG2000 Image Coding System: Part1-Core Coding System. ISO/IEC FCD15444-1. 2000:118-127.
12. R. C. Gonzalez & R. E. Woods, Digital Image Processing , Addison-Wesley Publishing Company, Inc., 1992.
13. Biggs, D.S.C. Acceleration of Iterative Image Restoration Algorithms. Applied Optics. Vol. 36. Number 8, 1997, pp. 1766–1775.
14. Hanisch, R.J., R.L. White, and R.L. Gilliland. Deconvolution of Hubble Space Telescope Images and Spectra. Deconvolution of Images and Spectra (P.A. Jansson, ed.). Boston, MA: Academic Press, 1997, pp. 310–356.
15. ZHOU Nan, QI Wenwen, CAO Shixiang, HE Hongyan, XING Kun, YUE Chunyu, Adaptive MTFC Restoration Algorithm of Remote Sensing Images Based on Spatial Domain , SPACECRAFT RECOVERY & REMOTE SENSING., 2015, 36(4):54-62.

Research on Sunglint Point Positioning Accuracy Based on Greenhouse Gas Detector

Jianjie Yin [a], Pengmei Xu, Lizhou Hou, Long Chen and Qian Cao

Beijing Institute of Space Mechanics and Electricity, Visible Light Camera Research Department, China

Abstract. Based on greenhouse gas detector of a satellite, the operation principle, system composition, and design scheme are introduced. It emphasizes on the method of the sunglint point positioning on the detector and particularly analyses two major factors: pointing accuracy of pointing mirror and calculation accuracy of sunglint position, which have an effect on sunglint point positioning accuracy. The theoretical analysis result of sunglint point positioning accuracy is less than 6.0km. That is in the allowed tolerant range. Physical verification system for sunglint point positioning accuracy is built by high precision two dimensional moving platform, point source and a visible light camera which has common-path with the main optical path in the detector. It shows that the physical verification result and the theoretical analysis result to be consistent. It provides a new verification idea for positioning accuracy analysis of similarity principle aim detection.

1 Introduction

The information of sunglint point on the ocean is helpful and important to analyses atmospheric composition over the ocean. Usually, the sun's reflection of surface is week. When the incident angle is equal to the reflection angle, mirror reflection is occurred on the surface and the reflected light becomes strong. The reflected point is called sunglint point [1].As to detect sunglint point, the input energy enhances. Calculation of atmospheric composition and retrieval of aerosol parameters will be more accurate[2-3]. So, accurate positioning of sunglint point is an important presupposition for enhance the accuracy of atmospheric remote detection.

In order to find the sunglint point accurately and quickly and detect and analyse atmosphere over the ocean, based on an onboard greenhouse gas detector, the paper presents a calculation method of the sunglint point position in a local orbit coordinate system and verifies the sunglint positioning accuracy by theoretical analysis and physical verification.

2 The introduction of greenhouse gas detector

2.1 Working principle

[a] Jianjie Yin : yinjjbjut@sina.com

Greenhouse gas detector is a time modulated Fouier transform interference spectrometer. It provides a atmospheric absorption spectrum of sunlight scattering by ground objects including the information of geographic and optical path to the user on orbit operation. The data array of atmospheric absorption spectrum can cover all over the world in required revisit time.

Figure.1 shows the working principle of Greenhouse gas detector. The aiming signal is introduced by tow-dimensional pointing mirror and then goes into the interferometer module to be modulated. The modulated signal is gathered to the detector by after-optical system and colour separation and converging optical system. The detector changes the modulated optical signal to interferential electrical signal. Interferential electrical signal is amplified, filtered, quantized and encoded using the analog and digital signal processing circuit. Meanwhile, there is a visible light camera in greenhouse gas detector. The visible light camera has a common optical path with the interferential signal. But the visible light field is greater than the interferential signal field.

Sunglint point observation is an observation mode of greenhouse gas detector for atmosphere above the ocean. The detector built a mathematical model of sunglint point according to the six orbital elements and local time. It finally realized the location of sunglint point precisely through coordinate transformation and geometric conversion.

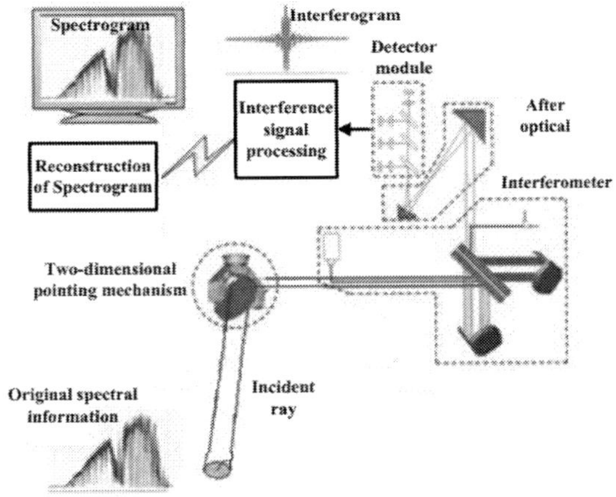

Figure 1. Schematic diagram of greenhouse gas detector.

2.2 The method of sunglint point location

The mathematical model of sunglint point location involves two kinds of coordinate: inertial system and orbit system [4-5]. Inertial system is that the earth center is the coordinate origin, the positive of X axis is from earth center point to vernal equinox, the Z axis coincides with the earth's rotation axis and north is positive, the Y axis is perpendicular to the XZ plane and constitutes the right-handed coordinate system. Orbit system is that the satellite centroid is the coordinate origin, the positive of X axis is the flight direction of satellite orbit, the Z axis coincides with the line between satellite centroid and the earth center and point to the earth center is positive, the Y axis is perpendicular to the XZ plane and constitutes the right-handed coordinate system.

Figure.2 shows the geometric relation of sun, satellite and sunglint point in inertial system. Point M is in the middle of the satellite vector $\boldsymbol{R}_{\text{sat}}$ and the sun vector $\boldsymbol{R}_{\text{sun}}$. The included angle between the two vectors is α . If α is divided into a value by sunglint vector $\boldsymbol{R}_{\text{M}}$, that time meet the conditions of the formation of sunglint point, and the point M is the sunglint point. In order to find the sunglint point M, we can assume the existence of point M and finally find the point M equal to the incident angle A_{sunM} and the reflection angle A_{satM} through repeated iteration and optimization of the relationship between A_{sunM} and A_{satM}. Then we can obtain the sunglint vector $\boldsymbol{R}_{\text{M}}$ in inertial system. After that, we can get the sunglint vector $\boldsymbol{S}_{\text{M}}$ in inertial system through coordinate transformation [6-7].

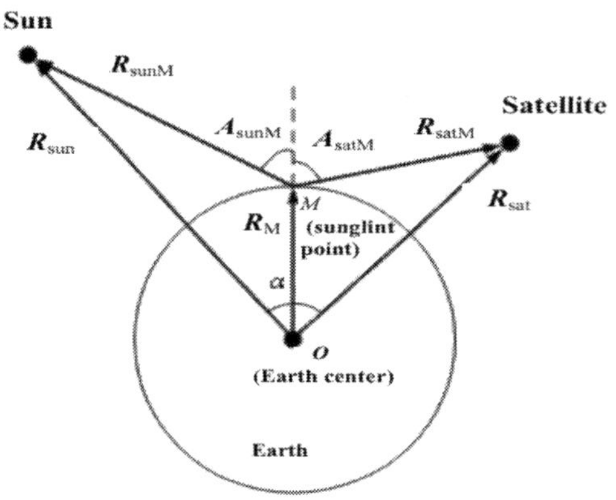

Figure.2. Geometric relation of sun, satelliate and sunglint in inertial system.

3 Analysis of the sunglint point positioning accuracy

3.1 Theoretical analysis

According to the working principle of greenhouse gas detector, the sunglint point positioning accuracy mainly consists of two parts: pointing accuracy of two-dimensional pointing mechanism and accuracy of sunglint positioning algorithm.

3.1.1 Analysis of pointing accuracy

The pointing accuracy of two-dimensional pointing mechanism is determined by the AT axis (along orbit direction) and CT axis (vertical orbit direction) motor and property of rotary transformer.

The AT axis is composed of a voice coil motor, pivot and rotary transformer.The step accuracy of voice coil motor is can be ignored. After quantification, the angle accuracy of the rotary transformer is 1.2'' and the electrical error of the rotary transformer is ±7.5''. The final calculation precision is ±7.5'' (±0.002°). So the pointing accuracy of AT axis is ±0.002°.

The CT axis is composed of a stepping motor, bearing and rotary transformer. The minimum output angle of the stepping motor is 0.036°. The angle accuracy of the rotary transformer is 1.2'' and the electrical error of the rotary transformer is ±40''. The final calculation precision is ±40'' (±0.01°). So the pointing accuracy of AT axis is ±0.028°.

3.1.2 Analysis of sunglint point positioning algorithm accuracy

Based on the STK model of satellite which carries with greenhouse gas detector, the sunglint point vector at 11 different time moments of one period in x, y and z direction of orbit coordinate system are extracted [8-9]. It's shows in Table.1. The orbit six elements of 11 groups at the same time are as input for the sunglint point positioning algorithm. The Matlab compile results of sunglint point positioning algorithm is showed in Table.1.

Table.1. Part of results of the sunglint vector in STK and sunglint calculation

time	The calculation results of sunglint vector exported from STK /km			The calculation results of sunglint vector used by sunglint point positioning algorithm in Matlab /km		
	x	*y*	*z*	*x*	*y*	*z*
2016-11-8 21:26:12	*212.97*	*-423.00*	*861.15*	*214.63*	*-434.73*	*895.26*

2016-11-8 21:27:12	*171.65*	*-417.35*	*858.24*	*172.76*	*-429.65*	*893.85*
2016-11-8 21:28:12	*131.05*	*-412.87*	*855.83*	*131.48*	*-425.59*	*892.67*
2016-11-8 21:29:12	*91.00*	*-409.53*	*853.78*	*90.71*	*-422.72*	*891.74*
2016-11-8 21:30:12	*51.34*	*-407.29*	*852.12*	*50.22*	*-420.74*	*890.99*
2016-11-8 21:31:12	*11.92*	*-406.15*	*850.85*	*9.96*	*-419.93*	*890.49*
2016-11-8 21:32:12	*-27.41*	*-406.10*	*849.99*	*-30.26*	*-420.04*	*890.16*
2016-11-8 21:33:12	*-66.83*	*-407.16*	*849.55*	*-70.59*	*-421.23*	*890.05*
2016-11-8 21:34:12	*-106.49*	*-409.32*	*849.50*	*-111.12*	*-423.34*	*890.13*
2016-11-8 21:35:12	*-146.55*	*-412.62*	*849.89*	*-152.00*	*-426.50*	*890.43*
2016-11-8 21:36:12	*-187.19*	*-417.10*	*850.74*	*-193.38*	*-430.74*	*890.96*

For comparison conveniently, the sunglint point vectors which are exported from STK and matlab at the same time are transferred to the pointing angle θ of greenhouse gas detector. The angle θ is the include angle between sunglint vector and z axis. Table.2 shows the caculation algorithm results error of sunglint point positioning is less than 0.37°.

Table.2. The contrast result between STK and sunglint calculation

Time	The angle calculated by sunglint vectors in STK / (°)	The angle calculated by sunglint vectors in sunglint point positioning algorithm/ (°)	error/ (°)
2016-11-8 21:26:12	*28.81*	*28.44*	*0.37*
2016-11-8 21:27:12	*27.73*	*27.39*	*0.35*
2016-11-8 21:28:12	*26.85*	*26.52*	*0.33*
2016-11-8 21:29:12	*26.17*	*25.87*	*0.30*
2016-11-8 21:30:12	*25.72*	*25.43*	*0.29*
2016-11-8 21:31:12	*25.53*	*25.25*	*0.27*
2016-11-8 21:32:12	*25.59*	*25.32*	*0.27*
2016-11-8 21:33:12	*25.91*	*25.63*	*0.27*
2016-11-8 21:34:12	*26.47*	*26.18*	*0.28*
2016-11-8 21:35:12	*27.26*	*26.95*	*0.31*
2016-11-8 21:36:12	*28.25*	*27.92*	*0.33*

3.1.3 The results of theoretical analysis

The ground distance of each degree of pointing mirror is 14.50km. So the ground field error of AT axis is 0.06km. The ground field error of CT axis is 0.81km. The ground field error caused by algorithm accuracy is 5.37km.Stated thus, the maximum ground field error of sunglint point detection of theoretical analysis is 6.18km. The sunglint point width is in 40~50km. this error has no effect for obtaining the sunglint point. In conclusion, the greenhouse gas detector could locate and track sunglint point effectively.

3.2 Physical verification

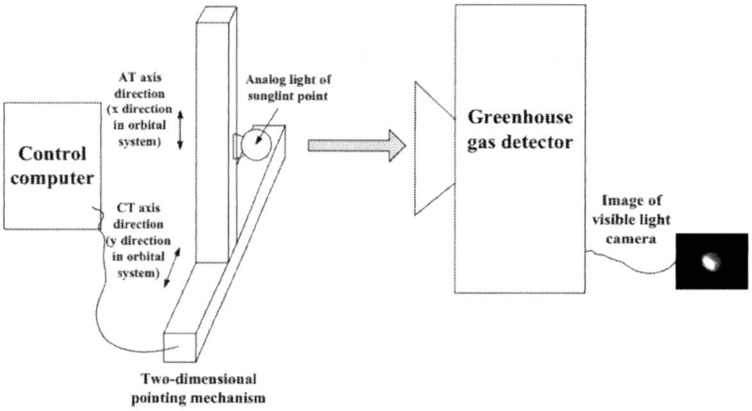

Figure.3. Schematic diagram of physical verification system.

Figure.3 shows the schematic diagram of physical verification system. Sunglint point position data which are the same period time with the orbit six elements are translated into the displacement data in two directions of two-dimensional moving platform [10]. After to ensure the first sunglint point position in the image center of visible light camera of greenhouse gas detector, two-dimensional moving platform and greenhouse gas detector are started at the same time. It is to collect the image information of visible light camera at this period. Through analysis of the change of centroid position of light source in the image of visible light camera, sunglint point positioning accuracy of greenhouse gas detector is obtained.

The detector in visible light camera consists of 640×512 pixels. The ground pixel resolution is 0.125km.Figure.4 shows all of the centroid position of light source in the image of visible light camera. The maximum distance from each centroid position of light source to the centroid position of light source of the first image is 49.20 pixels. The ground distance is 6.15km.

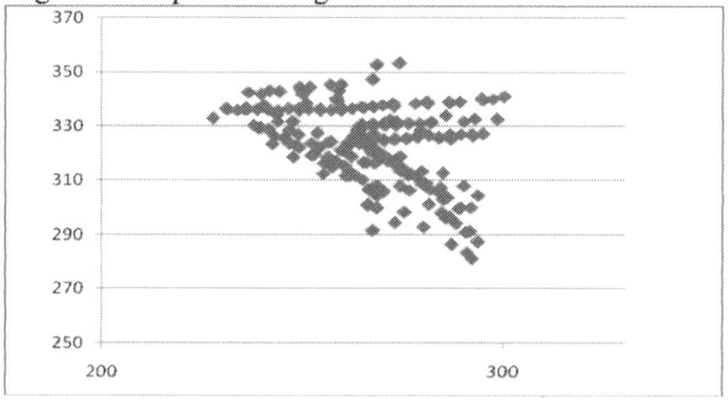

Figure.4. Distribution of light source centroid position in visible light image.

4 Conclusion

In conclusion, the maximum ground field error of sunglint point detection of physical verification is 6.15km. This result is basically consistent with the results (6.18km) of theoretical analysis. Both of those analysis methods provided in this paper verified the sunglint point positioning accuracy and also verified the analysis results each other.

References

1. G.N.PLASS, G.W.KATTAWAR, J.A.GUINN. Isophotes of Sunlight Glitter in a Wind-ruffled Sea. Applied Optics, 3,643-653 (1997)

2. SUTO.Hiroshi, KAWASHIMA.Takahiro, YOSHIDA.Jun, et al. The Pre-launch Performance Test and Calibration Results of Thermal and Near Infrared Sensor for Carbon Observation （TANSO） on GOSAT. Proceedings of SPIE - The International Society for Optical Engineering, 7106,71060L-1-71060L-10(2008)
3. G.INOUE, T.YOKOTA, H.OGUMA, et al. Overview of Greenhouse Gases Observing Satellite(GOSAT). American Geophysical Union Fall Meeting. San Francisco,(2004)
4. Qizhen.HU, Fang.YANG, An introduction to astronautics. Beijing,China Science and Technology Press，(2010)
5. Aoning.Xi, Wei.WANG, Near Earth Spacecraft Orbit Foundation. ChangSha,National University of Denfence Technology Press, (2003)
6. Ming.LUO, Fajie.DUAN, Xuejun.WANG, Development on Non - Contacting 3D Measurement System of Theodolites and It's Application in Aerospace. Shanghai Measurement and testing. (2009)
7. Yang.LIU, Dongyun.YI,Zhengming. WANG, Coordinate Transformation Methods from the Inertial System to the Centroid orbit System. Aerospace Control, 2,4-8(2007)
8. Suquan.DING, Bo.ZHANG, Shiyong.LIU, STK in Application of Spaceflight Simulation Analysis. BeiJing, National Defence Industry Press(2011)
9. Ying.YANG, Qi.WANG, STK in Application of Computer Simulation. BeiJing,National Defence Industry Press,(2005)
10. H.R GORDON, A.Y.MOREL, Remote Assessment of Ocean Color for Interpretation of Satellite Visible Imatery： A Review. New York,Springer-Verlag(1983)

MATEC Web of Conferences **114**, 04015 (2017) DOI: 10.1051/matecconf/201711404015
2MAE 2017

Mathematical Modeling and Kinematics Analysis of Double Spherical Shell Rotary Docking Skirt

Haixia Gong[1], Zhe Liu[1,a] and Liquan Wang[1]

[1]*School of Mechanical and Electrical, Harbin Engineering University, 150001, China*

Abstract. In order to solve the problem of large trim and heel angles of the wrecked submarine, the double spherical shell rotating docking skirt is studied. According to the working principle of the rotating docking skirt, and the fixed skirt, the directional skirt, the angle skirt are simplified as the connecting rod. Therefore, the posture equation and kinematics model of the docking skirt are deduced, and according to the kinematics model, the angle of rotation of the directional skirt and the angle skirt is obtained when the wrecked submarine is in different trim and heel angles. Through the directional skirt and angle skirt with the matching rotation can make docking skirt interface in the $0°\sim2\gamma$ range within the rotation, to complete the docking skirt and the wrecked submarine docking. The MATLAB software is used to visualize the rotation angle of fixed skirt and directional skirt, which lays a good foundation for the development of the control of the double spherical shell rotating docking skirt in future.

1 Introduction

At present, underwater space station docking, marine life-saving are widely used underwater docking technology [1] . Underwater rescue mainly relies on life-saving vehicle equipped with docking skirt active docking wrecked submarine. Docking skirt becomes bridge of connection life rescue vehicle and the wrecked submarine. Most of the initial life-saving vehicle take a fixed docking skirt [2-3]. In recent years, foreign countries has developed rotating docking skirt [4]. Based on the working principle of rotating docking skirt. In this paper, a mathematical model is established and a kinematic model is derived, and finally verifies the correctness of the theoretical analysis by MATLAB.

2 The working principle of rotating docking skirt

Double spherical shell rotating skirt consists of three parts: fixed skirt, directional skirt and angle skirt [5], as is shown in Figure 1. Fixed skirt and life-saving vehicle is fixed connection, play the role of connecting directional skirt and life-saving vehicle. The directional skirt and angle skirt are respectively the upper and lower part of the spherical shell, the angle between the rotation axis I of the directional skirt and the rotation axis II of the angle skirt is γ, in which the directional skirt can revolve around shaft I, angle skirt can revolve around shaft II. Through the directional skirt and angle skirt with the matching rotation can make docking skirt interface in the $0° \sim 2\gamma$ range within the rotation, as shown in Figure 2.

[a] Corresponding author : 350762275@qq.com

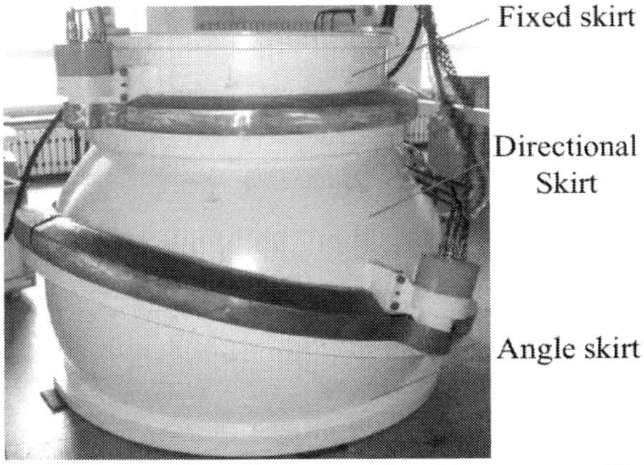

Figure 1. Composition of rotary docking skirt

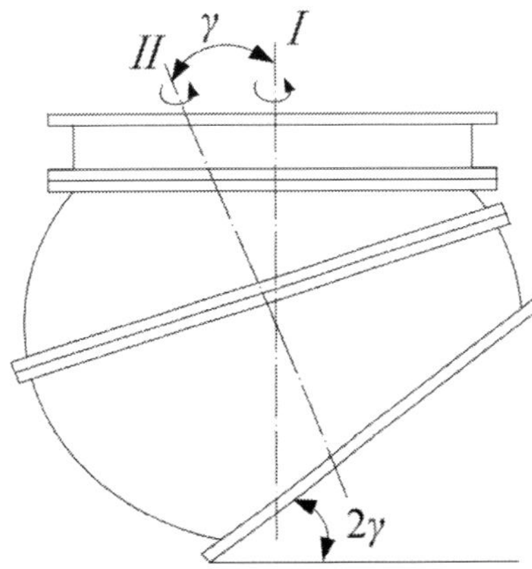

Figure 2. Rotation condition of the rotary docking skirt

3 The establishment of the position matrix of the rotating docking skirt

In order to facilitate deduction, this paper establishes a unified system. The midpoint of the contact plane of the fixed skirt and the directional skirt is O_1, Y_1's forward direction is forward direction of the lifeboat, Z_1's forward direction is vertical downward, X_1's forward direction is perpendicular to the plane of the Y1Z1 and points to the inside. In order to facilitate the subsequent calculation, here making the global coordinate system $OXYZ$ and coordinate system $O_1X_1Y_1Z_1$ coincidence. The intersection of the axis Z_1 and the lower plane of the directional skirt is O_2, the intersection of the axis II and the lower plane of angle skirt is O_3, the intersection of the axis I and the axis II is O_4, the intersection of the axis II and the lower plane of the angle skirt is O_5, the center of docking skirt interface is O_6; We postulate the length of O_1O_2 is D_1 and O_2O_3 is D_2, and also postulate the length of O_3O_4 is D_3 and O_5O_6 is D_4. The interface center of wrecked submarine is O_n, as is shown in Figure 3.

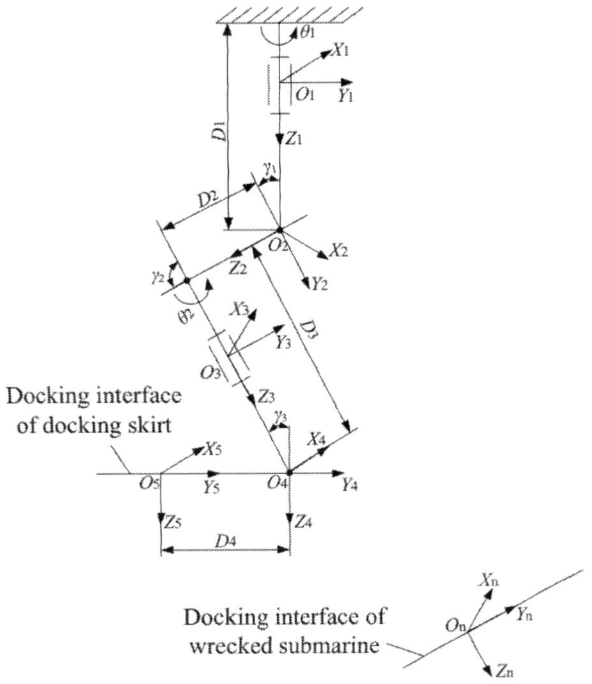

Figure 3. Rotary docking skirt docking motion coordinate system

Here is the position transformation matrix of coordinate system $O_1X_1Y_1Z_1$ to $O_6X_6Y_6Z_6$ [6-8], the parameters are shown in Table 1.

$$\begin{aligned} {}^0_6T = & Rot(Z,\theta_1)Trans(Z,D_1)Rot(X,-\gamma_1)Trans(Z,D_2)Rot(X,\gamma_2)\times \\ & Trans(Z,D_3)Rot(Z,\theta_2)Trans(Z,D_4)Rot(X,-\gamma_3)Trans(Y,-D_5) \end{aligned} \tag{1}$$

$${}^0_6T = {}^0_1T\,{}^1_2T\,{}^2_3T\,{}^3_4T\,{}^4_5T\,{}^5_6T = \begin{bmatrix} n_x & o_x & a_x & p_x \\ n_y & o_y & a_y & p_y \\ n_z & o_z & a_z & p_z \\ 0 & 0 & 0 & 1 \end{bmatrix} \tag{2}$$

Table 1. Parameters in the coordinate system

serial number of the link	γ_i/(°)	θ_i/(°)	D_i/mm
1	0	θ_1	D_1
2	$-\gamma_1$	0	D_2
3	γ_2	0	D_3
4	$-\gamma_3$	θ_2	D_4
5	0	0	D_5

In equation:

$$\begin{aligned} a_x = & (-\cos\theta_1\sin\theta_2 - \sin\theta_1\cos\gamma_1\cos\gamma_2\cos\theta_2 - \sin\theta_1\sin\gamma_1\sin\gamma_2\cos\theta_2)\sin\gamma_3 \\ & + (\sin\theta_1\cos\gamma_1\sin\gamma_2 - \sin\theta_1\sin\gamma_1\cos\gamma_2)\cos\gamma_3 \end{aligned} \tag{3}$$

$$\begin{aligned} a_y = & (-\sin\theta_1\sin\theta_2 + \cos\theta_1\cos\gamma_1\cos\gamma_2\cos\theta_2 + \cos\theta_1\sin\gamma_1\sin\gamma_2\cos\theta_2)\sin\gamma_3 \\ & + (-\cos\theta_1\cos\gamma_1\sin\gamma_2 + \cos\theta_1\sin\gamma_1\cos\gamma_2)\cos\gamma_3 \end{aligned} \tag{4}$$

$$a_z = (-\sin\gamma_1\cos\gamma_2\cos\theta_2 + \cos\gamma_1\sin\gamma_2\cos\theta_2)\sin\gamma_3 + (\sin\gamma_1\sin\gamma_2 + \cos\gamma_1\cos\gamma_2)\cos\gamma_3 \tag{5}$$

$$\begin{aligned} p_x = & -[(-\cos\theta_1\sin\theta_2 - \sin\theta_1\cos\gamma_1\cos\gamma_2\cos\theta_2 - \sin\theta_1\sin\gamma_1\sin\gamma_2\cos\theta_2)\cos\gamma_3 \\ & - \sin\theta_1\cos\gamma_1\sin\gamma_2\sin\gamma_3 + \sin\theta_1\sin\gamma_1\cos\gamma_2\sin\gamma_3)]D_5 + (\sin\theta_1\cos\gamma_1\sin\gamma_2 - \sin\theta_1\sin\gamma_1\cos\gamma_2)D_4 \\ & + (\sin\theta_1\cos\gamma_1\sin\gamma_2 - \sin\theta_1\sin\gamma_1\cos\gamma_2)D_3 - \sin\theta_1\sin\gamma_1 D_2 \end{aligned} \tag{6}$$

$$\begin{aligned} p_y = & -[(-\sin\theta_1\sin\theta_2 + \cos\theta_1\cos\gamma_1\cos\gamma_2\cos\theta_2 + \cos\theta_1\sin\gamma_1\sin\gamma_2\cos\theta_2)\cos\gamma_3 \\ & - (-\cos\theta_1\cos\gamma_1\sin\gamma_2 + \cos\theta_1\sin\gamma_1\cos\gamma_2)\sin\gamma_3]D_5 + (-\cos\theta_1\cos\gamma_1\sin\gamma_2 + \cos\theta_1\sin\gamma_1\cos\gamma_2)D_4 \\ & + (-\cos\theta_1\cos\gamma_1\sin\gamma_2 + \cos\theta_1\sin\gamma_1\cos\gamma_2)D_3 + \cos\theta_1\sin\gamma_1 D_2 \end{aligned} \tag{7}$$

$$\begin{aligned} p_z = & -[(-\sin\gamma_1\cos\gamma_2\cos\theta_2\cos\gamma_3 + \cos\gamma_1\sin\gamma_2\cos\theta_2\cos\gamma_3) - (\sin\gamma_1\sin\gamma_2\sin\gamma_3 + \cos\gamma_1\cos\gamma_2\sin\gamma_3)]D_5 \\ & + (\sin\gamma_1\sin\gamma_2 + \cos\gamma_1\cos\gamma_2)D_4 + (\sin\gamma_1\sin\gamma_2 + \cos\gamma_1\cos\gamma_2)D_3 + \cos\gamma_1 D_2 + D_1 \end{aligned} \tag{8}$$

4 Analysis on the Movement of Rotary Docking Skirt

In this article, γ_2=90°. so a_x, a_y and a_z can be simplified as:

$$a_x = -\cos\theta_1\sin\theta_2\sin\gamma_3 - \sin\theta_1\sin\gamma_1\cos\theta_2\sin\gamma_3 + \sin\theta_1\cos\gamma_1\cos\gamma_3 \tag{9}$$

$$a_y = -\sin\theta_1\sin\theta_2\sin\gamma_3 + \cos\theta_1\sin\gamma_1\cos\theta_2\sin\gamma_3 - \cos\theta_1\cos\gamma_1\cos\gamma_3 \tag{10}$$

$$a_z = \cos\theta_2\cos\gamma_1\sin\gamma_3 + \sin\gamma_1\cos\gamma_3 \tag{11}$$

We postulate the docking interface center of wrecked submarine is coordinate system $O_nX_nY_nZ_n$, the position matrix N of the wrecked submarine's docking interface center can be regarded as $O_1X_1Y_1Z_1$ which is respectively moving X_n along the axis X_1, moving Y_n along the axis Y_1 and moving Z_n along the axis Z_1, rotate β_2 around the axis Y_1, rotate the β_1 around the axis X_1.

$$\begin{aligned} N &= Trans(X,X_n)Trans(Y,Y_n)Trans(Z,Z_n)Rot(Y,\beta_2)Rot(X,\beta_1) \\ &= \begin{bmatrix} \cos\beta_2 & \sin\beta_1\sin\beta_2 & \sin\beta_2\cos\beta_1 & X_n \\ 0 & \cos\beta_1 & -\sin\beta_1 & Y_n \\ -\sin\beta_2 & \sin\beta_1\cos\beta_2 & \cos\beta_1\cos\beta_2 & Z_n \\ 0 & 0 & 0 & 1 \end{bmatrix} \end{aligned} \tag{12}$$

By the above formula, the direction cosine of the axis Z unit principal vector of the coordinate system $O_nX_nY_nZ_n$ relative to the coordinate system $O_1X_1Y_1Z_1$ is:

$$Z_N = \begin{bmatrix} \sin\beta_2\cos\beta_1 & -\sin\beta_2 & \cos\beta_1\cos\beta_2 \end{bmatrix} \tag{13}$$

The direction cosine of the axis Z unit principal vector of the coordinate system $O_6X_6Y_6Z_6$ relative to the coordinate system $O_1X_1Y_1Z_1$ is:

$$^0_6a = \begin{bmatrix} a_x & a_y & a_z \end{bmatrix} \tag{14}$$

In order to ensure docking skirt interface and wrecked submarine interface parallel docking , should make:

$$^0_6a = Z_N \tag{15}$$

We can obtain directional skirt angle θ_1 and angle skirt angle θ_2 from equation (15):

$$\cos\theta_1 = \frac{\sin\beta_1\cos\gamma_1\cos\gamma_3 - \sin\beta_1\cos\theta_2\sin\gamma_1\sin\gamma_3 - \sin\beta_2\cos\beta_1\sin\theta_2\sin\gamma_3}{(\sin\theta_2\sin\gamma_3)^2 + (\cos\theta_2\sin\gamma_1\sin\gamma_3 - \cos\gamma_1\cos\gamma_3)^2} \tag{16}$$

$$\cos\theta_2 = \frac{\cos\beta_1\cos\beta_2 - \sin\gamma_1\cos\gamma_3}{\cos\gamma_1\sin\gamma_3} \tag{17}$$

In order to ensure the docking skirt interface and the wrecked submarine interface coincidence, the life-saving vehicle needs to travel distance is:

$$L_x = X_n - P_x,\ L_y = Y_n - P_y,\ L_z = Z_n - P_z \tag{18}$$

The optimal distance is obtained, as shown in Figure 4:

$$L_{\min} = \sqrt{(X_n - P_x)^2 + (Y_n - P_y)^2 + (Z_n - P_z)^2} \tag{19}$$

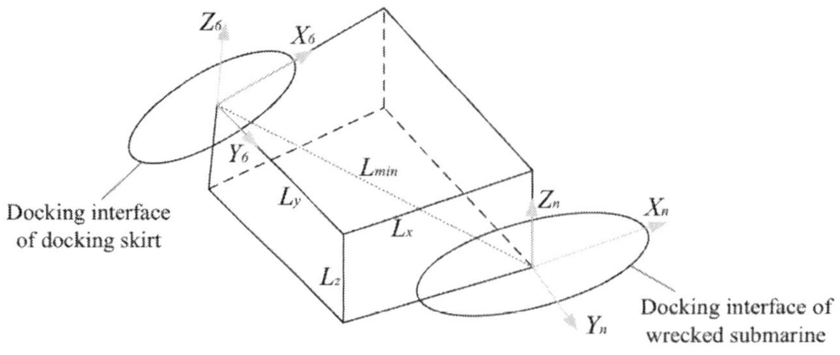

Figure 4. The best route for a lifeboat

5 Determination of the range of rotary docking operations

Because of the position of the wrecked submarine is random in the seabed. Therefore, it is need to get the rotation angle of the directional skirt and angle skirt when the wrecked submarine is in different heel angle and trim angle. It is assumed that the range of the heel angle and trim angle of the wrecked submarine is all 0°~60°, and the samples were sampled at intervals of 10°. In this article, γ_1=60°, γ_2=90° and γ_3=30°. Based on the range of angle cosine of the directional skirt and angle skirt in the kinematics model above, using Matlab to program can get the angle of directional skirt and angle skirt , as shown in Figure 5 and Figure 6.

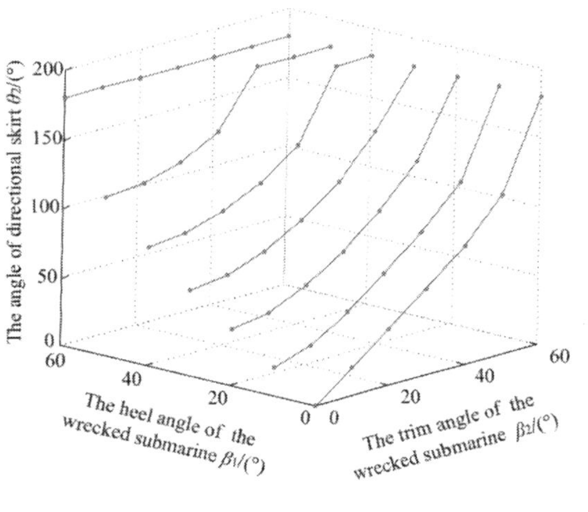

Figure 5. The angle of rotation by angle skirt

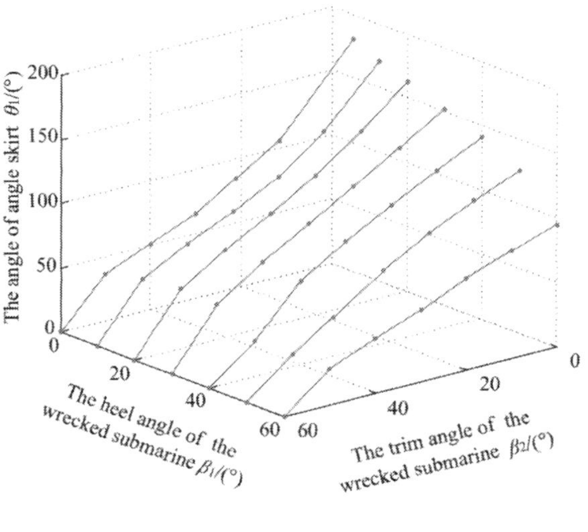

Figure 6. The angle of rotation by directional skirt

6 Conclusion

Through the study of the double spherical shell rotating docking skirt. And the fixed skirt, the directional skirt, the angle skirt are simplified as the connecting rod. We get the pose equation and the kinematic model by analyzing the working principle of the rotating docking skirt. The angle of rotation of the directional skirt and the angle skirt is obtained when the wrecked submarine is in different trim and heel angles. Through the directional skirt and angle skirt with the matching rotation can make docking skirt interface in the 0°~2γ range within the rotation, to complete the docking skirt and the wrecked submarine docking. The MATLAB software is used to visualize the rotation angle of fixed skirt and directional skirt, and the correctness of the theoretical analysis of the rotary skirt docking is proved, which lays a good foundation for the development of the control of the double spherical shell rotating docking skirt in future.

References

1. L. Q. WANG, D. D. TANG, J. R. Wu, Z. L. Zhang, Research on Structure and Kinematics of a Novel Mating Skirt on the Underwater Vehicle [J]. China Mechanical Engineering, **19**, 2814-2818 (2008)
2. X. Y. Meng, W. Tian. Comments on the nuclear submarine escape and rescue system development of foreign countries [J]. Ship Science and Technology, **34**, 137-143 (2012)
3. B. G. Fu, Q. X. Meng, H. M. Liu, Current situation and development trend of deep submergence rescue vehicles [J]. The Ocean Engineering, **25**, 120-126 (2007)
4. Owen F. Submarine rescue with an Australian twang, indepth rescue [J]. A Newsletter for Submarine Rescue Professionals, **1**, 1-6 (2004)
5. D. D. TANG, L. Q. WANG, Q. X. Meng, J. R. Wu, Z. L. Zhang, Structure design and carrying capacity analysis of mating skirt on the DSRV [J], **31**, 1-4 (2009)
6. X. Zhang, Z. L. Zheng, Y. Qi, Parameter Identification and Calibration of D-H Model for 6-DOF Serial Robots [J]. Robot, **38**, 360-370 (2016)
7. Y. Gan, J. L. Wang, F. J. Sun, Optimal Design of D-H Parameters of a 6R Robot for a Prescribed Workspace [J]. China Mechanical Engineering, **25**, 3003-3007+3011 (2014)
8. F. Y. Guo, T. Mei, J. H. Zhao, Problems and Improvement of D-H Method for Establishing Connecting Rod Coordinate System [J]. China Mechanical Engineering, **25**, 2710-2714 (2014)

MATEC Web of Conferences **114**, 04016 (2017) DOI: 10.1051/matecconf/201711404016
2MAE 2017

Revisiting the two-side optimization problem in satellite pursuit-evasion

KeKe Zhang[1,2,3], ZhongCheng Mu[3, a] , Dong Ye[4],JinLing Qi[3]

[1]*ShangHai Institute of Microsystem and Information Technology,ShangHai 200050, China*
[2]*University of Chinese Academy of Sciences, BeiJing 100049, China*
[3]*ShangHai Engineering Center for Microsatellite, ShangHai 201210, China*
[4]*Harbin Institute of Technology, Harbin 150001, China*

Abstract. In the present paper two-side optimization problem where the intercepting satellite pursuits the target satellite was studied. In the pursuit-evasion game the continuous thruster and the simple central gravity with J2 perturbation were involved for two satellites. With applying the functional extremum condition, the optimal control outputs of both side satellites were obtained by building the systems Hamilton function. Further, the two side optimal problem was transformed to the TPBVP, which contributed the solution with the mixed numerical method. Finally, the proposed method was validated by the numerical simulation, where the optimal interception and escaping trajectories were illustrated.

1 Introduction

In previous studies, for the satellite interception question, the targets were assumed generally to be static without any motion ability and moved in the pre-assigned orbits. Thus the solver only needs to design a maneuvering trajectory for the attacking satellite in solving the interception which could be regarded as single side optimization question. The design and optimization method was different as the propulsive force exerting on the interceptor. Luo, etc. studied the multiple impulsive fuel optimal orbit rendezvous problem under the linear [1] and nonlinear [2] dynamics, they directly applied the simplex, quadratic, genetic [3] and simulated annealing algorithms [4] to optimize the impulse vectors. Simulation results show that combining with the meta optimization method and the SQP method can effectively solve the problem [5]. Lawden, etc. proposed the principal vector theory to determine the trajectory optimal property [6] and added the coast stage or middle impulse to achieve the fuel optimal trajectory [7]. For the continuous thruster, the indirect method was applied based on the calculus of variation and maximum principle, causing that the interception question was transformed to the two point boundary value problem (TPBVP). Greenwood solved the transformed problem by the shooting method and applied it to study the small thrust fuel optimal transferring problem in the co-plane orbit [8]. Unfortunately, the initial value of the indirect method is difficult to estimate and is sensitive for the result achievement. The direct method scatters the total transferring procedure to small steps and transfers the dynamical optimization to the constrained minimum solving problem. Hargraves, etc. put forward the collocation method to solve the time optimal interception climbs problem [9]. Enright [10]

[a] Corresponding author : muzhongcheng@aliyun.com

improved the method and reduced the computing scale by developing the nonlinear programming with the parallel shooting frame. Herman, etc. [11] proposed the high order collocation method which was widely used to solve various earth transferring optimal orbits.

Contrarily, for the scenarios with having the orbit maneuvering ability, the method described above may become infeasible since both the interceptor and the target are in the non-stationary state. Considering both sides optimal features, the designer needs to give the control method, where the target satellite could escape from the interceptor when it discovered the later attacking intention. Actually, it is a classical pursuit-evasion problem and can be coped with the game theory. Iaacs firstly concentrated on this problem and defined the two side optimal solution as the saddle solution [12]. For the realistic situation, the saddle solution is difficult to get the solution since the system dynamics cover two players and the dynamics often are highly nonlinear and complex. Anderson [13] solved the coplanar spacecraft pursuit-evasion with large thrust by finding the difference between the current state and the pre-calculating state to update the conjugate. Menon [14] obtained the feedback control strategy for the spacecraft with saturated output by applying the back-step method and the quadratic objective function. Mauro Pontani [15, 16] transferred the original interception problem to the TPBVP and proposed the semi-direct method to solve it. This method was used in intercepting an extra-atmosphere missile and two spacecrafts attack-escape. Stupik [17] studied satellite combat based on the CW equation with the help of Kriging method. Great many standard solutions with various initial conditions solved with the former method were interpolated to obtain the suboptimal feedback solution when initial states domain and the disturbance were finite.

On the other hand, earth non-spherical perturbation has a great influence on the low orbital satellite motion. No matter how accurate the initial trajectory was designed, the satellite can still drifted an intolerable distance. Thus the present paper mainly studied the pursuit-evasion game of two satellites with J2 perturbation effect, where both satellites carried the continuous thruster with limited magnitude to perform orbital maneuver.

2 Pursuit and Evasion Dynamics with J2 Perturbation

This intercepting event ends when the pursuer catches the target, namely both satellites arrives at the same location in the space. During the catching and escaping process, the pursuer adopts the control strategy to catch the target as soon as possible while the later tries its best to postpone the intercept terminal time. In order to make the question simple, it was assumed that both players can completely acquire the counter dynamics information.

Firstly, aiming to complete the interception, it is required that UP (the specific thrusts of the interceptor) is bigger than UE (the specific thrusts of the escaper). Mauro [16] used a group of new state variables to describe the spacecraft's motion to make the simulation more real. Thus the same variables were adopted in the paper. The state variables of the pursuer and evader can be defined as $(r_i, \xi_i, \varphi_i, \gamma_i, v_i, \zeta_i)$ with representing respectively the satellite's geocentric distance, geometrical longitude, latitude, path angle, speed and azimuth angle, and $i = P, E$ stand for the two sides in the interception, respectively.

The simple two-body gravity exerts a centripetal force on the satellite which produces an acceleration vectors can been represent in the geodetic coordinate system as following

$$\boldsymbol{a}_\mu = \begin{bmatrix} -\frac{\mu}{r^2} & 0 & 0 \end{bmatrix}^T \tag{1}$$

where μ represents the gravitational constant of the earth.

The acceleration arose by the J2 non-spherical perturbation can been decomposed in the same coordinate as below

$$\boldsymbol{a}_{J_2} = -\frac{3J_2\mu R_e^2}{2r_4}\begin{bmatrix} 1-3\sin^2\varphi \\ 2\sin\varphi\cos\varphi \\ 0 \end{bmatrix} \tag{2}$$

where $J_2 = 1.018\times10^{-3}$ is the J2 perturbation constant and $R_e = 6378.165$ km is the mean earth radius. Using the coordinates transformation relation, the dynamics of the satellite can be given as equation (3), where $G_{J2} = J_2\mu R_e^2$.

$$\begin{aligned}
\dot{r} &= v\sin\gamma \\
\dot{\xi} &= \frac{v\cos\gamma\cos\zeta}{r\cos\varphi} \\
\dot{\varphi} &= \frac{v\cos\gamma\sin\zeta}{r} \\
\dot{\gamma} &= \frac{v}{r}\cos\gamma + \frac{U\cos\beta\sin\alpha}{v} - \frac{\mu\sin\gamma}{r^2 v} - \frac{3G_{J2}}{4r^4 v}\{\sin\gamma[3\cos(2\varphi)-1] + 2\cos\gamma\sin\zeta\sin(2\varphi)\} \\
\dot{v} &= U\cos\beta\cos\alpha - \frac{\mu\sin\gamma}{r^2} - \frac{3G_{J2}}{4r^4}\{\cos\gamma[3\cos(2\varphi)-1] + 2\sin\gamma\sin\zeta\sin(2\varphi)\} \\
\dot{\zeta} &= \frac{U\sin\beta}{v\cos\gamma} - \frac{v\tan\varphi\cos\gamma\cos\zeta}{r} - \frac{3G_{J2}}{2r^4 v\cos\gamma}\cos\zeta\sin(2\varphi)
\end{aligned} \tag{3}$$

Let the states of the interceptor (P) and target (E) be

$$\begin{aligned}
\boldsymbol{x}_P &= [x_1 \quad x_2 \quad x_3 \quad x_4 \quad x_5 \quad x_6]^T \\
&= [r_P \quad \xi_P \quad \varphi_P \quad \gamma_P \quad v_P \quad \zeta_P]^T \\
\boldsymbol{x}_E &= [x_7 \quad x_8 \quad x_9 \quad x_{10} \quad x_{11} \quad x_{12}]^T \\
&= [r_E \quad \xi_E \quad \varphi_E \quad \gamma_E \quad v_E \quad \zeta_E]^T
\end{aligned} \tag{4}$$

Obviously the pursuer and evader optimize the angle of the thruster to compete for the interception ending time, thus the control variables of satellites are given as below

$$\begin{aligned}
\boldsymbol{u}_P &= [u_1 \quad u_2]^T = [\alpha_P \quad \beta_P]^T \\
\boldsymbol{u}_E &= [u_3 \quad u_4]^T = [\alpha_E \quad \beta_E]^T
\end{aligned} \tag{5}$$

Further, the dynamics of both sides can be rewritten as below

$$\begin{aligned}
\dot{\boldsymbol{x}}_P &= \boldsymbol{f}_P(\boldsymbol{x}_P, \boldsymbol{u}_P, t) \\
\dot{\boldsymbol{x}}_E &= \boldsymbol{f}_E(\boldsymbol{x}_E, \boldsymbol{u}_E, t)
\end{aligned} \tag{6}$$

Interception ending requires that the location coordinate of both satellites is equal, namely the following conditions are satisfied at the terminal time t_f

$$\begin{aligned}
r_P = r_E|_{t=t_f} &\Rightarrow x_1(t_f) - x_7(t_f) = 0 \\
\xi_P = \xi_E|_{t=t_f} &\Rightarrow x_2(t_f) - x_8(t_f) = 0 \\
\varphi_P = \varphi_E|_{t=t_f} &\Rightarrow x_3(t_f) - x_9(t_f) = 0
\end{aligned} \tag{7}$$

Rewritten them to a compact vector form as following

$$N(x,t)\big|_{t=t_f}=0 \tag{8}$$

Finally, the orbital elements can be transformed to the state variables according to the relation equation in reference [13].

3 Saddel Solution of the Pursuit-evasion Game

In terms of the pursuer, the interception ending time is expected to become short as little as possible, while the counterpart escapes the catch to the best of its ability. Thus the ending time can be chosen as the objective namely

$$J=t_f \tag{9}$$

This interception is a traditional zero sum differential game. When the control strategies variables pair (U_P,U_E) satisfies the condition below, it is defined as the saddle solution (U_P^*,U_E^*)

$$J(U_P^*,U_E)\le J(U_P^*,U_E^*)\le J(U_P,U_E^*) \quad \forall U_P,U_E \tag{10}$$

which means that if only the purser adopts the optimal control strategy, the objective will be better than that they both make use the optimal control strategies, and the interception ending time will shorten. Otherwise, the time will increase.

The open loop solution of the differential game is obtained based on the initial states x_0 of both sides and the current time, whilst the close loop solution which is more flexible to the real situation can be solved according to the initial states, current time and the current states. Conclusion has been made that if the value function is continuous in the admissible domain of control strategies, the open loop solution and close loop solution are identical [18]. And for the system dynamics with high complexity and nonlinearity, it is almost impossible to analytically solve the saddle solution. Thus obtaining the close loop solution is unpractical currently, the optional way is to solve the open loop control strategy and optimal trajectory relevant.

For the two-side dynamical optimization problem with system (5), Maya type objective (8) and terminal condition collection (7), the Hamiltonian function can be defined as the optimal control theory

$$H=\lambda_P^T f_P+\lambda_E^T f_E=H_P+H_E \tag{11}$$

where λ_P and λ_E stand for the conjugates of the pursuer and evader's state variables, respectively. H_P is given as equation (12)

$$\begin{aligned}H_P&=\lambda_1 v_P\sin\gamma_P+\lambda_2\frac{v_i\cos\gamma_P\cos\xi_i}{r_P}+\lambda_3\frac{v_P\cos\gamma_P\sin\xi_P}{r_P\cos\varphi_P}+\lambda_4\left(\frac{v_P\cos\gamma_P}{r_P}-\frac{\mu\sin\gamma_P}{r_P^2 v_P}\right.\\&\left.-\frac{3G_{J2}}{4r_P^4 v_P}\left\{\sin\gamma_P\left[3\cos(2\varphi_P)-1\right]+2\cos\gamma_P\sin\zeta_P\sin(2\varphi_P)\right\}+\frac{U_P\cos\beta_P\sin\alpha_P}{v_P}\right)\\&+\lambda_5\left(U_P\cos\beta_P\cos\alpha_P-\frac{3G_{J2}}{4r_P^4}\left\{\cos\gamma_P\left[3\cos(2\varphi_P)-1\right]+2\sin\gamma_P\sin\zeta_P\sin(2\varphi_P)\right\}\right.\\&\left.-\frac{\mu\sin\gamma_P}{r_P^2}\right)+\lambda_6\left(\frac{U_P\sin\beta_P}{v_P\cos\gamma_P}-\frac{3G_{J2}\cos\zeta_P\sin(2\varphi_P)}{2r_P^4 v_P\cos\gamma_P}-\frac{v_P\tan\varphi_P\cos\gamma_P\cos\zeta_P}{r_P}\right)\end{aligned} \tag{12}$$

H_E is like H_P with the subscript substituting to be E and the conjugates changing from $\lambda_i, i = 1,2,\ldots,6$ to λ_{i+6}.

If defining the co-state variables of terminal conditions (7) as $\boldsymbol{\upsilon} = \begin{bmatrix} \upsilon_1 & \upsilon_2 & \upsilon_3 \end{bmatrix}^T$, the conjugates at terminal time can be obtained with satisfying the conditions as equations (13)

$$\begin{aligned} \boldsymbol{\lambda}_P^T &= \frac{\partial \varphi\left(t_f, \boldsymbol{x}\left(t_f\right)\right)}{\partial \boldsymbol{x}_P\left(t_f\right)} + \left.\frac{\partial \boldsymbol{N}^T}{\boldsymbol{x}_P}\right|_{t=t_f} \\ \boldsymbol{\lambda}_E^T &= \frac{\partial \varphi\left(t_f, \boldsymbol{x}\left(t_f\right)\right)}{\partial \boldsymbol{x}_E\left(t_f\right)} + \left.\frac{\partial \boldsymbol{N}^T}{\boldsymbol{x}_E}\right|_{t=t_f} \end{aligned} \tag{13}$$

Substituting the correlative variables and equations, $\boldsymbol{\lambda}\left(t_f\right)$ is given below

$$\begin{aligned} &\begin{cases} \lambda_1\left(t_f\right) = \upsilon_1 \\ \lambda_2\left(t_f\right) = \upsilon_2 \\ \lambda_3\left(t_f\right) = \upsilon_3 \\ \lambda_4\left(t_f\right) = \lambda_5\left(t_f\right) = \lambda_6\left(t_f\right) = 0 \end{cases} \\ &\begin{cases} \lambda_7\left(t_f\right) = -\upsilon_1 \\ \lambda_8\left(t_f\right) = -\upsilon_2 \\ \lambda_9\left(t_f\right) = -\upsilon_3 \\ \lambda_{10}\left(t_f\right) = \lambda_{11}\left(t_f\right) = \lambda_{12}\left(t_f\right) = 0 \end{cases} \end{aligned} \tag{14}$$

And collecting the equations, and the conjugates satisfy the following conditions

$$\begin{aligned} \lambda_4\left(t_f\right) = \lambda_5\left(t_f\right) = \lambda_6\left(t_f\right) = \lambda_{10}\left(t_f\right) \\ = \lambda_{11}\left(t_f\right) = \lambda_{12}\left(t_f\right) = 0 \\ \lambda_1\left(t_f\right) + \lambda_7\left(t_f\right) = 0 \\ \lambda_2\left(t_f\right) + \lambda_8\left(t_f\right) = 0 \\ \lambda_3\left(t_f\right) + \lambda_9\left(t_f\right) = 0 \end{aligned} \tag{15}$$

Then applying the function extremum theory to system Hamiltonian function H_P, the co-state equations are derived as the following

$$\begin{aligned} \dot{\boldsymbol{\lambda}}_P &= -\left[H_{\boldsymbol{x}_P}\right]^T = -\left[\frac{\partial H_P}{\partial \boldsymbol{x}_P}\right]^T = -\left[\frac{\partial \boldsymbol{f}_P}{\partial \boldsymbol{x}_P}\right]^T \boldsymbol{\lambda}_P \\ \dot{\boldsymbol{\lambda}}_E &= -\left[H_{\boldsymbol{x}_E}\right]^T = -\left[\frac{\partial H_E}{\partial \boldsymbol{x}_E}\right]^T = -\left[\frac{\partial \boldsymbol{f}_E}{\partial \boldsymbol{x}_E}\right]^T \boldsymbol{\lambda}_E \end{aligned} \tag{16}$$

And using $\mathrm{s}(\cdot)$ and $\mathrm{c}(\cdot)$ with replacing $\sin(\cdot)$ and $\cos(\cdot)$ respectively and substituting $\boldsymbol{f}_P$ as equations (6), then the co-state equations above can be written as equation (17).

$$\begin{aligned}
\dot{\lambda}_1 &= -\lambda_2 \frac{x_5 \,\mathrm{c}\, x_4 \,\mathrm{c}\, x_6}{x_1^2 \,\mathrm{c}\, x_3} - \lambda_3 \frac{x_5 \,\mathrm{c}\, x_4 \,\mathrm{s}\, x_6}{x_1^2} + \lambda_4 \left(\frac{3G_{J2}}{x_1^5 x_5} \left\{ \mathrm{s}\, x_4 \left[3\mathrm{c}(2x_3) - 2 \right] + 2\mathrm{s}(2x_3) \mathrm{c}\, x_5 \,\mathrm{s}\, x_6 \right\} \right. \\
&\quad \left. \frac{2\mu \,\mathrm{c}\, x_4}{x_1^3 x_5} - \frac{x_5 \,\mathrm{c}\, x_4}{x_1^2} \right) + \lambda_5 \left(\frac{3G_{J2}}{x_1^5} \left\{ \mathrm{s}\, x_4 \left[3\mathrm{c}(2x_3) - 2 \right] + 2\mathrm{s}(2x_3) \mathrm{c}\, x_5 \,\mathrm{s}\, x_6 \right\} + \frac{2\mu \,\mathrm{s}\, x_4}{x_1^3} \right) \\
&\quad + \lambda_6 \left[\frac{x_5 \,\mathrm{c}\, x_4 \,\mathrm{c}\, x_6 \,\mathrm{s}\, x_3}{x_1^2 \,\mathrm{c}\, x_3} + \frac{6G_{J2} \mathrm{s}(2x_3) \mathrm{c}\, x_6}{x_1^5 x_5 \,\mathrm{c}\, x_4} \right] \\
\dot{\lambda}_2 &= 0 \\
\dot{\lambda}_3 &= \lambda_2 \frac{x_5 \,\mathrm{c}\, x_4 \,\mathrm{s}\, x_3}{x_1 \,\mathrm{c}^2\, x_3} + \lambda_4 \frac{3G_{J2}}{4x_1^4 x_5} \left[6\mathrm{s}(2x_3) \mathrm{c}\, x_4 + 4\mathrm{c}(2x_3) \mathrm{s}\, x_4 \,\mathrm{s}\, x_6 \right] + \lambda_5 \frac{3G_{J2}}{4x_1^4} \left[6\mathrm{s}(2x_3) \mathrm{s}\, x_4 \right. \\
&\quad \left. -4\mathrm{c}(2x_3) \mathrm{c}\, x_5 \,\mathrm{s}\, x_6 \right] - \lambda_6 \left[\frac{x_5 \,\mathrm{c}\, x_4 \,\mathrm{c}\, x_6}{x_1} \left(1 + \frac{\mathrm{s}^2\, x_3}{\mathrm{c}^2\, x_3} \right) + \frac{3G_{J2} \mathrm{c}(2x_3) \mathrm{c}\, x_6}{x_1^4 x_5 \,\mathrm{c}\, x_4} \right] \\
\dot{\lambda}_4 &= \lambda_1 x_5 \,\mathrm{c}\, x_4 - \lambda_2 \frac{x_5 \,\mathrm{c}\, x_6 \,\mathrm{s}\, x_4}{x_1 \,\mathrm{c}\, x_3} - \lambda_3 \frac{x_5}{x_1} \mathrm{s}\, x_4 \,\mathrm{s}\, x_6 + \lambda_4 \left(\frac{\mu \,\mathrm{s}\, x_4}{x_1^2 x_5} + \frac{3G_{J2}}{4x_1^4 x_5} \left\{ \mathrm{s}\, x_4 \left[3\mathrm{c}(2x_3) - 1 \right] \right. \right. \\
&\quad \left. \left. + 2\mathrm{s}(2x_3) \mathrm{c}\, x_4 \,\mathrm{s}\, x_6 \right) - \frac{x_5 \,\mathrm{s}\, x_4}{x_1} \right) - \lambda_5 \left(\frac{3G_{J2}}{4x_1^4} \mathrm{c}\, x_4 \left[3\mathrm{c}(2x_3) - 1 \right] + \frac{\mu \,\mathrm{c}\, x_4}{x_1^2} \right) \\
&\quad + \lambda_6 \left[\frac{U_E \,\mathrm{s}\, \beta_P \,\mathrm{s}\, x_4}{x_5 \,\mathrm{c}^2\, x_4} + \frac{x_5 \,\mathrm{c}\, x_6 \,\mathrm{s}\, x_3 \,\mathrm{s}\, x_4}{x_1 \,\mathrm{c}\, x_3} - \frac{3G_{J2} \mathrm{s}(2x_3) \mathrm{c}\, x_6 \,\mathrm{s}\, x_4}{2x_1^4 x_5 \,\mathrm{c}^2\, x_4} \right] \\
\dot{\lambda}_5 &= \lambda_1 \,\mathrm{s}\, x_4 + \lambda_2 \frac{\mathrm{c}\, x_4 \,\mathrm{c}\, x_6}{x_1 \,\mathrm{c}\, x_3} + \lambda_3 \frac{\mathrm{c}\, x_4 \,\mathrm{s}\, x_6}{x_1} + \lambda_4 \left(\frac{\mathrm{c}\, x_4}{x_1} - \frac{U_P \,\mathrm{s}\, \alpha_P \,\mathrm{c}\, \beta_P}{x_5^2} + \frac{3G_{J2}}{4x_1^4 x_5^2} \left\{ -2\mathrm{s}(2x_3) \mathrm{c}\, x_4 \,\mathrm{s}\, x_6 \right\} + \frac{\mu \,\mathrm{c}\, x_4}{x_1^4 x_5^2} \right) \\
&\quad -\lambda_5 \frac{3G_{J2}}{2x_1^4} \mathrm{s}(2x_3) \mathrm{s}\, x_4 \,\mathrm{s}\, x_6 - \lambda_6 \left[\frac{U_P \,\mathrm{s}\, \beta_P}{x_5^2 \,\mathrm{c}\, x_4} + \frac{\mathrm{c}\, x_4 \,\mathrm{c}\, x_6 \,\mathrm{s}\, x_3}{x_1 \,\mathrm{c}\, x_3} - \frac{3G_{J2} \mathrm{s}(2x_3) \mathrm{c}\, x_6 \,\mathrm{s}\, x_4}{2x_1^4 x_5^2 \,\mathrm{c}\, x_4} \right] \\
\dot{\lambda}_6 &= -\lambda_2 \frac{x_5 \,\mathrm{c}\, x_4 \,\mathrm{s}\, x_6}{x_1 \,\mathrm{c}\, x_3} + \lambda_3 \frac{x_5 \,\mathrm{c}\, x_4 \,\mathrm{c}\, x_6}{x_1} + \lambda_4 \frac{3G_{J2} \mathrm{s}(2x_3) \mathrm{s}\, x_4 \,\mathrm{c}\, x_6}{2x_1^4 x_5} \\
&\quad -\lambda_5 \frac{3G_{J2} \mathrm{s}(2x_3) \mathrm{c}\, x_4 \,\mathrm{c}\, x_6}{2x_1^4} + \lambda_6 \left[\frac{x_5 \,\mathrm{c}\, x_4 \,\mathrm{s}\, x_6 \,\mathrm{s}\, x_3}{x_1 \,\mathrm{c}\, x_3} + \frac{3G_{J2} \mathrm{s}(2x_3) \mathrm{s}\, x_6}{2x_1^4 x_5 \,\mathrm{c}\, x_4} \right]
\end{aligned} \tag{17}$$

The terminal time of interception is uncertain, thus the system Hamilton satisfies the following terminal condition

$$H\big|_{t=t_f} + \left[\frac{\partial \varphi\left(t_f, \boldsymbol{x}\left(t_f\right)\right)}{\partial t_f} + \boldsymbol{\upsilon}^T \frac{\partial \boldsymbol{N}}{\partial t} \right] = H\big|_{t=t_f} + 1 = 0 \tag{18}$$

The Hamilton function can be divided to two parts H_P and H_E, so applying the min-max condition to H is equal to deriving the minimum of H_P and maximum of H_E. Then the optimal control strategies of the pursuer and evader satisfy the equations as below

$$\begin{aligned}
\boldsymbol{u}_P^* &= \arg\min_{\boldsymbol{u}_P} H = \arg\min_{\boldsymbol{u}_P} H_P \\
&= \arg\min_{[\alpha_P, \beta_P]} \left[\begin{array}{l} \lambda_4 \dfrac{\sin\alpha_P \cos\beta_P}{v_P} + \lambda_6 \dfrac{\sin\beta_P}{v_P \cos\gamma_P} \\ +\lambda_5 \cos\alpha_P \cos\beta_P + (\cdot)_P \end{array} \right]
\end{aligned} \tag{19}$$

$$\begin{aligned}
\boldsymbol{u}_E^* &= \arg\max_{\boldsymbol{u}_E} H = \arg\max_{\boldsymbol{u}_E} H_E \\
&= \arg\max_{[\alpha_E, \beta_E]} \left[\begin{array}{l} \lambda_{10} \dfrac{\sin\alpha_E \cos\beta_E}{v_E} + \lambda_{12} \dfrac{\sin\beta_E}{v_E \cos\gamma_E} \\ +\lambda_{11} \cos\alpha_E \cos\beta_E + (\cdot)_E \end{array} \right]
\end{aligned} \tag{20}$$

where $(\cdot)_P$ and $(\cdot)_E$ represent the remainders which is irrelevant to the control angles of the pursuer and evader, respectively. Continuing the derivation process, the optimal controls of both sides $\left(\left[\alpha_P^*,\beta_P^*\right],\left[\alpha_E^*,\beta_E^*\right]\right)$ can be obtained as equation (22).

$$\begin{cases} \beta_P^* = -\arcsin\dfrac{\lambda_6}{v_P\cos\gamma_P}\left[\left(\dfrac{\lambda_4}{v_P}\right)^2+\lambda_5^2+\left(\dfrac{\lambda_6}{v_P\cos\gamma_P}\right)^2\right]^{-\frac{1}{2}} \\ \cos\alpha_P^* = -\dfrac{\lambda_5}{\cos\beta_P^*}\left[\left(\dfrac{\lambda_4}{v_P}\right)^2+\lambda_5^2+\left(\dfrac{\lambda_6}{v_P\cos\gamma_P}\right)^2\right]^{-\frac{1}{2}} \\ \sin\alpha_P^* = -\dfrac{\lambda_4}{v_P\cos\beta_P^*}\left[\left(\dfrac{\lambda_4}{v_P}\right)^2+\lambda_5^2+\left(\dfrac{\lambda_6}{v_P\cos\gamma_P}\right)^2\right]^{-\frac{1}{2}} \end{cases}, \quad \begin{cases} \beta_E^* = \arcsin\dfrac{\lambda_{12}}{v_E\cos\gamma_E}\left[\left(\dfrac{\lambda_{10}}{v_E}\right)^2+\lambda_{11}^2+\left(\dfrac{\lambda_{12}}{v_E\cos\gamma_E}\right)^2\right]^{-\frac{1}{2}} \\ \cos\alpha_E^* = \dfrac{\lambda_{11}}{\cos\beta_E^*}\left[\left(\dfrac{\lambda_{10}}{v_E}\right)^2+\lambda_{11}^2+\left(\dfrac{\lambda_{12}}{v_E\cos\gamma_E}\right)^2\right]^{-\frac{1}{2}} \\ \sin\alpha_E^* = \dfrac{\lambda_{10}}{v_E\cos\beta_E^*}\left[\left(\dfrac{\lambda_{10}}{v_E}\right)^2+\lambda_{11}^2+\left(\dfrac{\lambda_{12}}{v_E\cos\gamma_E}\right)^2\right]^{-\frac{1}{2}} \end{cases} \tag{21}$$

4 Satellites Pursuit and Evasion solving based on Mixed Optimization

Obviously if the conjugates at any time are known, the control angles can be derived from equation (22). However the former needs to be integrated from the starting of interception with known initials. In order to solve this problem, it was performed to collect the system (6) and conjugate (17) to a set of differential equations and combine them with equations (15), (17) and (18) to form a two point boundary value problem. Thus the problem has 13 equality constraints and 13 knowns with including 12 initial conjugates and the interception ending time. Normally, the TPBVP is difficult to solve analytically for the nonlinearity of system dynamics. Numerical methods like shooting and difference are usually used to derive the solution. And good initial guess is required to guarantee the convergence and correctness of the solving process. In order to avoid these faults, the TPBVP was transformed to an unconstrained optimization problem where the design variables are the 13 knowns and the objective was constructed by minimize the distance between the function values formed from the equalities constraints (15), (17) and (18) and the original. The distance driving to zero in the optimize process equals to the satisfaction of the equalities constraints.

A mixed optimization method consisting of two stages was used in solving the optimization problem described above. Firstly, we make use of heuristic optimization method to search the approximated solution in the globally optimum's convergent region. In the second step, the sequential quadratic programming was employed to accurately refine the approximated solution. Based on the second order Taylor approximation, the SQP algorithm calculates the gradients and Hessian matrix of the objective function sequentially, and transform the preview complex constrained optimization problem to a series linear or quadratic regulating problems which satisfy the convex optimization theory. Solution of the original question was corrected after this sub-optimization problem solved.

Heuristic optimization imitates the natural evolving process, including genetic algorithm, simulated annealing algorithm, swarm algorithm, cuckoo search, etc. This paper employed the genetic algorithm as reference [15]. So the objective construction was same. But in the second step, the semi-direct method was changed by directly optimize the constructed objective in the first stage with the safe altitude constraints on the attacking-escaping satellites like the below

$$\begin{aligned} \left(r_e + h_{safe}\right) - r_P \le 0 \\ \left(r_e + h_{safe}\right) - r_E \le 0 \end{aligned} \tag{22}$$

where h_{safe} is the safe altitude. The inequalities above require the radial distance of both player should not be less than a predefine value to keep them away from intersecting the atmosphere layer.

5 Simulations and Two Side Optimal Trajectories

The optimization method described in section 4 was applied to solve the pursuit-evasion game of satellites with J2 non-spherical perturbation. Dimensional normalization was implemented as the reference [18] where the normalized length and time were given as $\mathrm{DU} = 1R_E$ and TU=806.8s respectively. Then the normalized acceleration derived from the DU and TU would be $1\mathrm{AU}=9.798\cdot10^{-3}\mathrm{km/s^2}$. h_{safe} equals to 100km. Satellites orbital elements of the pursuer and evader were given in table 1.

Table 1. Orbital elements of the purser and evader

Orbital elements	a / (km)	e	i / (°)	Ω / (°)	M / (°)
pursuer	**6978.165**	**0**	**40**	**0**	**310**
evader	**6978.165**	**0**	**60**	**30**	**290**

The intercepting and target satellites moved in the low orbits with different inclinations. The accelerations of them were given as $U_P = 0.02\mathrm{AU}, U_E = 0.01\mathrm{AU}$. The genetic algorithm was embodied by the ga function in Matlab. The inherit-selection process continues 2000 generations and each generation has 100 individuals. Also cross and mutation probability were set to be 0.8 and 0.2 respectively. The SQP accurate solution was achieved by the Matlab function *fmincon* . The genetic search process goes 17 minutes in total. Lastly the optimization gave the solution with normalized interception ending time being 2.482. The final optimized costate variables were given in table 2.

Table 2. Costate variables

λ_1	λ_2	λ_3	λ_4	λ_5	λ_6	λ_7	λ_8	λ_9	λ_{10}	λ_{11}	λ_{12}
2.948	**-0.206**	**-0.316**	**-2.366**	**3.420**	**-2.152**	**-4.116**	**0.061**	**0.147**	**0.562**	**3.603**	**-4.997**

Substituting the initials of co-states and system states to the integration of the conjugate and dynamic equation, the states variations of the pursuer and evader were obtained, as shown in figure 1.

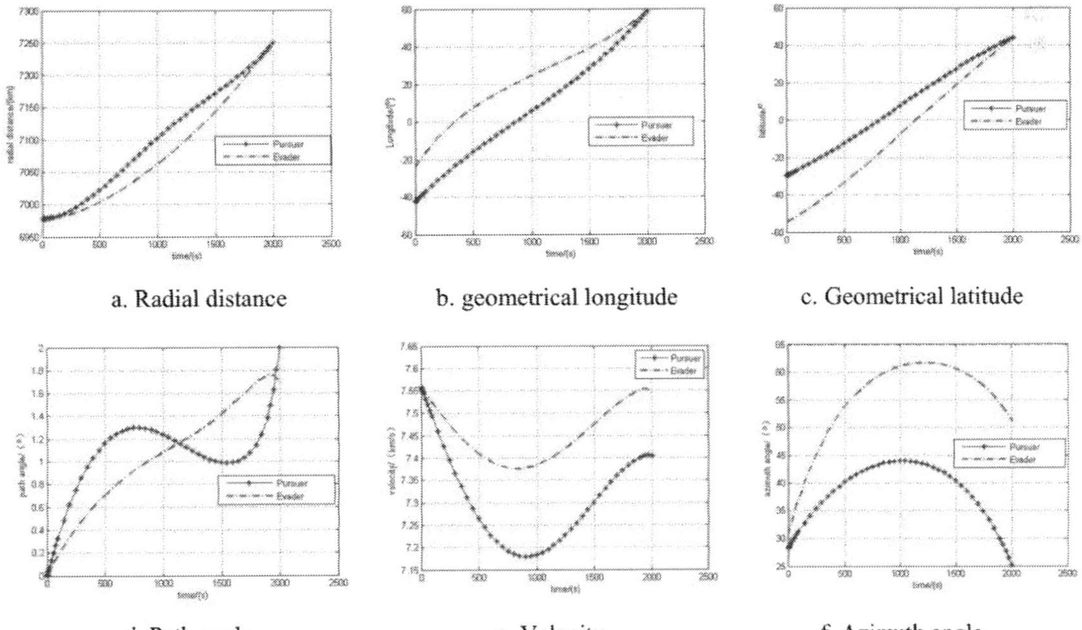

a. Radial distance b. geometrical longitude c. Geometrical latitude

d. Path angle e. Velocity f. Azimuth angle

Figure 1. States of pursuer and evader

From the figures, it was found that the pursuer successfully intercept the evader when the orbital attitude get 872km, geometrical longitude and latitude arrive at 60°and 42°after 2002 seconds. The optimal trajectories of the interceptor and escaper are given in figure 2.

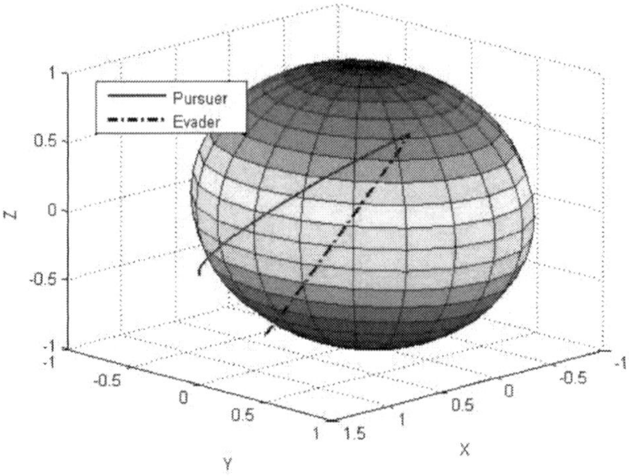

Figure 2. Trajectories of the pursuer and evader

6 Conclusion

In the present paper the typical satellite interception problem was revised, where both the pursuer and the evader can apply the continuous thrust to perform orbital maneuver in the earth central force with the J2 non-spherical perturbation. The two-side optimization was involved since the interceptor and the escaper will try their best to compete in the attacking-defense scenario. The satellite dynamic with J2 perturbation was derived. The function extremum theory was used to obtain the necessary condition of the Hamilton system and the optimal control strategies of both sides when the pursuer and evader game have got its saddle solution. Then the pursuit-evasion game was turned into a two point boundary value problem and a further constrained optimization problem. Genetic algorithm was employed to search the approximation of the global optimal solution. The accurate solution was achieved by application of the SQP method to re-optimize the objective with the approximating solution as starting initials. Lastly, the two side optimal control angles and the optimal trajectories of the pursuer and evader were simulated and calculated to validate the efficiency of the method.

References

1. Luo, Y. Z., Tang, G. J., & Li, H. Y. Optimization of multiple-impulse minimum-time rendezvous with impulse constraints using a hybrid genetic algorithm [J]. Aerospace Science and Technology, 2006, 10(6): 534-540.
2. Hughes, S. P., Mailhe, L. M., & Guzman, J. J. A comparison of trajectory optimization methods for the impulsive minimum fuel rendezvous problem[C]. In 26th Rocky Mountain Guidance and Control Conference. 2003: 3-6.
3. Luo, Y. Z., Li, H. Y., & Tang, G. J. Hybrid approach to optimize a rendezvous phasing strategy[J]. Journal of Guidance, Control and Dynamics, 2007, 30(1): 185-191.
4. Luo, Y. Z., & Tang, G. J. Spacecraft optimal rendezvous controller design using simulated annealing[J]. Aerospace Science and Technology, 2005, 9(8): 732-737.

5. Luo, Y. Z., Tang, G. J., Li, Y. J., & Li, H. Y. Optimization of multiple-impulse, multiple-revolution, rendezvous-phasing maneuvers[J]. Journal of Guidance, Control, and Dynamics, 2007, 30(4): 946-952.
6. F. Lawden. Optimal Trajectories for Space Navigation[J]. London Butter Worths. 1963
7. P. M. Lion, M. Handelsman.Primer Vector on Fixed-Time Impulsive Trajectories [J].AIAA Journal. 1968, 6(1): 127-132.
8. Greenwood D T, Kern E A. Minimum-fuel thrust limited transfer trajectories between coplanar elliptic orbits[J]. AIAA Journal, 1970, 8(10): 1772-1779.
9. Hargraves C R, Paris S W. Direct trajectory optimization using nonlinear pro-gramming and collocation[J]. Journal of Guidance, Control, and Dynamics, 1987, 10(4): 338-342.
10. Enright P J, Conway B A. Discrete approximations to optimal trajectories using direct transcription and nonlinear programming[J]. Journal of Guidance, Control, and Dynamics, 1992, 15(4): 994-1002.
11. Herman A L, Spencer D B. Optimal, low-thrust earth-orbit transfers using higher- order collocation methods[J]. Journal of Guidance, Control, and Dynamics, 2002, 25(1): 40-47.
12. Isaacs R. Differential games: a mathematical theory with applications to warfare and pursuit, control and optimization[M]. Courier Dover Publications, 1999. 3-43
13. Anderson G M. Feedback control for a pursuing spacecraft using differential dynamic programming[J]. AIAA Journal, 1977, 15(8): 1084-1088.
14. Menon P K A, Calise A J, Leung S K M. Guidance laws for spacecraft pursuit-evasion and rendezvous[C]//AIAA Guidance Navigation and Control. 1988, 668-697.
15. Mauro Pontani, Bruce A. Conway. Optimal Interception of Evasive Missile Warheads: Numerical Solution of the Differential Game[J]. Journal of Guidance, Control, and Dynamics, 2008, 31(4): 1111-1122.
16. Mauro Pontani, Bruce A. Conway. Numerical Solution of the Three-Dimensional Orbital Pursuit–Evasion Game Orbital Pursuit–Evasion Game[J]. Journal of Guidance, Control, and Dynamics, 2009, 32(2): 474-487.
17. Stupik J, Pontani M, Conway B. Optimal pursuit-evasion spacecraft trajectories in the hill reference frame[C]//AIAA/AAS Astrodynamics Specialist Conference. Reston, VA: AIAA, 2012: 4882.
18. Pontani M. Numerical Solution of Orbital Combat Games Involving Missiles and Spacecraft[J]. Dynamic Games and Applications, 2011, 1(4): 534-557.

An adaptive filtering method based on EMD for X-ray pulsar navigation with uncertain measurement noise

N. Li[1], Z. W. Kang[1], J. Liu[2] and X. M. Xu[1]

[1]*College of Information Science and Engineering, Hunan University, Changsha 410082, China*
[2]*College of Information Science and Engineering, Wuhan University of Science and Technology, Wuhan 430081, China*

Abstract: Affected by the unstable pulse radiation and the pulsar directional errors, the statistical characteristics of the pulsar measurement noise may vary with time slowly and cannot be accurately determined, which cause the filtering accuracy of the extended Kalman filter(EKF) in pulsar navigation positioning system decline sharply or even diverge. To solve this problem, an adaptive extended Kalman filtering algorithm based on the empirical mode decomposition(EMD) is proposed. In this method, the high frequency noise is separated from measurement information of pulsar by the method of EMD, and the noise variance can be estimated to update the parameters of EKF. The simulation results demonstrate that compared with conventional EKF, the proposed method can adaptively track the change of the measurement noise, and still keeps high estimation accuracy with unknown measurement noise, the positioning accuracy of the pulsar navigation is improved simultaneously.

1 Introduction

X-ray pulsar navigation is an emerging autonomous navigation technology for spacecraft, which can provide high precision navigation information[1-3]. X-ray pulsars are neutron stars that spin rapidly and emit stable pulsed radiation signal. Based on the cumulative processing of the pulsed radiation signal, the pulse time of arrival (TOA) is obtained and be utilized as the basic measurement of the pulsar-based navigation[4-6]. Combining the measurement information with orbital dynamics equation, the accurate positioning of the spacecraft can be achieved by the method of optimal estimation[7].

At present, the EKF is the most widely used filtering method in the study of autonomous navigation of spacecraft based on X-ray pulsars[8-9]. when a prior information is given and the model parameters are known, EKF can get better positioning accuracy[10]. However, affected by the unstable pulse radiation, the pulsar background noise and the pulsar directional errors, the statistical characteristics of the pulsar measurement noise may vary with time slowly and cannot be accurately determined limited to the current measurement equipment [11-12]. Scince EKF can not adjust to the change of the measurement noise, then the filtering accuracy of EKF will decline or even diverge. To this end, many scholars have proposed a variety of adaptive Kalman filtering algorithm which mainly concentrated in two aspects: adaptive filtering based on innovation and multiple model

Corresponding author: Z. W. Kang <hn_zwkang@126.com>

estimation(MME)[13-17]. The former can adjust the covariance matrix of measurement noise by introducing the adjustment factor, and the calculation is simple but the filtering accuracy is limited. The MME uses multiple filters with different parameters simultaneously, although the accuracy has been improved, but the calculation is rather complex.

Based on the previous studies, combing the EMD with the EKF, an adaptive filtering method based on EMD_EKF is proposed in this paper. The pulsar measurement model with unknown parameters is established and the noise component can be extracted from the measurement information by meas of EMD. Then, the noise variance is estimated and the relevant parameter of the EKF algorithm is modified in real time to realize the state estimation of the spacecraft.

2 Pulsar navigation model

2.1 The obit dynamic model

This paper utilizes the transfer orbit to illustrate, assuming the position vector of the spacecraft $\boldsymbol{r}=[x,y,z]^T$ and the velocity vector $\boldsymbol{v}=[v_x,v_y,v_z]^T$, the orbit dynamic equation can be suggested as:

$$\begin{bmatrix} \dot{\boldsymbol{r}} \\ \dot{\boldsymbol{v}} \end{bmatrix} = \begin{bmatrix} \boldsymbol{v} \\ \boldsymbol{a} \end{bmatrix} + \begin{bmatrix} \boldsymbol{w}_r \\ \boldsymbol{w}_v \end{bmatrix} \tag{1}$$

Assuming system state variable $\mathbf{x}=\left[x,y,z,v_x,v_y,v_z\right]^T$, noise variable $\boldsymbol{w}(t)=[0,0,0,\Delta f_x,\Delta f_y,\Delta f_z]^T$

Then (1) can be written as:

$$\dot{\boldsymbol{X}} = \begin{bmatrix} \boldsymbol{v} \\ \boldsymbol{a} \end{bmatrix} = \mathrm{f}(\boldsymbol{X},\mathrm{t}) + \boldsymbol{w}(\mathrm{t}) \tag{2}$$

$\boldsymbol{a}$ is acceleration vector of spacecraft, since the spacecraft is mainly affected by the gravity of the Sun, Mars, Earth and Jupiter in the transfer orbit, the specific forms of $\boldsymbol{a}$ can be written as:

$$\dot{\boldsymbol{v}} = \boldsymbol{a} = -u_s \frac{\boldsymbol{r}}{r_{ps}^3} - u_m \left[\frac{\boldsymbol{r}-\boldsymbol{r}_1}{r_{pm}^3} + \frac{\boldsymbol{r}_1}{r_{sm}^3} \right] - u_e \left[\frac{\boldsymbol{r}-\boldsymbol{r}_2}{r_{pe}^3} + \frac{\boldsymbol{r}_2}{r_{se}^3} \right] - u_j \left[\frac{\boldsymbol{r}-\boldsymbol{r}_3}{r_{pj}^3} + \frac{\boldsymbol{r}_3}{r_{sj}^3} \right] + \boldsymbol{\Delta f} \tag{3}$$

Specific parameters refer to the reference[18].

2.2 The measurement model of X-ray pulsar with uncertain noise

The basic measurement of X-ray pulse navigation is the pulse TOA. The measurement model based on the pulse TOA deviation can be written as:

$$\Delta t = t_{ssb} - t_b = \frac{1}{c} \boldsymbol{n} \cdot \boldsymbol{r} \tag{4}$$

Where, t_{ssb} and t_b respect the pulse TOA at the spacecraft and SSB . $\boldsymbol{n}$ and $\boldsymbol{r}$ are the pulsar direction vector and the spacecraft position vector respectively, and c represents the light speed. Assuming the pulsar observation vector as $z=[c\cdot\Delta t]$, then the pulsar navigation measurement equation can be presented as follows:

$$z = h(\boldsymbol{X}(t), t) + v \tag{5}$$

Here, v is the pulsar measurement noise with the standard deviation of σ which can be calculated according to Taylor formula[19] .

$$h[X(t), t] = \boldsymbol{n} \cdot \boldsymbol{r} \tag{6}$$

The navigation accuracy of the pulsar navigation system is depends on the accuracy of the TOA, but due to exists of the unstable pulse radiation, the background noise and the pulsar directional errors the pulsar measurement will be inevitably disturbed. The interference mentioned above can be classified as the pulsar measurement noise which is regrade as white Gaussian noise, and the noise variance can not be calculated limited to the current measurement equipment and technology. The measurement equation is:

$$z = h(\boldsymbol{X}(t), t) + v' \tag{7}$$

$$v' = v + \xi \tag{8}$$

Assuming that the variance of v' is σ', then σ' is the unknown parameter of pulsar measurement model. This paper employs three pulsars as navigation pulsar and the observation vector and measurement model of pulsar can be recorded as:

$$\boldsymbol{Z} = [z_1, z_2, z_3]^T \tag{9}$$

$$\boldsymbol{Z}(t) = h(\boldsymbol{X}, t) + \boldsymbol{V}' \tag{10}$$

where

$$h(\boldsymbol{X}, t) = [h_1[\boldsymbol{X}(t), t], h_2[\boldsymbol{X}(t), t], h_3[\boldsymbol{X}(t), t]]^T \tag{11}$$

$$\boldsymbol{V}' = [v_1', v_2', v_3']^T \tag{12}$$

Here, $\boldsymbol{V}'$ is the system measurement noise and the corresponding covariance matrix $\boldsymbol{R}'$ which is the system parameter to be estimated can be expressed as :

$$\boldsymbol{R}' = [\sigma_1', \sigma_2', \sigma_3']^T \tag{13}$$

3 Adaptive filtering algorithm based on empirical mode decomposition

As the models of the navigation system are nonlinear, the EKF is adopted as the navigation filter to realize continuous orbit determination of the spacecraft[8,20]. However, when the statistical characteristics of the measurement noise varies, the traditional EKF can not adjust to the change, which may cause serious estimation bias. This paper introduce the EMD to the the EKF, the high-frequency noise component can be extracted from the pulsar measurement. Then the noise variance is estimated to adjust the relevant parameters in EKF constantly to achieve the optimal state estimation of the spacecraft.

3.1 The empirical mode decomposition

The NAE member N.E. Huang has proposed the empirical mode decomposition method(EMD),which is a new signal processing method that apply for nonlinear processes[21]. According to its own characteristics, the signal is decomposed into several intrinsic mode functions (IMF) which represent different frequency components of the signal. In addition, Wu and Huang has found that each IMF component obtained by decomposing white noise via empirical mode decomposition satisfy the normal distribution. Therefore, the EMD method can be used to decompose the noisy signal with white noise.

After decomposed by EMD, the noisy signal can be resolved into several IMF components, which can be expressed as[22]:

$$x(t)=\sum_{i=1}^{n} c_i(t) \tag{14}$$

Where $x(t)$ represent the noisy signal, $c_i(t)$ is the IMF components which represent different frequency components of the signal and n is the number of the IMF. According to the correlation of white noise, we can determine whether the component is the desired noise component by calculating the auto-correlation or cross-correlation of these IMF components, and then noise variance can be calculated.

3.2 The filtering algorithm of EMD_EKF

Discretizing (2) and (10), the discrete system model is obtained as follows:

$$\boldsymbol{X}(k+1)=\boldsymbol{F}(k+1,k)\boldsymbol{X}(k)+\boldsymbol{W}(k) \tag{15}$$

$$\boldsymbol{Z}(k+1)=\boldsymbol{H}(k+1)\boldsymbol{X}(k+1)+\boldsymbol{V}'(k+1) \tag{16}$$

Where $\boldsymbol{X}$ is the state vector of the system and $\boldsymbol{Z}$ is the measurement vector of the pulsar, $\boldsymbol{F}$ is the system state transition matrix and $\boldsymbol{H}$ is the observation matrix. $\boldsymbol{W}$ is the state noise matrix with variance of $\boldsymbol{Q}$ and $\boldsymbol{V}'$ is the observation noise matrix with unknown variance $\boldsymbol{R}'$. Because the measurement noise variance is unknown, so the observed signal should be deposed by EMD before using the EKF to update the state of the spacecraft. The specific filtering process is as follows:

1) Calculating the one step prediction value $\hat{\boldsymbol{X}}_{k+1,k}$ of the state and error variance matrix $P_{k+1,k}$

2) Calculating one step prediction of measurement: $\hat{\boldsymbol{Z}}_{k+1,k}=\boldsymbol{H}_{k+1}\hat{\boldsymbol{X}}_{k+1,k}$

3) Calculating one step prediction of covariance matrix: $\boldsymbol{P}_{k+1,k}=\boldsymbol{F}_{k+1,k}\boldsymbol{P}_{k,k}\boldsymbol{F}_{k+1,k}{}^{T}+\boldsymbol{Q}_{k+1}$

4) Deposing the measurement by the method of EMD, assuming that the first K (K<N) term is the noise component, the noise variance is calculated as: $\boldsymbol{R}'_{k+1}=\sum_{i=1}^{K} IMF_i$

5) Calculating the gain matrix: $\boldsymbol{K}_{k+1}=\boldsymbol{P}_{k+1,k}\boldsymbol{H}_{k+1}\left[\boldsymbol{H}_{k+1}\boldsymbol{P}_{k+1,k}\boldsymbol{H}_{k+1}{}^{T}+\boldsymbol{R}'_{k+1}\right]^{-1}$

6) Updating state estimation: $\hat{\boldsymbol{X}}_{k+1}=\hat{\boldsymbol{X}}_{k+1,k}+\boldsymbol{K}_{k+1}\left(\boldsymbol{Z}_{k+1}-\hat{\boldsymbol{Z}}_{k+1,k}\right)$

7) Updating covariance matrix: $\boldsymbol{P}_{k+1}=\left[\boldsymbol{I}-\boldsymbol{K}_{k+1}\boldsymbol{H}_{k+1}\right]\cdot\boldsymbol{P}_{k+1,k}$

4 Simulation results and analysis

The Mars explorer orbit data of the "Pathfinder Mars" is adopted in this paper, the specific simulation conditions refer to the reference[5].There are two kinds of measurement noise used in simulation:

DOI: 10.1051/matecconf/201711404017

white Gaussian noise with fixed variance σ and time-varying noise whose variance intensity changed on the basis of the fixed variance. In the above two kinds of noise, the EKF and EMD_EKF method are used to estimate the state of the spacecraft, and the navigation performance is analyzed.

Figure 1 shows the simulation results of EKF based on those two noises described above. As can be seen from Figure 1, EKF can achieve a good filtering effect with fixed noise. But when the variance of measurement noise vary with time, the EKF cannot track and adjust the changes, which result in a sharp decline in the navigation accuracy or even lead to filter divergence.

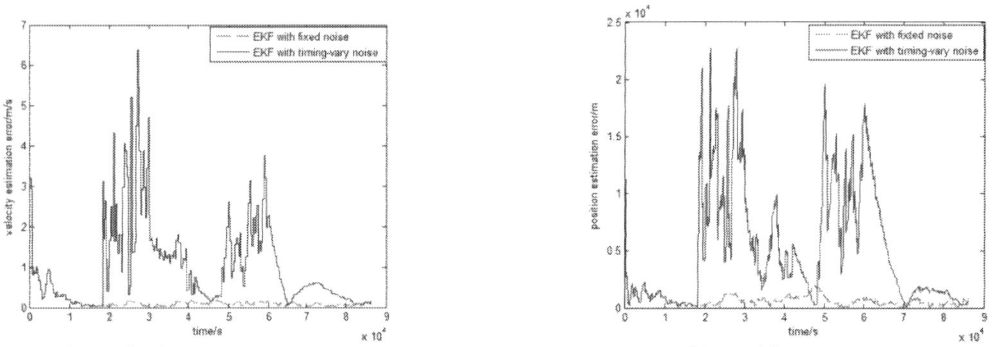

(a) velocity error (b) position error

Figure 1. Simulation results of conventional EKF under two kinds of noise

Figure 2 shows the simulation results based on EKF and EMD_EKF with timing-vary noise. From Figure 2, we can see that when the variance of measurement noise varies, the positioning error of navigation based on EKF increases sharply because the variance of measurement noise adapted in EKF is inconsistent with the true value. The proposed adaptive filtering based on EMD_EKF adjust the noise variance adaptively and still maintain high accuracy of positioning when the noise variance is unknown. A further explanation of Figure 2 is given by table 1 , We can see that the positioning error of EKF algorithm has achieved to 3000m and that error of this method is only 500m, the positioning accuracy is 5 times higher than that of EKF.

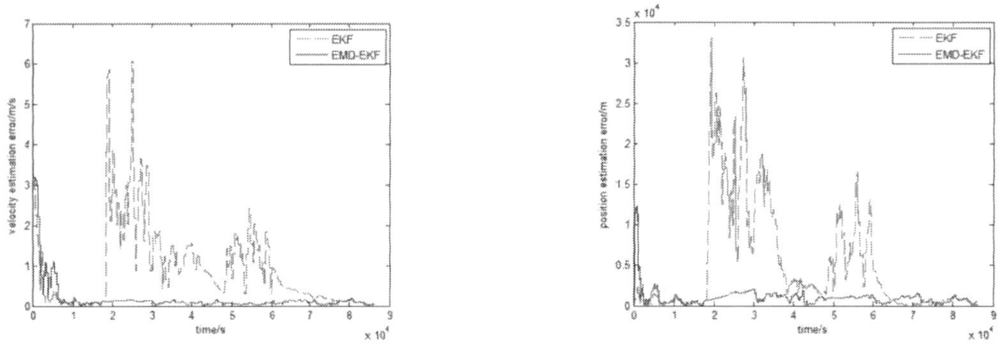

(a) velocity error (b) position error

Figure 2. Filtering results of two methods with timing-vary noise

Table 1. Performance comparison of two methods with unknown noise variance

Filter method	position error	velocity error
EKF	3538 m	0.625 m/s
EMD_EKF	526 m	0.0726 m/s

Table 2. Performance comparison of two methods with fixed noise variance

Filter method	position error	velocity error
EKF	464 m	0.0701 m/s
EMD_EKF	467 m	0.0714 m/s

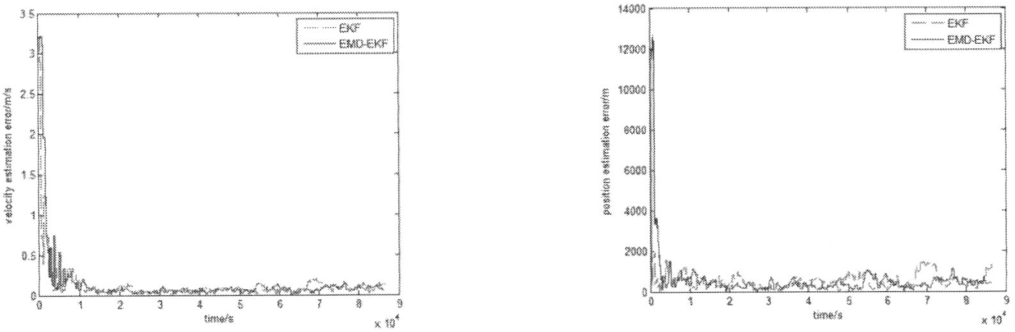

(a) velocity error   (b) position error

Figure 3. Filtering results of two methods with fixed noise

It can be seen from Figure 3 that the navigation accuracy of the two algorithms with fixed noise is consistent basically. It shows that the proposed method can obtain high positioning accuracy both in the fixed noise and unknown noise which can restrain the influence of unknown disturbance on navigation accuracy. Table 2 shows the performance comparison of these two methods.

Figure 4 shows the comparison between the reconstructed noise after EMD decomposition and the original noise with fixed variance of pulsar B0531+21. We can see that the high-frequency part extracted from the pulsar measurement by means of EMD can reconstruct the original noise primely. Figure 5 suggests that the proposed method can adaptively track the change when the noise variance varies with time and ensure that the noise variance used in the EKF is consistent with the real value through estimating the noise variance in real time.

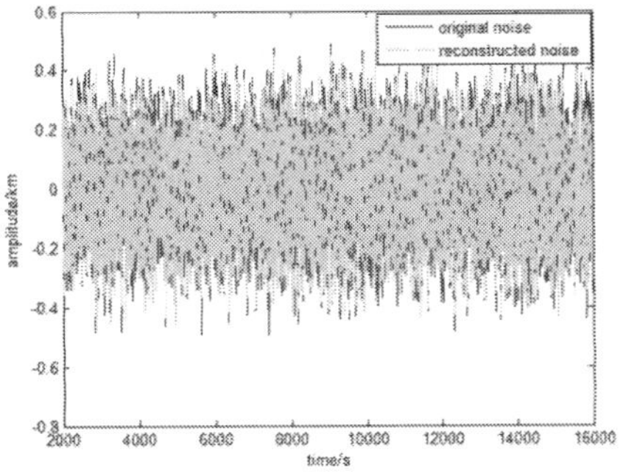

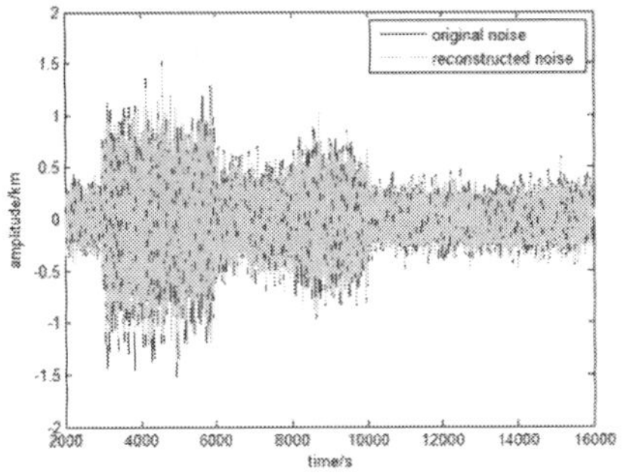

Figure 4. Comparison of two noise with fixed variance

Figure 5. Comparison of two noise with timing-vary variance

5 Conclusion

In this paper, an adaptive extended Kalman filtering method based on empirical mode decomposition is studied to solve the filtering problem of the X-ray pulsar-based navigation system with uncertain measurement noise. This method does not need to determine the statistical properties of the measurement noise in advance and can track the varies of the noise adaptively to modify the relevant parameters of EKF in real time. Compared with the conventional EKF, the proposed method can still maintain a high estimation accuracy with timing-vary and uncertain measurement noise. Thus it is an adaptive Kalman filtering algorithm that is more suitable for the spacecraft navigation in deep space environment in which the measurement noise is difficult to determine.

References

1. S.I. Sheikh, D.J. Pines, P.S. Ray, The use of X-ray pulsars for spacecraft navigation, Adv. Astronaut. Sci, **119** (2005) 105-119.
2. S.I. Sheikh, D.J. Pines, Recursive Estimation of Spacecraft Position and Velocity Using X-ray Pulsar Time of Arrival Measurements, J. Inst. Navig, **53** (2006) 149 - 166.
3. K. Xiong, C.L. Wei, L.D. Liu, Research on the autonomous navigation of satellite constellation using pulsars, J. Astronaut , **29** (2008) 545-566.
4. B.J. Fei, G.Z. Yao, J. Du, Some problem of research on X-ray pulsar-based navigation., J. Aca. Armor. Fro. Eng, **27** (2013) 1-8.
5. J. Liu, J.C. Fang, Z.H. Yang, X-ray pulsar/Doppler difference integrated navigation for deep space exploration with unstable solar spectrum, Aerosp. Sci. Technol., **41** (2015) 144~150.
6. P.Y. Cui, S. Wang, A. Gao, et al, X-ray pulsars/Doppler integrated navigation for Mars final approach, Adv. Space. Res. , **57** (2016) 1889-1900.
7. Y.D. Wang, W. Zheng, S.M. Sun, L. Li, X-ray pulsar-based navigation system with the errors in the planetary ephemerides for Earth-orbiting satellite, Adv. Space Res. **51** (2013) 2394–2404.
8. M.Y. Fu, Z.H. Deng, L.P. Yan, Kalman filtering theory and application in navigation system, *Science Publishing House*, 2010.
9. X.L. Ning, X. Ma, C. Peng, J.C. Fang, Analysis of filtering methods for satellite autonomous orbit determination using celestial and geomagnetic measurement, Math. Probl. Eng., **2** (2012) 75-78.
10. J. Liu, Z.W. Kang, P. White, J. Ma, J.W. Tian, Doppler/XNAV integrated navigation system using small-area X-ray sensor, IET Radar Sonar Navig, **5** (2011) 1010~1017.
11. J. Liu, J. Ma, J.W. Tian, Z.W. Kang, P. White, X-ray pulsar navigation method for spacecraft with pulsar direction error, Adv. Space. Res., **46** (2010) 1409~1417.
12. X.L. Ning, Z. Li, P.P. Huang, Y.Q. Yang, An adaptive Kalman filter for Mars spacecraft acquisition phase, J. Deep. Space. Explor., **3** (2016)237-245.
13. M.J. Yu, INS/GPS integration system using adaptive filter for estimating measurement noise variance, IEEE Trans. Aerosp. Electron. Syst., **48** (2012) 1786-1792.
14. B.N. Alsuwaidan, J.L. Crassidis, Y. Cheng, Generalized multiple-model adaptive estimation using an auto correlation approach, IEEE Trans. Aerosp. Electron. Syst., **47** (2010) 2138–2152.
15. K. Xiong, T. Liang, Y.J. Lei, Multiple model Kalman filter for attitude determination of precision pointing spacecraft, Acta Astronau., **68** (2011) 843–852 .
16. K. Xiong, C.L. Wei, L.D. Liu, Robust multiple model adaptive estimation for spacecraft autonomous navigation, Aerosp. Sci. Technol., **42** (2015) 249~258.
17. F. Yang, Z. Wang, Y.S. Hung, Robust Kalman filtering for discrete time-varying uncertain systems with multiplicative noises, IEEE Trans. Autom. Control, **47** (2002) 1179–1183.
18. J. Liu, J.C. Fang, Z.W. Kang, J. Wu, X.L. Ning, Novel Algorithm for X-Ray Pulsar Navigation Against Doppler Effects, IEEE Trans. Aerosp. Electron. Syst., **51** (2015) 228~241.
19. J.H. Taylor, Pulsar timing and relativistic gravity, Philo-sophical Trans, R Soc. London, **341** (1992) 117-134.

 DOI: 10.1051/matecconf/201711404017

20. K. Xiong, C.L. Wei, L.D. Liu, Robust Kalman filtering for discrete-time nonlinear systems with parameter uncertainties, Aerosp. Sci. Technol. **18** (2012) 15–24.
21. N.E. Huang , S. Zheng, R. Steven, The empirical mode decomposition and the Hilbert spectrum for nonlinear non-stationary time Series Analysis, *P. Math. Phys. Eng. Sci, london,* **454** (1998) 903-995.
22. X.Y. Li, J. Jin, Y. Shen, Y.P. Liu, Noise level estimation method with application to EMD_based signal decomposing, J. Syst. Eng. Electron., **27** (2016) 763-771.

AUTHOR INDEX

AUTHOR INDEX

AUTHOR INDEX